Advances in Experimental Medicine and Biology

Volume 870

Advances in Experimental Medicine and Biology presents multidisciplinary and dynamic findings in the broad fields of experimental medicine and biology. The wide variety in topics it presents offers readers multiple perspectives on a variety of disciplines including neuroscience, microbiology, immunology, biochemistry, biomedical engineering and cancer research. *Advances in Experimental Medicine and Biology* has been publishing exceptional works in the field for over 30 years and is indexed in Medline, Scopus, EMBASE, BIOSIS, Biological Abstracts, CSA, Biological Sciences and Living Resources (ASFA-1), and Biological Sciences. The series also provides scientists with up to date information on emerging topics and techniques.

2014 Impact Factor: 1.958

More information about this series at http://www.springer.com/series/5584

Isabella C. Felli • Roberta Pierattelli
Editors

Intrinsically Disordered Proteins Studied by NMR Spectroscopy

Springer

Editors
Isabella C. Felli
CERM and Department of Chemistry
"Ugo Schiff" University of Florence
Florence
Italy

Roberta Pierattelli
CERM and Department of Chemistry
"Ugo Schiff" University of Florence
Florence
Italy

ISSN 0065-2598 ISSN 2214-8019 (electronic)
Advances in Experimental Medicine and Biology
ISBN 978-3-319-20163-4 ISBN 978-3-319-20164-1 (eBook)
DOI 10.1007/978-3-319-20164-1

Library of Congress Control Number: 2015948151

Springer Cham Heidelberg New York Dordrecht London

Printed on acid-free paper

Springer International Publishing AG Switzerland is part of Springer Science+Business Media (www.springer.com)

Preface

Ignored for a long time in high-resolution studies of proteins, intrinsic protein disorder is now recognized as one of the key features for a large variety of cellular functions, where structural flexibility presents a functional advantage in terms of binding plasticity and promiscuity. The properties of intrinsically disordered proteins (IDPs) and protein regions (IDPRs) are highly complementary to those deriving from the presence of a unique and well-defined three-dimensional fold. Structural order, well characterized through the vast number of protein 3D structures present in the Protein Data Bank (www.pdb.org), leading to highly organized protein machines with well-defined binding pockets, lies at the heart of structural biology. However, the functional importance of conformational flexibility, and the characterization of residual transient structure in these protein regions, has only recently attracted the attention of the scientific community and has immediately found widespread interest in molecular biology research.

The tendency of a protein to adopt a stable globular structure such as the number of folds we find in the Protein Data Bank or to be highly flexible and able to sample many different conformations, directly results from the amino acid primary sequence. It is not so surprising that a wide range of properties is needed for proteins to carry out a broad range of functions and the idea that a high extent of dynamics and flexibility provides important functional features nowadays appears quite obvious. Structural disorder is abundant in higher multi-cellular organisms, in particular in regulatory protein regions that orchestrate dynamic cellular functions relying on spatial and temporal malleability. As a consequence, IDPs and IDPRs are overrepresented in key functions of higher eukaryotes, i.e. transcriptional and translational regulation, intracellular signaling, protein homeostasis, inter-cellular communication, cell-fate decisions, and many more.

In this frame nuclear magnetic resonance (NMR) spectroscopy is the unique technique able to provide high resolution information. However the peculiar properties of IDPs do influence the NMR observables raising also several critical questions that should be considered in the design of optimal NMR experiments and in the interpretation of the data in terms of protein's structural and dynamic properties.

Recent progress in the field has radically changed our perspective to study IDPs through NMR: increasingly complex IDPs can now be characterized, a wide range of observables can be determined reporting on their structural and dynamic properties, computational methods to describe the structure and dynamics are in continuous development and IDPs can be studied in environments as complex as whole cells.

Therefore we felt timely to convey these exciting recent developments in a book and to do this in close interactions with pioneers in the field of IDPs as well as with newcomers able to bring fresh energy and enthusiasm. We hope to be able to communicate the new exciting possibilities offered by NMR and to present open questions to foster further developments.

After an introductory chapter by Dunker and Oldfield describing the key steps opening the field of IDPs, the book focuses on the many aspects that make NMR a unique technique to study IDPs (Chaps. 2–5). Contributions discuss different aspects starting from the first principles of NMR spectroscopy, including hardware requirements, to the design and application of complex NMR experiments, to data interpretation in terms of structural and dynamic properties, to the best ways to achieve snapshots of IDPs in cells. Key to NMR analysis of complex proteins is the possibility to have efficient heterologous protein expression, enabling stable isotope incorporation. The tricks useful for IDPs samples preparation are reported in Chap. 6.

Information at atomic resolution derived from NMR needs to be complemented by information achieved through a variety of different biophysical methods reporting on different properties of IDPs, as discussed in Chaps. 7–8. The use of predictors, which may represent a preliminary step for any investigation, is reported in Chap. 9 while the perspectives for in-cell NMR in this field of research are discussed in Chap. 10. Several examples of recent investigations and open questions are reported in the final part of the book (Chaps. 12–14).

A separated chapter (Chap. 11) is dedicated to the PED, the database that has been designed to deposit experimental data and calculate structural ensembles in order to render these data feely accessible to the scientific community and stimulate discussion and progress in the field.

With this book we hope to provide a useful guide to students and post-docs approaching the field and willing to contribute to the characterization of IDPs through NMR. Recent progress in NMR instrumentation, combined with the development of new methods, has radically changed the perspective of the kind of molecules we can study, the amount of information that we can achieve as well as the time needed to complete an NMR characterization. Therefore we hope that the new NMR tools developed will be increasingly used and contribute a wealth of experimental information on IDPs. Speculating on more long-term perspectives, the development of improved NMR methods to study IDPs is expected to provide a large amount of experimental data on them, contributing to our understanding of the molecular basis responsible for their function and filling a gap of about 50 years with respect to our knowledge on the structural and dynamic behaviour of folded proteins. This is expected to reveal a much larger number of ways in which proteins communicate in the cell. Other expected outcomes of NMR experimental data on

IDPs include the improvement of prediction tools, which still suffer from the bias that they are derived from the missing information in the electron density maps in X-ray crystallography data!

We hope you will enjoy the book, and even more studying IDPs through NMR, as much as we do! A great thanks to all the Authors that we had the luck to work with and to the IDPbyNMR EC Marie Curie Initial Training Network which contributed to make it possible.

Contents

Contributors

Matilde Beltrandi Aix-Marseille Université; CNRS, Marseille, France

Pallab Bhowmick VIB Department of Structural Biology, Vrije Universiteit Brussel (VUB), Brussels, Belgium

Christophe Bignon Aix-Marseille Université; CNRS, Marseille, France

Martin Blackledge IBS, University Grenoble Alpes; IBS, CNRS; IBS, CEA, Grenoble, France

David Blocquel Aix-Marseille Université; CNRS, Marseille, France

Bernhard Brutscher Institut de Biologie Structurale, Université Grenoble 1, CNRS, CEA, Grenoble Cedex 9, France

Eduardo O. Calçada Magnetic Resonance Center (CERM), University of Florence, Sesto Fiorentino, Italy

Cesyen Cedeño VIB Department of Structural Biology, Vrije Universiteit Brussel, Brussels, Belgium

Nicolas Coudevylle Department of Computational and Structural Biology, Max F. Perutz Laboratories, University of Vienna, Vienna, Austria

András Dinnyés BioTalentum Ltd, Godollo, Hungary

Marion Dosnon Aix-Marseille Université; CNRS, Marseille, France

A. Keith Dunker Center for Computational Biology and Bioinformatics, Indiana University School of Medicine, Indianapolis, IN, USA

Jenny Erales Aix-Marseille Université; CNRS, Marseille, France

Isabella C. Felli CERM and Department of Chemistry "Ugo Schiff", University of Florence, Sesto Fiorentino, Florence, Italy

Biao Fu Department of Chemistry, University of Cambridge, Cambridge, UK

Sergio Gil-Caballero Bruker BioSpin AG, Industriestrasse 26, Fällanden, Switzerland

Antoine Gruet Aix-Marseille Université; CNRS, Marseille, France

Mainak Guharoy VIB Department of Structural Biology, Vrije Universiteit Brussel (VUB), Brussels, Belgium

Johnny Habchi Aix-Marseille Université; CNRS, Marseille, France

Tomáš Hošek CERM and Department of Chemistry "Ugo Schiff", University of Florence, Sesto Fiorentino, Florence, Italy

Malene Ringkjøbing Jensen IBS, University Grenoble Alpes; IBS, CNRS, Grenoble, France

Priyanka Joshi Department of Chemistry, University of Cambridge, Cambridge, UK

Michael Kachala Hamburg Outstation, European Molecular Biology Laboratory; Department of Chemistry, Hamburg University, Hamburg, Germany

Robert Konrat Department of Computational and Structural Biology, Max F. Perutz Laboratories, University of Vienna, Vienna, Austria

Georg Kontaxis Department of Computational and Structural Biology, Max F. Perutz Laboratories, University of Vienna, Vienna, Austria

Magdalena Korsak Giotto Biotech, Sesto Fiorentino, Italy

Tatiana Kozyreva Giotto Biotech, Sesto Fiorentino, Italy

Jaka Kragelj IBS, University Grenoble Alpes; IBS, CNRS; IBS, CEA, Grenoble, France

Rainer Kümmerle Bruker BioSpin AG, Industriestrasse 26, Fällanden, Switzerland

Dennis Kurzbach Department of Computational and Structural Biology, Max F. Perutz Laboratories, University of Vienna, Vienna, Austria

Sonia Longhi Aix-Marseille Université; CNRS, Marseille, France

Christopher J. Oldfield Center for Computational Biology and Bioinformatics, Indiana University School of Medicine, Indianapolis, IN, USA

Alessandro Piai CERM and Department of Chemistry "Ugo Schiff", University of Florence, Sesto Fiorentino, Florence, Italy

Roberta Pierattelli CERM and Department of Chemistry "Ugo Schiff", University of Florence, Sesto Fiorentino, Florence, Italy

Hadas Raveh-Amit BioTalentum Ltd, Godollo, Hungary

Zsófia Sólyom Institut de Biologie Structurale, Université Grenoble 1, CNRS, CEA, Grenoble Cedex 9, France

Dmitri I. Svergun Hamburg Outstation, European Molecular Biology Laboratory, Hamburg, Germany

Peter Tompa VIB Department of Structural Biology, Vrije Universiteit Brussel (VUB), Brussels, Belgium

Universiteit Brussel (VUB), Structural Biology Brussel (SBB), Brussels, Belgium

Institute of Enzymology, Research Center of Natural Sciences, Hungarian Academy of Sciences, Budapest, Hungary

Vladimir N. Uversky Department of Molecular Medicine and USF Health Byrd Alzheimer's Research Institute, Morsani College of Medicine, University of South Florida, Tampa, FL, USA

Biology Department, Faculty of Science, King Abdulaziz University, Jeddah, Kingdom of Saudi Arabia

Institute for Biological Instrumentation, Russian Academy of Sciences, Pushchino, Moscow Region, Russia

Erica Valentini Hamburg Outstation, European Molecular Biology Laboratory; Department of Chemistry, Hamburg University, Hamburg, Germany

Mihaly Varadi Department of Structural Biology, Vlaams Institute voor Biotechnologie (VIB); Vrije Universiteit Brussel (VUB), Structural Biology Brussel (SBB), Brussels, Belgium

Michele Vendruscolo Department of Chemistry, University of Cambridge, Cambridge, UK

Chapter 1
Back to the Future: Nuclear Magnetic Resonance and Bioinformatics Studies on Intrinsically Disordered Proteins

A. Keith Dunker and Christopher J. Oldfield

Abstract From the 1970s to the present, regions of missing electron density in protein structures determined by X-ray diffraction and the characterization of the functions of these regions have suggested that not all protein regions depend on prior 3D structure to carry out function. Motivated by these observations, in early 1996 we began to use bioinformatics approaches to study these intrinsically disordered proteins (IDPs) and IDP regions. At just about the same time, several laboratory groups began to study a collection of IDPs and IDP regions using nuclear magnetic resonance. The temporal overlap of the bioinformatics and NMR studies played a significant role in the development of our understanding of IDPs. Here the goal is to recount some of this history and to project from this experience possible directions for future work.

Keywords Protein structure and function

1 Introduction

The mainstream view of the relationship between protein structure and function is that the amino acid sequence of each protein contains the information needed to fold into a specific 3D structure. Upon folding, a set of amino acid side chains, which are typically separated along the linear sequence, become organized into a spatially co-localized region called the active site. This site carries out the function of the protein, typically enzyme catalysis or the binding of small molecules.

A. K. Dunker (✉) · C. J. Oldfield
Center for Computational Biology and Bioinformatics,
Indiana University School of Medicine, Indianapolis, IN 46202, USA
e-mail: kedunker@iu.edu

C. J. Oldfield
e-mail: cjoldfie@iu.edu

I. C. Felli, R. Pierattelli (eds.), *Intrinsically Disordered Proteins Studied by NMR Spectroscopy,* Advances in Experimental Medicine and Biology,
DOI 10.1007/978-3-319-20164-1_1

This sequence → structure → function point of view has dominated thinking about proteins since the lock-and-key mechanism to explain enzyme function was first presented in the 1890s (Fischer 1894; translated and discussed in Lemieux and Spohr 1994), and since the protein denaturation was ascribed to the loss of specific structure in the 1930s (Wu 1931; Mirsky and Pauling 1936). By the 1930s protein denaturation by acid, base, elevated temperature, and urea had been shown for multiple proteins, and protein renaturation (e.g., protein refolding) had even been accomplished by that time for a few proteins (Mirsky and Anson 1930; Anson and Mirsky 1931b; Anson and Mirsky 1931a).

The much later studies on the refolding of ribonuclease (Sela et al. 1957), which led to the Nobel Prize, were carried out in the context of the idea that information flows from DNA sequence → RNA sequence → amino acid sequence. This work focused on understanding the connection between the amino acid sequence and the biologically active conformation. Furthermore, this Nobel prize was shared with researchers who determined the amino acid sequence of ribonuclease (Hirs et al. 1960), which was the first enzyme to be sequenced. The folding of a protein (typically an enzyme) into its active conformation was regarded to be so important that this process was considered to be "the second half of the genetic code" (Kolata 1986; Gierasch and King 1990), leading eventually to the structural genomics initiative (Burley 2000).

This concept that prior formation of structure is required for a protein to carry out function has deep roots in the theory of chemical structure. A famous example is the structure of benzene in which its chemical formula, C_6H_6, could be reconciled with the known chemical bonding properties of carbon by a structure having a 6 membered ring, like a snake biting its tail (Read 1957), with alternating single and double bonds (Kekule 1865). Of course, the concept of alternating double and single bonds was modified into one type of partial double bond all around the ring by the theory of resonance (Pauling 1932). Similarly, knowledge of the chemical structure of the peptide bond leads to the conclusion that the nitrogen electron pair forms a partial double bond by resonance interaction with the π electrons of the C=O moiety, thus stabilizing the peptide bond into a planar structure, with of course the potential for strain to twist this moiety out of its planar configuration (Corey and Pauling 1953; Edison 2001). Indeed, thinking about the chemistry of any particular molecule is typically dominated by concepts taken from the theory of chemical structure. Thus, the sequence → structure → function paradigm for proteins can be thought of as an extension of fundamental principles that arise from the structural features of molecules.

Given this background, it is no wonder that it has taken a very long time for intrinsically disordered proteins (IDPs) and IDP regions to be recognized as important for biological function. Here we will provide some of the history of how IDPs came to be recognized.

2 Discovery and Initial Characterization of Intrinsically Disordered Proteins (IDPs)

Here we present some very early theoretical conjectures suggesting the existence of unstructured proteins and their role in an important biological process. These early theoretical conjectures were followed up fairly recently with experimental studies, but our view is that the experimental work gives only limited support to the original conjectures. Next we present several important early experiments showing the existence of IDPs and IDP regions. These data show that experimental support for IDPs has existed for a long time prior to the recent flurry of work on these proteins.

2.1 Early Theoretical Work on IDPs and Recent Follow-Up

In the 1930s it was thought that a single antibody molecule is able to recognize multiple antigen molecules having different shapes and different chemical properties (Rothen and Landsteiner 1939). To explain how one antibody could bind to differently shaped antigens, it was suggested that antibodies initially exist in an unfolded state and then undergo antigen-directed folding (Rothen and Landsteiner 1939; Pauling 1940). In this model, the antibody molecule undergoes a disorder-to-order transition as it binds to each antigen, with differently folded antibody structures occurring when the same antibody binds to differently shaped antigens. More than 60 years after the initial proposal, X-ray crystal structures of one antibody molecule bound to different antigen partners were purported to support the antigen-dependent folding model (James et al. 2003; James and Tawfik 2003a). If true, this early theoretical work combined with the more recent experimental validation would represent the first discovery of an IDP-based mechanism for biological function.

Close examination of these antibody-antigen complexes, however, reveals that sidechain rearrangements—not alternatively folded backbones—account for the observed structural changes induced by the association of the same antibody molecule with different antigens. Also, changes in hydrogen bonding appear to be more important than changes in hydrophobic contacts (James and Tawfik 2003b). These structural differences truly represent antigen-induced changes in folding, because, like the backbones, side chains also have multiple conformational choices. However, in our view, this result differs from the original proposal in which the entire proteins, both side chains and backbones, were represented as unfolded before association with their cognate antigens.

As the immune response progresses, B cells produce antibodies with increasing affinity for the antigen via a process involving hypermutation and clonal selection (Teng and Papavasiliou 2007). Molecular dynamics studies have been carried out on a set of homology models developed from antibody sequences obtained from different stages in this evolutionary process (Zimmermann et al. 2006; Thorpe and

Brooks 2007). The naïve, less specific, more weakly binding antibodies were found to contain less well packed, much more dynamic antigen binding sites. Such sites are therefore able to alter their structures upon binding to differently shaped antigens. As the evolutionary process continues, the antibody binding sites become tighter with better fit, and antigen binding becomes more specific (Zimmermann et al. 2006; Thorpe and Brooks 2007). It is as if antibodies evolve from induced fit to lock and key binding mechanisms. Prior to this selection process, the ability of the circulating antibodies to bind to differently shaped antigens provides the basis for a very broad immuno-surveillance.

We wonder whether there are some antibodies with binding sites that are truly disordered rather than merely flexible. Such antibodies, if they exist, would fulfill the original hypothesis of antigen-directed folding, and thereby push back the start-date for the study of IDP regions. On the other hand, there are many examples of IDPs and IDP regions that bind to different partners and fold differently when they bind to the different partners (Oldfield et al. 2008; Hsu et al. 2013) just as suggested earlier (Rothen and Landsteiner 1939; Pauling 1940). Thus, the authors of the 1939 and 1940 papers certainly presented an insightful model for protein-partner interactions that turns out to be correct, but this model might not apply to antigen binding by antibodies as they originally proposed.

A comment is necessary on the nomenclature and mechanisms used to describe partner binding by IDPs. Induced fit (Koshland 1958; Koshland 1959) has frequently been suggested to be analogous to the structural adjustments made by both the glove and hand as the hand is inserted (Koshland 2004). This description implies that the enzyme is folded but still flexible and so can adjust its structure to fit the substrate, and that the substrate also has flexibility that comes into play upon binding. Thus, in our view, the interaction between an IDP or IDP region and its partner should not be described as induced fit as is very often done, but rather as a disorder-to-order transition (Schulz 1979) or as coupled binding and folding (Spolar and Record 1994). As for the mechanism of binding, two models have been suggested as the extremes on a continuum, namely conformational selection, for which the partner selects out members of the ensemble that are already in the same configuration as when bound (Burgen et al. 1975), and coupled binding and folding, for which the partner binds to a local site on the IDP followed by concomitant folding and binding (Spolar and Record 1994; Sugase et al. 2007). Complex formation between an IDP and its cognate partner may also involve a mixed mechanism, with conformational selection for part of the interface followed by coupled folding and binding for the remainder (Espinoza-Fonseca 2009).

2.2 *Early Experimental Characterization of IDPs and Recent Follow-Up*

In 1950 and 1958 the milk protein casein and the egg protein phosvitin, respectively, were reported to lack 3D structure in vitro under physiological conditions.

The lack of 3D structure in casein was determined by single wave length optical rotation using the sodium D line (McMeekin 1952), which had been used since the 1930s to follow protein denaturation induced by acid, base, and heat (Herriott 1938). The lack of 3D structure in phosvitin was shown by optical rotatory dispersion (ORD) (Jirgensons 1958), which is the multi-wave length extension of optical rotation. Phosvitin was later confirmed to lack structure by circular dichroism (CD) spectroscopy (Grizzuti and Perlmann 1970). Note that CD and ORD depend on the same physical principle, namely the differential absorption of left and right circularly polarized light by optically active molecules, so these two experiments are equivalent to each other. Finally, phosvitin was also shown to be disordered by NMR spectroscopy (Vogel 1983).

Following application of ORD to a collection of proteins, in 1966 the suggestion was made to classify proteins according to their conformation (Jirgensons 1966). This suggestion predated the Structural Classification of Proteins (SCOP) (Murzin et al. 1995) and the Class, Architecture, Topology, and Homology (CATH) (Orengo et al. 1997) databases by about 30 years. Unlike SCOP and CATH, however, this classification scheme contained a category called "disordered", based mainly on phosvitin but also on the histones, which had been shown at the time by ORD to lack structure at low salt concentration (Jirgensons 1966). Indeed, CD has been used to show that histones 1, 3 and 4 exhibit much less helix content in low ionic strength buffers (Smerdon and Isenberg 1976; Feldman et al. 1980).

CD and ORD are both effective in identifying whole protein structure and disorder, especially when used in combination with intrinsic viscosity (Jirgensons 1958) or size exclusion chromatography (Uversky et al. 2002) to confirm a nonglobular, extended structure. However, neither CD nor ORD can be used to identify localized regions of disorder. The first method used to identify local regions of intrinsic disorder was X-ray crystallography.

In 1971 two regions of missing electron density in staphylococcus nuclease were identified and suggested to be due to lack 3D structure. Indeed, the regions corresponding to missing electron density were called "disordered" (Arnone et al. 1971). Such local regions of missing electron density in protein crystal structures can arise either from static disorder, in which the missing region adopts multiple fixed positions, or from dynamic disorder, in which the missing region remains mobile even in the confines of the crystal lattice (Bode et al. 1976; Ringe and Petsko 1986). Such collections of multiple, alternative conformations, whether static or dynamic, lack regular, repeated spacing and therefore fail to scatter X-rays coherently, thus leading to regions of missing electron density. One way to distinguish static from dynamic disorder is to repeat the structure determination at lower temperatures. Dynamic disorder tends to be reduced as the temperature is lowered because a single, low energy conformation tends to be favored, whereas static disorder remains basically unchanged (Ringe and Petsko 1986).

Over the years, most protein crystal structures have been observed to have regions of missing electron density. Indeed, in one analysis of the Protein Data Bank (PDB), only ~7 % of the structures determined by X-ray crystallography were observed to be complete with no regions of missing electron density and only ~25 %

of the proteins in the PDB were reported to have >95 % of their lengths observed in the corresponding crystal structures (Le Gall et al. 2007). However, regions of missing electron density in protein crystal structures are mostly short. Thus, even though the large majority of protein crystal structures in the PDB contain disorder, only ~3 % of the residues in the PDB structures are consistently missing and an additional ~2 % of the residues are observed in some structures and unobserved in other structures of the same protein.

These "ambiguous" residues, which are observed in some structures but not in others of the same protein (Le Gall et al. 2007), have been studied in more detail. Regions containing such residues were named "dual personality fragments" (Zhang et al. 2007). Protein regions with predictions intermediate between structure and disorder, called "semi-disorder" (Zhang et al. 2007; Zhang et al. 2013), are in many instances equivalent to dual personality or ambiguous regions. Basically these segments undergo order ←→ disorder transitions depending on the crystallization conditions. In some cases, a protein containing a region with ambiguous residues crystallizes into two different lattices, with the IDP region facing open space in one crystal lattice and with the same region forming structure as the result of being part of a crystal contact in an alternative lattice. Other causes of dual personality can be differences in pH or other solvent conditions between the two crystals, with the altered solvent either causing the disorder or failing to promote structure (Mohan et al. 2009). We speculate that the buffers found to be successful for protein crystallization have the tendency to stabilize or induce structure in proteins, thereby reducing the overall content of IDP residues.

About 25 % of unrelated or distantly related proteins in the PDB have IDP regions of 10 or more residues in length, while an additional 15 % have ambiguous segments in this length range. For IDPs or for ambiguous segments of 20 or more residues in length, these percentages drop to about 13 and 5 % of these PDB proteins, respectively, with further decreased percentages as the length range of the IDP regions becomes longer (Le Gall et al. 2007).

Sufficiently long regions of missing electron density can arise from structured domains that move as rigid bodies that assume multiple conformations relative to the remainder of the protein due to a flexible hinge. Alternatively, the missing electron density can arise from long regions of intrinsically disordered protein (IDP) in which the backbone assumes multiple conformations. Several examples of mobile, structured domains that lead to missing electron density have been observed (Bennett and Huber 1984), so one needs to be cautious about assigning disordered status to large regions of missing electron density. Thus, disorder in protein crystal structures is not a direct indication of a disordered backbone, but rather a disordered backbone is inferred from a region of missing electron density.

A few proteins in the PDB apparently have very long regions of disorder that comprise more than 50 % of the protein asymmetric unit in the crystal. However, a careful analysis suggests that such apparently large fractions of disorder are generally due to annotation errors in the PDB. In such examples, only a fragment of a given protein was typically isolated and crystallized, but the entire sequence is given in the PDB. This leads to a large over-estimate of the amount of missing electron density (Oldfield et al. 2013). The removed segments that are mistakenly annotated as being present sometimes turn out to be disordered, which would suggest that they had to be removed for crystallization to occur, but such a result can't be assumed without detailed study.

The largest region of confirmed disorder so far observed in a protein structure determined by X-ray crystallography is a 273-residue segment at the C-terminus of the α chain of the $\alpha_2\beta_2\gamma_2$ fibrinogen hexamer. This segment comprises 28 % of the asymmetric unit (Oldfield et al. 2013). While short segments within this region apparently exhibit preference for specific structure, the rapid hydrogen exchange of the large majority of this 273-residue segment, even when in the context of the intact $\alpha_2\beta_2\gamma_2$ hexamer, confirms that this region is indeed mostly an IDP (Marsh et al. 2013).

3 Examples of IDP-Based Mechanisms

As indicated above, a large fraction of the protein crystal structures in the PDB contain regions of missing electron density, regions that, with some exceptions, are IDPs. Furthermore, a significant number of these IDP regions have been associated with various biological functions (Dunker et al. 2002). Two fairly early examples are discussed below.

3.1 Tobacco Mosaic Virus

In 1976 the tobacco mosaic virus (TMV) coat protein crystal structure was shown to contain a 25 residue, lysine-rich loop region of missing electron density (Champness et al. 1976). In solution and in the crystal, this coat protein forms a double-disc that contains 34 chains with a mass over 500,000 Da (Champness et al. 1976; Bloomer et al. 1978). This double disc is suggested to be an intermediate in TMV assembly by means of a mechanism in which the TMV RNA molecule threads into interface between the two layers via a central hole. During RNA encapsulation, the 25 residue, lysine-rich flexible region undergoes a disorder-order transition as it binds to and largely surrounds the singled stranded RNA molecule (Stubbs et al. 1977; Namba and Stubbs 1986). It has been suggested that the flexibility of this 25-residue IDP segment is crucial for TMV assembly because without this IDP region the central hole would be too small for the RNA to make its way into its binding groove (Holmes 1983).

This work suggesting a disorder-to-order transition upon RNA binding for the TMV coat protein stimulated thinking about the possible advantages of IDPs as compared to structure for partner binding. This led to an important paper in which it was pointed out that, for disorder-to-order transitions upon binding, some of the binding energy is used for folding the disordered region rather than for increasing the affinity (Schulz 1979). Thus, compared to a comparable-sized, otherwise similar interface between two structured moieties, an IDP-based interaction would be expected to have comparable specificity but lower affinity. It was pointed out that such a combination would be very useful for biological signaling and regulation.

3.2 *Trypsinogen*

Trypsin is a protease that cleaves peptide bonds following a lysine or arginine amino acid residue. This protein is synthesized as an inactive precursor called trypsinogen, which is exported from the cell and then becomes activated via extracellular proteolysis. The bovine trypsinogen structure determined in 1976 reveals a 15-residue disordered amino terminus and a three additional regions of disorder called the activation domain (Bode et al. 1976), shown in Fig. 1.1a. These segments are associated with the correct formation of the binding pocket for lysine or arginine. Lowering the temperature reduces the structural variation of the 15-residue amino terminus, suggesting dynamic disorder for this region. On the other hand, lowering the temperature has little effect on the residues in the activation domain, suggesting static disorder for this region (Bode et al. 1976).

The amino acid sequence of bovine trypsinogen's 15 residue amino terminal IDP region is VDDDDKIVGGYTCGA. Cleavage after the lysine group at residue 6 by enteroprotease exposes a new amino terminus at the end of the IDP region. This new terminus, which starts with the IV dipeptide, is quite hydrophobic. The conformation of active trypsin (Fig. 1.1b) is formed when this IV hydrophobic terminus docks into its binding site in the activation domain by means of a tethered search over the allowed conformations of the several IDP residues that remain after the cleavage. This docking reaction is accompanied by a disorder-to-order transition for the activation domain, due to direct interaction between the IV terminus and the interaction domain (Fig. 1.1c), which enables full expression of trypsin's protease activity (Bode et al. 1976). Additional experiments (Bode and Huber 1976), including molecular dynamics simulations (Brünger et al. 1987), support this overall mechanism.

Finally, Fig. 1.1d compares disorder prediction, the specific amino acid residues that are in contact with the IV residues at the amino terminus, and the regions of missing electron density. Higher values for the predictor indicate greater likelihood for the presence of disorder. Note that the regions of missing electron density approximately align with higher values of the disorder prediction, and note also that the IV amino terminal dipeptide of trypsin interacts directly with the regions that were disordered in the trypsinogen structure. These observations agree with the model that the docking of the IV peptide induces disorder-to-order transitions via direct interaction with three distinct disordered regions. Thus, the IV peptide prob-

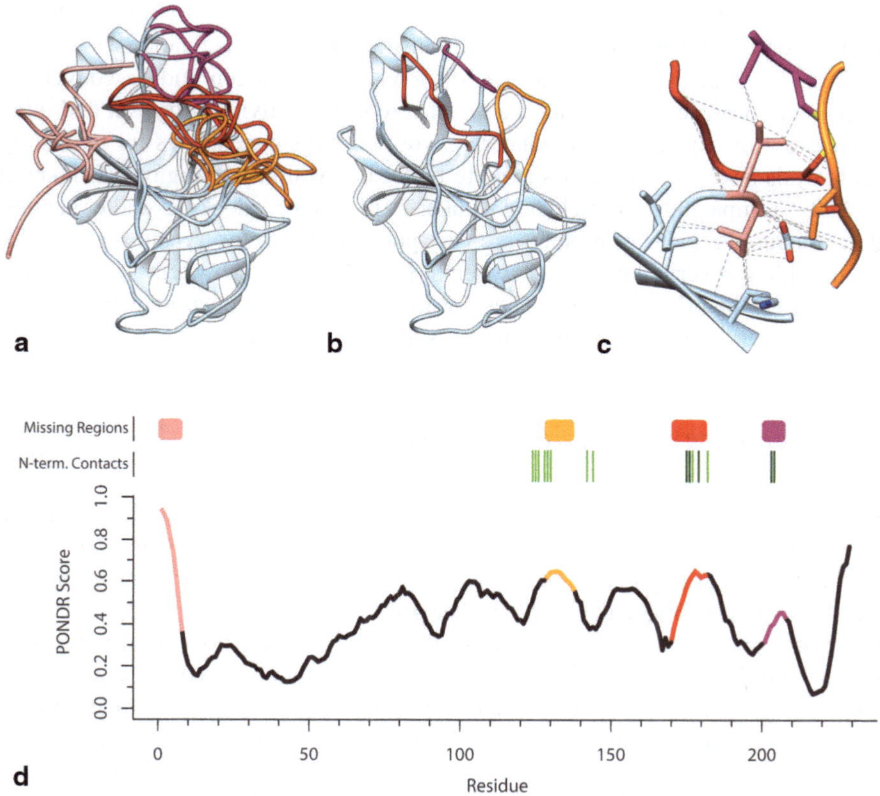

Fig. 1.1 Trypsin, trypsinogen, and intrinsic disorder. **a** Structure of trypsinogen (PDB ID 2TGT) (Walter et al. 1982) with the four regions of missing density (colored, from N- to C-terminus: pink, orange, red, and purple) modeled with 5 different random conformations (note that three of these regions are modeled in the PDB structure without supporting density). **b** Structure of trypsin (PDB ID 2PTN) (Walter et al. 1982). **c** Illustration of contacts within trypsin between the N-terminal Ile-Val and other regions of trypsin. Dotted lines indicate atomic contacts between atoms within 0.4 Å of their Van der Waals radii. **d** Comparison between missing density regions (boxes, with colors corresponding to (**a**)), contacts made by the N-terminal Ile-Val (light green and dark green ticks for contacts with Ile and Val, respectively), and PONDR VSL2B predictions of intrinsic disorder (black line with colored segments corresponding to (**a**)). Prediction values greater than 0.5 indicate residues likely to be disordered, and values less than 0.5 indicate residues likely to be structured

ably does not bind into a pre-existing cavity, but rather likely uses a coupled binding and folding mechanism to induce structure formation.

Every biochemistry text we have examined discusses the importance of synthesizing trypsin as an inactive precursor followed by its secretion and activation by proteolysis. Indeed, inappropriate activation of trypsinogen inside of cells results in a very serious disease, pancreatitis (Lerch and Gorelick 2000), which generally leads to hospitalization and can lead to death (Mayer et al. 1999). Yet none of the biochemistry texts that we have examined mention the very interesting disorder-based mechanism for keeping trypsin inactive nor do they discuss the fascinating proteolysis-dependent disorder-to-order transition leading to activation.

In addition to the role of intrinsic disorder in trypsinogen and its conversion to trypsin, intrinsic disorder plays a direct role in the activity of trypsin and other proteases. Subsequent to this pioneering work on trypsinogen, sites for regulatory proteolysis have often been observed to be located within IDP regions (Dunker et al. 2002; Tompa 2002; Oldfield and Dunker 2014). Indeed, docking studies suggest that structured regions of proteins are very poor substrates for trypsin such that local unfolding is very likely needed when trypsin digestion does occur in a region of structure (Hubbard et al. 1991). Thus, locating regulatory protease cut sites in IDP regions is very important for trypsinogen's activation mechanism.

3.3 Canonization of IDP-Based Mechanisms

By the early 1980s, many more IDP examples had been discovered (McMeekin 1952; Jirgensons 1958; Doolittle 1973; Manalan and Klee 1983; Sigler 1988) and the likely general importance of IDP or highly flexible regions for protein function had been proposed (Schulz 1979; Huber 1979; Holmes 1983). Additionally, the number of publications that characterize IDPs and their functions is currently skyrocketing (Oldfield and Dunker 2014). However, despite these early examples and current popularity, IDPs have not been emphasized in current general biochemistry and structural biology monographs, and only recently have very brief sections on IDPs been added to textbooks (Voet and Voet 2010; Tymoczko et al. 2011; Nelson and Cox 2012; Voet et al. 2012; McKee and McKee 2013; Pratt and Cornely 2013). The slow adoption of IDP-based mechanisms as part of the educational cannon is possibly due to the lack of succinct descriptions of IDP mechanisms; there is no simple analogy to the "lock-and-key" or "molecular machine" description often invoked when introducing the functions of ordered proteins. Possibly, the closest generalized description of IDP function come from analogies to protein folding, e.g. "binding-and-folding" or "conformational selection". The obvious weakness of such descriptions is that they do not leverage concepts familiar to a general audience. Furthermore, these descriptions do not encompass the functions of IDPs, completely neglecting entropic functions, e.g. neurofilament H (Brown and Hoh 1997) and Sic1 (Borg et al. 2007), for which there seems to be no suitable simplified descriptions.

We speculate that the concurrence of two lines of study is finally beginning to breakdown the resistance to the acceptance of IDP-based protein mechanisms, especially regulatory mechanisms, underlying protein function. These two lines of study are the application of nuclear magnetic resonance (NMR) to the study of protein structure and dynamics and the study of IDPs using computational or bioinformatics methods.

4 Computational Investigations of IDPs

4.1 Initial Development of Disorder Prediction

At the time we first encountered IDP examples, both in our own laboratory work (Dunker et al. 1991) and the literature (Bode et al. 1976; Bloomer et al. 1978; Holmes 1983; Kissinger et al. 1995), we were just starting to study protein structure and function by computational and bioinformatics approaches (Arnold et al. 1992). Given the challenges of predicting protein secondary and tertiary structures from their amino acid sequences, and given our awareness of IDP examples, we decided to attempt to use amino acid sequence information to predict whether proteins or regions of protein were structured or disordered.

Studies using highly simplified (e.g. lattice) models of protein folding had already suggested that highly polar amino acid sequences would simply fail to fold (Shakhnovich and Gutin 1993). Other work suggested that amino acid composition rather than the details of the sequence could be used to predict the folding class of a protein such as all α, all β, α/β and so on (Nishikawa and Ooi 1982). Furthermore, encouraged by the latter work, we developed a new method based on conditional probabilities to investigate the relationships between specific amino acid compositional features and secondary structure types (Arnold et al. 1992). An interesting finding of this work was that, compared to many indicators of helicity, the helical hydrophobic moment (Eisenberg et al. 1982) gave the strongest indication that a local region is likely to be helical.

From the aforementioned studies by us and others on the relationship between composition and structure, we asked whether structure and disorder could be distinguished by amino acid compositions over windows of 21 amino acids using our conditional probability method. The results showed clearly that structure or disorder are determined by features of the amino acid composition, with high net charge, low hydrophobicity, low aromaticity, and high proline content being among those features that favor disorder over structure (Xie et al. 1998). While the signal that associated proline with IDP regions was evident but not particularly strong in this first study, later work with much larger datasets showed proline content to be a very strongly correlated with the presence of IDP regions (Campen et al. 2008; Theillet et al. 2013).

4.2 Application of Disorder Prediction to Whole Proteomes

Next we used these promising results to develop predictions of structure versus disorder for sliding window-based algorithms that gave structure versus disorder predictions significantly better than expected by chance (Romero et al. 2001, 1997a, 1997b). These first-generation, composition-based algorithms were used to predict the disorder content of amino acid sequence databases (Romero et al. 1998) and then of whole genomes (Dunker et al. 2000). Both of these studies suggested that disorder is very common. An especially important finding, which has held up in subsequent studies by different research groups using improved disorder predictors, is that eukaryotes are much richer in IDP residues than are either bacteria or archaea (Dunker et al. 2000; Ward et al. 2004; Xue et al. 2012).

One example of such a study, which also included disorder predictions on viral proteomes (Xue et al. 2012), is given in Fig. 1.2. These data clearly show that, overall, archaea have a range of predicted amounts of disorder that highly overlaps the range observed for bacteria, while eukaryotes have a range that is clearly larger than the amounts observed for either archaea or bacteria. Another finding is that the viruses have the widest range of predicted disorder.

A few archaea have especially high amounts of predicted disorder (Fig. 1.2). These are all halophiles (Xue et al. 2010). Halophiles live in very high salt environments, and, despite the pumps that lower their internal salt concentrations, their intracellular environments still contain on the order of 2 M salt (Lanyi 1974). Compared to proteins from non-halophiles, the halophile proteins have fewer hydrophobic side chains and are enriched in negatively charged amino acids (which interact with surrounding cations to stabilize the proteins) (Fukuchi et al. 2003; Allers 2010). Reduced hydrophobic residue content and increased net charge likely causes the predictors to identify these proteins as being disordered (Oldfield and Dunker

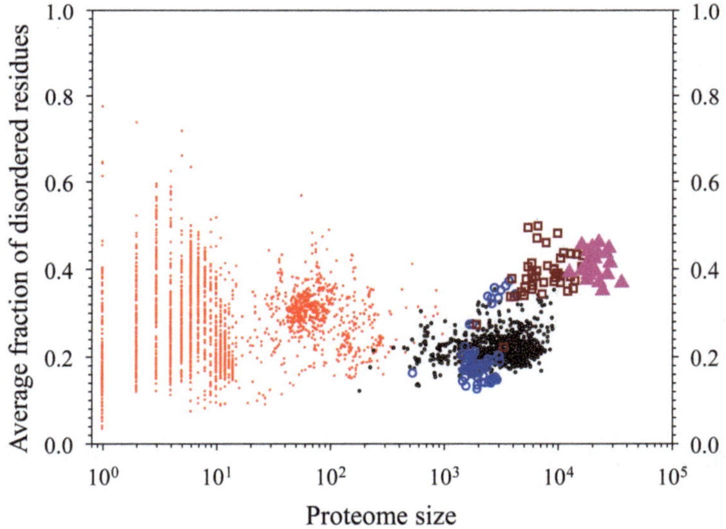

Fig. 1.2 Disorder content for 3484 species across all domains of life. The number of proteins in each species is plotted against the average fraction of predicted disordered per species. Five different groups of species are plotted with different symbols and colors: multicellular eukaryotes (*pink triangle*), unicellular eukaryotes (*red squares*), archaea (*blue circles*), bacteria (*green circles*), and viruses (*red circles*). Note that viruses expressing a single polyprotein precursor are given a proteome length of one. Reproduced from (Xue et al. 2012) with permission

2014). Indeed halophile proteins are observed to be disordered when expressed in *E. coli* or when the salt concentration is reduced to the physiological range for non-halophilic organisms (Allers 2010). In summary, the halophile proteins are structured, but they are incorrectly predicted to be rich in disorder due to their atypical amino acid compositional features that have evolved to maintain structure in their high salt environments.

4.3 Evaluation of Disorder Predictors

By now more than 50 disorder predictors have been published (He et al. 2009). Links for many of these predictors are available at www.disprot.org or at http://labs.cas.usf.edu/bioinfo/IDPgurus.html. Many of these predictors were developed in order to participate in the Critical Assessment of (Protein) Structure Prediction (CASP) experiments. For more information on CASP, visit http://predictioncenter.org/. Evaluation of the performance of disorder prediction was included as part of the CASP experiment from CASP 5 (2002) to CASP 10 (2012). These experiments provided objective evaluation of blind predictions of protein disorder. Shown in Fig. 1.3 are the five top ranking predictions as estimated by the area under the receiver operating characteristic curve (AUC) for CASP 6 through CASP 10 (Melamud and Moult 2003; Jin and Dunbrack 2005; Noivirt-Brik et al. 2009; Monastyrskyy et al. 2011; Monastyrskyy et al. 2014).The AUC data was not given for the disorder prediction evaluation for CASP 5. While these data show that disorder is

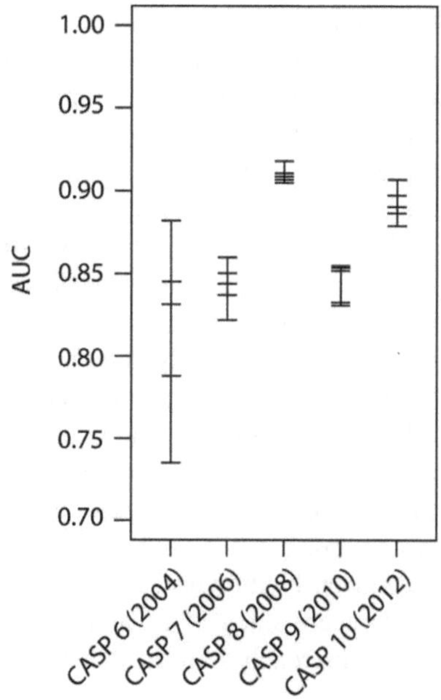

Fig. 1.3 Performance of the top five predictors of intrinsic disorder as evaluated in Critical Assessment of Structure Prediction (CASP) events (Melamud and Moult 2003; Jin and Dunbrack 2005; Noivirt-Brik et al. 2009; Monastyrskyy et al. 2011; Monastyrskyy et al. 2014). Accuracies were evaluated by the area under the curve (AUC) method, where a value of 1.0 indicates perfect prediction. Vertical lines represent the range of AUC values for the top five predctors, with horizontal lines indicate the accuracy of individual predictors

fairly well predicted (e.g. an AUC near 0.9), no clear improvement has been observed over time.

One severe weakness of this exercise has been the small number of IDP residues in each CASP experiment. The variability of the prediction accuracies likely results from differences in the degree of difficulty presented by the targets each year. For example, we know that many experimentally characterized regions of disorder often contain localized regions that are predicted to be structured, and indeed, in many cases these regions with increased structural propensities correlated with protein partner binding sites located within longer segments of disorder (Garner et al. 1999; Oldfield et al. 2005a). Changing amounts of such regions could make a set of disordered regions more or less easy to predict.

A more comprehensive, but less unbiased, evaluation of disordered predictors was recently performed. This evaluation of 19 well known disorder predictors applied these predictors to 514 chains that contained IDP regions of various lengths. In this experiment the values for the top five predictors applied to all lengths of disorder gave AUC values that ranged from 0.781 to 0.821. When the predictors were applied to disordered regions of 30 residues or longer, the AUC values for the top five predictors ranged from 0.828 to 0.869 (Peng and Kurgan 2012). These values are slightly smaller than the values observed in the CASP experiments described above, suggesting that CASP targets may not be representative of all IDPs.

4.4 Combining Disorder and Structure Prediction

An especially interesting computational development is the coupling of structure prediction with disorder prediction (Fukuchi et al. 2009; Fukuchi et al. 2011). When the prediction of structure is based on homology to all currently known protein structures, the results give fairly good coverage of each sequence in a given proteome, but often with one or more gaps. Such a gap can arise because of a domain that doesn't resemble any of the currently known protein structural domains. Alternatively, such a gap can be due to a disordered region. Disorder prediction can then indicate which of these two alternatives is more likely. If a region that is homologous to a protein of known structure and if that same region is also predicted to be structured by an order/disorder predictor, then the two completely different prediction methods support each other. On the other hand, if a region lacks resemblance to any currently known structure and if that same region is also predicted to be disordered, then again two completely different types of prediction support each other. Some regions exhibit contradictions between the two methods. For example, a gap between regions homologous to known structures can sometimes be predicted to be structured. In this case, the likely explanation is that such a region is indeed structured, but it is nonhomologous to (or is too distantly related to match) any of the currently known structures.

A recent study of 1,765 proteomes from 1,256 distinct species using this combined approach showed that the large majority of residues demonstrated mutual agreement between the two types of predictions (Oates et al. 2013). This "Database of disordered protein prediction" or, D^2P^2, is a very important computational resource that can be found at d2p2.pro.

5 Nuclear Magnetic Resonance (NMR) Determination of Protein Structure and Disorder

5.1 Early Studies of Protein Structure Determination by NMR

NMR determination of protein structure provides an important alternative to structure determination by X-ray crystallography. The single most important difference from X-ray diffraction methods is that structure determination by NMR does not require crystallization. This means that NMR-determined structures can contain significant fractions of IDP regions. Indeed, NMR structure determination can even indicate that the entire protein lacks specific structure, existing instead as an IDP. A second major difference between X-ray-based and NMR-based structure determination is that the former method does give signals from the disorder but rather disorder is inferred from the absence of signals, i.e. inferred from regions of missing electron density (Bode et al. 1976; Bloomer et al. 1978; Ringe and Petsko 1986), whereas the latter method includes relaxation protocols that give signals which distinguish local motion from overall motion, with IDP regions indicated as regions having motions faster than the global estimates (Muchmore et al. 1996; Daughdrill et al. 1997; Sprangers et al. 2000; Pawley et al. 2001; Larsson et al. 2003).

Just as we were beginning our computational studies of IDPs in the spring of 1996, we became aware of several NMR investigations of these IDPs and IDP regions. These NMR investigations gave compelling data in support of the lack of structure for these proteins or for localized regions within these structures, and, in several cases, these studies confirmed the importance of disorder for protein function. Thus, these NMR determinations of disorder were crucial for providing confidence that the development of the computational methods would be potentially worth all the effort. Here we provide a brief summary of these early NMR investigations of IDPs and IDP regions.

The second protein structure to be determined by NMR, in 1989, was the DNA binding region of the Antennapedia homeodomain transcription factor from drosophila (Qian et al. 1989). In addition to the three-helix DNA binding domain revealed by NMR analysis, this protein contains an ill-defined extension that was interpreted to be disordered as shown in Fig. 1.4a. Upon binding to DNA, one of the three helices nestles into the major groove while the ill-defined extension adopts structure (Otting et al. 1990) as it binds within the minor groove (Fig. 1.4b).

Other DNA-binding IDPs have been shown by NMR, and in some cases by X-ray diffraction as well, to bind into the minor grooves of DNA (Evans et al. 1995;

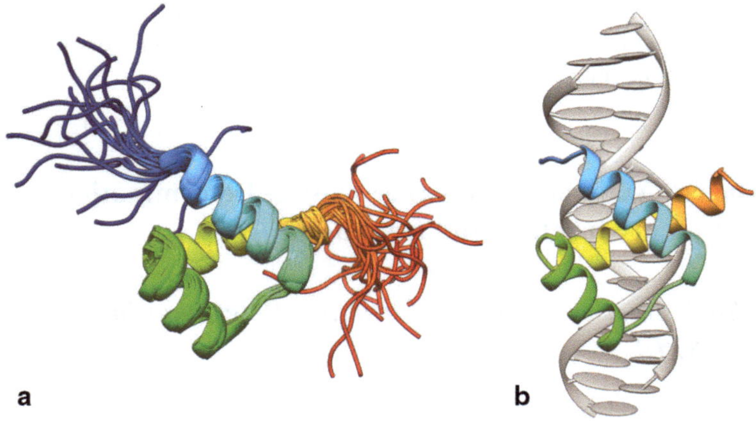

Fig. 1.4 Structures of (**a**) free and (**b**) DNA-bound antennapedia (PDB IDs 1HOM and 9ANT, respectively).The structure of free antennapedia was determined by NMR (Billeter et al. 1990) and the bound structure was determined by X-ray crystallography (Fraenkel and Pabo 1998). Several disordered residues at the N- and C- termini show multiple conformations in the NMR structure (**a**) and are absent in the bound structure (**b**)

Singh et al. 2006; Fonfría-Subirós et al. 2012). What is biologically interesting for these examples, including the Antennapedia homeodomain, is that the minor-groove binding by IDP regions is accompanied by direct interactions between their amino acid side chains and the DNA bases. Thus, such IDP-based interactions often play crucial roles in the recognition of sequence-specific binding sites by DNA binding proteins (Otting et al. 1990; Evans et al. 1995; Fonfría-Subirós et al. 2012).

A very interesting collection of proteins regulate the cell cycle by binding to and inhibiting several cyclin dependent kinases (CDKs) complexed with their activating cyclins, including $p21^{Waf1/Cip1/Sdi}$, $p27^{Kip1}$, and $p57^{Kip2}$ (Lee et al. 1995), herein called p21, p27, and p57, respectively. By binding to specific CDKs and their activating cyclins, these proteins bring about cell cycle arrest. By reducing the production of these inhibitors and by concomitantly removing the existing molecules by protease digestion, the arrest of the cell-cycle is overturned and cell division continues. The protease digestion step depends on a multistep process, which evidently provides a means to integrate different signals (Lee et al. 1995; Kriwacki et al. 1996; Galea et al. 2008a; Dunker and Uversky 2008; Galea et al. 2008b).

In 1996, NMR spectroscopy was used to determine the structural basis of cell-cycle regulation by one of these proteins, p21, and its binding to the CDK2 (Kriwacki et al. 1996). By labeling p21 with ^{15}N, its resonances were distinguishable from those of the CDK2. As shown in Fig. 1.5a, in the absence of CDK2, the ^{1}H-^{15}N heteronucear single quantum correlation (HSQC) NMR spectrum of p21 is poorly dispersed and exhibits overlapping peaks. This spectrum exhibits only subtle changes in resonance dispersion in 6 M urea, showing that p21 is indeed largely unstructured. When p21 is mixed with CDK2, the NMR spectrum becomes significantly dispersed (Fig. 1.5b) suggesting that p21 undergoes a disorder-to-order

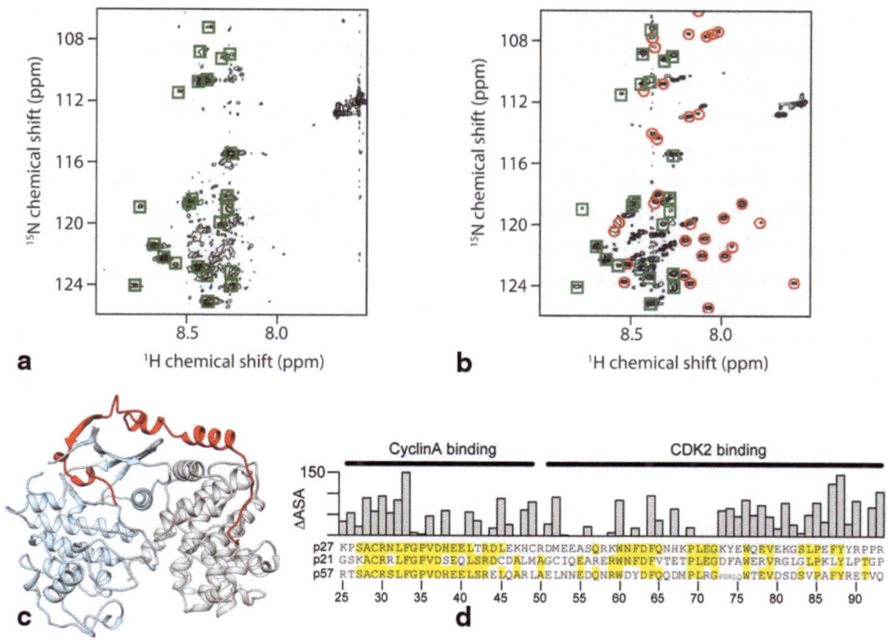

Fig. 1.5 Folding-and-binding in p21/p27 recognition of CDK-cyclin. The ^{1}H-^{15}N HSQC spectra of (**a**) free and (**b**) CDK2-bound p21 is shown with common (*green boxes*) and unique (*red circles*) resonances highlighted. **c** The structure (PDB ID 1JSU) of p27 (*red*) bound to CDK2 (*blue*) and cyclinA (*grey*). **d** Buried surface area (ΔASA) of p27 in the p27-CDK2/cyclinA complex and sequence conservation among p27, p21, and p57, where identical residues are highlighted yellow. **a** and **b** reproduced from (Kriwacki et al. 1996); **c** and **d** modified from (Russo et al. 1996)

transition upon binding to CDK2. Also reported in 1996, the X-ray crystal structure of p27 bound to CDK2-Cyclin A shows how this rather long IDP binds to both the kinase and its associated cyclin by wrapping around the heterodimeric complex (Fig. 1.5c). Finally, the sequences of p21, p27, and p57 for the binding region are aligned and the buried surface estimated for p27 when it binds to the CDK2-Cyclin A complex is also shown (Fig. 1.5d), showing that in most cases fairly high sequence similarity occurs for the residues involved in the binding interface between p27 and the CDK2-Cyclin A complex.

The proteolytic digestion of p27 bound to CDK2-cyclinA likely involves localized "breathing" of the p27 to thereby make Y88 accessible for phosphorylation. Given the extended nature of the CDK2-cyclin A-p27 complex (Fig. 1.5c), it seems reasonable to suppose that some localized regions will bind more weakly than others, thus leading to preferential "breathing" of certain sites along the binding interface. Once exposed, the Y88 moiety becomes phosphorylated by a non-receptor tyrosine kinase. Upon phosphorylation of Y88, the CDK2 kinase active site becomes constitutively exposed, thus speeding up the unimolecular phosphosphorylation of T187. This second phosphorylation in turn signals ubiquitination, following by digestion of p27 by the proteasome. This multistep-step "signaling conduit"provides a means to integrate multiple signals so that cell cycle progression proceeds only when each of these steps have been completed (Galea et al. 2008a; Dunker and Uversky 2008; Galea et al. 2008b; Follis et al. 2012).

Also reported in 1996 was the study of Bcl-x_L by both NMR and X-ray diffraction (Muchmore et al. 1996). Bcl-x_L includes disordered regions at the N- and C- termini as well as a large disordered loop (Fig. 1.6a). One interesting feature of the Muchmore et al. study is that this loop maps to missing electron density in the X-ray determined-structure and also has sparse data in the NMR experiment, and so was represented as an extended loop (Fig. 1.6b). Such regions of sparse data can arise from an IDP region or they can arise from structured regions that have structures that simply fail to give NMR signals. A second interesting feature of this study was that NMR relaxation experiments were carried out to distinguish between these two possibilities. For these relaxation experiments, Bcl-x_L was labeled with ^{15}N and

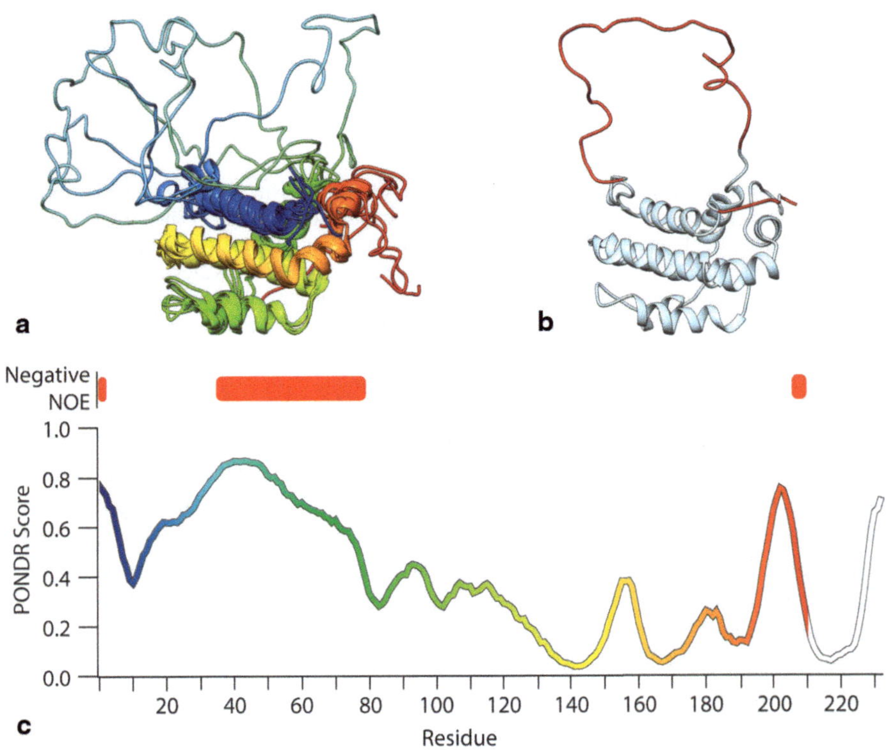

Fig. 1.6 Characterization of intrinsic disorder in BCL-xL by measurement of the heteronuclear NOE. **a** Ribbon rendering of multiple conformations of BCL-xL as determined by NMR (PDB ID 2ME9) (Follis et al. 2014), with ribbons colored with a gradient from blue at the N-terminus to red at the C-terminus. **b** Heteronuclear NOE measurements (Muchmore et al. 1996) mapped to the structure of BCL-xL (PDB ID 1LXL), where residues with negative (*red*) and positive (*blue*) heteronuclear NOE values are highlighted. **c** Comparison of intrinsic disordered prediction (gradient colored line, with colors corresponding to (**a**)) with regions that have negative heteronuclear NOE values (*red boxes*)

the values of nuclear Overhauser effect (NOE) between the peptide nitrogen and the bound proton were determined. The negative ^{15}N (^{1}H) NOE values correspond closely to the regions that have sparse NMR signals (Fig. 1.6b), which in turn closely match the regions of missing electron density. As discussed below, the ^{15}N (^{1}H) NOE experiments suggest that the loop is moving faster than the remainder of the protein and is therefore more likely to be disordered rather than to be structured and to be giving sparse data. Additionally, these regions correspond closely to regions predicted to be intrinsically disordered (Fig. 1.6c).

Several measurable and derived parameters reflect the dynamics of each residue in a protein's backbone at various time scales (Chaps. 3 and 5). Longitudinal and transverse ^{15}N relaxation rates and ^{15}N (^{1}H) heteronuclear NOE values (Muchmore et al. 1996), provide information on the fluctuations of the peptide nitrogen-hydrogen bond (Chap. 3).

These relaxation parameters are often interpreted through the model-free formalism (Lipari and Szabo 1982), so called because it does not depend on a specific model of internal molecular motion, but rather uses data-fitting to estimate the magnitude (S^2) and time scale (τ_e) that characterizes the internal motion. A complication in application of the model-free formalism to IDPs is that this approach assumes that internal and overall motion are independent, i.e. that the protein does not change shape (Peng 2012). Generalizations of the model-free formalism and alternative formalisms that overcome this limitation have been developed (reviewed in Peng 2012).

At least for an initial analysis, measurement of the ^{15}N (^{1}H) NOE values can provide an excellent characterization of residues as intrinsically ordered or intrinsically disordered (Chap. 3). Theoretically possible values of the heteronuclear NOE range from +0.82 to −3.6 (at 11.7 T) for residues with a well-ordered and mobile backbone, respectively (Kay et al. 1989). This method is very sensitive to intrinsically disordered regions because dynamics of the peptide bond are extremely different in ordered regions, in which motion of the peptide bond is tightly coupled to motion of the entire protein, and disordered regions, in which motion of the peptide bond is independent of and much faster than the entire protein. In fact, it is common to consider residues with only slightly depressed heteronuclear NOE values, in the range of 0.6 to 0.7, as being highly mobile (Sprangers et al. 2000; Pawley et al. 2001; Larsson et al. 2003). A limitation of the ^{15}N (^{1}H) NOE values is that quantitative analysis, e.g. relative dynamics, is difficult due to confounding factors (Kay et al. 1989), but can be accomplished with careful analysis and supporting data (Lacy et al. 2004).

Another study that strongly motivated our computational studies on IDPs was the 1997 NMR–based report on the interaction between the FlgM and σ^{28} proteins from E.coli (Daughdrill et al. 1997), a study that stimulated a commentary on the importance of being unfolded (Plaxco and Gross 1997). FlgM and σ^{28} regulate the morphogenesis of the flagellar organelle, which contains a transmembrane basal body motor, a flexible hook, and the flagellum. Each flagellum contains ~20,000 copies of a protein called flagellin and accounts for about 8 % of the total cell protein. The σ^{28} transcription factor promotes the transcription of the flagellin gene, whereas FlgM binds to σ^{28} and inhibits its activity. The basal body contains a central pore that forms the basis of the type III export apparatus (Chilcott and Hughes 1998; Minamino and MacNab 2000; Calvo and Kearns 2015). Once the assembly of the basal body is completed, FlgM leaks out through this central pore. Loss of FlgM leads to activation of σ^{28} and thus to the synthesis of flagellin. Using the NMR methods described in both the p21 and Bcl-x_L papers (Muchmore et al. 1996; Kriwacki et al. 1996), FlgM was shown to be an IDP, but to gain structure in its C-terminal half upon binding to σ^{28} (Daughdrill et al. 1997). FlgM's IDP status almost certainly helps to increase its rate of transport through the somewhat narrow central pore of the basal body. On the other hand, if the basal body is improperly assembled either from mutation or environmental stress, then inhibition of σ^{28} by FlgM prevents the wasteful synthesis of flagellin.

These NMR studies on p21, Bcl-xL, and FlgM all appeared during 1996 and 1997, just when we were constructing our first algorithms to predict IDPs and IDP regions from their amino acid sequence. Consider that a region of missing electron density in the X-ray-determined structure of Bcl-x_L matched a region of high variability in the NMR-determined-structure of this same protein, and furthermore that this same region was shown by its low and negative ^{15}N (^1H) NOE values to be faster moving than the structured parts of the same molecule. Thus, Bcl-x_L was shown to contain an IDP region that was indicated to lack structure by three independent experimental methods. This strongly reinforced our confidence that these IDP regions were likely to be real and not experimental artifacts. Also, all three of these proteins exhibited extremely interesting and very important biological functions. Thus, these examples plus the ones we already knew about suggested that these proteins carry out such important functions that understanding their (lack of) structure-function relationships might lead to a change in the way we think about proteins.

5.2 Using NMR to Identify IDP Regions

As published previously and discussed above, a collapsed HSQC NMR spectrum indicates a protein that is very likely an IDP throughout (Chap. 3). On the other hand, dispersed spectra indicate a structured protein, and if various conditions are right, the protein structure can be determined by a collection of NMR measurements coupled with molecular modeling. NMR-determined protein structures are generally represented as ensembles of structures. Conformational variability between structures within an ensemble indicates uncertainty in the protein structure, where well-ordered regions will show little variability and flexible regions may show high variability. In the current PDB, there are 9501 protein structures determined by NMR methods. The number of ensemble members range from 1 to 640, with 20 members being by far the most commonly reported number and with very few having more than 30, including just 39 structures with more than 60 ensemble members.

A likely IDP region can be identified in such an ensemble by considering the pairwise structural deviations along the chain, with IDP regions giving much larger deviations as compared to the structured regions. While such high deviation regions typically arise from IDP regions, these high deviation regions can also arise from a region of structure that fails to give sufficient data. As mentioned above, these two alternatives can be distinguished by ^{15}N (^{1}H) NOE values, which are estimated as described above. Also as described above, for values from about 0.7–0.6 down to −3.6 are generally taken to be disordered (Sprangers et al. 2000; Pawley et al. 2001; Larsson et al. 2003).

Systematically determination of regions of NMR structures that show high deviations is useful as a way of identifying potential IDPs regions. Two different methods for systematically identifying regions of high deviation have been reported (Di Domenico et al. 2012; Ota et al. 2013; Potenza et al. 2015). In one of the methods, IDP regions are described as being mobile rather than disordered.

The method that identifies IDP regions as being mobile combines two measures of disorder, namely structural superposition and changes in torsion angles (Di Domenico et al. 2012; Potenza et al. 2015). The superposition measure is based on the TM-score (Zhang and Skolnick 2005), which was developed to compare two different structures formed by the same amino acid sequence, such as comparing a predicted 3D structure of a protein with its actual structure. The torsion angles are compared by their averages and standard deviations over the various models in the ensemble. These two measures then lead to the identification of each residue as being mobile or not mobile. An alternative comparison is to use the DSSP (Define Secondary Structure of Proteins) method (Kabsch and Sander 1983) to assign secondary structure. In this case residues are identified as mobile if the secondary structure changes from model to model. The mobile versus not mobile assignment from this criterion overrides that from combining superposition and changes in torsion angles. Finally, one or two not mobile amino acids flanked by mobile amino acids are reassigned as mobile and one or two mobile amino acids flanked by not mobile amino acids are reassigned as not mobile. These NMR assignments of mobility (or disorder) are available in a database called MobiDB, http://mobidb.bio.unipd.it/.

A much simpler method for identifying IDP regions from NMR-determined structures was developed in a study aimed at comparing regions of missing electron density in X-ray-determined structures with regions showing high deviation in NMR ensembles of the same proteins or in NMR ensembles of closely related proteins (Ota et al. 2013). The root-mean-square deviation (RMSD) was used as a measure of variation between the equivalent Cα atoms in an NMR ensemble. For a given threshold, residues above the threshold were considered to be disordered, and those below the threshold, structured. These threshold-based structure-disorder assignments were compared with the X-ray assignments using the Matthews Correlation Coefficient (MCC) (Matthews 1975), which measures the quality of the agreement between two binary classifications. This comparison was repeated for different RMSD threshold values in order to find the threshold that gave the best agreement between the two assignments, i.e. the maximum value for the MCC. In this study, an RMSD value of 3.2 Å was found to give the best correlation between disorder determined from divergence in NMR ensembles and disorder determined by regions of missing electron density. One complication is that, in many cases, the structure of the same protein has been determined multiple times by X-ray diffraction and, as described above, often with differences in the regions observed to be disordered. A reasonable approach here is to simply use the structure showing the largest region of missing electron density.

Another feature of this study was that the results of structure and disorder determined by the 3.2 Å value for the RMSD were compared with structure and disorder determined by ^{15}N (^{1}H) NOE values (Ota et al. 2013). Of the 55 proteins for which both NMR–determined and X-ray–determined structures were found, only 24 of these could be associated with relaxation studies using ^{15}N (^{1}H) NOE measurements. For regions identified as structured (less than 3.2 Å RMSD), a total of 1773 ^{15}N (^{1}H) NOE signals were found. Of these, just 130 (~7 %) exhibited ^{15}N (^{1}H) NOE peak values smaller than 0.5, indicating that the RMSD threshold method and the relaxation method give >90 % coincidence in their identification of structured regions. The identification of disorder by the two methods is not quite so similar. For this set of 24 proteins, about 400 residues were identified as being in IDP regions by virtue of RMSD values greater than 3.2 Å. Of these residues, 320 (80 %) displayed ^{15}N (^{1}H) NOE peak values less than 0.5, giving 80 such residues having ^{15}N (^{1}H) NOE peak values greater than this threshold. These 80 residues were mostly in loops in the protein structure.

In addition to the interpretation of NMR ensembles, bioinformatics methods can be applied directly to chemical shift and ^{15}N (^{1}H) NOE data. The latter approach has the advantages that NMR parameters directly characterize mobility and flexibility—as opposed to indirectly through ensemble RMSD—and that no structured region is required—as is required for calculation of RMSD—allowing for inclusion of highly disordered proteins. Chemical shifts are a very rich source of structural and dynamic information (Chap. 3) and readily indicate regions of the polypeptide chains that are characterized by partially populated secondary structural elements (helices, sheets, etc). Many proteins which are highly disordered and flexible have been characterized through NMR and their chemical shifts and ^{15}N (^{1}H) NOE have been determined and deposited in the Biological Magnetic Resonance Bank (BMRB) (Markley et al. 2008). Therefore chemical shifts of IDPs offer a rich dataset to be further explored and used to develop prediction methods. One such method, DynaMine (Cilia et al. 2013), is a linear model that predicts chemical shift-based order parameters directly from amino acid sequence and has a disorder prediction performance on par with more complex predictors trained from order-disorder datasets.

6 Future Directions

Above we pointed out the usefulness of characterizing the same IDP or IDP region by multiple methods. As we look to the future, we suggest that it will be important to expand the number of IDPs and IDP regions characterized by multiple methods. A second major thrust should be to improve current methods and to develop new methods for characterizing IDPs and IDP regions by NMR. A third area is application and improvement of in-cell NMR techniques to support and extend in vitro studies to in vivo and in situ environments.

6.1 Accumulating IDPs and IDP Regions Characterized by Multiple Methods

With regard to experimental determination of IDPs and IDP regions by multiple methods, an obvious need is for the characterization of many more proteins by X-ray–determined structures, by NMR–determined structures, and by NMR relaxation methods such as was done for Bcl-x_L. In our attempts to compare these three types of results (Ota et al. 2013), we could find ^{15}N (^{1}H) NOE relaxation data for just 24 of the 55 proteins that had both X-ray– and NMR–determined structures and that met our other criteria. Furthermore, just 1 of the 24 instances of relaxation data was found to be deposited in the BMRB (Markley et al. 2008); the other 23 instances were found by manual literature searches. To give another comparison, as indicated above, the current PDB contains 9501 structures determined by NMR methods, while the BMRB currently contains just 212 proteins with ^{15}N (^{1}H) NOE relaxation data. Thus, in the future we would like to see a push for more proteins to be characterized by all three of these approaches, and for a much larger fraction of the NMR relaxation data to be deposited into a single location, the BMRB. This would facilitate systematic comparisons of the three types of data for the identification of IDP regions. This in turn would lead to a larger collection of proteins with well characterized IDP regions. Additionally, the information provided by ^{15}N (^{1}H) NOEs, chemical shifts, and other NMR observables are useful for identification of IDP regions even if no structure has been deposited in the PDB.

It would be helpful to expand the number of IDPs and IDP regions that are characterized by methods in addition to those based on X-ray and NMR. To give one example, the protein calcineurin contains a long IDP region of about 150 residues that features a calmodulin binding site and an autoinhibitory domain. This region has been characterized as disordered by regions of missing electron density located on both sides of the bound autoinhibitory domain as observed in the X-ray–determined structure (Kissinger et al. 1995). In addition, this same region has been indicated to be disordered by protease digestion, for which multiple sites are cleaved more or less simultaneously (Manalan and Klee 1983), and also by H/D exchange with proteolysis coupled with mass spectrometry being used to identify the specific regions that are undergoing rapid exchange (Rumi-Masante et al. 2012). To give another example, a segment within the axin scaffold protein was shown to be disordered because this segment exhibited a collapsed HSQC NMR spectrum that showed little change upon the addition of 4 M urea, because this segment failed to show evidence of unfolding upon heating, and because this segment migrated in size exclusion chromatography as a protein with a much larger mass. Despite its disordered status as indicated by these multiple methods, this axin segment exhibited its biological function of accelerating phosphorylation when added to a mixture containing a kinase and its substrate, both of which bind to this same segment (Noutsou et al. 2011; Xue et al. 2013).

Perhaps the community needs to form an "unstructural genomics initiative." In the structural genomics initiative (SGI) (Burley 2000), many IDPs and IDP regions were found (Oldfield et al. 2005b; Johnson et al. 2012; Oldfield et al. 2013). Thus, one approach would be to evaluate proteins originally studied in the SGI by multiple methods to reveal IDPs and IDP regions. The goal of the original SGI was to find representatives of the various types of protein structures found in nature (e.g. to find different types of protein folds), which was to be used in parallel with studies of their functions thereby to expand our understanding of the sequence → structure → function paradigm. In a similar fashion, the goal of the unstructured genomics initiative would be to expand our understanding of the newly emerging sequence → IDP ensemble → function paradigm.

6.2 Developing Improved Methods for the Study of IDPs by NMR

In some respects IDPs are ideal candidates for NMR relaxation experiments, particularly due to sharp peaks due to their relaxation properties. However, IDPs can present some substantial challenges. High solvent exposure leads to similar local environments for all residues, leading to collapsed spectra with overlapping peaks. Additionally, the generally lower complexity of IDPs, relative to ordered proteins, leads to additional environmental symmetry between residues and further collapse of spectra (Felli and Pierattelli 2014; Nováček et al. 2014).

Several approaches have been developed to overcome the high signal overlap of many IDPs (Nováček et al. 2014). One of these approaches is sampling of spectra at a higher rate to resolve spectra overlap. However, optimal sampling is generally impractical due to long experiment times that would be required (Felli and Pierattelli 2014; Nováček et al. 2014), and instead non-uniform sampling methods are employed to approximate the optimal sampling schedule (Mobli et al. 2012). Lower complexity sequences are typically problematic because only neighboring residues are encoded in spectra, which leads to collapse due to degenerate sequence positions. One method to combat this collapse is to encode longer range sequence into the spectra (Nováček et al. 2014), where the resulting high dimensional spectra are approximated with non-uniform sampling methods. Another method to combat collapsed spectra is the use of ^{13}C detection instead of ^{1}H detection (Bermel et al. 2006). Detection of ^{13}C can separate resonances by a factor of 3 compared to detection of ^{1}H. Additionally, ^{13}C is not sensitive to exchange broadening or as sensitive to salt concentration as ^{1}H, which can further improve spectra taken near physiological conditions (Gil et al. 2013; Felli and Pierattelli 2014; Nováček et al. 2014; Chap. 3).

6.3 In-Cell NMR

By over-expressing one specific protein in *E. coli* in the presence of ^{15}N or ^{13}C enriched compounds, it becomes possible to carry out NMR studies of proteins inside of cells (Serber et al. 2001; Serber and Dötsch 2001). Subsequently, other methods such as injection of frog oocytes or adding penetration signals such as polyR have been used to introduce labeled proteins for in-cell NMR studies (Ito and Selenko 2010). This body of work appeared after we had already published our first predictors of IDPs and IDP regions (Romero et al. 1997a, 1997b, 2001). Furthermore, a much earlier NMR experiment showing the disorder status of the protein chromogranin A in slices of rat adrenal medulla (Daniels et al. 1976), which was a harbinger of the ability to observe IDPs inside cells by NMR, was simply unknown by us. Thus, these in-cell NMR experiments had no influence regarding our decision to develop predictors of IDPs and IDP regions. However, these and many subsequent in-cell NMR experiments have had substantial effect on the entire IDP field.

One suggestion has been that IDPs don't really exist in the crowded environment of the cell. Instead, molecular crowding has been suggested to cause proteins to collapse and fold. To test this suggestion, acid-unfolded cytochrome *c* and the prototypical IDP, α synuclein, were studied by both NMR and CD with and without the addition of 1 M glucose as a small-molecule crowding agent. Glucose caused acid-unfolded cytochrome *c* to collapse to the size and structure closely approximating its native state. On the other hand, α-synuclein did collapse but failed to gain significant secondary structure (Morar et al. 2001), suggesting to us that it remains disordered.

Several in-cell NMR experiments confirm that α-synuclein remains disordered even when in such highly crowded conditions as the inside of cells (McNulty et al. 2006; Barnes and Pielak 2011; Fauvet et al. 2012; Binolfi et al. 2012; Smith et al. 2015). It is worth noting that a soluble construct of α-synuclein forms a dynamic tetrameric helical bundle (Wang et al. 2011b). This finding has caused some discussion regarding the IDP status of this protein (reviewed in (Alderson and Markley 2013)). Our view is that such a result is common and does not negate its IDP status because many IDPs self-associate or form complexes with other proteins. Furthermore, the equilibrium between the IDP monomeric state and the structured tetramer might be related to its biological function.

For typical IDPs, crosspeaks in the ^{1}H-^{15}N HSQC spectra are very similar whether obtained in vitro or via in-cell experiments (Dedmon et al. 2002; McNulty et al.

2006). Amazingly, recent H/D exchange experiments for FlgM and α-synuclein yielded nearly identical results for in vitro and in-cell experiments, indicating that true disorder can persist inside *E. coli* despite the crowded environment and further indicating that the interactions inside the cell don't appreciably affect the exchange rates (Smith et al. 2015). In contrast, in-cell NMR experiments on structured proteins show that the crowded environment can have very significant effects on the resulting spectra (Li et al. 2008). These large effects arise from hindered rotational motions that result from interactions with the other macromolecules in the crowded intra-cellular environment, which for *E. coli* reach levels near 300 to 400 g/L (Zimmerman and Trach 1991; Ellis 2001). In some cases the hindered motions result from hydrophobic interactions (Wang et al. 2011a), and in other cases they result from electrostatic interactions, either with other proteins (Crowley et al. 2011) or with RNA (Kyne et al. 2015).

An especially interesting discovery is that crowding can lead to protein destabilization or even unfolding (Miklos et al. 2011; Harada et al. 2013). Evidently, at high levels of crowding, the sum of the energies of even weak interactions can overcome the intramolecular folding energies and thereby destabilize a structured protein. Using in-cell NMR coupled with H/D exchange, the degree of instability can be estimated (Monteith and Pielak 2014). Using this approach, a surface mutation in a structured protein was found to be 10-fold more destabilizing in cells than in vitro (Monteith et al. 2015), a truly remarkable result.

The unfolding of structured proteins by crowding raises several very fundamental questions for the IDP field. Should such proteins be considered IDPs? Do cells actively modulate such destabilization by altering the relative amounts of stabilizing and destabilizing macromolecules in the intra-cellular milieu? Are such coupled crowding and unfolding events involved in particular biological functions? If so, which functions? We look forward to future NMR and related experiments aimed at answering these and other questions that come out of these recent findings.

Acknowledgements Gary Pielak, Roberta Pierattelli, and Isabella Felli are thanked for their helpful suggestions for improving the NMR sections of this paper. Ya-Yue Van and Molecular Kinetics are thanked for funding.

References

Alderson TR, Markley JL (2013) Biophysical characterization of α-synuclein and its controversial structure. Intrinsically Disord Proteins 1:18–39. doi:10.4161/idp.26255

Allers T (2010) Overexpression and purification of halophilic proteins in Haloferax volcanii. Bioeng Bugs 1:288–290. doi:10.4161/bbug.1.4.11794

Anson ML, Mirsky AE (1931a) Protein coagulation and its reversal: serum albumin. J Gen Physiol 14:725–732

Anson ML, Mirsky AE (1931b) Protein coagulation and its reversal: globin. J Gen Physiol 14:605–609

Arnold GE, Dunker AK, Johns SJ et al (1992) Use of conditional probabilities for determining relationships between amino acid sequence and protein secondary structure. Proteins 12:382–399. doi:10.1002/prot.340120410

Arnone A, Bier CJ, Cotton FA et al (1971) A high resolution structure of an inhibitor complex of the extracellular nuclease of *Staphylococcus aureus*. I. Experimental procedures and chain tracing. J Biol Chem 246:2302–2316
Barnes CO, Pielak GJ (2011) In-cell protein NMR and protein leakage. Proteins: Struct Funct Bioinform 79:347–351. doi:10.1002/prot.22906
Bennett WS, Huber R (1984) Structural and functional aspects of domain motions in proteins. CRC Crit Rev Biochem 15:291–384
Bermel W, Bertini I, Felli IC et al (2006) ^{13}C-detected protonless NMR spectroscopy of proteins in solution. Prog Nucl Magn Reson Spectros 48:25–45. doi:10.1016/j.pnmrs.2005.09.002
Billeter M, Qian Y, Otting G et al (1990) Determination of the three-dimensional structure of the Antennapedia homeodomain from Drosophila in solution by ^{1}H nuclear magnetic resonance spectroscopy. J Mol Biol 214:183–197
Binolfi A, Theillet F-X, Selenko P (2012) Bacterial in-cell NMR of human α-synuclein: a disordered monomer by nature? Biochem Soc Trans 40:950–954. doi:10.1042/BST20120096
Bloomer AC, Champness JN, Bricogne G et al (1978) Protein disk of tobacco mosaic virus at 2.8 Å resolution showing the interactions within and between subunits. Nature 276:362–368
Bode W, Huber R (1976) Induction of the bovine trypsinogen-trypsin transition by peptides sequentially similar to the N-terminus of trypsin. FEBS Lett 68:231–236
Bode W, Fehlhammer H, Huber R (1976) Crystal structure of bovine trypsinogen at 1–8 Å resolution. I. Data collection, application of patterson search techniques and preliminary structural interpretation. J Mol Biol 106:325–335
Borg M, Mittag T, Pawson T et al (2007) Polyelectrostatic interactions of disordered ligands suggest a physical basis for ultrasensitivity. Proc Natl Acad Sci U S A 104:9650–9655. doi:10.1073/pnas.0702580104
Brown HG, Hoh JH (1997) Entropic exclusion by neurofilament sidearms: A mechanism for maintaining interfilament spacing. Biochemistry 36:15035–15040. doi:10.1021/bi9721748
Brünger AT, Huber R, Karplus M (1987) Trypsinogen-trypsin transition: a molecular dynamics study of induced conformational change in the activation domain. Biochemistry 26:5153–5162
Burgen AS, Roberts GC, Feeney J (1975) Binding of flexible ligands to macromolecules. Nature 253:753–755
Burley SK (2000) An overview of structural genomics. Nat Struct Biol 7(Suppl):932–934. doi:10.1038/80697
Calvo RA, Kearns DB (2015) FlgM is secreted by the flagellar *export apparatus in Bacillus subtilis*. J Bacteriol 197:81–91. doi:10.1128/JB.02324-14
Campen A, Williams RM, Brown CJ et al (2008) TOP-IDP-scale: a new amino acid scale measuring propensity for intrinsic disorder. Protein Pept Lett 15:956–963
Champness JN, Bloomer AC, Bricogne G et al (1976) The structure of the protein disk of tobacco mosaic virus to 5Å resolution. Nature 259:20–24
Chilcott GS, Hughes KT (1998) The type III secretion determinants of the flagellar anti-transcription factor, FlgM, extend from the amino-terminus into the anti-sigma28 domain. Mol Microbiol 30:1029–1040
Cilia E, Pancsa R, Tompa P et al (2013) From protein sequence to dynamics and disorder with DynaMine. Nat Commun. doi:10.1038/ncomms3741
Corey RB, Pauling L (1953) Fundamental dimensions of polypeptide chains. Proc R Soc Lond B Biol Sci 141:10–20
Crowley PB, Chow E, Papkovskaia T (2011) Protein interactions in the Escherichia coli cytosol: an impediment to in-cell NMR spectroscopy. ChemBioChem 12:1043–1048. doi:10.1002/cbic.201100063
Daniels A, Williams RJ, Wright PE (1976) Nuclear magnetic resonance studies of the adrenal gland and some other organs. Nature 261:321–323
Daughdrill GW, Chadsey MS, Karlinsey JE et al (1997) The C-terminal half of the anti-sigma factor, FlgM, becomes structured when bound to its target, σ28. Nat Struct Mol Biol 4:285–291. doi:10.1038/nsb0497-285
Dedmon MM, Patel CN, Young GB et al (2002) FlgM gains structure in living cells. Proc Natl Acad Sci U S A 99:12681–12684. doi:10.1073/pnas.202331299

Di Domenico T, Walsh I, Martin AJM et al (2012) MobiDB: a comprehensive database of intrinsic protein disorder annotations. Bioinformatics 28:2080–2081. doi:10.1093/bioinformatics/bts327

Doolittle RF (1973) Structural aspects of the fibrinogen to fibrin conversion. Adv Protein Chem 27:1–109

Dunker AK, Uversky VN (2008) Signal transduction via unstructured protein conduits. Nat Chem Biol 4:229–230. doi:10.1038/nchembio0408-229

Dunker AK, Ensign LD, Arnold GE et al (1991) Proposed molten globule intermediates in fd phage penetration and assembly. FEBS Lett 292:275–278

Dunker AK, Obradovic Z, Romero P et al (2000) Intrinsic protein disorder in complete genomes. Genome Inform Ser Workshop Genome Inform 11:161–171

Dunker AK, Brown CJ, Lawson JD et al (2002) Intrinsic disorder and protein function. Biochemistry 41:6573–6582

Edison AS (2001) Linus Pauling and the planar peptide bond. Nat Struct Biol 8:201–202. doi:10.1038/84921

Eisenberg D, Weiss RM, Terwilliger TC (1982) The helical hydrophobic moment: a measure of the amphiphilicity of a helix. Nature 299:371–374

Ellis RJ (2001) Macromolecular crowding: an important but neglected aspect of the intracellular environment. Curr Opin Struct Biol 11:114–119

Espinoza-Fonseca LM (2009) Reconciling binding mechanisms of intrinsically disordered proteins. Biochem Biophys Res Commun 382:479–482. doi:10.1016/j.bbrc.2009.02.151

Evans JN, Zajicek J, Nissen MS et al (1995) ^{1}H and ^{13}C NMR assignments and molecular modelling of a minor groove DNA-binding peptide from the HMG-I protein. Int J Pept Protein Res 45:554–560

Fauvet B, Fares M-B, Samuel F et al (2012) Characterization of semisynthetic and naturally Nα-acetylated α-synuclein in vitro and in intact cells: implications for aggregation and cellular properties of α-synuclein. J Biol Chem 287:28243–28262. doi:10.1074/jbc.M112.383711

Feldman L, Beaudette NV, Stollar BD et al (1980) Conformational changes in the H3. H4 histone complex. Serological and circular dichroism studies. J Biol Chem 255:7059–7062

Felli IC, Pierattelli R (2014) Novel methods based on ^{13}C detection to study intrinsically disordered proteins. J Mag Reson 241:115–125. doi:10.1016/j.jmr.2013.10.020

Fischer E (1894) Einfluss der configuration auf die wirkung der enzyme. Ber Dt Chem Ges 27:2985–2993

Follis AV, Galea CA, Kriwacki RW (2012) Intrinsic protein flexibility in regulation of cell proliferation: advantages for signaling and opportunities for novel therapeutics. Adv Exp Med Biol 725:27–49. doi:10.1007/978-1-4614-0659-4_3

Follis AV, Llambi F, Ou L et al (2014) The DNA-binding domain mediates both nuclear and cytosolic functions of p53. Nat Struct Mol Biol 21:535–543. doi:10.1038/nsmb.2829

Fonfría-Subirós E, Acosta-Reyes F, Saperas N et al (2012) Crystal structure of a complex of DNA with one AT-hook of HMGA1. PLoS ONE 7:e37120. doi:10.1371/journal.pone.0037120

Fraenkel E, Pabo CO (1998) Comparison of X-ray and NMR structures for the Antennapedia homeodomain-DNA complex. Nat Struct Biol 5:692–697. doi:10.1038/1382

Fukuchi S, Yoshimune K, Wakayama M et al (2003) Unique amino acid composition of proteins in halophilic bacteria. J Mol Biol 327:347–357

Fukuchi S, Homma K, Minezaki Y et al (2009) Development of an accurate classification system of proteins into structured and unstructured regions that uncovers novel structural domains: its application to human transcription factors. BMC Struct Biol 9:26. doi:10.1186/1472-6807-9-26

Fukuchi S, Hosoda K, Homma K et al (2011) Binary classification of protein molecules into intrinsically disordered and ordered segments. BMC Struct Biol 11:29. doi:10.1186/1472-6807-11-29

Galea CA, Nourse A, Wang Y et al (2008a) Role of intrinsic flexibility in signal transduction mediated by the cell cycle regulator, p27 Kip1. J Mol Biol 376:827–838. doi:10.1016/j.jmb.2007.12.016

Galea CA, Wang Y, Sivakolundu SG et al (2008b) Regulation of cell division by intrinsically unstructured proteins: intrinsic flexibility, modularity, and signaling conduits. Biochemistry 47:7598–7609. doi:10.1021/bi8006803

Garner E, Romero P, Dunker A et al (1999) Predicting binding regions within disordered proteins. Genome Inform Ser Workshop Genome Inform 10:41–50

Gierasch LM, King J (1990) Protein folding: deciphering the second half of the genetic code. American Association for the Advancement of Science, Washington, D.C.

Gil S, Hošek T, Solyom Z et al (2013) NMR Spectroscopic studies of intrinsically disordered proteins at near-physiological conditions. Angew Chem Int Ed 52:11808–11812. doi:10.1002/anie.201304272

Grizzuti K, Perlmann GE (1970) Conformation of the phosphoprotein, phosvitin. J Biol Chem 245:2573–2578

Harada R, Tochio N, Kigawa T et al (2013) Reduced native state stability in crowded cellular environment due to protein-protein interactions. J Am Chem Soc 135:3696–3701. doi:10.1021/ja3126992

He B, Wang K, Liu Y et al (2009) Predicting intrinsic disorder in proteins: an overview. Cell Res 19:929–949. doi:10.1038/cr.2009.87

Herriott RM (1938) Isolation, crystallization, and properties of swine pepsinogen. J Gen Physiol 21:501–540

Hirs CH, Moore S, Stein WH (1960) The sequence of the amino acid residues in performic acid-oxidized ribonuclease. J Biol Chem 235:633–647

Holmes KC (1983) Flexibility in tobacco mosaic virus. Ciba Found Symp 93:116–138

Hsu W-L, Oldfield CJ, Xue B et al (2013) Exploring the binding diversity of intrinsically disordered proteins involved in one-to-many binding. Protein Sci 22:258–273. doi:10.1002/pro.2207

Hubbard SJ, Campbell SF, Thornton JM (1991) Molecular recognition. Conformational analysis of limited proteolytic sites and serine proteinase protein inhibitors. J Mol Biol 220:507–530

Huber R (1979) Conformational flexibility in protein molecules. Nature 280:538–539

Ito Y, Selenko P (2010) Cellular structural biology. Curr Opin Struct Biol 20:640–648. doi:10.1016/j.sbi.2010.07.006

James LC, Tawfik DS (2003a) Conformational diversity and protein evolution—a 60-year-old hypothesis revisited. Trends Biochem Sci 28:361–368. doi:10.1016/S0968-0004(03)00135-X

James LC, Tawfik DS (2003b) The specificity of cross-reactivity: promiscuous antibody binding involves specific hydrogen bonds rather than nonspecific hydrophobic stickiness. Protein Sci 12:2183–2193. doi:10.1110/ps.03172703

James LC, Roversi P, Tawfik DS (2003) Antibody multispecificity mediated by conformational diversity. Science 299:1362–1367. doi:10.1126/science.1079731

Jin Y, Dunbrack RL (2005) Assessment of disorder predictions in CASP6. Proteins 61(Suppl 7):167–175. doi:10.1002/prot.20734

Jirgensons B (1958) Optical rotation and viscosity of native and denatured proteins. X. Further studies on optical rotatory dispersion. Arch Biochem Biophys 74:57–69

Jirgensons B (1966) Classification of proteins according to conformation. Die Makromolekulare Chemie 91:74–86. doi:10.1002/macp.1966.020910105

Johnson DE, Xue B, Sickmeier MD et al (2012) High-throughput characterization of intrinsic disorder in proteins from the Protein Structure Initiative. J Struct Biol 180:201–215. doi:10.1016/j.jsb.2012.05.013

Kabsch W, Sander C (1983) Dictionary of protein secondary structure: Pattern recognition of hydrogen-bonded and geometrical features. Biopolymers 22:2577–2637. doi:10.1002/bip.360221211

Kay LE, Torchia DA, Bax A (1989) Backbone dynamics of proteins as studied by ^{15}N inverse detected heteronuclear NMR spectroscopy: application to staphylococcal nuclease. Biochemistry 28:8972–8979

Kekule A (1865) Sur la constitution des substances aromatiques. Bull Soc Chim Paris 3:98–110

Kissinger CR, Parge HE, Knighton DR et al (1995) Crystal structures of human calcineurin and the human FKBP12-FK506-calcineurin complex. Nature 378:641–644. doi:10.1038/378641a0
Kolata G (1986) Trying to crack the second half of the genetic code. Science 233:1037–1039
Koshland DE (1958) Application of a theory of enzyme specificity to protein synthesis. Proc Natl Acad Sci U S A 44:98–104
Koshland DE Jr (1959) Enzyme flexibility and enzyme action. J Cell Comp Physiol 54:245–258
Koshland DE (2004) Crazy, but correct. Nature 432:447–447. doi:10.1038/432447a
Kriwacki RW, Hengst L, Tennant L et al (1996) Structural studies of p21Waf1/Cip1/Sdi1 in the free and Cdk2-bound state: conformational disorder mediates binding diversity. Proc Natl Acad Sci U S A 93:11504–11509
Kyne C, Ruhle B, Gautier VW et al (2015) Specific ion effects on macromolecular interactions in *Escherichia coli* extracts. Protein Sci 24:310–318. doi:10.1002/pro.2615
Lacy ER, Filippov I, Lewis WS et al (2004) p27 binds cyclin–CDK complexes through a sequential mechanism involving binding-induced protein folding. Nat Struct Mol Biol 11:358–364. doi:10.1038/nsmb746
Lanyi JK (1974) Salt-dependent properties of proteins from extremely halophilic bacteria. Bacteriol Rev 38:272–290
Larsson G, Martinez G, Schleucher J et al (2003) Detection of nano-second internal motion and determination of overall tumbling times independent of the time scale of internal motion in proteins from NMR relaxation data. J Biomol NMR 27:291–312
Lee MH, Reynisdóttir I, Massagué J (1995) Cloning of p57KIP2, a cyclin-dependent kinase inhibitor with unique domain structure and tissue distribution. Genes Dev 9:639–649
Le Gall T, Romero PR, Cortese MS et al (2007) Intrinsic disorder in the Protein Data Bank. J Biomol Struct Dyn 24:325–342. doi:10.1080/07391102.2007.10507123
Lemieux RU, Spohr U (1994) How Emil Fischer was led to the lock and key concept for enzyme specificity. Adv Carbohydr Chem Biochem 50:1–20
Lerch MM, Gorelick FS (2000) Early trypsinogen activation in acute pancreatitis. Med Clin North Am 84:549–563, viii
Li C, Charlton LM, Lakkavaram A et al (2008) Differential dynamical effects of macromolecular crowding on an intrinsically disordered protein and a globular protein: implications for in-cell NMR spectroscopy. J Am Chem Soc 130:6310–6311. doi:10.1021/ja801020z
Lipari G, Szabo A (1982) Model-free approach to the interpretation of nuclear magnetic resonance relaxation in macromolecules. 1. Theory and range of validity. J Am Chem Soc 104:4546–4559. doi:10.1021/ja00381a009
Manalan AS, Klee CB (1983) Activation of calcineurin by limited proteolysis. Proc Natl Acad Sci U S A 80:4291–4295
Markley JL, Ulrich EL, Berman HM et al (2008) BioMagResBank (BMRB) as a partner in the Worldwide Protein Data Bank (wwPDB): new policies affecting biomolecular NMR depositions. J Biomol NMR 40:153–155. doi:10.1007/s10858-008-9221-y
Marsh JJ, Guan HS, Li S et al (2013) Structural insights into fibrinogen dynamics using amide hydrogen/deuterium exchange mass spectrometry. Biochemistry 52:5491–5502. doi:10.1021/bi4007995
Matthews BW (1975) Comparison of the predicted and observed secondary structure of T4 phage lysozyme. Biochim Biophys Acta 405:442–451
Mayer J, Rau B, Schoenberg MH et al (1999) Mechanism and role of trypsinogen activation in acute pancreatitis. Hepatogastroenterology 46:2757–2763
McKee T, McKee JR (2013) Biochemistry: the molecular basis of life, 5th edn. Oxford University Press, UK
McMeekin TL (1952) Milk proteins. J Food Prot 15:57–63
McNulty BC, Young GB, Pielak GJ (2006) Macromolecular crowding in the *Escherichia coli* periplasm maintains α-synuclein disorder. J Mol Biol 355:893–897. doi:10.1016/j.jmb.2005.11.033
Melamud E, Moult J (2003) Evaluation of disorder predictions in CASP5. Proteins 53(Suppl 6):561–565. doi:10.1002/prot.10533
Miklos AC, Sarkar M, Wang Y et al (2011) Protein crowding tunes protein stability. J Am Chem Soc 133:7116–7120. doi:10.1021/ja200067p

Minamino T, MacNab RM (2000) Interactions among components of the Salmonella flagellar export apparatus and its substrates. Mol Microbiol 35:1052–1064

Mirsky AE, Anson ML (1930) Protein coagulation and its reversal: improved methods for the reversal of the coagulation of hemoglobin. J Gen Physiol 13:477–481

Mirsky AE, Pauling L (1936) On the structure of native, denatured, and coagulated proteins. Proc Natl Acad Sci U S A 22:439–447

Mobli M, Maciejewski MW, Schuyler AD et al (2012) Sparse sampling methods in multidimensional NMR. Phys Chem Chem Phys 14:10835–10843. doi:10.1039/C2CP40174F

Mohan A, Uversky VN, Radivojac P (2009) Influence of sequence changes and environment on intrinsically disordered proteins. PLoS Comput Biol 5:e1000497. doi:10.1371/journal.pcbi.1000497

Monastyrskyy B, Fidelis K, Moult J et al (2011) Evaluation of disorder predictions in CASP9. Proteins 79(Suppl 10):107–118. doi:10.1002/prot.23161

Monastyrskyy B, Kryshtafovych A, Moult J et al (2014) Assessment of protein disorder region predictions in CASP10. Proteins 82(Suppl 2):127–137. doi:10.1002/prot.24391

Monteith WB, Pielak GJ (2014) Residue level quantification of protein stability in living cells. Proc Natl Acad Sci U S A 111:11335–11340. doi:10.1073/pnas.1406845111

Monteith WB, Cohen RD, Smith AE et al (2015) Quinary structure modulates protein stability in cells. Proc Natl Acad Sci U S A 112:1739–1742. doi:10.1073/pnas.1417415112

Morar AS, Olteanu A, Young GB et al (2001) Solvent-induced collapse of alpha-synuclein and acid-denatured cytochrome c. Protein Sci 10:2195–2199. doi:10.1110/ps.24301

Muchmore SW, Sattler M, Liang H et al (1996) X-ray and NMR structure of human Bcl-xL, an inhibitor of programmed cell death. Nature 381:335–341. doi:10.1038/381335a0

Murzin AG, Brenner SE, Hubbard T et al (1995) SCOP: a structural classification of proteins database for the investigation of sequences and structures. J Mol Biol 247:536–540. doi:10.1006/jmbi.1995.0159

Namba K, Stubbs G (1986) Structure of tobacco mosaic virus at 3.6 Å resolution: implications for assembly. Science 231:1401–1406

Nelson DL, Cox MM (2012) Lehninger Principles of biochemistry, 6th edn. W.H. Freeman, New York

Nishikawa K, Ooi T (1982) Correlation of the amino acid composition of a protein to its structural and biological characters. J Biochem 91:1821–1824

Noivirt-Brik O, Prilusky J, Sussman JL (2009) Assessment of disorder predictions in CASP8. Proteins 77(Suppl 9):210–216. doi: 10.1002/prot.22586

Noutsou M, Duarte AMS, Anvarian Z et al (2011) Critical scaffolding regions of the tumor suppressor Axin1 are natively unfolded. J Mol Biol 405:773–786. doi:10.1016/j.jmb.2010.11.013

Nováček J, Žídek L, Sklenář V (2014) Toward optimal-resolution NMR of intrinsically disordered proteins. J Magn Reson 241:41–52. doi:10.1016/j.jmr.2013.12.008

Oates ME, Romero P, Ishida T et al (2013) D^2P^2: database of disordered protein predictions. Nucleic Acids Res 41:D508–D516. doi:10.1093/nar/gks1226

Oldfield CJ, Dunker AK (2014) Intrinsically disordered proteins and intrinsically disordered protein regions. Annu Rev Biochem 83:553–584. doi:10.1146/annurev-biochem-072711-164947

Oldfield CJ, Cheng Y, Cortese MS et al (2005a) Coupled folding and binding with alpha-helix-forming molecular recognition elements. Biochemistry 44:12454–12470. doi:10.1021/bi050736e

Oldfield CJ, Ulrich EL, Cheng Y et al (2005b) Addressing the intrinsic disorder bottleneck in structural proteomics. Proteins 59:444–453. doi:10.1002/prot.20446

Oldfield CJ, Meng J, Yang JY et al (2008) Flexible nets: disorder and induced fit in the associations of p53 and 14-3-3 with their partners. BMC Genomics 9(Suppl 1):S1. doi:10.1186/1471-2164-9-S1-S1

Oldfield CJ, Xue B, Van Y-Y et al (2013) Utilization of protein intrinsic disorder knowledge in structural proteomics. Biochim Biophys Acta 1834:487–498. doi:10.1016/j.bbapap.2012.12.003

Orengo CA, Michie AD, Jones S et al (1997) CATH—a hierarchic classification of protein domain structures. Structure 5:1093–1108

Ota M, Koike R, Amemiya T et al (2013) An assignment of intrinsically disordered regions of proteins based on NMR structures. J Struct Biol 181:29–36. doi:10.1016/j.jsb.2012.10.017
Otting G, Qian YQ, Billeter M et al (1990) Protein–DNA contacts in the structure of a homeodomain–DNA complex determined by nuclear magnetic resonance spectroscopy in solution. EMBO J 9:3085–3092
Pauling L (1932) Interatomic distances in covalent molecules and resonance between two or more Lewis electronic structures. Proc Natl Acad Sci U S A 18:293–297
Pauling L (1940) A theory of the structure and process of formation of antibodies. J Am Chem Soc 62:2643–2657
Pawley NH, Wang C, Koide S et al (2001) An improved method for distinguishing between anisotropic tumbling and chemical exchange in analysis of ^{15}N relaxation parameters. J Biomol NMR 20:149–165
Peng JW (2012) Exposing the moving parts of proteins with NMR spectroscopy. J Phys Chem Lett 3:1039–1051. doi:10.1021/jz3002103
Peng Z-L, Kurgan L (2012) Comprehensive comparative assessment of in-silico predictors of disordered regions. Curr Protein Pept Sci 13:6–18
Plaxco KW, Gross M (1997) Cell biology. The importance of being unfolded. Nature 386:657, 659. doi:10.1038/386657a0
Potenza E, Domenico TD, Walsh I et al (2015) MobiDB 2.0: an improved database of intrinsically disordered and mobile proteins. Nucleic Acids Res 43:D315–320. doi:10.1093/nar/gku982
Pratt CW, Cornely K (2013) Essential biochemistry, 3rd edn. Wiley, Hoboken
Qian YQ, Billeter M, Otting G et al (1989) The structure of the Antennapedia homeodomain determined by NMR spectroscopy in solution: comparison with prokaryotic repressors. Cell 59:573–580
Read J (1957) From alchemy to chemistry. Courier Dover Publications, New York
Ringe D, Petsko GA (1986) Study of protein dynamics by X-ray diffraction. Meth Enzymol 131:389–433
Romero P, Obradovic Z, Dunker AK (1997a) Sequence data analysis for long disordered regions prediction in the calcineurin family. Genome Inform Ser Workshop Genome Inform 8:110–124
Romero P, Obradovic Z, Kissinger CR et al (1997b) Identifying disordered regions in proteins from amino acid sequence. Int Conf Neural Netw 1:90–95. doi:10.1109/ICNN.1997.611643
Romero P, Obradovic Z, Kissinger CR et al (1998) Thousands of proteins likely to have long disordered regions. Pac Symp Biocomput 3:437–448
Romero P, Obradovic Z, Li X et al (2001) Sequence complexity of disordered protein. Proteins 42:38–48
Rothen A, Landsteiner K (1939) Absorption of antibodies by egg albumin films. Science 90:65–66. doi:10.1126/science.90.2325.65-a
Rumi-Masante J, Rusinga FI, Lester TE et al (2012) Structural basis for activation of calcineurin by calmodulin. J Mol Biol 415:307–317. doi:10.1016/j.jmb.2011.11.008
Russo AA, Jeffrey PD, Patten AK et al (1996) Crystal structure of the p27Kip1 cyclin-dependent-kinase inhibitor bound to the cyclin A-Cdk2 complex. Nature 382:325–331. doi:10.1038/382325a0
Schulz GE (1979) Nucleotide binding proteins. In: Balaban M (ed) Moledular mechanisms of biological recognition. Elsevier/North-Holland Biomedical Press, Amsterdam, pp 79–94
Sela M, White FH, Anfinsen CB (1957) Reductive cleavage of disulfide bridges in ribonuclease. Science 125:691–692
Serber Z, Dötsch V (2001) In-cell NMR spectroscopy. Biochemistry 40:14317–14323
Serber Z, Keatinge-Clay AT, Ledwidge R et al (2001) High-resolution macromolecular NMR spectroscopy inside living cells. J Am Chem Soc 123:2446–2447
Shakhnovich EI, Gutin AM (1993) Engineering of stable and fast-folding sequences of model proteins. Proc Natl Acad Sci U S A 90:7195–7199
Sigler PB (1988) Transcriptional activation. Acid blobs and negative noodles. Nature 333:210–212. doi:10.1038/333210a0

Singh M, D'Silva L, Holak TA (2006) DNA-binding properties of the recombinant high-mobility-group-like AT-hook-containing region from human BRG1 protein. Biol Chem 387:1469–1478. doi:10.1515/BC.2006.184

Smerdon MJ, Isenberg I (1976) Conformational changes in subfractions of calf thymus histone H1. Biochemistry 15:4233–4242

Smith AE, Zhou Z, Pielak GJ (2015) Hydrogen exchange of disordered proteins in *Escherichia coli*. Protein Sci. doi:10.1002/pro.2643

Spolar RS, Record MT Jr (1994) Coupling of local folding to site-specific binding of proteins to DNA. Science 263:777–784

Sprangers R, Bottomley MJ, Linge JP et al (2000) Refinement of the protein backbone angle psi in NMR structure calculations. J Biomol NMR 16:47–58

Stubbs G, Warren S, Holmes K (1977) Structure of RNA and RNA binding site in tobacco mosaic virus from 4-A map calculated from X-ray fibre diagrams. Nature 267:216–221

Sugase K, Dyson HJ, Wright PE (2007) Mechanism of coupled folding and binding of an intrinsically disordered protein. Nature 447:1021–1025. doi:10.1038/nature05858

Teng G, Papavasiliou FN (2007) Immunoglobulin somatic hypermutation. Annu Rev Genet 41:107–120. doi:10.1146/annurev.genet.41.110306.130340

Theillet FX, Kalmar L, Tompa P et al (2013) The alphabet of intrinsic disorder: 1. Act like a Pro: on the abundance and roles of proline residues in intrinsically disordered regions. Intrinsically Disord Proteins 1:e24360

Thorpe IF, Brooks CL (2007) Molecular evolution of affinity and flexibility in the immune system. Proc Natl Acad Sci U S A 104:8821–8826. doi:10.1073/pnas.0610064104

Tompa P (2002) Intrinsically unstructured proteins. Trends Biochem Sci 27:527–533

Tymoczko JL, Berg JM, Stryer L (2011) Biochemistry: a short course, 2nd edn. W.H. Freeman, New York

Uversky VN, Li J, Souillac P et al (2002) Biophysical properties of the synucleins and their propensities to fibrillate: inhibition of α-synuclein assembly by β- and γ-synucleins. J Biol Chem 277:11970–11978. doi:10.1074/jbc.M109541200

Voet D, Voet JG (2010) Biochemistry, 4th edn. Wiley, Hoboken

Voet D, Voet JG, Pratt CW (2012) Fundamentals of biochemistry: life at the molecular level, 4th edn. Wiley, Hoboken

Vogel HJ (1983) Structure of hen phosvitin: A ^{31}P NMR, ^{1}H NMR, and laser photochemically induced dynamic nuclear polarization ^{1}H NMR study. Biochemistry 22:668–674

Walter J, Steigemann W, Singh TP et al (1982) On the disordered activation domain in trypsinogen: chemical labelling and low-temperature crystallography. Acta Cryst B 38:1462–1472. doi:10.1107/S0567740882006153

Wang Q, Zhuravleva A, Gierasch LM (2011a) Exploring weak, transient protein–protein interactions in crowded in vivo environments by in-cell nuclear magnetic resonance spectroscopy. Biochemistry 50:9225–9236. doi:10.1021/bi201287e

Wang W, Perovic I, Chittuluru J et al (2011b) A soluble α-synuclein construct forms a dynamic tetramer. Proc Natl Acad Sci U S A 108:17797–17802. doi:10.1073/pnas.1113260108

Ward JJ, Sodhi JS, McGuffin LJ et al (2004) Prediction and functional analysis of native disorder in proteins from the three kingdoms of life. J Mol Biol 337:635–645. doi:10.1016/j.jmb.2004.02.002

Wu H (1931) Studies on denaturation of proteins XIII. A theory of denaturation. Chin J Physiol 5:321–344

Xie Q, Arnold G, Romero P et al (1998) The sequence attribute method for determining relationships between sequence and protein disorder. Genome Inform Ser Workshop Genome Inform 9:193–200

Xue B, Williams RW, Oldfield CJ et al (2010) Archaic chaos: intrinsically disordered proteins in Archaea. BMC Syst Biol 4(Suppl 1):S1. doi:10.1186/1752-0509-4-S1-S1

Xue B, Dunker AK, Uversky VN (2012) Orderly order in protein intrinsic disorder distribution: disorder in 3500 proteomes from viruses and the three domains of life. J Biomol Struct Dyn 30:137–149. doi:10.1080/07391102.2012.675145

Xue B, Romero PR, Noutsou M et al (2013) Stochastic machines as a colocalization mechanism for scaffold protein function. FEBS Lett 587:1587–1591. doi:10.1016/j.febslet.2013.04.006

Zhang Y, Skolnick J (2005) TM-align: a protein structure alignment algorithm based on the TM-score. Nucleic Acids Res 33:2302–2309. doi:10.1093/nar/gki524

Zhang Y, Stec B, Godzik A (2007) Between order and disorder in protein structures: analysis of "dual personality" fragments in proteins. Structure 15:1141–1147. doi:10.1016/j.str.2007.07.012

Zhang T, Faraggi E, Li Z, Zhou Y (2013) Intrinsically semi-disordered state and its role in induced folding and protein aggregation. Cell Biochem Biophys. doi:10.1007/s12013-013-9638-0

Zimmerman SB, Trach SO (1991) Estimation of macromolecule concentrations and excluded volume effects for the cytoplasm of *Escherichia coli*. J Mol Biol 222:599–620

Zimmermann J, Oakman EL, Thorpe IF et al (2006) Antibody evolution constrains conformational heterogeneity by tailoring protein dynamics. Proc Natl Acad Sci U S A 103:13722–13727. doi:10.1073/pnas.0603282103

Chapter 2
Structure and Dynamics of Intrinsically Disordered Proteins

Biao Fu and Michele Vendruscolo

Abstract Intrinsically disordered proteins (IDPs) are involved in a wide range of essential biological processes, including in particular signalling and regulation. We are only beginning, however, to develop a detailed knowledge of the structure and dynamics of these proteins. It is becoming increasingly clear that, as IDPs populate highly heterogeneous states, they should be described in terms of conformational ensembles rather than as individual structures, as is instead most often the case for the native states of globular proteins. Within this context, in this chapter we describe the conceptual tools and methodological aspects associated with the description of the structure and dynamics of IDPs in terms of conformational ensembles. A major emphasis is given to methods in which molecular simulations are used in combination with experimental nuclear magnetic resonance (NMR) measurements, as they are emerging as a powerful route to achieve an accurate determination of the conformational properties of IDPs.

Keywords Structure · Dynamics · Molecular dynamics · Conformational ensembles

1 Introduction: From Average Structures to Conformational Ensembles

Intrinsically disordered proteins (IDPs) play crucial roles in many aspects of molecular and cell biology, as these proteins are involved in a variety of signalling and regulation processes as well as being implicated in a range of neurodegenerative and systemic disorders such as Alzheimer's and Parkinson's diseases, and type II diabetes (Dyson and Wright 2005; Knowles et al. 2014; Uversky 2013). From the point of view of structural biology, IDPs pose formidable challenges since they are conformationally highly heterogeneous (Fig. 2.1) and are thus not readily amenable to the standard approaches for structure determination that have been developed for folded proteins.

M. Vendruscolo (✉) · B. Fu
Department of Chemistry, University of Cambridge, Cambridge CB2 1EW, UK
e-mail: mv245@cam.ac.uk

I. C. Felli, R. Pierattelli (eds.), *Intrinsically Disordered Proteins Studied by NMR Spectroscopy,* Advances in Experimental Medicine and Biology,
DOI 10.1007/978-3-319-20164-1_2

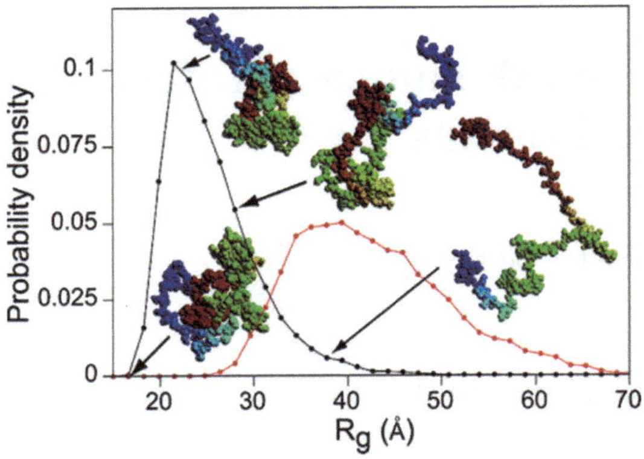

Fig. 2.1 IDPs are conformationally highly heterogeneous. This fundamental aspect of the nature of IDPs is illustrated here by the probability distribution of the radius of gyration, *Rg*, of α-synuclein (*black line*), an IDP associated with Parkinson's disease (Dedmon et al. 2005). The values of *Rg* range from about 18 Å to more than 40 Å. For comparison the probability distribution of a polypeptide chain with the same length as α-synuclein in a random coil state is also shown (*red line*) (Dedmon et al. 2005)

Native states are also undergoing structural fluctuations, the dynamics of which are important for enzymatic catalysis, ligand binding and the formation of biomolecular complexes (Frauenfelder et al. 1991; Fersht 1999; Karplus and Kuriyan 2005; Mittermaier and Kay 2006; Vendruscolo and Dobson 2006; Boehr et al. 2009). The dynamics of native states are usually represented by a conformational variability around a well-defined structure, and powerful techniques are available to calculate them and their related conformational fluctuations (Brooks et al. 1983; Brunger et al. 1998; Schwieters et al. 2006). This type of description, however, is not suitable in the case of highly heterogeneous states because in such states, in the absence of a specific reference structure, an IDP populates a wide range of conformations having very dissimilar structures (Varadi et al. 2014).

The characterization of the behaviour of IDPs requires novel approaches with respect to standard protein structure determination procedures. The gold standard for the determination of the structures of native states is represented by X-ray crystallography, a technique that allows the positions of all the atoms to be identified with great accuracy through the mapping of the corresponding electron densities (Blundell and Johnson 1976). Nuclear magnetic resonance (NMR) spectroscopy can also achieve this type of accurate positioning of the atoms making up a protein molecule through the measurement of inter-proton distances by exploiting nuclear Overhauser effects (Wüthrich 1986). In this context, the problem of protein structure determination is solved by acquiring an amount of experimental information sufficient to determine essentially all the degrees of freedom of a protein molecule once its sequence and covalent bond topology are known. In the case of IDPs, by contrast, this approach is not possible, since the presence of a wide variety of different conformations prevents the definition of the structural properties of proteins by providing a single list of three-dimensional atomic coordinates.

A powerful conceptual framework in this case is that of statistical mechanics (Chandler 1987; van Kampen 1992). In this type of description the objective is to determine a range of representative conformations populated by IDPs together with their statistical weights. In other words, the aim is to characterise the Boltzmann distributions of IDPs. The reason for adopting this approach is that if one calculates the number of possible states of an IDP, one realizes that no experiment will ever be able to provide sufficient information to determine the atomic coordinates of the exceedingly large number of different conformations that it can explore. To obtain an insight into this issue, one can consider a most common textbook example, in which the velocities of the particles of an ideal gas in a box are provided in terms of a well-defined probability distribution, the Maxwell-Boltzmann distribution (Chandler 1987). The knowledge of such a distribution enables a great variety of properties of the ideal gas to be calculated, and these calculations provide accurate predictions for experimental measurements that can be performed on rarefied weakly-interacting gasses. In this view, the goal of measuring the positions of all the atoms in the myriad different conformations of a protein is not only practically impossible to achieve, but also essentially irrelevant, since one can perform accurate predictions of many aspects of its behaviour even without such knowledge.

For a given IDP, in order to generate an ensemble of structures according to their Boltzmann probabilities, or statistical weights, the availability of only sparse experimental measurements for structure determination can be complemented with the use of *a priori* information, including about covalent bond lengths, dihedral angles and rotameric states of side chains. This type of information can be provided through the use of force fields in molecular dynamics (Brooks et al. 1983; Hornak et al. 2006; Lindorff-Larsen et al. 2012a) or through effective potentials derived from protein structure databases (Das and Baker 2008). In this approach, a computational model of the conformational space populated by IDPs is combined with the information provided by the experimental measurements in order to achieve a description of the structure and dynamics of IDPs simultaneously consistent with the overall theoretical knowledge of the behaviour of these proteins and with the specific observation made about specific systems. As we will describe in the following, a range of different methods have been proposed to combine theoretical knowledge about IDPs and the experimental measurements on them. Before coming to that, however, we address the two major, and in many ways complementary, problems that should be considered in the determination of the structure and dynamics of IDPs.

2 The Two Fundamental Problems in the Computational Study of IDPs

The strategy in which experimental data are combined with a theoretical modelling of IDPs requires an ability to generate a relatively accurate sampling of their conformational space. A powerful approach to achieve this result is provided, for example, by all-atom molecular dynamics simulations (Karplus and Kuriyan 2005; Shaw et al. 2010; Best 2012). In these simulations, the conformational space of a protein is sampled by integrating the equations of motion for a time interval sufficiently long to enable the relevant regions to be explored. There are, however, two major challenges in the implementation of this approach. The first is the 'force field problem' and the second is 'the conformational sampling problem'. We should also note that although we describe these two problems here in the case of all-atom molecular dynamics simulations, they are common to essentially any scheme to sample the conformational space of proteins, as one needs always to evaluate the energy of a given protein conformation and to explore the range of its available conformations.

2.1 The Force Field Problem

One of the most fundamental aspects of any theoretical method to describe the behaviour of proteins concerns the ability to associate an energy to a given conformation. In molecular dynamics simulations, the function that associates an energy to a given conformation is called a 'force field' (although rather than a force it is actually an energy, or more precisely, a potential energy). The most common force fields are based on molecular mechanics, in which classical mechanics is used to describe the behaviour of proteins and the interactions are provided in a classical framework, involving a combination of terms describing the covalent bond distances and angles ('bonded terms') and of terms describing other interactions, including van der Waals and Coulomb interactions, between atoms ('non-bonded terms') (Brooks et al. 1983; Hornak et al. 2006; Lindorff-Larsen et al. 2012a).

These energy terms, however, represent only an approximate model of the actual interactions between atoms. Although better representations of these interactions are possible in principle (e.g. through the use of quantum mechanics), they become computationally more expensive and as a consequence they are more seriously affected by the conformational sampling problem (see Sect. 1.2.2) (Brooks et al. 1983; Hornak et al. 2006; Lindorff-Larsen et al. 2012a; Baker and Best 2013). The energies that can be associated with given conformations, therefore, can only be of limited accuracy, and the corresponding exploration of the conformational space is carried out with inaccurate statistical weights. Despite a range of significant recent advances in the improvement of force fields (Lindorff-Larsen et al. 2012a; Bottaro et al. 2013; Baker and Best 2013; Piana et al. 2014), one should thus bear in mind that force fields are not exact. Having said that, the use of molecular dynamics simulations provides a range of opportunities that have been explored in a series of recent studies that are beginning to provide descriptions of the structure and dynamics of IDPs and of the disordered states of other proteins (Lindorff-Larsen et al. 2012b; Camilloni and Vendruscolo 2014; Knott and Best 2012; Krzeminski et al. 2013; Varadi et al. 2014).

2.2 The Conformational Sampling Problem

As mentioned above, the number of possible conformations of a protein molecule is enormous. It is thus out of the question to enumerate all such possible conformations using a computer, since it would require an essentially infinite amount of time and memory. In statistical mechanics, however, it is relevant to sample the conformational space only in the regions where the statistical weights are non-negligible. For folded states, this means that only a relatively small number of conformations need to be considered, and indeed single X-ray structures represent the state of a protein quite faithfully. By contrast, many more conformations should be explored for IDPs, as the statistical weights are significantly different from zero for a wide range of different structures.

In molecular dynamics the speed at which the conformational space can be explored is inherently limited by the step of integration of the equations of motion, which is typically of 1 to 2 femtoseconds. Even with the most powerful supercomputers, trajectories can currently be followed up to the millisecond timescale—a feat that involves something like a trillion integration steps! (Shaw et al. 2010; Vendruscolo and Dobson 2011). As IDPs tend to explore their relevant conformational space on longer timescales (e.g. seconds and beyond), one should bear in mind that the sampling will necessarily be incomplete.

Several methods have been proposed to enhance the sampling efficiency. For example, one of the most common ones involves the 'coarse-graining' of the conformational degrees of freedom (Tozzini 2005; Monticelli et al. 2008). In this approach, rather than representing a protein molecule by providing a list of the three-dimensional coordinates of all its atoms, one simplifies the representation by specifying only the most relevant degrees of freedom, such as for instance only the position of the Cα atoms. In coarse-grained approaches, while the integration step becomes much less expensive, the force field becomes less accurate because in eliminating some of the atoms of a protein the corresponding interactions should be incorporated in some averaged manner in the force field, and such averaging is inherently approximate. There is therefore a trade-off between speed in the sampling and accuracy in the energy estimation.

In other approaches, the all-atom representation is maintained but the force field is modified in a controlled manner to bias the sampling towards the relevant regions of the conformational space. One of the first methods proposed for this purpose is that of 'umbrella sampling', in which a weighting function is introduced in the force field to prevent the sampling of structures outside a given region of the conformational space (Chandler 1987). This bias is then removed in order to reweight the conformations and obtain their correct statistical weights. Series of umbrella sampling simulations can be then analysed using the weighted histogram analysis method (WHAM) or its generalizations (Kumar et al. 1992; Hub et al. 2010; Zhu and Hummer 2012). A related method is that of accelerated molecular dynamics, in which the sampling of the conformational space is enhanced by reducing the energy barriers separating the different states populated by a protein (Board et al. 1992;

Markwick et al. 2007). This method modifies the potential energy landscape by raising the energy wells below a given threshold level, while leaving those above this level unaffected. As a result, the barriers between neighbouring energy basins are reduced, allowing the protein to sample regions of the conformational space that cannot be easily accessed in conventional molecular dynamics simulations.

A particularly effective method that is becoming increasingly adopted in IDP simulations is that of metadynamics (Laio and Parrinello 2002; Laio and Gervasio 2008). In this method one assumes that the behaviour of a protein can be described accurately through a small number of collective variables. The basic idea of the method is that the protein is discouraged from returning to the proximity of the conformations that it has already visited by 'remembering' their positions. This idea is implemented by calculating the position of the protein in terms of the collective variables during the simulation and by adding a Gaussian function in this position to the energy landscape of the protein itself. As the simulation progresses, the Gaussian functions accumulate preferentially in the energy minima until the free energy eventually becomes a constant as a function of the collective variables. The three main parameters that control the convergence of the simulations are the time between the addition of Gaussian functions and the height and width of the Gaussian functions themselves.

3 Combining Experiments with Simulations Using the Maximum Entropy Principle

As mentioned above, several approaches have been proposed for characterizing non-native states. These approaches differ in the particular way in which the system-dependent experimental measurements are combined with the system-independent theoretical information provided by the force field. A general framework for carrying out this plan is provided by the maximum entropy principle (Pitera and Chodera 2012; Cavalli et al. 2013; Roux and Weare 2013; Boomsma et al. 2014). According to this principle, the conformational space populated by a protein should be the largest possible one compatible with the information available. In the context of molecular dynamics simulations, the incorporation of the information provided by a given set of experimental data to a force field should be carried out in a manner that maximizes the number of conformations that are sampled, with the only requirement that they be compatible with the experimental data. In this sense, the maximum entropy principle provides the opposite prescription to the 'Occam's razor', according to which the minimal number of structures should be determined to generate a set consistent with the available experimental data.

The maximum entropy principle only provides a guideline about how to combine experiments with simulations, and there are many possible alternatives for its practical implementation. For example, in an approach often used for the characterization of the behaviour of IDPs, the experimental information is used to filter out conformations in disagreement with the observations from a previously generated ensemble of conformations (Choy and Forman-Kay 2001; Bernadó et al. 2005; Heise et al. 2005). The success of this approach relies on the ability of the conformational sampling to explore regions that are populated with significant probability by IDPs, as otherwise it becomes impossible to select conformations consistent with the experimental data. When this condition is met, the maximum entropy principle framework offers a highly effective way to carry out the selection.

An approach that has been investigated extensively in recent years consists of extending the methods of structure determination that have been developed for native states to highly heterogeneous states (Bonvin et al. 1994; Bonvin and Brunger 1995; Burgi et al. 2001; Constantine et al. 1995; Fennen et al. 1995; Kemmink and Scheek 1995; Kessler et al. 1988; Torda et al. 1989; Clore and Schwieters 2004; Lindorff-Larsen et al. 2005). In this approach the experimental information is used to construct structural restraints to be used in molecular simulations. In this case the sampling is biased to take place in regions of conformational space that satisfy the available experimental information. It has been shown that the addition of the bias can be carried out in a manner compatible with the maximum entropy principle (Pitera and Chodera 2012; Cavalli et al. 2013; Roux and Weare 2013; Boomsma et al. 2014). In this context, if the experimental restraints are imposed as averages over a number N of replicas of the protein molecule, the sampling is carried out according to the maximum entropy principle in the limit of large values of N and large values of the force constant in front of the energy restraint term. In practice, it has also been shown that the number N of replicas can be relatively small, ranging from 2 to 16 (Cavalli et al. 2013; Roux and Weare 2013; Boomsma et al. 2014).

By building on these advances, the recently proposed replica-averaged metadynamics (RAM) method (Camilloni et al. 2013; Camilloni and Vendruscolo 2014) combines the advantages of advanced sampling techniques (in this case metadynamics) to improve the conformational sampling problem with the use of experimentally-driven energy biases in the molecular dynamics simulations to improve on the force field problem. In RAM simulations, the replicas needed for the maximum entropy principle implementation of the experimental restraints are also exploited opportunistically to speed up the sampling of conformational space as they are used as collective variables.

4 Free Energy Representations of Conformational Ensembles

In order to define the specific conformation of a protein one can provide a list of the coordinates of all its atoms (e.g. by the analysis of electron density maps obtained by X-ray crystallography). In NMR spectroscopy, alternative options include the specification of a large set of distances between atom pairs (e.g. by using NOE-derived interproton distance information), or of orientations of interatomic vectors either with respect to each other (e.g. by considering J couplings) or relative to an external direction (e.g. by the use of residual dipolar couplings (RDCs)). This information is then readily translated, usually in an unequivocal manner, into atomic coordinates using standard computational methods. This approach, however, as noted above, is unsuitable for IDPs, as these proteins populate a vast number of different conformations, so that conformational ensembles, which include a variety of structures together with their statistical weights, should be specified.

A very effective way to represent conformational ensembles is through the use of free energy landscapes (Boehr et al. 2009; Frauenfelder et al. 1991; Vendruscolo and Dobson 2006). A free energy landscape represents the probability of observing a given value of a given parameter of a system (Fig. 2.2). For example, the free energy landscape $F(R_g)$ as a function of the radius of gyration R_g can be calculated as

$$F(R_g) = -\alpha \log P(R_g) \tag{2.1}$$

where α is a proportionality constant. This free energy landscape is thus proportional to the negative of the logarithm of the probability distribution $P(R_g)$ of the radius of gyration R_g. This probability distribution can be calculated from a simulation as

$$P(R_g) = \frac{N(R_g)}{N} \tag{2.2}$$

where $N(R_g)$ is the number of times that the trajectory has visited a conformation with the value R_g and N is the total number of conformations generated during the trajectory. In many cases, the free energy landscape can be calculated as a function of multiple parameters, although when more than two parameters are used the graphical representation becomes less intuitive.

The major advantage of working with free energy landscapes is that they readily give insights about a number of essential properties of IDPs, including: (1) the list

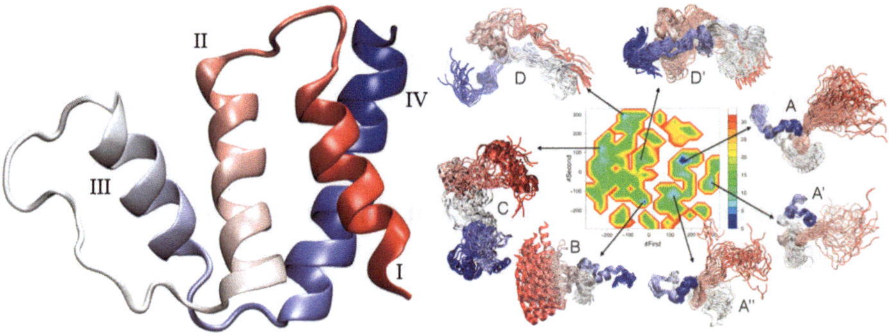

Fig. 2.2 The free energy landscape of a disordered protein is characterized by the presence of a large number of local minima. This feature is illustrated for the case of the low pH state of acyl-CoA-binding protein (*ACBP*) (Camilloni and Vendruscolo 2014), a four helix bundle protein (*left panel*) that populates a variety of conformationally distinct substates under acidic conditions (*right panel*). The characterization of highly heterogeneous conformational states of proteins in terms of free energy landscapes provides a concise and comprehensive overview of the nature and properties of such states

of their conformational states, (2) the structural features of these states, (3) the extent of their dynamics (in the sense of their equilibrium structural fluctuations), (4) their populations (i.e. their statistical weights), and (5) their mechanisms of function. More specifically, one can use the free energy landscape of a protein to find the number of its states by counting the number of minima, as such minima correspond to regions of high occupation probability, as specified by Eq. (2.1), even if sometimes in disordered states, such as those populated by IDPs, the number of minima can be very large and their populations very small. Furthermore, the extension of the basin around a given free energy minimum provides information about the overall size of the conformational ensemble corresponding to that state, as wide basins will correspond to conformational fluctuations of larger amplitude and hence to larger conformational ensembles.

Most importantly, the knowledge of the different states accessible to a protein is crucial in providing insights into the molecular basis of its function. A very common example is that of the description of the molecular recognition process between two proteins in terms of the 'conformational selection' model (Lange et al. 2008; Boehr et al. 2009). In this model, bound-like conformations are explored by the unbound protein, which then recognises its partner preferentially by binding it in one of these bound-like structures. The characterisation of the free energy landscapes of IDPs can provide a compelling demonstration of this principle (Knott and Best 2012).

5 Validation Methods for Conformational Ensembles

As the translation of the experimental measurements into structural restraints and their use in computational methods require a range of assumptions, the resulting structures should be critically assessed in order to establish whether they are correct or not. Ultimately, a powerful guiding principle is that a given conformational ensemble should enable successful predictions to be made about the outcome of the measurements of a variety of different properties of an IDP. In this case such an ensemble represents a comprehensive description of this protein within a statistical mechanics framework. When this happens, one should conclude that the conformations that have been determined, together with their statistical weights, provide a satisfactory representation of the state of a protein as their validity can be tested extensively. Suitable types of experimental parameters available for validating non-native states include fluorescence resonance energy transfer (FRET) derived distances (Haas 2005; Schuler et al. 2002; Sherman and Haran 2006; Moglich et al. 2006) and several NMR observables such as RDCs (Bernadó et al. 2005), paramagnetic relaxation enhancement (PRE) derived distances (Francis et al. 2006; Dedmon et al. 2005; Lindorff-Larsen et al. 2004), J-couplings (Smith et al. 1996), chemical shifts (Korzhnev et al. 2004; Camilloni and Vendruscolo 2014), R_2 values (Klein-Seetharaman et al. 2002) and protection factors from hydrogen exchange (Gsponer et al. 2006). The exploitation of these techniques will undoubtedly direct future efforts for increasing the resolution of IDP structures.

Several other methods of validation have been considered in the context of protein structure determination, many of which can be extended readily to IDPs. The internal consistency of a structural determination procedure can be verified by using only a subset of restraints and by testing whether the remaining ones are reproduced (cross-validation) (Spronk et al. 2004). The use of cross-validation, however, is potentially prone to error, especially in the case of highly heterogeneous ensembles of structures (Francis et al. 2006; Dedmon et al. 2005; Lindorff-Larsen et al. 2004). If for example several average inter-atomic distances are imposed on a single molecule, the only conformations compatible with this type of restraint may be compact ones. As a consequence of the time and ensemble averaging during the acquisition of NMR spectra, however, not all of the inter-atomic contacts detected experimentally need to be simultaneously present in any given conformation. For instance, the Δ131Δ fragment of staphylococcal nuclease was represented as a rather compact and native-like ensemble by imposing PRE-derived distances on a single molecule in the simulations (Gillespie and Shortle 1997). When instead the experimental distances were imposed on the average over many molecules, a much more expanded ensemble of conformations was obtained, in which states with an overall native-like topology were present but with very low statistical weights (Francis et al. 2006, Vendruscolo 2007).

In alternative validation methods, the statistical properties of the conformations obtained can be compared with those in the protein structure databases. These methods have become highly sophisticated for native states (Grishaev and Bax 2004; Spronk et al. 2004), and it will become increasingly possible to apply them to non-native states, since large repositories of high resolution structures are beginning to be available (Varadi et al. 2014).

Another highly effective strategy to assess the quality and performance of different structure determination methods is to directly compare their performances on a set of common targets. In the case of structure prediction, this community-wide strategy has been implemented and optimised in the series of Critical Assessment of Protein Structure Prediction (CASP)[1] exercises, which have run every 2 years since 1994 (Moult et al. 2014). In these assessments, experimental groups provide a set of sequences for which they have predicted the structures, and the various computational groups submit their predicted structures within a given deadline. The structures are then released publicly after the completion of the exercise and the performance of the various prediction methods is assessed. For protein structure determination methods this strategy has recently been extended to NMR spectroscopy methods with the Critical Assessment of Automated Structure Determination by NMR (CASD-NMR) assessment (Rosato et al. 2009; Rosato et al. 2012). Within the IDPbyNMR[2] initiative, there is now a plan to extend this assessment to IDPs

[1] http://predictioncenter.org/.

[2] IDPbyNMR (High resolution tools to understand the functional role of protein intrinsic disorder) is a Marie Curie activity funded under the FP7 people programme, project number 264257; http://www.idpbynmr.eu/home/.

within the broader Protein Ensemble Database (pE-DB)[3] initiative (Varadi et al. 2014), which is described in more detail in Chap. 11.

6 Conclusions

The characterization at high resolution of the conformationally heterogeneous states of IDPs is challenging because the dynamics of these states make it difficult both to obtain accurate experimental measurements and to translate them into a source of structural information. These states have structural features that are difficult to extract by extending experimental and theoretical techniques developed for describing native states (for which an average structure is well-defined) or highly denatured states (for which random coil models are often well-suited), and they require the development of novel computational methods for determining ensembles of structures. In this chapter we have discussed methods to incorporate experimental measurements into molecular simulations, as these methods represent a promising approach that is beginning to provide highly accurate conformational ensembles of IDPs.

References

Baker CM, Best RB (2013) Matching of additive and polarizable force fields for multiscale condensed phase simulations. J Chem Theor Comp 9(6):2826–2837

Bernadó P, Bertoncini CW, Griesinger C et al (2005) Defining long-range order and local disorder in native alpha-synuclein using residual dipolar couplings. J Am Chem Soc 127(51):17968–17969

Best RB (2012) Atomistic molecular simulations of protein folding. Curr Op Struct Biol 22(1):52–61

Blundell TL, Johnson LN (1976) Protein crystallography. Academic Press, New York

Board JA Jr, Causey JW, Leathrum JF Jr et al (1992) Accelerated molecular dynamics simulation with the parallel fast multipole algorithm. Chem Phys Lett 198(1):89–94

Boehr DD, Nussinov R, Wright PE (2009) The role of dynamic conformational ensembles in biomolecular recognition. Nat Chem Biol 5(11):789–796

Bonvin A, Brunger AT (1995) Conformational variability of solution nuclear-magnetic-resonance structures. J Mol Biol 250(1):80–93. doi:10.1006/jmbi.1995.0360

Bonvin A, Boelens R, Kaptein R (1994) Time-averaged and ensemble-averaged direct NOE restraints. J Biomol NMR 4(1):143–149

Boomsma W, Ferkinghoff-Borg J, Lindorff-Larsen K (2014) Combining experiments and simulations using the maximum entropy principle. PLoS Comp Biol 10(2):e1003406

Bottaro S, Lindorff-Larsen K, Best RB (2013) Variational optimization of an all-atom implicit solvent force field to match explicit solvent simulation data. J Chem Theor Comp 9(12):5641–5652

Brooks BR, Bruccoleri RE, Olafson BD et al (1983) CHARMM—a program for macromolecular energy, minimization, and dynamics calculations. J Comp Chem 4(2):187–217

Brunger AT, Adams PD, Clore GM et al (1998) Crystallography & NMR system: a new software suite for macromolecular structure determination. Acta Crystallogr D 54:905–921

[3] http://pedb.vib.be/.

Burgi R, Pitera J, van Gunsteren WF (2001) Assessing the effect of conformational averaging on the measured values of observables. J Biomol NMR 19(4):305–320. doi:10.1023/a:1011295422203
Camilloni C, Vendruscolo M (2014) Statistical mechanics of the denatured state of a protein using replica-averaged metadynamics. J Am Chem Soc 136:8982–8991
Camilloni C, Cavalli A, Vendruscolo M (2013) Replica-averaged metadynamics. J Chem Theor Comp 9(12):5610–5617
Cavalli A, Camilloni C, Vendruscolo M (2013) Molecular dynamics simulations with replica-averaged structural restraints generate structural ensembles according to the maximum entropy principle. J Chem Phys 138(9):094112
Chandler D (1987) Introduction to modern statistical mechanics. Oxford University Press, New York
Choy WY, Forman-Kay JD (2001) Calculation of ensembles of structures representing the unfolded state of an SH3 domain. J Mol Biol 308(5):1011–1032
Clore GM, Schwieters CD (2004) How much backbone motion in ubiquitin is required to account for dipolar coupling data measured in multiple alignment media as assessed by independent cross-validation? J Am Chem Soc 126(9):2923–2938
Constantine KL, Mueller L, Andersen NH et al (1995) Structural and dynamic properties of a beta-hairpin-forming linear peptide.1. Modeling using ensemble-averaged constraints. J Am Chem Soc 117(44):10841–10854. doi:10.1021/ja00149a007
Das R, Baker D (2008) Macromolecular modeling with Rosetta. Annu Rev Biochem 77:363–382
Dedmon MM, Lindorff-Larsen K, Christodoulou J et al (2005) Mapping long-range interactions in alpha-synuclein using spin-label NMR and ensemble molecular dynamics simulations. J Am Chem Soc 127(2):476–477
Dyson HJ, Wright PE (2005) Intrinsically unstructured proteins and their functions. Nat Rev Mol Cell Biol 6(3):197–208
Fennen J, Torda AE, van Gunsteren WF (1995) Structure refinement with molecular-dynamics and a Boltzmann-weighted ensemble. J Biomol NMR 6(2):163–170
Fersht AR (1999) Structure and mechanism in protein science: a guide to enzyme catalysis and protein folding. W. H. Freeman, New York
Francis CJ, Lindorff-Larsen K, Best RB et al (2006) Characterization of the residual structure in the unfolded state of the Δ131Δ fragment of staphylococcal nuclease. Proteins 65(1):145–152
Frauenfelder H, Sligar SG, Wolynes PG (1991) The energy landscapes and motions of proteins. Science 254(5038):1598–1603
Gillespie JR, Shortle D (1997) Characterization of long-range structure in the denatured state of staphylococcal nuclease. 2. Distance restraints from paramagnetic relaxation and calculation of an ensemble of structures. J Mol Biol 268(1):170–184
Grishaev A, Bax A (2004) An empirical backbone-backbone hydrogen-bonding potential in proteins and its applications to NMR structure refinement and validation. J Am Chem Soc 126(23):7281–7292
Gsponer J, Hopearuoho H, Whittaker SBM et al (2006) Determination of an ensemble of structures representing the intermediate state of the bacterial immunity protein Im7. Proc Natl Acad Sci U S A 103(1):99–104
Haas E (2005) The study of protein folding and dynamics by determination of intramolecular distance distributions and their fluctuations using ensemble and single-molecule FRET measurements. ChemPhysChem 6(5):858–870
Heise H, Luca S, de Groot BL et al (2005) Probing conformational disorder in neurotensin by two-dimensional solid-state NMR and comparison to molecular dynamics simulations. Bioph J 89(3):2113–2120
Hornak V, Abel R, Okur A et al (2006) Comparison of multiple amber force fields and development of improved protein backbone parameters. Proteins 65(3):712–725
Hub JS, De Groot BL, Van Der Spoel D (2010) g_wham—a free weighted histogram analysis implementation including robust error and autocorrelation estimates. J Chem Theor Comp 6(12):3713–3720
Karplus M, Kuriyan J (2005) Molecular dynamics and protein function. Proc Natl Acad Sci U S A 102(19):6679–6685

Kemmink J, Scheek RM (1995) Dynamic modeling of a helical peptide in solution using NMR data—multiple conformations and multi-spin effects. J Biomol NMR 6(1):33–40. doi:10.1007/bf00417489

Kessler H, Griesinger C, Lautz J et al (1988) Conformational dynamics detected by nuclear magnetic-resonance NOE values and J-coupling constants. J Am Chem Soc 110(11):3393–3396. doi:10.1021/ja00219a008

Klein-Seetharaman J, Oikawa M, Grimshaw SB et al (2002) Long-range interactions within a nonnative protein. Science 295(5560):1719–1722

Knott M, Best RB (2012) A preformed binding interface in the unbound ensemble of an intrinsically disordered protein: evidence from molecular simulations. PLoS Comp Biol 8(7):e1002605

Knowles TP, Vendruscolo M, Dobson CM (2014) The amyloid state and its association with protein misfolding diseases. Nat Rev Mol Cell Biol 15(6):384–396

Korzhnev DM, Salvatella X, Vendruscolo M et al (2004) Low-populated folding intermediates of Fyn SH3 characterized by relaxation dispersion NMR. Nature 430(6999):586–590

Krzeminski M, Marsh JA, Neale C et al (2013) Characterization of disordered proteins with ensemble. Bioinformatics 29(3):398–399

Kumar S, Rosenberg JM, Bouzida D et al (1992) The weighted histogram analysis method for free-energy calculations on biomolecules. I. The method. J Comp Chem 13(8):1011–1021

Laio A, Gervasio FL (2008) Metadynamics: a method to simulate rare events and reconstruct the free energy in biophysics, chemistry and material science. Rep Prog Phys 71(12):126601

Laio A, Parrinello M (2002) Escaping free-energy minima. Proc Natl Acad Sci U S A 99(20): 12562–12566

Lange OF, Lakomek N-A, Farès C et al (2008) Recognition dynamics up to microseconds revealed from an RDC-derived ubiquitin ensemble in solution. Science 320(5882):1471–1475

Lindorff-Larsen K, Kristjansdottir S, Teilum K et al (2004) Determination of an ensemble of structures representing the denatured state of the bovine acyl-coenzyme A binding protein. J Am Chem Soc 126(10):3291–3299

Lindorff-Larsen K, Best RB, DePristo MA et al (2005) Simultaneous determination of protein structure and dynamics. Nature 433(7022):128–132

Lindorff-Larsen K, Maragakis P, Piana S et al (2012a) Systematic validation of protein force fields against experimental data. PLoS ONE 7(2):e32131

Lindorff-Larsen K, Trbovic N, Maragakis P et al (2012b) Structure and dynamics of an unfolded protein examined by molecular dynamics simulation. J Am Chem Soc 134(8):3787–3791

Markwick PR, Bouvignies G, Blackledge M (2007) Exploring multiple timescale motions in protein GB3 using accelerated molecular dynamics and NMR spectroscopy. J Am Chem Soc 129(15):4724–4730

Mittermaier A, Kay LE (2006) New tools provide new insights in NMR studies of protein dynamics. Science 312(5771):224–228

Moglich A, Joder K, Kiefhaber T (2006) End-to-end distance distributions and intrachain diffusion constants in unfolded polypeptide chains indicate intramolecular hydrogen bond formation. Proc Natl Acad Sci U S A 103(33):12394–12399

Monticelli L, Kandasamy SK, Periole X et al (2008) The Martini coarse-grained force field: extension to proteins. J Chem Theor Comp 4(5):819–834

Moult J, Fidelis K, Kryshtafovych A et al (2014) Critical assessment of methods of protein structure prediction (CASP)—round X. Proteins 82(S2):1–6

Piana S, Klepeis JL, Shaw DE (2014) Assessing the accuracy of physical models used in protein-folding simulations: quantitative evidence from long molecular dynamics simulations. Curr Op Struct Biol 24:98–105

Pitera JW, Chodera JD (2012) On the use of experimental observations to bias simulated ensembles. J Chem Theor Comp 8(10):3445–3451

Rosato A, Bagaria A, Baker D et al (2009) CASD-NMR: critical assessment of automated structure determination by NMR. Nat Methods 6(9):625–626

Rosato A, Aramini JM, Arrowsmith C et al (2012) Blind testing of routine, fully automated determination of protein structures from NMR data. Structure 20(2):227–236

Roux B, Weare J (2013) On the statistical equivalence of restrained-ensemble simulations with the maximum entropy method. J Chem Phys 138(8):084107
Schuler B, Lipman EA, Eaton WA (2002) Probing the free-energy surface for protein folding with single-molecule fluorescence spectroscopy. Nature 419(6908):743–747
Schwieters CD, Kuszewski JJ, Clore GM (2006) Using Xplor-NIH for NMR molecular structure determination. Prog Nucl Mag Res Spectrosc 48(1):47–62
Shaw DE, Maragakis P, Lindorff-Larsen K et al (2010) Atomic-level characterization of the structural dynamics of proteins. Science 330(6002):341–346
Sherman E, Haran G (2006) Coil-globule transition in the denatured state of a small protein. Proc Natl Acad Sci U S A 103(31):11539–11543
Smith LJ, Bolin KA, Schwalbe H et al (1996) Analysis of main chain torsion angles in proteins: Prediction of NMR coupling constants for native and random coil conformations. J Mol Biol 255(3):494–506
Spronk C, Nabuurs SB, Krieger E et al (2004) Validation of protein structures derived by NMR spectroscopy. Prog Nucl Mag Res Spectrosc 45(3–4):315–337
Torda AE, Scheek RM, van Gunsteren WF (1989) Time-dependent distance restraints in molecular-dynamics simulations. Chem Phys Lett 157(4):289–294. doi:10.1016/0009-2614(89)87249-5
Tozzini V (2005) Coarse-grained models for proteins. Curr Op Struct Biol 15(2):144–150
Uversky VN (2013) A decade and a half of protein intrinsic disorder: Biology still waits for Physics. Protein Sci 22(6):693–724
van Kampen NG (1992) Stochastic processes in physics and chemistry. North-Holland, Amsterdam, New York
Varadi M, Kosol S, Lebrun P et al (2014) pE-DB: A database of structural ensembles of intrinsically disordered and of unfolded proteins. Nucl Acids Res 42(D1):D326–D335
Vendruscolo M, Dobson CM (2006) Dynamic visions of enzymatic reactions. Science 313(5793):1586–1587
Vendruscolo M (2007) Structure determination of highly heterogenous states of proteins. Curr Op Struct Biol 17:15–20
Vendruscolo M, Dobson CM (2011) Protein dynamics: Moore's law in molecular biology. Curr Biol 21(2):R68–R70
Wüthrich K (1986) NMR of proteins and nucleic acids. Wiley, New York
Zhu F, Hummer G (2012) Convergence and error estimation in free energy calculations using the weighted histogram analysis method. J Comp Chem 33(4):453–465

Chapter 3
NMR Methods for the Study of Instrinsically Disordered Proteins Structure, Dynamics, and Interactions: General Overview and Practical Guidelines

Bernhard Brutscher, Isabella C. Felli, Sergio Gil-Caballero, Tomáš Hošek, Rainer Kümmerle, Alessandro Piai, Roberta Pierattelli and Zsófia Sólyom

Abstract Thanks to recent improvements in NMR instrumentation, pulse sequence design, and sample preparation, a panoply of new NMR tools has become available for atomic resolution characterization of intrinsically disordered proteins (IDPs) that are optimized for the particular chemical and spectroscopic properties of these molecules. A wide range of NMR observables can now be measured on increasingly complex IDPs that report on their structural and dynamic properties in isolation, as part of a larger complex, or even inside an entire living cell. Herein we present basic NMR concepts, as well as optimised tools available for the study of IDPs in solution. In particular, the following sections are discussed hereafter: a short introduction to NMR spectroscopy and instrumentation (Sect. 3.1), the effect of order and disorder on NMR observables (Sect. 3.2), particular challenges and bottlenecks for NMR studies of IDPs (Sect. 3.3), 2D HN and CON NMR experiments: the fingerprint of an IDP (Sect. 3.4), tools for overcoming major bottlenecks of IDP NMR studies (Sect. 3.5), ^{13}C detected experiments (Sect. 3.6), from 2D to 3D: from simple snapshots to site-resolved characterization of IDPs (Sect. 3.7), sequential NMR assignment: 3D experiments (Sect. 3.8), high-dimensional NMR experiments (nD, with n > 3) (Sect. 3.9) and conclusions and perspectives (Sect. 3.10).

B. Brutscher (✉) · Z. Sólyom
Institut de Biologie Structurale, Université Grenoble 1, CNRS, CEA, 71 avenue des Martyrs, 38044 Grenoble Cedex 9, France
e-mail: bernhard.brutscher@ibs.fr

I. C. Felli (✉)· T. Hošek · A. Piai · R. Pierattelli (✉)
CERM and Department of Chemistry "Ugo Schiff", University of Florence, Via Luigi Sacconi 6, 50019 Sesto Fiorentino, Florence, Italy
e-mail: felli@cerm.unifi.it

R. Pierattelli
e-mail: pierattelli@cerm.unifi.it

R. Kümmerle · S. Gil-Caballero
Bruker BioSpin AG, Industriestrasse 26, 8117 Fällanden, Switzerland

I. C. Felli, R. Pierattelli (eds.), *Intrinsically Disordered Proteins Studied by NMR Spectroscopy*, Advances in Experimental Medicine and Biology,
DOI 10.1007/978-3-319-20164-1_3

Keywords NMR basics · pulse sequences · NMR instrumentation · Sequential assignment · BEST · ^{13}C detection · high dimensional NMR

1 A Short Introduction to NMR Spectroscopy and Instrumentation

The possibilities offered by modern nuclear magnetic resonance (NMR) spectroscopy are enormous and cover a wide range of applications in physics, chemistry, biology, medicine and material sciences. Past and present progress in the field is mainly based on technical improvements of the NMR spectrometer (magnetic field strength, NMR electronics, probes, etc.) and on the development of a variety of pulse sequences that exploit the basic principles of NMR spectroscopy in an ingenious way in order to obtain the desired information on a particular sample. The first part of this chapter will therefore be dedicated to the most recent key developments in NMR instrumentation, mainly stimulated by emerging scientific challenges, and to the basic principles of NMR spectroscopy.

The phenomenon of nuclear magnetic resonance was discovered by Purcell et al. (1946) and Bloch (1946); shortly afterwards they were awarded the Nobel Prize in Physics *"for their development of new methods for nuclear magnetic precision measurements and discoveries in connection therewith"*. Since its discovery, NMR spectroscopy, further to its fundamental role in physics, has become a very powerful tool in chemistry and biology for structural studies of small organic and inorganic compounds as well as large systems, including polymers and biomolecules. Besides X-ray crystallography and electron microscopy, NMR is the only method capable of providing atomic resolution information on the structure of biological macromolecules. Compared to X-ray crystallography, it allows studies of biological macromolecules in the liquid state—meaning that crystallization of the molecule is not required. This makes structural studies of highly dynamic systems such as intrinsically disordered proteins (IDPs) possible. NMR spectroscopy can also be employed for studies of interactions of biomolecules with small ligands, metal ions and other biological macromolecules. Recently, the importance of regarding biomolecules as dynamic ensembles instead of a single static entity has become widely recognized. NMR spectroscopy provides a unique tool to access dynamics information at atomic resolution, from the picosecond timescale to slow exchange processes on the second (or even slower) timescale. This is particularly important for intrinsically disordered proteins, which are often characterized by heterogeneous dynamic properties along the polypeptide chain.

1.1 The Basic Principles of NMR

NMR spectroscopy studies the interaction of matter with radiofrequency electromagnetic waves that excite magnetic transitions of the atomic nuclear spins. Indeed, an atomic nucleus, in order to be observable by NMR spectroscopy, has to possess

Table 3.1 Gyromagnetic ratios and natural abundance of nuclei important for NMR studies of proteins

Isotope	Spin I	Gyromagnetic ratio γ_n (10^6 rad s^{-1} T^{-1})	Gyromagnetic ratio/2π $\gamma_n/2\pi$ (MHz T^{-1})	Natural abundance isotope (%)
^1H	½	267.513	42.576	99.98
^{13}C	½	67.262	10.705	1.108
^{15}N	½	−27.116	−4.316	0.37

a nonzero spin quantum number I. The relevant nuclear isotopes for NMR studies of proteins are ^1H, ^{13}C, and ^{15}N; all of them have a spin quantum number of $I=½$. When the nuclei are placed in a static magnetic field B_0, which is by convention aligned along the z-axis, the nuclear magnetic momentum $\boldsymbol{\mu}$ of the spins interact with the magnetic field leading to a splitting of the energy levels (Zeeman splitting) according to Eq. 3.1

$$E = -\boldsymbol{\mu}\cdot\boldsymbol{B} = -\mu_Z B_0 = -\gamma I_Z B_0 = -\gamma m\hbar B_0 \tag{3.1}$$

where γ is the gyromagnetic ratio and $\hbar$ is the reduced Planck constant. For spin ½ nuclei the magnetic quantum number m can take the values of $m=+½$ and $m=-½$. The gyromagnetic ratios and natural abundance of the nuclear isotopes important in biomolecular NMR are summarized in Table 3.1.

Equation 3.1 is of the utmost importance for the sensitivity of NMR experiments. The energy difference between the two Zeeman energy levels is given by:

$$\Delta E = -\gamma\hbar B_0 = \hbar\omega_0 \tag{3.2}$$

where ω_0 is the characteristic NMR frequency (Larmor frequency) of a nuclear spin at a given magnetic field strength.

At thermal equilibrium and temperature T, the ratio of nuclear spins in the lower (E_α) and higher (E_β) energy state can be calculated from the Boltzmann distribution:

$$\frac{N_\beta}{N_\alpha} = e^{-\frac{\Delta E}{kT}} \tag{3.3}$$

where k is the Boltzmann constant.

The NMR signal is proportional to the magnetization, which is in turn dependent on the spin polarization P given by the population difference between the two states divided by the total number of spins:

$$P = \frac{N_\alpha - N_\beta}{N_\alpha + N_\beta} \tag{3.4}$$

In the case of ^1H nuclei, which possess the highest gyromagnetic ratio among the spin ½ nuclei in proteins, the ratio of spins in the upper energy state versus the lower energy state is 0.999872 at room temperature on an 800 MHz magnet ($B_0=18.8$ T). This means that only a very small fraction of the spins present in the sample contributes to the observed NMR signal. Furthermore, in the case of ^{13}C and ^{15}N nuclei (often called heteronuclei), the natural abundance of the NMR active nuclei is very low, as shown in Table 3.1. Isotope enrichment techniques have therefore been developed to enhance the sensitivity of NMR techniques involving such heteronuclei.

Sensitivity and resolution are the two most important factors influencing the outcome of NMR experiments. NMR sensitivity, which is defined as the signal to noise ratio (SNR) obtained in a fixed amount of time, can be described by the following equation:

$$SNR \propto \frac{N}{V} \gamma_{exc} \gamma_{det}^{2.5} B_0^{2.5} n_{scan}^{0.5} \frac{1}{\sqrt{R_s(T_a + T_s) + R_c(T_a + T_c)}} \tag{3.5}$$

where N is the number of spins, V the active volume, γ_{exc} the gyromagnetic ratio of the excited nuclei and γ_{det} that of the detected nuclei, B_0 the static magnetic field, n_{scan} the number of scans (experimental repetitions), R_S the resistance of the sample and R_c that of the coil, T_a the temperature of the preamplifier and T_s and T_c those of the sample and of the coil, respectively (Hoult and Richards 1976; Kovacs et al. 2005). Other factors that depend on the properties of the sample under investigation and on the type of pulse sequence used contribute to the sensitivity of an NMR experiment, and therefore to the possibility of accessing the desired information. These will be discussed in detail in the following paragraphs.

Some general conclusions for improving the sensitivity of an NMR experiment can be derived from Eq. 3.5. The NMR signal is proportional to the amount of spins present in the sample; it is therefore desirable to use highly concentrated samples for NMR spectroscopy. However, in practice, the maximum concentration of a protein sample is often limited by protein solubility and obtaining large quantities of isotopically labelled proteins can be expensive and time consuming. Most protein NMR experiments use proton excitation and detection because of the high gyromagnetic ratio of these nuclei; nevertheless, ^{13}C direct detection also provides a valuable tool for biomolecular NMR applications (see Sect. 3.6). The linear dependence on B_0 explains the on-going efforts of NMR manufacturers to develop magnets with higher magnetic fields. To date, the highest magnetic field of a commercial magnet corresponds to 1 GHz proton Larmor frequency (23.5 T), and 1.2 GHz magnets (28.2 T) are under development. The SNR increases with the square root of the number of scans. Increasing the overall measurement time thus represents a common method for spectral improvement. The last contribution to NMR sensitivity comes from the electronic detection circuit. The SNR of NMR increases with a lowering of the temperature of the preamplifier (T_a), the sample (T_s), and the coil (T_c), as well as the resistance of the sample (R_s) and the coil (R_c). A recent development that greatly enhanced the sensitivity of NMR spectroscopy was the introduction of probes with cryogenically cooled (to about 20 K) detection circuits. However, protein samples with high ionic strengths degrade the beneficial effects of cryogenically cooled probes.

Resolution is the other important factor to extract atomic resolution information from NMR spectra. It depends on chemical shift dispersion and signal linewidths. Two peaks can be resolved when their difference in frequency is larger with respect to their linewidths. Nuclear spins that are characterized by favourable chemical shift dispersion and small linewidths for the system investigated should thus be exploited to improve the resolution of the spectra. The possibility of introducing additional indirect dimensions in NMR experiments provides the other invaluable tool for enhancing the resolution by spreading signals in additional dimensions and thus reducing the possibility of accidental cross-peak overlap.

1.2 NMR Instrumentation and Recent Improvements

NMR spectrometers are composed of three main components, plus some optional accessories, such as automatic sample changers or variable temperature (VT) gas preconditioning devices:

1. The superconducting magnet generating the static magnetic field B_0, with the superconducting coil immersed in liquid helium at a temperature of 4.2 K (or about 2 K for subcooled "pumped" magnets). Commercially available superconducting NMR magnets have a freely accessible vertical bore with a diameter of 50–54 mm (standard bore, used mainly for liquid state and solid state applications), 89 mm (wide bore, used mainly for solid state and micro-imaging applications) or 154 mm (super wide bore, mostly used for micro-imaging applications). The temperature inside the magnet bore can be adjusted independently.
2. The probe, positioned in the bore of the superconducting magnet, is equipped with radiofrequency (RF) coils/antennas emitting RF pulses and detecting with the same RF coils the voltage induced by the precessing nuclear spin magnetization in the NMR sample. One can distinguish between three major fields of NMR applications, requiring different types of probes: imaging (or micro-imaging), magic-angle spinning (MAS) solid-state applications, and high-resolution liquid-state applications. In the following, we will focus on NMR probes for liquid-state applications. The standard sample tube diameter is 5 mm, but optimized probes are available for sample diameters in the range 1 to 10 mm. Modern probes are equipped with temperature control, a field lock channel (e.g. ^{2}H), an actively shielded coil enabling pulsed field gradients of ~5 mT along one or several axis by applying DC currents (~10 A), and the capability to optimize the resonance circuit tuning and matching via software control. Most liquid state probes are built with two distinct RF coils to accommodate 2, 3 or 4 frequency circuits besides the field lock. Consequently, they are called double, triple or quadruple resonance probes. A further distinction or classification can be done depending on whether the inner RF coil (closer to the sample) is tuned to ^{1}H ("inverse probe", TXI) or to a X nucleus ("observe probe", TXO). Historically, the inner coil of probes was tuned to observe the less sensitive X nuclei. With the introduction of pulse sequence elements like INEPT, the advantage of detecting X nuclei via the more sensitive ^{1}H nucleus has led to the development of inverse probes that provide increased sensitivity.
3. The spectrometer console has all the electronic components required for executing complex multi-channel, multi-nuclear pulse experiments, as well as for detection of the electronic signal induced in the RF coils of the probe. In particular, RF pulses are generated at the Larmor frequencies of the nuclear spins

that need to be manipulated during the experiment. The console ensures their correct timing, phase and amplitude modulation, and signal amplification. The spectrometer electronics is typically characterized by the number of independent radiofrequency (RF) channels, e.g. the number of different nuclei being addressable within one NMR experiment. A further distinction is the peak power delivered by the RF amplifier, the requirements strongly depending on the probe used for the application. For liquid state NMR, high-band frequency (^{3}H, ^{1}H, ^{19}F) excitation typically needs less than 50 W, whereas for low γ nuclei (often also called X nuclei) RF pulses may require more than 500 W at high B_0 fields. The console also hosts an equipment, called spectrometer or field lock, that serves to maintain the magnetic field strength seen by the spins throughout the whole experiment duration, resulting in field variations of less than ~0.1 Hz (^{1}H Larmor frequency). This corresponds to a precision of $\sim 10^{-10}$ of the static B_0 field, maintained for up to several days (in the case of long biomolecular experiments). The B_0 field is homogenized over the sample volume by means of a shim unit that generates additional small magnetic fields (so-called shims), again resulting in spatial B_0 field variation of less than 10^{-9}. Last but not least, a variable temperature unit serves to measure and regulate the NMR sample temperature throughout the experiment.

Over the past years all these spectrometer components have seen major technological improvements, so that besides ease of use, both performance and stability of modern NMR spectrometers have been significantly improved.

Improved Magnet Technology. Using NbTi and Nb_3Sn as superconducting material, combined with state-of-the art wire technology, the highest currently achievable magnetic field strength is 23.5 T, equivalent to a ^{1}H Larmor frequency of 1000 MHz. Increasing the magnetic field strength further will require the use of new types of superconducting wire, so-called high temperature superconducting tapes presenting substantially higher critical currents I_C compared to conventional low temperature superconductors in the presence of magnetic field. The first next generation magnets with a magnetic field strength of ~1.2 GHz are expected to become available within the next 3 years. For magnetic field strengths up to 21.1 T (900 MHz) incremental improvements of low temperature superconducting wire have been made over the past years to design more compact and better shielded magnets. These compact magnets have, at identical field strength, smaller dimensions and less weight with substantially reduced stray fields, thus making the laboratory and on-site space requirements for an NMR spectrometer less critical and less costly. As an example, the weight of an 18.8 T magnet (800 MHz) could be cut in half with two design steps over the last decade.

In addition, modern NMR magnets have a significantly reduced consumption of liquid helium. This not only translates into reduced operational costs, but increased liquid helium hold times may become critical during periods of helium shortage, as experienced at the end of 2012/beginning of 2013. Recently also "cryogen-free" magnets became available, where an active refrigeration technology, mainly consisting in two pulse tubes, allows to re-liquefy evaporating helium gas (zero boil-off), and to completely avoid the outer liquid nitrogen dewar. Such "cryogen-free" magnets are currently available for magnetic resonance imaging (MRI) systems, and for gyrotron magnets of solids DNP NMR spectrometers. However, they are not (yet) suitable for high-resolution NMR applications as required for the study of IDPs.

Improved Probe Technology. Experimental sensitivity at constant mass or constant sample concentration has always been a critical factor for NMR probe design. A first possibility to improve the probe performance is to enhance the efficiency of the RF coil, e.g. to achieve higher B_1 fields in the sample with a given coil current I (B_1/I). This can be achieved for smaller sample geometries, so that for example the mass sensitivity of a 1 mm probe is roughly 4-fold higher compared to a 5 mm inverse probe. A second possibility to enhance sensitivity for a given probe type is to reduce the contribution of the thermal noise originating from the RF coil, RF resonance circuit as well as the pre-amplification of the observed NMR signal (CPTXI, CPTXO). Cooling these parts down from room temperature to roughly −200 °C (using liquid nitrogen as cooling medium) leads to sensitivity enhancements of a factor of 2 to 3. Using helium gas (instead of nitrogen) to cool the pre-amplification stages allows to further decrease the temperature to roughly −260 °C, resulting in sensitivity gains of a factor 4–5 compared to an equivalent conventional probe. Historically, such helium cooled probes were actually available before the nitrogen-cooled ones. A drawback of these helium-cooled cryoprobes is that they require a more complex and costly infrastructure (helium compressor, cooling of the helium compressor) compared to the liquid nitrogen cooled probe, resulting in higher running costs. Also note that these maximum sensitivity gains stated above for cryogenically cooled probes only apply to samples of low electronic conductivity. For salty samples the gain in sensitivity is less. However, the situation may be improved by using NMR sample tubes of smaller diameter or of optimized geometry.

The impressive improvements in sensitivity achieved with cryogenic technology have also enabled the design of probes optimized for ^{13}C direct detection, stimulating the design of a variety of new experiments optimized for IDPs, as described later in this chapter.

Improved NMR Electronics. Modern console electronics typically offer fully digital RF frequency generation with a timing resolution of a few tens of nanoseconds to enable the user fast switching of frequency, amplitude and phase as required for modern NMR experiments. The timing as well as the amplitude of the delivered RF and field gradient pulses need to be stable over the full length of the NMR experiment, e.g. up to several days. Another critical parameter for high-resolution NMR spectroscopy is the uniformity and stability of the NMR sample temperature. Most NMR samples are positioned in a stream of temperature-controlled gas, and the sensors are placed outside the NMR sample volume in the gas flow. This setup allows a high level of temperature stability and uniformity over the sample volume. However, the RF pulses in the NMR pulse sequence may heat up the sample, an effect that is not detected by the external temperature sensors. As a result the actual sample temperature may deviate by several degrees from the temperature measured close to the sample. This problem is most critical for high frequency nuclei. It can be reduced by optimizing the RF probe coil design to limit the presence of electrical fields in the sample area, or by reducing sample conductivity if possible. An alternative is the measurement of the actual temperature during an experiment from the ^2H resonance of spins within the NMR sample. Such an "NMR thermometer" requires the presence of two ^2H spin species with temperature-dependent chemical shifts. The measurement of the chemical shift difference of the two ^2H signals, via the lock channel, allows to regulate permanently the temperature within the NMR sample and to correct for RF heating. A major advantage of this direct measurement of the sample temperature via the nuclear spins is the possibility of recording data on different NMR spectrometers (e.g. field-dependent relaxation studies) at identical absolute temperature.

2 The Effect of Order and Disorder on NMR Observables

In the following sections we will introduce the main NMR observables that are used to derive structural and dynamic information on proteins, with a particular emphasis on how the particular features of IDPs affect these NMR observables.

2.1 Structured Versus Disordered Proteins

In order to better understand the impact of structural order and disorder on NMR experiments and spectral parameters, it is important to recognize the principal features that distinguish a well-folded globular protein from a largely unfolded, highly flexible IDP. These are illustrated in Fig. 3.1.

Hydrophobic interactions and the formation of hydrogen bonds are the main driving forces for the folding process of a polypeptide chain into a globular 3D structure and for the stabilization of such a compact fold. The main consequences of the presence of a stable and well-defined structure that are of importance for the NMR properties of the molecule are a high proton density, the exclusion (to a large extent) of water from the interior of the protein and, to a first crude approximation, the possibility of describing its rotational motion with an overall rotational correlation time. The high proton density and large rotational correlation time make ^1H-^1H nuclear Overhauser effects (NOEs) the main source of structural information for globular proteins, whereas they are only of little use in IDP research, where the effect is mainly restricted to proton pairs that are close in the primary sequence and is on average smaller than that for globular proteins. Indeed, the dynamic behaviour of IDPs is significantly different from that of structured proteins as a result of the small energy difference between conformers that is responsible for the high flexibility typical of IDPs and allows easy inter-conversion between many different conformations. Therefore, even to a first approximation, a single overall rotational correlation time cannot be defined for IDPs and local effective correlation times are on average smaller with respect to those used to describe structured proteins of similar size. The differences in ^1H density and molecular dynamics between IDPs and globular proteins also have a strong influence on the NMR signal chemical shifts and the relaxation parameters. Conformational averaging drastically reduces contributions to chemical shifts deriving from the local environment and causes severe resonance

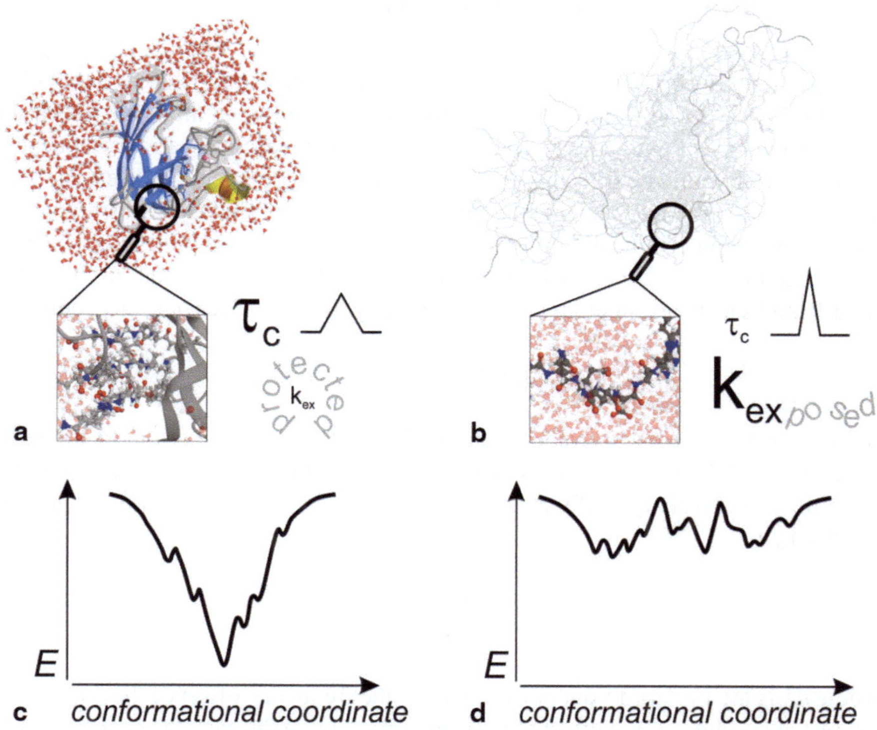

Fig. 3.1 Schematic illustrations of differences between structured (**a**) and intrinsically disordered (**b**) proteins (for reasons of clarity the molecules of water were omitted in the ensemble of IDP conformers), and energy landscapes of well-folded proteins (**c**) and IDPs (**d**) in the native state

overlap. The high flexibility of IDPs has a strong impact on nuclear relaxation rates, and thus on NMR linewidths. Depending on the time scale of motions, several situations can be encountered: on one hand, conformational exchange processes can result in extensive line broadening, and in the extreme case, the absence of any detected NMR signal. On the other hand fast motions result in narrow NMR lines, a property that makes highly flexible IDPs particularly amenable to NMR characterization. A variety of complex NMR experiments can be conceived for such highly flexible IDPs, where the favourable relaxation properties allow for multiple transfer steps and the narrow linewidths contribute to increase the resolution in the resulting spectra. Finally, IDPs are also characterized by non-compact conformations and the exposure of labile protons, e.g. amide protons, to the solvent results in high exchange rates. In fact, the measurement of solvent exchange rates is used to distinguish between highly structured and disordered regions of proteins by identifying solvent exposed and solvent protected amide sites (Schanda et al. 2006a). Approaching physiological pH and temperature, chemical exchange of solvent exposed amide protons may broaden resonances beyond detection and in this case alternative nuclear spins, such as aliphatic ^{1}H or ^{13}C, should be detected to access information on the IDP through NMR (Gil et al. 2013).

2.2 NMR Peak Positions and Chemical Shifts

Peak positions are dependent on the resonance frequencies of the observed nuclei. Based on Eq. 3.2 one would expect to observe a single line in the NMR spectrum for each nucleus of a given type (^{1}H, ^{13}C, ^{15}N) at the respective Larmor frequency. However, this would be true only in the hypothetical experiment of observing "naked" nuclei. In practice, nuclei of the same type resonate at slightly distinct frequencies if they are in chemically different environments within the molecule. The reason is that the nuclei experience a net magnetic field B that is the sum of the static field B_0 and secondary shielding fields induced by the local electronic environment:

$$B = (1 - \sigma) B_0 \tag{3.6}$$

where σ is the isotropic average shielding factor. This is at the origin of the term "chemical shift" used to measure resonance positions in NMR spectroscopy. Chemical shift values δ are typically measured relative to the chemical shift of a standard according to:

$$\delta = \frac{\Omega - \Omega_{ref}}{\omega_0} \cdot 10^6 \tag{3.7}$$

where ω_0 is the Larmor frequency in MHz, Ω the frequency offset of the nucleus of interest in Hz and Ω_{ref} the offset of the standard in Hz. In this way the chemical shift becomes a field-independent quantity, expressed in parts per million (ppm). The chemical shift makes NMR an atomic resolution technique: once the resonances are assigned to their respective nuclei, their response to further manipulations can be followed.

Chemical shifts can be very characteristic for nuclei in different chemical moieties such as the different amino acids constituting a protein, and they provide the first useful information that can be obtained from the NMR spectrum. The stable local three-dimensional structure, or partially populated secondary structural conformations, create a unique local electronic environment and thus the contribution to chemical shift coming from isotropic shielding is different for each nuclear spin. In IDPs, the lack of a stable structure results in averaging of a large part of the contributions to the chemical shift coming from the local chemical environment, which is the reason for the low chemical shift dispersion observed in the NMR spectra (Fig. 3.2).

In addition, peak intensities (areas or volumes) can be easily determined and may provide useful atomic resolution information. Peak intensities in the NMR spectra are proportional to the number of nuclear spins giving rise to the peaks. They also report on differential signal loss during an NMR experiment due to relaxation processes. They therefore provide a first useful indication of the heterogeneous structural and dynamic properties of a protein. Peak linewidths would also provide useful information but they are more difficult to measure as they are influenced by different factors (see Sect. 3.2.5). In many cases, for simplicity, the determination of peak heights is used.

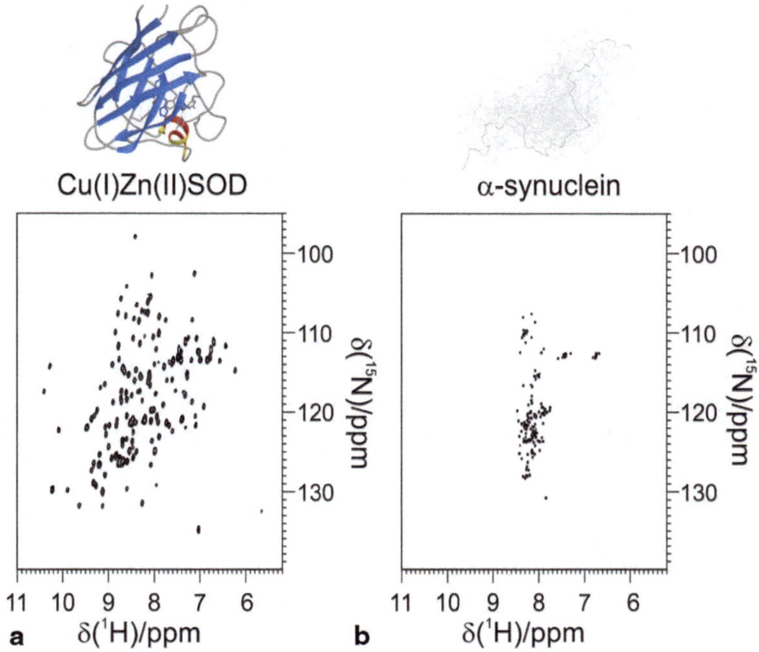

Fig. 3.2 The different chemical shift dispersion of the structured and intrinsically disordered proteins is demonstrated by 2D ^{1}H-^{15}N correlation spectra acquired on two proteins of similar size, but characterized by different structural properties. **a** The HSQC spectrum of structured monomeric Cu(I)Zn(II) superoxide dismutase (1.5 mM sample in 20 mM phosphate buffer, pH 5.0, at 298 K; PDB code: 1BA9). **b** The HSQC spectrum of intrinsically disordered α-synuclein (1.0 mM sample in 20 mM phosphate, pH 6.4, 0.5 mM EDTA, 200 mM NaCl, at 285.5 K). The experiments were acquired on a 700 MHz Bruker AVANCE spectrometer equipped with a CPTXI probe

2.3 Secondary Chemical Shifts

The chemical shifts of backbone nuclei are sensitive to the local backbone geometry, and therefore provide useful information on the occurrence and propensity of secondary structural elements along the protein backbone (Spera and Bax 1991; Wishart et al. 1991). For a ^{13}C, ^{15}N enriched protein, a large number of chemical shifts ($^{1}H^{N}$, ^{15}N, $^{13}C'$, $^{13}C^{\alpha}$, $^{13}C^{\beta}$) can be measured, all of which are reporters of secondary structure even if to a different extent. In order to extract this information, a so-called secondary chemical shift is computed as the difference of the measured chemical shift and a predicted random coil value for each nucleus (Wishart et al. 1992; Wishart and Sykes 1994; Schwarzinger et al. 2001; Tamiola et al. 2010; Kjaergaard et al. 2011; Kjaergaard and Poulsen 2012). Random coil chemical shifts are the theoretical chemical shifts of a polypeptide of the same amino acid sequence characterized by lack of long-range order and secondary structure. However, the conformational sampling of a polypeptide is never completely random, in the sense that all dihedral angles are sampled with equal probability because of steric hindering and chemical interactions between neighbouring side-chains. The distribution of the sampled dihedral angles along the protein backbone is determined by the conformational Gibbs free energy, which in turn depends on temperature and solvent effects. The accuracy of secondary chemical shifts thus depends on the quality of the predicted random coil chemical shift values. This aspect becomes even more crucial for IDPs, since the measured chemical shifts are usually close to the corresponding random coil values. Figure 3.3 illustrates how C^{α} and C^{β} secondary chemical shifts can be used to identify α-helical and β-sheet regions and highly unstructured segments in globular proteins, as well as residual secondary structures in IDPs.

Different tables of ^{13}C and ^{15}N random coil values for different amino acids, taking into account next and previous neighbours, pH and solvent effects can be found in the literature (Schwarzinger et al. 2001; Zhang et al. 2003; Tamiola et al. 2010; Kjaergaard et al. 2011; Kjaergaard and Poulsen 2012). One of the most commonly used random coil chemical shift data sets, based on chemical shift measurements on small poly-glycine peptides, is the one reported by Wishart et al. (Wishart et al. 1995). Recently, temperature and pH correction factors have been introduced from a peptide-based study employing glutamine peptides, as their conformational sampling is considered to be more representative (Kjaergaard et al. 2011). The other

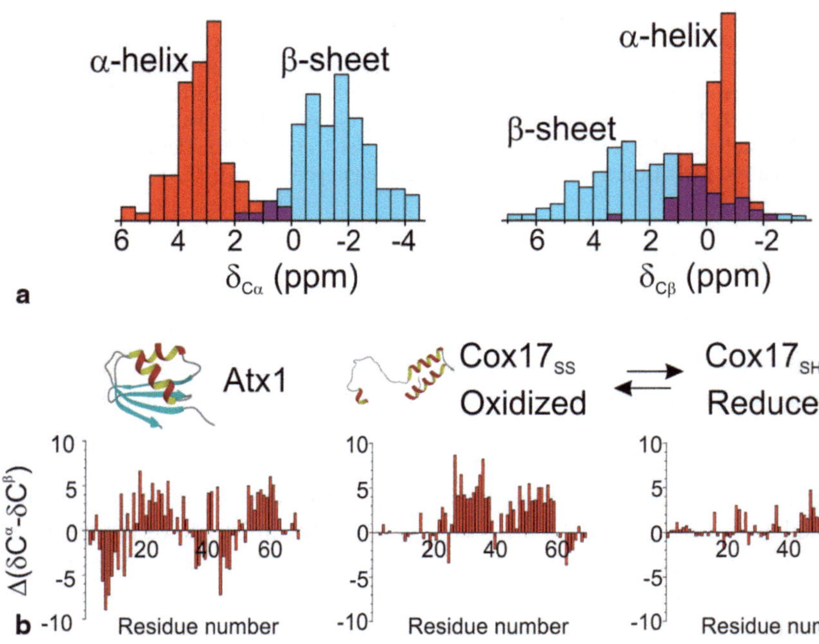

Fig. 3.3 **a** Histograms of the C^{α} and C^{β} secondary chemical shift distribution in an α-helix and a β-sheet (adapted from Spera and Bax 1991). **b** The identification of the secondary structure elements based on the secondary chemical shifts is shown on the examples of the structured metallochaperone Atx1 (Arnesano et al. 2001), the partially structured oxidized form of the copper chaperone Cox17 (Arnesano et al. 2005) and of its unfolded reduced form (Bertini et al. 2011a). (Adapted from Felli et al. 2012)

type of reference data can be obtained from protein chemical shift databases. This approach has the drawback that sample conditions such as pH and temperature vary between different entries. Furthermore, neighbouring and other chemical effects will be biased depending on the composition of the entries. The most commonly used database is refDB by Zhang et al. (Zhang et al. 2003). An IDP-based random coil chemical shift database has recently been introduced (Tamiola et al. 2010). Neighbour corrections are usually required to account for changes in conformational sampling because of steric clashes and other chemical effects such as electrostatics or ring current shifts (Wishart et al. 1995; Schwarzinger et al. 2001; Kjaergaard and Poulsen 2011).

2.4 Line Splittings and Spin Coupling Constants

Nuclear spins are never completely isolated; mutual interactions between neighbouring nuclear spins are therefore always present and give rise to couplings. The coupling derives from two main contributions: scalar (through bonds) and dipolar (through space). While the latter is averaged out in isotropic solutions, the former, mediated by the electrons in covalent bonds (scalar or ^{n}J-coupling, where n indicates the number of covalent bonds separating the two nuclei involved in the coupling), causes splitting of the NMR signal. The frequency differences between the multiplet components of an NMR signal measured in Hz then reflect the strength of the scalar couplings. Scalar couplings involving ^{1}H, ^{13}C and ^{15}N are exploited for coherence transfer in the majority of multidimensional NMR experiments discussed in the following paragraphs. Scalar couplings, in particular ^{3}J-couplings in a polypeptide, also contain valuable structural information as they depend on the intervening dihedral angle according to the Karplus relation (Karplus 1959). Inter-conversion between different conformers typical of IDPs results in averaged scalar couplings; while this has little impact on coherence transfer in multidimensional NMR experiments, it does influence the interpretation of ^{3}J-couplings in terms of intervening dihedral angles. For example, $^{3}J_{HH}$ couplings are sometimes used to confirm the presence of partially populated secondary structural elements (Billeter et al. 1992; Vuister and Bax 1993; Case 2000; Otten et al. 2009).

The magnetic moments of two nuclear spins interact "through space" via the dipolar mechanism with an interaction strength that is inversely proportional to r^{3}, with r being the inter-nuclear distance. The dipolar interaction has a particular angular dependence with respect to the static magnetic B_0 field and, as a consequence, in isotropic solutions where all orientations are sampled with equivalent probability this interaction averages to zero and does not give rise to line splittings. However, it is possible to reintroduce the dipolar coupling by either exploiting the natural magnetic anisotropy of a molecule, resulting in slightly unequal sampling of the molecular orientations in a strong static magnetic field due to partial molecular alignment (Tolman et al. 1995; Tjandra et al. 1996; Banci et al. 1998), or by dissolving the protein in an anisotropic "alignment" medium, e.g. a liquid crystalline solution (Bax and Grishaev 2005). This allows residual dipolar coupling (RDC) to be measured as a line splitting, similarly to what is done for the measurement of *J*-coupling constants. In fact, the measured couplings in a partially aligned sample are the sum of the scalar and residual dipolar couplings. RDCs provide important information on local and global structure and dynamics in the molecule as explained in more detail in Chaps. 4 and 5.

2.5 NMR Spin Relaxation, Line Widths and Intensities

Relaxation of nuclear spins after excitation by a radio-frequency is caused by time-dependent local magnetic fields induced by the molecular tumbling and local rotational fluctuations that modulate the anisotropic spin interactions, the chemical shielding anisotropy (CSA) and the dipolar (DD) interaction. Additional contributions to spin relaxation arise from exchange processes, e.g. chemical exchange between labile protein and solvent protons, or exchange between different molecular conformations. In a simplified representation of nuclear spin relaxation, the relaxation process is described by two time constants. The longitudinal relaxation time constant T_1 accounts for the return of the spin system to thermal equilibrium associated with a loss of energy, while the transverse relaxation time constant T_2 describes the dephasing of coherence. A main consequence of spin relaxation for a particular NMR experiment is the loss of signal during the various transfer steps and chemical shift evolution delays. In addition, the longitudinal relaxation properties of the excited spin species determine the rate at which a pulse sequence can be repeated, as will be explained in more detail in Sect. 3.5.5, while transverse relaxation properties of nuclear spins determine their NMR linewidth.

The NMR linewidth is given by the transverse relaxation rate R_2 (reciprocal of T_2) of the detected nuclear spin, $\Delta\nu = R_2/\pi$, plus inhomogeneous contributions arising from the experimental setup (sample heterogeneities, B_0 inhomogeneity, temperature gradients, etc.). IDPs, in particular highly flexible ones, are generally characterized by large transverse relaxation time constants T_2, which lead to narrow peaks in the NMR spectra in comparison to the ones observed for globular proteins of comparable size. On the other hand, chemical and conformational exchange processes can be highly pronounced in IDPs, causing broadening of the peaks, especially in the spectra exploiting labile amide protons or in the case of structurally and dynamically heterogeneous proteins. Comparisons of cross-peak intensities (areas or volumes) thus report on differential intensity loss during the pulse sequence, while comparisons of cross-peak heights also report on differences in linewidth. The NMR analysis software often determines the peak heights in the spectrum during the peak picking procedure. When the peak is picked the program fits a function, such as a Lorentzian or Gaussian, and the maximum gives the peak height. For IDPs, peak heights in simple 2D spectra like the 2D HN and 2D CON (see Sect. 3.4.2) are often heterogeneous because relaxation properties of the spins vary depending on the extent of transient structure and on differences in local mobility. Therefore they provide a first indication of the different structural and dynamic properties of the protein.

Relaxation effects, including auto-correlated as well as cross-correlated relaxation rates, can in many cases be accurately quantified and used to extract structural and dynamic information. In this chapter we mainly focus on the determination of ^{15}N relaxation rates, one of the major tools to obtain information on the local dynamics of different parts of protein backbones (Barbato et al. 1992; Peng and Wagner 1994). Other interesting relaxation rates that can be determined and exploited to achieve structural and dynamic information on IDPs are instead discussed in Chap. 5 (paramagnetic relaxation enhancements, cross correlation rates). The determination of ^{1}H-^{1}H NOEs is instead not discussed in detail as it plays a minor role in the study of IDPs and many excellent books and reviews are available (Neuhaus and Williamson 1989; Cavanagh et al. 2007). However, the major contributions to ^{1}H relaxation are discussed in Sect. 3.5.5 because they are at the basis of the longitudinal relaxation enhancement effects that provide a valuable tool for the design of NMR experiments.

2.6 ^{15}N Relaxation Parameters

The quantification of nuclear spin relaxation effects provides a valuable tool to characterize local and global molecular motions. ^{15}N relaxation values, notably T_1, T_2 and the ^{1}H-^{15}N heteronuclear NOE (HETNOE), are the most commonly determined to characterize the dynamic behaviour of proteins, for both globular proteins and IDPs (Kay et al. 1989; Peng and Wagner 1992; Palmer 2004). The longitudinal relaxation time T_1 (or relaxation rate $R_1 = 1/T_1$) measures the decay of ^{15}N polarization, while the transverse relaxation time T_2 (or relaxation rate $R_2 = 1/T_2$) accounts for the loss of spin coherence. The HETNOE quantifies the polarization transfer from an amide ^{1}H to its attached ^{15}N.

Indeed, the major contributions to ^{15}N relaxation of a backbone amide ^{15}N in a protein are the ^{15}N chemical shift anisotropy (CSA) and the ^{15}N-^{1}H dipolar interaction (DD) with the directly bound proton, which to a good approximation, can be considered of equal magnitude throughout the protein backbone. This means that variations in ^{15}N relaxation properties of backbone amide ^{15}N spins can be interpreted in terms of differences in local motions (Peng and Wagner 1994). The measured ^{15}N relaxation parameters provide quantitative information on the amplitudes and the time scales of motions experienced by the ^{15}N-^{1}H amide group of a specific residue. ^{15}N relaxation (T_1, T_2 and HETNOE) is sensitive to motion occurring on the pico- to nanosecond time scale. As an example, Fig. 3.4 shows ^{1}H-^{15}N heteronuclear NOE values measured for individual amide sites along the backbones of different proteins; being especially sensitive to fast local motions, these observables allow protein regions characterized by different levels of local mobility to be easily identified.

The ^{15}N T_2 relaxation times are in addition also sensitive to slower conformational exchange processes in the micro- to millisecond time range that induce a change in the isotropic chemical shift. This exchange contribution (R_{ex}) adds to the relaxation induced by time modulation of the CSA and dipolar interactions ($1/T_2 = 1/T_2(CSA) + 1/T_2(DD) + R_{ex}$). In order to separate the exchange contribution, different experimental approaches have been proposed: repeating the relaxation measurements (i) at several static magnetic field strengths B_0, or (ii) by varying the applied radio-frequency field amplitude during the relaxation period. For more details, we

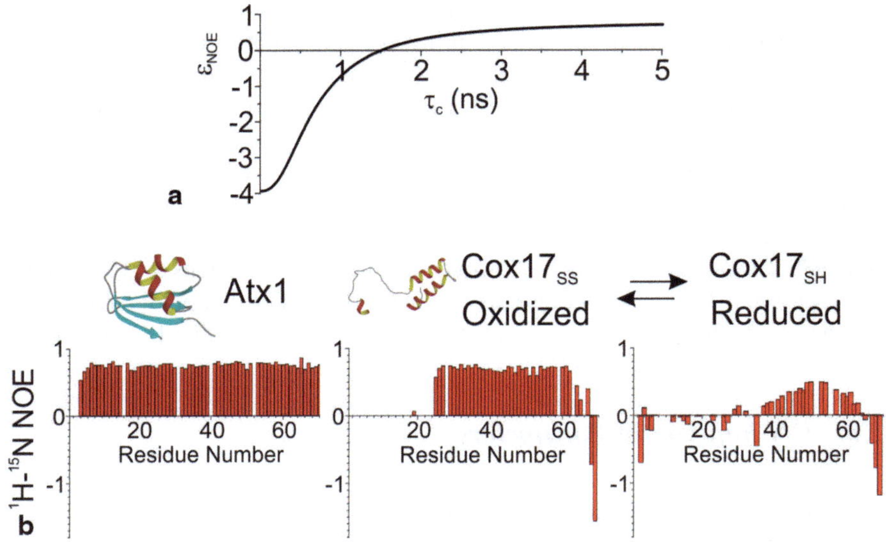

Fig. 3.4 **a** Variations of the steady-state NOE with the correlation time for the ^{1}H-^{15}N pair in a field of 11.74 T. **b** The graphs show the ^{1}H-^{15}N heteronuclear NOE for the majority of the residues of structured Atx1 protein (Arnesano et al. 2001), and of the partially structured oxidized form of the copper chaperone Cox17 (Arnesano et al. 2005) and of its unfolded reduced form (Bertini et al. 2011a). The experiments were acquired on a 500 MHz Bruker AVANCE spectrometer equipped with a CPTXI probe. Adapted from (Felli et al. 2012)

refer the reader to the scientific literature (Palmer et al. 2001; Tollinger et al. 2001; Palmer and Massi 2006).

As mentioned in the previous section, the oscillating magnetic fields induced by local and global rotational motions are causing spin relaxation in NMR. In order to account for the stochastic nature of these motions, they are best described by their frequency distribution, also called the power spectral density function $J(\omega)$. Furthermore, according to the standard relaxation theory for NMR spin relaxation, the so called BWR theory (Wangsness and Bloch 1953; Bloch 1956; Redfield 1957), spin relaxation is caused by the fluctuating magnetic fields created by molecular motions at the spin transition frequencies of the coupled ^{15}N-^{1}H spin system (ω_N, ω_H, $\omega_H - \omega_N$ and $\omega_H + \omega_N$) and at zero-frequency. The three relaxation parameters (T_1, T_2 and HETNOE) show a different dependence on the corresponding spectral density components ($J(0)$, $J(\omega_N)$, $J(\omega_H)$, $J(\omega_H - \omega_N)$ and $J(\omega_H + \omega_N)$) as described by the following equations:

$$\frac{1}{T_1} = \frac{d^2}{20}[J(\omega_H - \omega_N) + 3J(\omega_N) + 6J(\omega_H + \omega_N)] + \frac{c^2}{15} J(\omega_N)$$

$$\frac{1}{T_2}=\frac{d^2}{20}[4J(0)+J(\omega_H-\omega_N)+3J(\omega_N)+3J(\omega_H)+6J(\omega_H+\omega_N)]$$
$$+\frac{c^2}{15}[J(0)+3J(\omega_N)]+R_{ex} \tag{3.8}$$
$$HETNOE=1+\frac{d^2}{20}\frac{\gamma_H}{\gamma_N}[6J(\omega_H+\omega_N)]\cdot T_1$$

with $d=\frac{\mu_0}{4\pi}\gamma_H\gamma_N\hbar r_{NH}^{-3}$ the dipolar coefficient, $c=\omega_N\Delta\sigma$ the ^{15}N chemical shift anisotropy and R_{ex} the contribution from conformational exchange processes, if any.

In order to extract the desired dynamics information, the measured ^{15}N relaxation data (T_1, T_2 and HETNOE) can be analysed in several ways. A first quick analysis of the relaxation data, or some combinations thereof, already yields valuable qualitative information on the global and local conformational properties of the protein. For example, we have shown above that the HETNOE, which is only sensitive to high-frequency components of the spectral density (see Eq. 3.8), highlights highly-flexible protein segments (Fig. 3.4). Typical HETNOE values range from about $+0.9$ for amide groups in rigid protein fragments to largely negative values (HETNOE $<<0$) for highly flexible sites in disordered regions. Thus, the HETNOE provides a measure of the orientational degree of freedom of a particular amide group, a property that is also called an order parameter. In addition, the T_1/T_2 ratio can be computed as it provides information on the local protein rigidity (Kay et al. 1989). For globular proteins, T_1/T_2 is to a good approximation proportional to the protein's overall rotational correlation time. For IDPs, T_1/T_2 allows to distinguish peptide regions displaying significant secondary and tertiary structural propensities, characterized by longer effective correlation times, from segments lacking any residual structure, characterized by shorter effective correlation times.

Arguably, the most commonly used method for the analysis of ^{15}N relaxation data of globular proteins is the model-free formalism introduced by Lipari and Szabo where the molecular tumbling, described by a global correlation time, is separated from local motions (Lipari and Szabo 1982), characterized by site-specific correlation times and order parameters. This separation is justified by the difference in time scales of these motions in the case of globular proteins, but this model is not rigorous for random coil-like polymers and IDPs. However, a similar approach can still be applied for IDPs by replacing the global correlation time of the protein by an effective segmental correlation time that varies over the polypeptide chain, reporting on the persistence lengths of the segmental chain motions.

Another possibility for analysing ^{15}N relaxation data that applies also to IDPs is the reduced spectral density mapping approach, which provides information about the shape of the spectral density function at individual sites (Farrow et al. 1995). If we neglect for the moment conformational exchange (R_{ex}) contributions to ^{15}N relaxation, Equation 3.8 shows that the three measured rates depend on five different spectral density components. In order to reduce this number to three, and thus matching the number of NMR observables, the high-frequency spectral density components $J(\omega_H)$, $J(\omega_H-\omega_N)$ and $J(\omega_H+\omega_N)$ are replaced by an effective $J(0.87\cdot\omega_H)$ value.

This allows to solve Eq. 3.8 analytically. The validity of reduced spectral density mapping for very flexible IDPs has recently been investigated and a slightly different approach was proposed, including the removal of exchange contributions by the measurement of additional cross-correlated relaxation rates (Kadeřávek et al. 2014).

3 Particular Challenges and Bottlenecks for NMR Studies of IDPs

A wide range of NMR experiments has been developed throughout the years for the study of globular proteins and their interactions, with the major objective of providing high-resolution structural and dynamic information. These experiments are the natural starting point for the NMR investigation of IDPs. However, the peculiar properties of IDPs have a strong impact on NMR spectra, as already outlined in the previous section, and thus on NMR experiments. This means that conventional NMR experiments need to be tailored for the specific properties of highly disordered proteins in order to study IDPs of increasing size and complexity. The main bottlenecks for NMR studies of IDPs will be briefly reviewed in the following sections.

3.1 *Spectral Resolution*

In order to extract structural and dynamic information for single nuclear sites in an IDP, we need sufficient spectral resolution to distinguish individual resonances (or correlation peaks). As mentioned in the previous section the first consequence of the lack of a stable structure is the averaging of a large part of the contributions to the chemical shift deriving from the local chemical environment. This results in a drastic reduction of the chemical shift ranges for the different nuclear spin species (^{1}H, ^{13}C, ^{15}N) and thus in a problem of strong overlap in the corresponding NMR spectra (see Fig. 3.5).

This resolution problem is especially pronounced for ^{1}H NMR, as will be discussed in more detail later on. Identifying strategies to overcome the problem of resonance/cross-peak overlap is of key importance to be able to study IDPs of increasing size and complexity. Several strategies to address this critical point are discussed in the next sections, such as exploiting the favourable resolution in exclusively heteronuclear NMR experiments, correlating nuclei of neighbouring amino acids and exploiting amino acid selection to simplify spectra and identify residue types.

3.2 *Experimental Sensitivity*

We have already introduced the theoretical basis and the main determinants of NMR sensitivity in Sect. 3.1.2. Here we will focus on IDPs, which are by definition

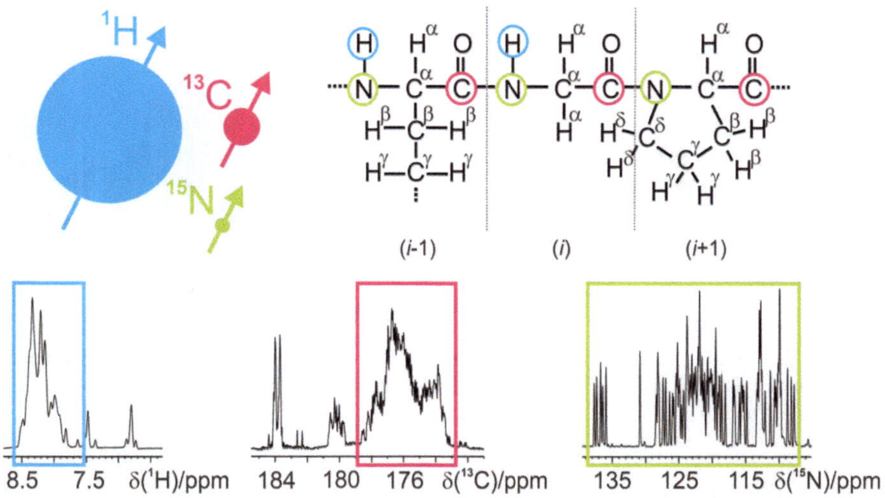

Fig. 3.5 The difference in the ^{1}H, ^{13}C and ^{15}N chemical shift dispersion of the IDPs is illustrated on the 1D spectra of α-synuclein. The experiments were acquired on a 700 MHz Bruker AVANCE spectrometer equipped with a CPTXO probe on a 1 mM α-synuclein sample (20 mM phosphate buffer, pH 6.4, 200 mM NaCl, 0.5 mM EDTA, at 285.5 K)

very flexible macromolecules. From an NMR sensitivity point of view, this flexibility has both advantages and disadvantages. The fast timescale molecular motions in IDPs are responsible for reduced effective rotational correlation times, a feature that in many cases contributes to long spin coherence lifetimes (long T_2) and narrow NMR lines. In principle, this enables the design of complex high-dimensional NMR pulse schemes to achieve the necessary resolution required for IDPs. It is interesting to note that for highly flexible IDPs the increase in molecular size does not have a major impact on relaxation times and linewidths, while it does for folded proteins. However, for protein regions with a significant amount of transient structure, this is no longer the case, as illustrated by amide ^{15}N T_2 relaxation time constants of the NS5A protein of HCV (Fig. 3.6). Some peptide segments have T_2 relaxation time constants that are four times shorter than those observed for other regions. This results in a large dynamic range of peak intensities observed in the NMR spectra. This feature becomes even more pronounced for complex NMR experiments involving an increasing number of transfer steps and frequency editing periods.

The effects of the presence of partially structured peptide regions and extensive conformational dynamics on the NMR spectra can become so pronounced that most of the nuclear spins are affected. No or only very limited solutions remain to study proteins characterized by strongly line-broadened resonances by NMR. Examples that often fall into this category are the so-called molten globule states that represent the "dark side" of biomolecular NMR, in between well-folded globular and highly disordered states.

Sensitivity can also be reduced because of protein aggregation, which is governed by the same principles as protein folding. Some IDPs, such as α-synuclein,

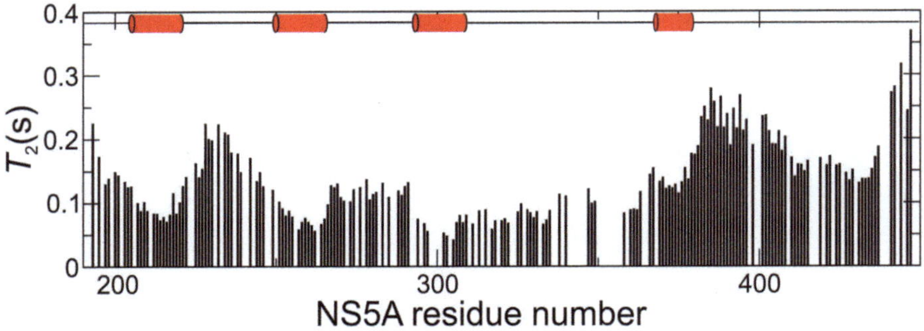

Fig. 3.6 ^{15}N amide T_2 relaxation time constants of NS5A measured as a function of the amino acid sequence at 278 K, at 18.8 T field. Red cylinders indicate regions adopting transient α-helical secondary structure

tau, or prion proteins, are known to cause neurodegenerative diseases in the aggregated form. Tendencies to aggregate can hamper NMR studies of these proteins. On the other hand, higher molecular mass aggregates become amenable to other NMR techniques, such as solid-state magic-angle spinning (MAS) NMR, which however is not addressed in this chapter (Tycko 2006; Chimon et al. 2007; Bertini et al. 2011c).

3.3 Experimental Time Requirements

Sample stability problems are often encountered because of aggregation, as mentioned before, or proteolytic degradation of IDPs. When the stability of the protein permits sample stabilization by boiling, the proteases can be deactivated by heat denaturation. However, in the case of significant residual structure, this is often not a valuable option as extensive heat treatment may result in irreversible changes of the structural features of the IDP (Chap. 6). In such cases the NMR spectra have to be acquired before sample degradation occurs. The requirements of short experimental times and at the same time high spectral resolution, achieved by long acquisition times in all dimensions of a high-dimensional (3D, 4D, etc.) experiment, seem to be contradictory, but they can be reconciled by the use of the fast NMR data acquisition techniques discussed in Sects. 3.5.3 and 3.5.5.

3.4 Sample Optimization

The conformational dynamics and transient structure of IDPs are highly sensitive to experimental conditions such as pH, buffer composition, salt concentration and temperature. Under certain conditions, some parts of the IDP may undergo conformational dynamics on a timescale that leads to extensive line broadening in the NMR spectra and thus missing correlation peaks. Optimization of the sample conditions is thus even more important than for globular proteins. On the other hand, because of the dependence of the structural features of the IDP on the experimental conditions, one may want to study the IDP under close to physiological conditions such as neutral pH and relatively elevated temperature. Studies of enzymatic reactions, for example the occurrence of post-translational modifications, also often require neutral pH and high temperature to ensure optimal enzyme activity. Finally, in-cell NMR can be used in order to experimentally show that IDPs remain flexible *in vivo* and that disorder is not just an artefact of the chosen sample conditions.

3.5 Prolines are Abundant in IDPs

Further problems can occur because of the typically large proline content in the amino acid sequence of IDPs (Tompa 2002; Theillet et al. 2013). As prolines do not have backbone amide protons, they are not detected in amide ^{1}H detected NMR spectra and therefore represent breakpoints in the sequential backbone resonance assignment strategy based on $^{1}H^{N}$ detected triple-resonance experiments. For the same reason, variants of 2D ^{1}H-^{15}N correlation experiments typically used to follow or monitor physiological processes, such as interactions or post-translational modifications, or to measure observables such as ^{15}N relaxation rates, paramagnetic relaxation enhancements or residual dipolar couplings, do not provide information about prolines.

4 2D HN and CON NMR Experiments: The Fingerprint of an IDP

Uniform ^{13}C and ^{15}N isotope labelling of proteins (and thus IDPs) in bacterial expression systems has become routinely used and is currently a necessary requirement to proceed with any high resolution NMR investigation. The nuclear spins we can deal with are therefore ^{1}H, ^{13}C and ^{15}N, meaning the vast majority of the nuclei in a protein.

The main considerations when designing or choosing an NMR experiment are its overall sensitivity, the resulting spectral resolution, the number of detected peaks and the information contained in the spectral parameters (peak positions and intensities). In order to optimize these parameters we can choose the source of spin polarization, the directly observed nuclei, the number and nature of the indirectly detected nuclei, the type of transfer steps used and the dimensionality of the experiment, to cite only the most important ingredients of an NMR experiment. The most useful 2D NMR experiments to characterize IDPs are introduced here.

4.1 NMR Properties of ^1H, ^{13}C and ^{15}N in IDPs

According to Eq. 3.5, the sensitivity of the NMR experiment strongly depends on the gyromagnetic ratio of the nuclear spins (see Table 3.1) directly detected at the end of the pulse sequence and those serving as polarization source at the beginning of the experiment. Because of the high gyromagnetic ratio of ^{1}H, most NMR experiments for protein applications are based on ^{1}H excitation and ^{1}H direct detection, while heteronuclear chemical shifts (^{13}C and ^{15}N) are exploited in indirect dimensions of multidimensional NMR experiments (Sattler et al. 1999; Lescop et al. 2007). However, ^{13}C detection has recently evolved into a useful tool to study biomolecules (Felli and Pierattelli 2014a), in particular thanks to the larger chemical shift dispersion of ^{13}C nuclei compared to protons and to other specific benefits of the technique that improve the sensitivity of the NMR experiment, thus compensating for the lower gyromagnetic ratio (Sect. 3.6). ^{15}N detection in protein NMR is rarely used, although some experiments have been proposed for specific applications (Vasos et al. 2006; Takeuchi et al. 2010; Gal et al. 2011). NMR sensitivity is also influenced by the specific properties of the investigated system, in particular by its relaxation properties as well as by homonuclear couplings, all contributions not explicitly considered in Eq. 3.5. These aspects should also be carefully considered in the choice of the most appropriate experimental strategy to access a specific kind of information, as discussed more in detail in the next sections considering applications to IDPs.

The second important point to consider is spectral resolution. It is well known from NMR textbooks that ^{13}C and ^{15}N nuclei show superior chemical shift dispersion compared to ^{1}H. This is definitely an interesting property provided it also holds for IDPs. As an example, Fig. 3.5 shows the chemical shift dispersion observed in α-synuclein, a well-characterized IDP, for three backbone nuclear spins, $^1H^N$, $^{13}C'$ and ^{15}N. It is clear that the chemical shift dispersion increases from $^1H^N$ to $^{13}C'$ to ^{15}N even in the absence of a stable structure. The same holds true for aliphatic and aromatic spin pairs, with ^{13}C yielding a higher resolution in the NMR spectrum than the attached ^{1}H. Therefore, the exploitation of the improved frequency resolution in the ^{13}C and ^{15}N dimensions of multidimensional NMR spectra is crucial for the study of IDPs.

For ^{1}H and ^{13}C resonances in the side-chains of IDPs we observe a larger overall dispersion (compared to backbone nuclei) due to the differences in chemical structure between different amino acids. However, side-chain resonances from the same amino acid type cluster in the same spectral region. Side chain resonances are therefore mainly used as indicators of the amino acid type of a given residue, while they are only of little use for the characterization of site-specific structure and dynamics in the IDP. As an example, the 2D ^{1}H-^{13}C (HC) correlation spectrum acquired on α-synuclein is shown in Fig. 3.7. Despite the high sensitivity of this experiment, the overlap of signals deriving from the same type of amino acid and the presence of resolved homonuclear couplings (see below) are responsible for the extensive overlap observed.

Since relatively narrow peaks are observed in the NMR spectra of IDPs in the absence of unfavourable dynamics, homonuclear *J*-couplings may contribute

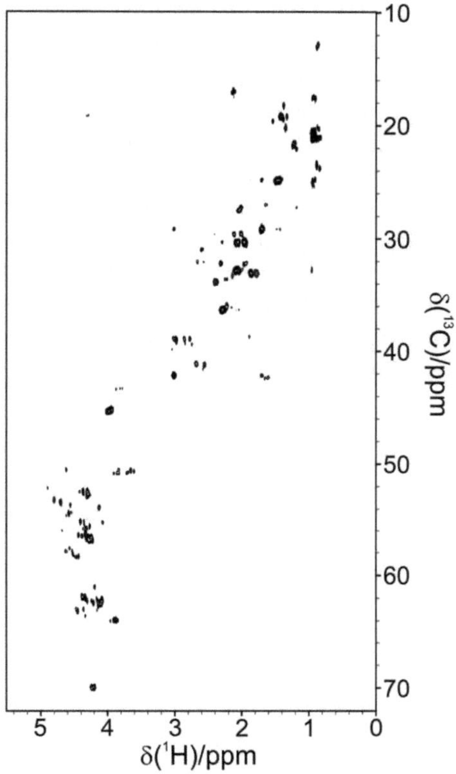

Fig. 3.7 The ^{1}H-^{13}C HSQC spectrum acquired on a 950 MHz Bruker AVANCE spectrometer equipped with a CPTCI probe on a 1 mM α-synuclein sample (20 mM phosphate buffer, pH 6.4, 200 mM NaCl, 0.5 mM EDTA, at 285.5 K)

significantly to the linewidth. Excluding for the moment the one-bond $^1J_{CC}$ couplings from the discussion, which definitely need to be suppressed experimentally to obtain well-resolved ^{13}C spectra (as explained in Sect. 3.6.1), the magnitude of ^2J and ^3J homonuclear coupling constants decreases passing from ^{1}H to ^{13}C to ^{15}N as an indirect consequence of the gyromagnetic ratios of these nuclei. This feature is thus again in favour of the detection of ^{13}C or ^{15}N instead of ^{1}H, unless the IDP has been perdeuterated and back-protonated at amide sites to suppress homonuclear ^{1}H-^{1}H couplings.

4.2 2D ^{1}H-^{15}N and ^{13}C′-^{15}N Correlation Fingerprint Spectra of IDPs

Recording simple 2D NMR "fingerprint" spectra of the IDP provides some useful information on the protein and allows the evaluation of the overall sample properties and the feasibility of a subsequent high resolution NMR study even in the absence of a sequence-specific resonance assignment. Indeed, the number of cross-peaks detected in the spectrum compared to the number of peaks expected from the primary sequence provides a first estimation of the amount of spectral overlap and indicates whether the totality of the peptide sequence or only a part of it is NMR-visible under the chosen sample conditions. The latter case occurs quite often as many proteins exhibit a significantly heterogeneous nature in terms of structural and dynamic properties leading to conformational exchange and/or aggregation induced line broadening. The observed chemical shift dispersion allows the identification of whether the protein is structured, partly structured, or highly unstructured.

For such a fingerprint spectrum, we want to detect a single correlation peak per residue with good dispersion in a reasonable amount of data acquisition time. The most common experiments used are therefore ^{1}H detected ^{1}H-^{15}N (HN) (Favier and Brutscher 2011) and ^{13}C detected $^{13}C'$-^{15}N (CON) (Bermel et al. 2006a) correlated spectra, as shown in Fig. 3.8 for human securin, a 200 amino acid IDP with more than 20 % prolines.

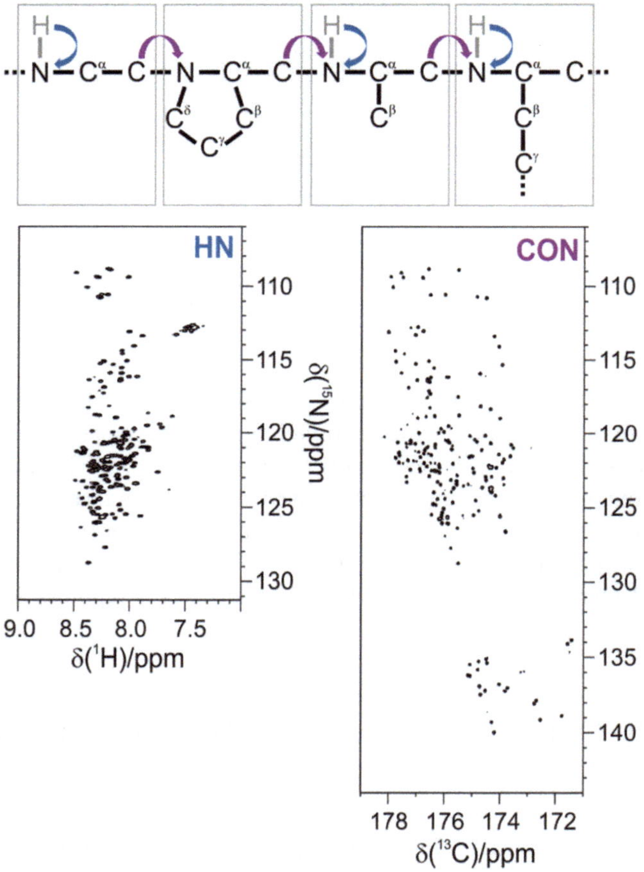

Fig. 3.8 The schematic illustration (*top*) of the correlations observed in the basic 2D ^{1}H-^{15}N HSQC and ^{13}C-^{15}N CON-IPAP experiments. The ^{1}H-^{15}N HSQC (*left*) and ^{13}C-^{15}N CON-IPAP (*right*) spectra were acquired on the ^{13}C, ^{15}N-labeled sample of the intrinsically disordered human securin protein (0.7 mM sample in 25 mM phosphate buffer, pH 7.2, 150 mM KCl, 10 mM 2-mercaptoethanol, at T 283 K) (Csizmok et al. 2008). The experiments were performed on a Bruker AVANCE 700 MHz spectrometer equipped with a CPTXO probe

While the HN experiment is much more sensitive and can be recorded in a significantly short time, prolines are only detected in the CON spectrum, which also shows a better spectral resolution. In addition, the CON spectrum does not suffer from hydrogen-exchange-induced line broadening and thus can still be recorded under conditions of high pH and temperature, where many HN peaks are no longer detectable (Gil et al. 2013); this is shown in Fig. 3.9, which reports the HN and CON spectra acquired on α-synuclein with increasing temperature. On the other hand, high-quality HN correlation spectra can be recorded on protein samples at concentrations of only a few μM on a spectrometer equipped with a cryogenic probe. The 2D HN and CON spectra are thus highly complementary both in terms of detectability and information content. These 2D correlation experiments can be acquired in different ways. The most appropriate variants for applications to IDPs are discussed in detail in the following section.

These 2D spectra can also be used to follow changes in the properties of the IDP upon changing the experimental conditions such as temperature, pH, ionic strength, buffer, reducing/oxidizing environment, or the addition of potential partners such as metal ions, small molecules, nucleic acid fragments and proteins. They also enable chemical reactions such as the occurrence of post-translational modifications to be followed (Selenko et al. 2008). Finally, they provide an invaluable tool to take a snapshot of a protein inside an entire cell (Serber et al. 2006; Selenko and Wagner 2007; Felli et al. 2014). As an example, the 2D HN and CON experiments acquired

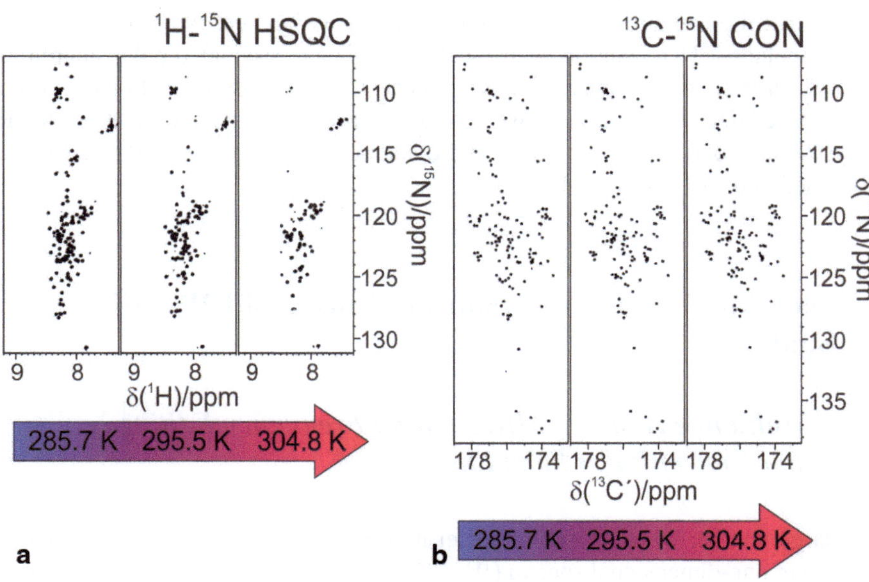

Fig. 3.9 2D ^{1}H-^{15}N HSQC (**a**) and ^{13}C-^{15}N CON-IPAP (**b**) spectra acquired on a 1 mM α-synuclein sample (20 mM phosphate buffer, pH 7.4, 200 mM NaCl, 0.5 mM EDTA) at different temperatures, from *left* to *right*: 285.7 K, 295.5 K and 304.8 K. Each spectrum was acquired with one scan per increment and with the same resolution (in Hz)

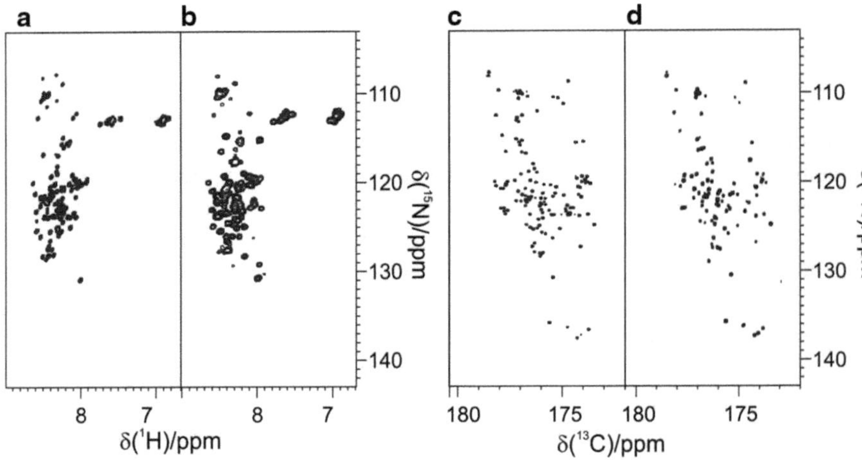

Fig. 3.10 The 2D ^{1}H-^{15}N SOFAST-HMQC (**a**, **b**) and ^{13}C-^{15}N $H^{\alpha\text{-flip}}$CON (**c**, **d**) spectra acquired on α-synuclein overexpressed in *E. coli* cells (**b**, **d**) and on cell lysates (**a**, **c**). The experiments were acquired on a Bruker AVANCE 700 MHz spectrometer equipped with a CPTXO probe; the acquisition time of SOFAST-HMQC was 13 min and of $H^{\alpha\text{-flip}}$CON 44 min

on α-synuclein in cell are compared with those acquired on the purified protein (Fig. 3.10), showing that they can be used for the investigation of IDPs in-cell.

Finally, modifications of these basic 2D experiments (HN and CON) enable the determination of a variety of observables that report on different properties of the IDP at atomic resolution, once sequence-specific assignment becomes available. These include ^{15}N relaxation rates, scalar couplings, residual dipolar couplings (RDCs), paramagnetic relaxation enhancements (PREs), cross-relaxation (σ_{HH}) and cross-correlation rates (CCR), as well as solvent exchange rates. All of these NMR observables report on the structural and dynamic features of the IDP and will be discussed in detail in Chaps. 4 and 5.

5 Tools for Overcoming Major Bottlenecks of IDP NMR Studies

5.1 Multidimensional NMR, Indirect Frequency Editing and Non-uniform Sampling

The large number of NMR-active nuclei in a protein results in severe resonance overlap in one-dimensional spectra (^{1}H, ^{13}C, ^{15}N), making it practically impossible to extract atom-resolved information from them. This problem is circumvented by using multidimensional (nD) NMR techniques that spread and correlate the signals of individual nuclear spins along different frequency dimensions (Ernst et al. 1987). Multidimensional NMR data are recorded by repeating the basic pulse sequence

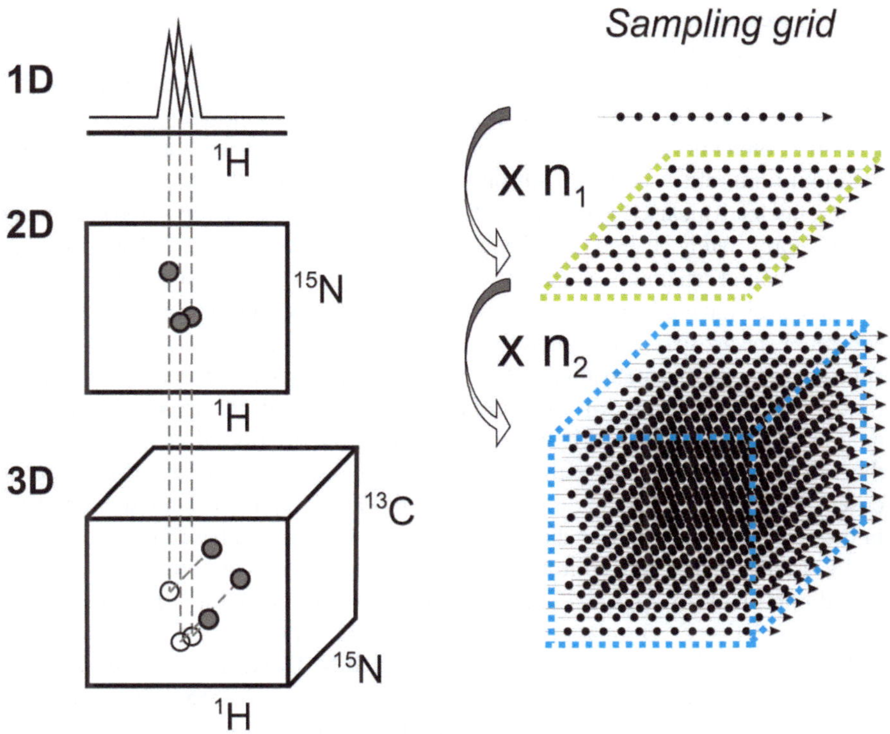

Fig. 3.11 Schematic representation of the multi-dimensional NMR spectra (*left*), and the increasing number of experimental repetitions required, resulting in longer acquisition times (*right*)

numerous times. For each repetition of the experiment (transient), the observable NMR signal is detected along one dimension, while additional 'indirect' dimensions are sampled by incrementing a time variable in the pulse sequence from one repetition to the next. Despite the great success of multidimensional NMR, an important drawback of this stepwise sampling procedure is the long experimental time, which is a direct result of the hundreds or even thousands of transients that are required for a single data set (Fig. 3.11).

In this section we will discuss different editing techniques to improve the spectral resolution in the indirect dimensions of nD NMR experiments and advanced non-uniform sampling (NUS) approaches that are indispensable for recording high-dimensional spectra in reasonable experimental times.

5.2 Real-time, Constant-time and Semi-constant Time Frequency Editing

Frequency editing of nuclear spins is the key step for introducing indirect dimensions in multidimensional experiments. In order to do so, a pulse sequence element that allows time evolution of the frequency-edited spins (*I*) while suppressing (refocusing) all other chemical shift and J-coupling evolutions is required. Figure 3.12 shows five common pulse sequence implementations of I-spin editing during an incremented time variable t_I: (a) conventional real-time, (b) optimized real-time, (c) and (d) constant-time (CT) and (e) semi-CT editing.

In the conventional real-time implementation, heteronuclear J_{IS} and J_{IK} coupling evolutions are refocused by a 180° pulse applied in the middle of t_I. In this context "heteronuclear" means that the particular spin species can be manipulated separately by an appropriately shaped radiofrequency pulse. As an example, $^{13}C'$ and $^{13}C^{\alpha}$ can be manipulated separately as they have well-separated chemical shift ranges, while $^{13}C^{\alpha}$ and $^{13}C^{\beta}$ have generally overlapping chemical shift ranges and thus cannot be separated spectroscopically. Therefore, the $^{13}C^{\alpha}$ -$^{13}C^{\beta}$ coupling gives rise to a peak splitting in the $^{13}C^{\alpha}$ dimension of a NMR spectrum recorded with conventional real-time editing as illustrated in Fig. 3.13a. A slightly optimized version of real-time editing is shown in Fig. 3.12b, where chemical shift evolution of the I-spins during the 180° S-spin decoupling pulses, as well as Bloch-Siegert phase shifts, are refocused. This allows recording a data set that does not require any phase correction in the corresponding indirect dimension (t_I) in order to avoid baseline distortions.

In order to avoid line splitting due to homonuclear J_{II}-coupling evolution, the CT editing blocks of Figs. 3.12c and 3.12d can be used, with the CT delay T set to $T = 1/J_{II}$. A $^{1}H^{N}$-$^{13}C^{\alpha}$ correlation spectrum recorded with the pulse sequence block of Fig. 3.12c is shown in Fig. 3.13b. The detected NMR signal in this spectrum is modulated by a factor $\cos^{n}(\pi J_{cc}T)$, where n is the number of carbon atoms attached. In addition, signals with different sign provide meaningful information about the spin-coupling topology; for example, correlation peaks of glycine residues are of opposite sign with respect to all others, as glycines have no $^{13}C^{\beta}$ attached to the $^{13}C^{\alpha}$.

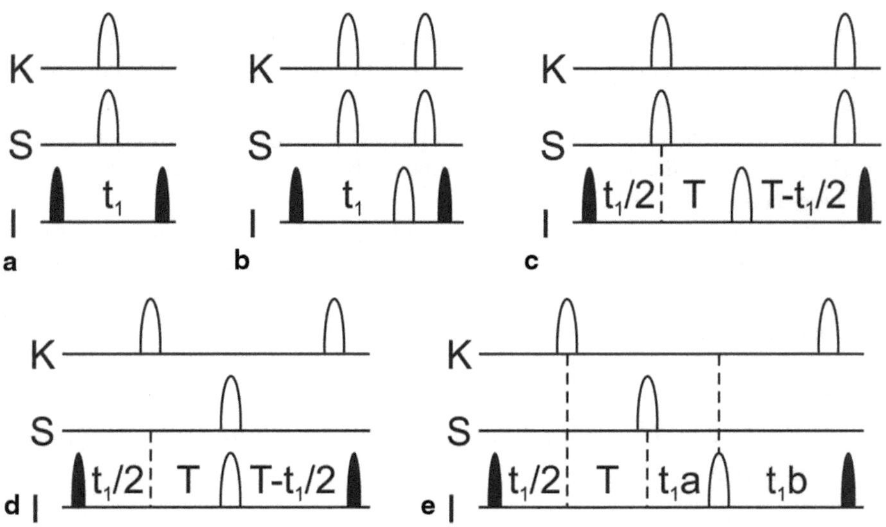

Fig. 3.12 Common pulse sequence implementations of I-spin editing during an incremented time variable t_I: **a** conventional real-time; **b** optimized real-time; **c** and **d** constant-time (CT), and **e** semi-CT editing

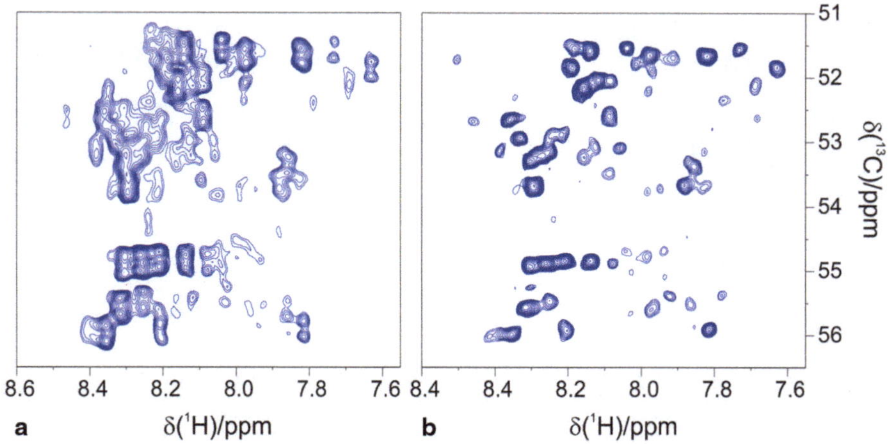

Fig. 3.13 Expansion of ^{1}H-$^{13}C^{\alpha}$ plane from a BEST-TROSY HN(CO)CA spectrum of NS5A recorded with real-time (**a**) and CT evolution (**b**). In both cases the acquisition time in the indirect $^{13}C^{\alpha}$ dimension was 20 ms with 200 points recorded and a spectral width of 30 ppm

Alternatively, CT frequency editing can be applied during an INEPT-type coherence transfer step (Morris and Freeman 1979). This is shown in Fig. 3.12d, where the *I*-spins are edited during an *I* to *S* transfer. Here, the CT delay has to be set to $T = 1/(2J_{IS})$ in order to achieve complete transfer.

A drawback of CT editing is that the maximum possible evolution time (t_1^{max}) is limited by the CT delay *T* so that $t_1^{max} \leq T$. This, of course, has an effect on the achievable spectral resolution, which is a crucial point for NMR studies of IDPs as discussed before. A possible solution could be to increase the CT delay to $n \cdot T$, where *n* is an integer number. However, introducing too long delays in the pulse sequences may cause pronounced relaxation losses. Semi-CT editing has therefore been proposed to enhance the spectral resolution while still exploiting spin evolution during a coherence transfer delay. The improvement achieved with semi-CT evolution in terms of spectral resolution in the ^{15}N dimension of an HNCA experiment is illustrated in Fig. 3.14.

5.3 *Strategies for Non-uniform Data Sampling in Multi-D NMR*

The amplitude of the NMR signal, or free induction decay (FID), is typically measured (sampled) at discrete, uniformly spaced time points. In a multidimensional NMR experiment, the same approach is generally employed in the indirect dimensions as well, since uniform sampling is required for data processing using the fast Fourier transform (FFT). The Nyquist theorem states that the sampling rate needs to be faster than twice the highest frequency expected or, in other words, determines a maximum value for the time interval between sampled points (dwell time) that has to be used to avoid spectral folding of the peaks. Hence, the sampled data points form a Cartesian grid with the spacing between time points in each dimension given by the inverse of the spectral width of the edited nuclei. As a result of this uniform

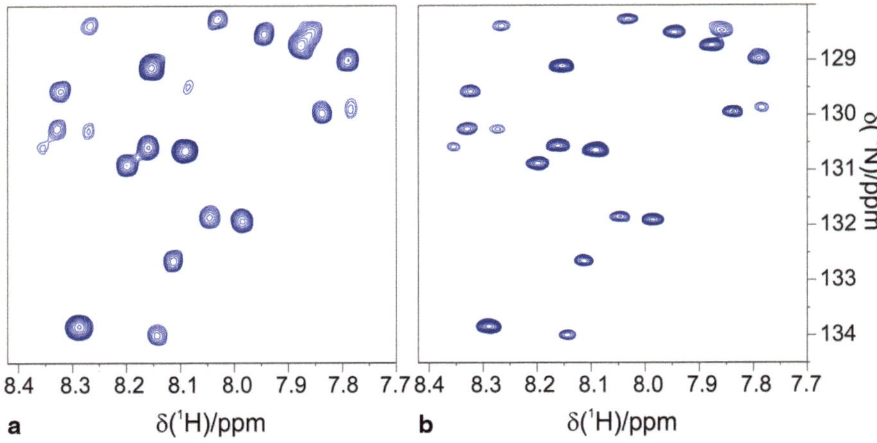

Fig. 3.14 Expansion of the ^{1}H-^{15}N plane from BEST-TROSY HNCA of NS5A acquired with **a** CT editing using the maximum number of increments (CT=24.5 ms) and **b** semi-CT evolution (CT=53 ms). In both cases the spectral width in the ^{15}N dimension was 30 ppm

data sampling procedure, the experimental time requirement increases roughly by about two orders of magnitude for each additional dimension, making high-dimensional experiments (>3D) impractical or requiring strong compromises on the spectral resolution in the indirect dimensions.

In order to overcome these limitations, alternative sampling strategies combined with appropriate processing tools have been developed over the past two decades (Hiller et al. 2005; Kazimierczuk et al. 2006; Mobli et al. 2006; Coggins and Zhou 2007; Kazimierczuk et al. 2007; Kazimierczuk et al. 2010a; Kazimierczuk et al. 2012; Yao et al. 2014). These sparse or non-uniform sampling (NUS) techniques rely on a reduction of the overall number of sampled time points by recording only a subset of the data points of the Cartesian grid. Examples of alternative sampling schemes are shown in Fig. 3.15.

Some of these sampling grids, e.g. linear under sampling or radial sampling, still yield data sets that can be processed using FFT (Szyperski et al. 1993b; Brutscher et al. 1995b; Kupce and Freeman 2003). The general NUS scheme, however, where a certain percentage of the grid points is randomly chosen, requires alternative processing tools. Several algorithms for processing non-uniformly sampled data are currently available and well-established, such as multidimensional decomposition (MDD) (Luan et al. 2005; Tugarinov et al. 2005), maximum entropy (MaxEnt) methods (Hoch and Stern 1996), compressed sensing (CS) (Holland et al. 2011; Kazimierczuk and Orekhov 2011), multidimensional Fourier transform (MFT) (Kazimierczuk et al. 2010a) and spectroscopy by integration of frequency and time domain information (SIFT) (Matsuki et al. 2009). The common feature of all these methods is that they aim to find the NMR spectrum that, when applying the inverse Fourier transform, best reproduces the measured time data points. As in the case of sparse data sampling, this is an underdetermined computational problem. Some additional assumptions must therefore be made in order to choose the most likely spectrum from

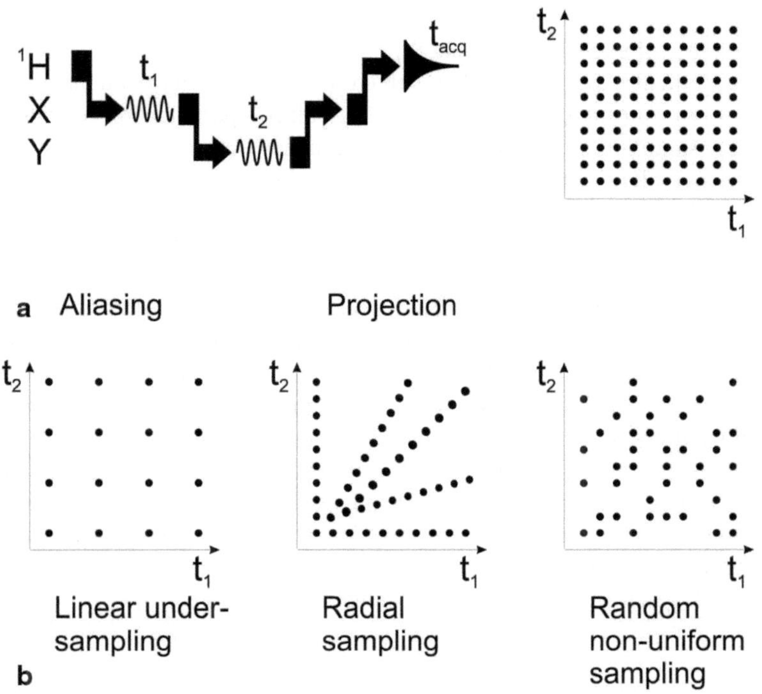

Fig. 3.15 Examples of different sampling schemes to speed up the acquisition of a 3D experiment. **a** Schematic illustration of a typical three-dimensional H-X-Y correlation experiment and the conventional time-domain sampling grid required to build the two indirect dimensions. Each point on the grid corresponds to a single repetition of the basic pulse sequence. **b** Some NUS schemes are shown. All the sampling patterns can be employed to reduce the experimental time, since a smaller number of points with respect to those of the conventional sampling grid are measured. The linear under sampling pattern can be used to fold chemical shifts in an advantageous manner, whereas radial sampling is generally employed in automated projection spectroscopy (APSY). Instead, random non-uniform sampling schemes are most commonly used when non-linear methods for spectral reconstruction such as MDD or MFT are used

among the possible solutions. For example, MDD exploits the prior knowledge that NMR signals are the direct product of Lorentzian (or Gaussian) line shapes in each frequency dimension. MaxEnt relies on the optimization of a penalty function in order to select the spectrum that has the highest entropy (minimal number of signals). Similarly, CS performs a l_1-norm minimization to recover the sparsest spectrum, while MFT performs a discrete multidimensional Fourier transform of the data using appropriate data weighting and filtering for the reduction of sampling noise. Finally, SIFT is an iterative method that replaces the missing time domain data points by zero for the first iteration, allowing classical FFT processing. In subsequent iterations, the missing data points are replaced by the result of the inverse FT of the obtained NMR spectrum after having set known empty spectral regions to zero.

At present, the question of the most appropriate sampling grid and processing technique for a given experiment is non-trivial, but a few general recommendations can be given: (i) the percentage of sampled data points can be decreased with increasing dimensionality of the experiment and sparseness of the final NMR spectrum (number of expected correlation peaks). Typically, in the case of a 4D backbone assignment experiment, a few percent of sampled data points are sufficient. (ii) A random distribution of the sampled data points is particularly recommended to spread the sampling noise over the entire frequency domain (with apparent reduction of its overall amount), preventing the clustering of artefacts in specific spectral regions. (iii) The choice of the processing algorithm mainly depends on the available software (degree of automation, required expertise for proper parameter adjustment) and the computer power (some algorithms require more computer power and memory than others). If possible, it is always a good option to test different processing tools on the same data set. Recent developments by spectrometer manufacturers have made it straightforward to set up a random sampling scheme and run experiments in NUS mode. Also, some of the non-linear processing routines, such as MDD and CS, have been interfaced with the NMR spectrometer software and are available for routine use. Alternatively, most of the processing algorithms can be accessed online and downloaded. These advanced data acquisition and processing tools should be largely exploited for NMR investigations of IDPs, as they allow the performance of experiments of high dimensionality in a reasonable amount of time at very high spectral resolution, as discussed in detail in Sect. 3.9.1.

5.4 Selecting ^{15}N Spin States with Favourable Transverse Relaxation Properties

Multi-dimensional NMR correlation experiments require a series of coherence transfer steps and frequency editing periods during which the detectable NMR signal, and thus the sensitivity of the experiment, decreases due to transverse spin relaxation. In order to improve the performance of such experiments and therefore their applicability to large IDPs and low sample concentrations, it is important to select the coherences with the longest transverse relaxation times for the transfer and chemical shift editing steps. Here we present two commonly used techniques for reducing signal losses during ^{15}N transverse evolution periods as required for all amide ^{1}H detected NMR experiments, and that are also of importance in ^{13}C detected experiments.

In a standard INEPT-type ^{15}N→^{13}C transfer step, ^{15}N single quantum coherence (SQC) oscillates back and forth between in-phase (N_x) and anti-phase ($2N_yH_z$) coherence, as a result of ^{15}N-^{1}H scalar coupling evolution. As a consequence, the effective relaxation during the transfer is given by the average of in-phase and anti-phase SQ relaxation rates. In uniformly ^{13}C, ^{15}N labelled proteins, relaxation of anti-phase SQC is about 30 % faster than relaxation of in-phase SQC. This difference is even more pronounced in the case of significant solvent exchange rates between the amide and water protons, further reducing the lifetime of anti-phase SQC. Improved transfer efficiency is thus achieved by ^{1}H composite pulse decoupling, which removes the ^{15}N-^{1}H scalar coupling evolution and maintains the in-phase SQC throughout the duration of the transfer. Common ^{1}H decoupling schemes are MLEV-16 (Levitt et al. 1982), DIPSI-2 (Shaka et al. 1988), WALTZ-16 (Shaka et al.

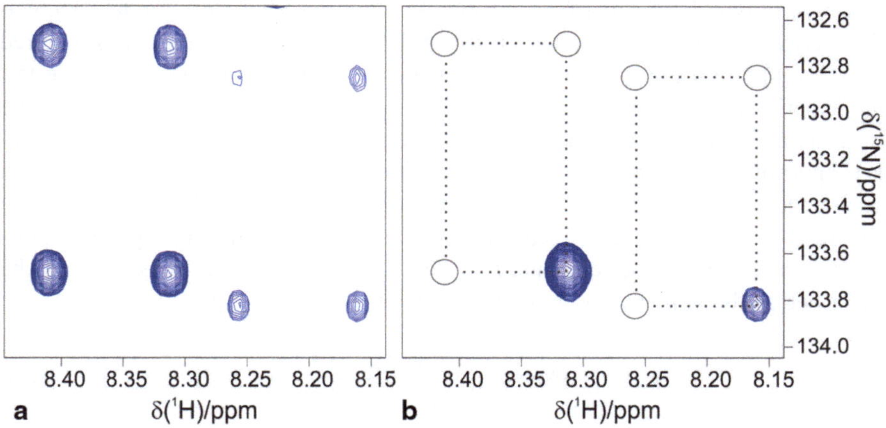

Fig. 3.16 Small spectral region of **a** ^{1}H-^{15}N coupled HSQC and **b** TROSY spectrum of NS5A recorded on a 950 MHz spectrometer. The spectral region displays correlation peaks for two NS5A residues. The TROSY effect is less pronounced for the left residue, which is located in a highly flexible region of the IDP, compared to the right one, which is part of a peptide segment with a high propensity to form an α-helical structure. In both cases the acquisition time in the indirect dimension was 44 ms with 256 points recorded and a spectral width of 30 ppm

1983a; Shaka et al. 1983b) and GARP-1 (Shaka et al. 1985). They are all composed of a basic pulse sequence element (R) that basically performs a broadband 180° spin inversion and is repeated by applying an additional phase cycle. They differ mainly by the B_1 field strength required to achieve a certain decoupling bandwidth.

A different approach to enhancing spectral resolution in ^{15}N spectra, especially at high magnetic field strength, is transverse relaxation-optimized spectroscopy (TROSY) introduced by Pervushin et al. in 1997 (Pervushin et al. 1997). The TROSY effect is based on the interference between two different spin relaxation mechanisms, e.g. the dipolar coupling (DD) and the chemical shift anisotropy (CSA), in a scalar coupled spin pair such as ^{1}H-^{15}N. This interference can be constructive (increasing the apparent relaxation rate) or destructive (decreasing the apparent relaxation rate) for different components of the peak multiplet (see Fig. 3.16). The TROSY pulse sequence allows the selection of the multiplet component (single-transition spin states) giving the sharpest lines (Pervushin et al. 1997). The TROSY effect for ^{1}H-^{15}N is highest at magnetic field strengths of about 1 GHz and increases with the effective tumbling correlation time. TROSY-based pulse sequences are thus especially useful to enhance the peak intensities and line shapes of the IDP regions that are involved in transient structure formation.

5.5 Longitudinal-relaxation Enhancement for Increased Sensitivity and Reduced Acquisition Times

In this section we discuss the dependence of the experimental sensitivity on the recovery delay T_{rec} (Schanda 2009). A schematic drawing of an NMR experiment is shown in Fig. 3.17, consisting of a pulse sequence of length t_{seq}, a data acquisition period t_{acq} and an additional inter-scan delay t_{rec}. A recovery delay is required for relaxation of the system in order to restore sufficient spin polarization to restart the experiment for a subsequent scan ($T_{rec}=t_{acq}+t_{rec}$). Its duration depends on the longitudinal relaxation time constant T_1. On one hand, the detected signal intensity is proportional to the amount of spin polarization available at the beginning of each scan under steady-state conditions. On the other hand, the SNR scales with the square root of the number of scans that can be performed during a given experimental time, and thus the scan time $T_{scan} = t_{seq} + t_{acq} + t_{rec}$. These two effects can be described analytically by the following equation:

$$SNR \propto \frac{1-exp(-T_{rec}/T_1)}{\sqrt{T_{scan}}} \tag{3.9}$$

Equation 3.9 implies that maximum sensitivity is obtained by adjusting the recycle delay T_{rec} to:

$$T_{rec}(SNR_{max}) \cong 1.25 \cdot T_1 \tag{3.10}$$

and thus the dependency of the experimental sensitivity on the longitudinal relaxation time T_1 is approximately given by:

$$SNR_{max} \propto \frac{1}{\sqrt{T_1}} \tag{3.11}$$

derived substituting T_{rec} and T_{scan} with T_1 in Equation 3.9.

Enhancing the longitudinal relaxation efficiency of the excited spins thus provides an interesting way to increase the experimental sensitivity and additionally to reduce the minimal data acquisition time. Here we will focus on ^1H excitation experiments. In order to understand the experimental schemes that have been proposed for longitudinal proton relaxation enhancement, we need to briefly discuss the spin interactions that govern proton relaxation. There are essentially two different mechanisms that are responsible for proton longitudinal relaxation in a protein: (i) ^1H-^1H dipolar interactions; (ii) hydrogen exchange processes between labile protein protons, e.g. amide and hydroxyl protons, and water solvent protons. The time evolution of the polarization of each proton spin in the molecule is given by the Solomon or Bloch-McConnell equations (Bloch 1946; Solomon 1955; McConnell 1958):

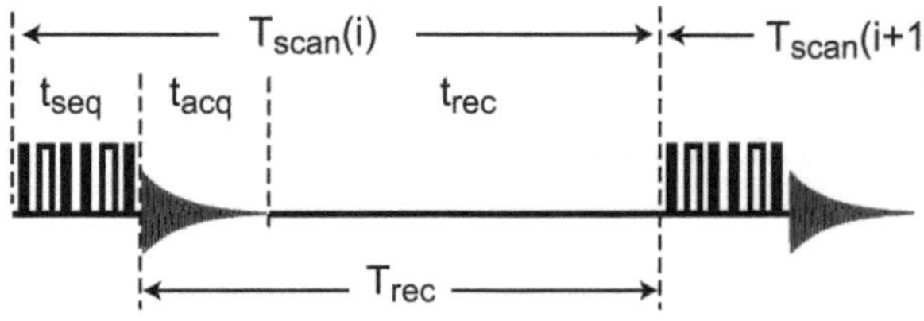

Fig. 3.17 A schematic drawing of an NMR experiment

$$-\frac{d}{dt}\begin{pmatrix} W_{1z}-W_{1z}^0 \\ H_{1z}-H_{1z}^0 \\ H_{2z}-H_{2z}^0 \\ \vdots \\ H_{nz}-H_{nz}^0 \end{pmatrix} = \begin{pmatrix} \rho_W & 0 & 0 & \cdots & 0 \\ 0 & \Sigma_j\rho_{1j} & \sigma_{12} & \cdots & \sigma_{1n} \\ 0 & \sigma_{21} & \Sigma_j\rho_{2j} & \cdots & \sigma_{2n} \\ \vdots & \vdots & \vdots & \ddots & \vdots \\ 0 & \sigma_{n1} & \sigma_{n2} & \cdots & \Sigma_j\rho_{nj} \end{pmatrix} \begin{pmatrix} W_{1z}-W_{1z}^0 \\ H_{1z}-H_{1z}^0 \\ H_{2z}-H_{2z}^0 \\ \vdots \\ H_{nz}-H_{nz}^0 \end{pmatrix} + \begin{pmatrix} 0 & 0 & 0 & \cdots & 0 \\ -k_{ex,1} & k_{ex,1} & 0 & \cdots & 0 \\ -k_{ex,2} & 0 & k_{ex,2} & \cdots & 0 \\ \vdots & \vdots & \vdots & \ddots & \vdots \\ -k_{ex,n} & 0 & 0 & \cdots & k_{ex,n} \end{pmatrix} \begin{pmatrix} W_{1z} \\ H_{1z} \\ H_{2z} \\ \vdots \\ H_{nz} \end{pmatrix} \quad (3.12)$$

where H_{iz} denotes the z-component of the polarization of proton i and H_{iz}^{0} is its thermal equilibrium value. The different ρ and σ terms stand for auto- and cross-relaxation rate constants, with values depending on the distance separating the two protons involved as well as the global and local dynamics of the protein experienced at the sites of the two protons. W_z stands for the bulk water polarization and $k_{ex,\,i}$ are the hydrogen exchange rates for individual protons with the water.

Equation 3.12 indicates that the relaxation of an individual proton spin depends on the spin state of all the other protons in the protein, as well as the bulk water, at the start of the recovery time. The selective spin manipulation of a subset of protons, e.g. amides, while leaving all other protein and water proton spins unperturbed, thus provides an efficient spectroscopic tool for enhancing longitudinal proton spin relaxation, as will be shown in the following section.

The contributions of solvent exchange and dipolar cross-relaxation of amide protons with unperturbed aliphatic proton spins depend on the size and residual structure of the IDP and the sample conditions. Figure 3.18 shows apparent amide ^{1}H T_1 values measured for different IDP samples upon selective or non-selective inversion of different sets of proton spins (Gil et al. 2013; Solyom et al. 2013). For NS5A and BASP1, studied at low temperature and pH, the major relaxation enhancement mechanisms are dipolar interactions, while this situation changes as solvent exchange becomes more efficient, as shown for α-synuclein and PV core protein, which were studied at higher temperature and pH. The average T_1 values measured for the different IDPs shown in Fig. 3.18, as well as the range of predicted exchange rates using the SPHERE program are given in Table 3.2 (Bai et al. 1993; Zhang 1995).

When dipolar interactions provide the dominant relaxation mechanism, non-selective ^{1}H T_1 values are on the order of 900 ms, while selective spin manipulation allows them to be reduced by a factor of three to four, reaching values of 200–300 ms (Schanda and Brutscher 2005; Lescop et al. 2007). This difference becomes even more pronounced in the case of fast hydrogen exchange. Under these conditions, the non-selective ^{1}H T_1 approach the T_1 of bulk water (~3 s at 25 °C), while the selective T_1 become as short as 60 ms (Gil et al. 2013). For α-synuclein at pH 7.4 and 15 °C, the average amide ^{1}H T_1 is reduced by a factor of 38, resulting

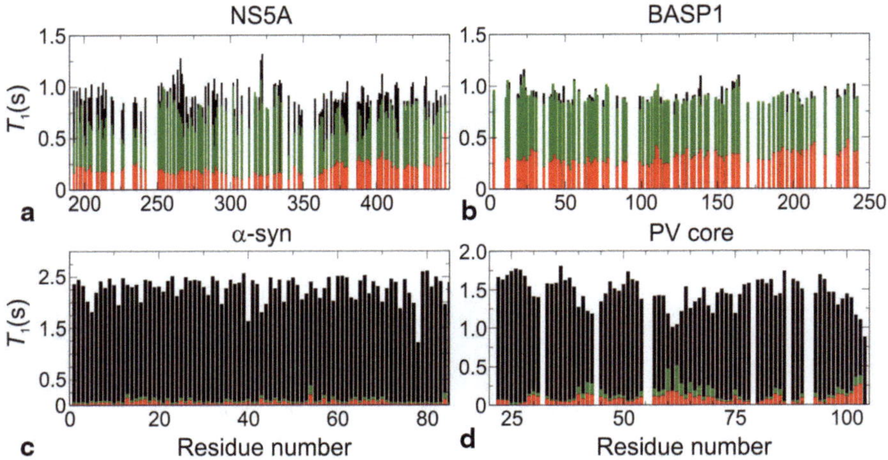

Fig. 3.18 Apparent ^{1}H T_1 relaxation time constants measured by inversion recovery experiments. Amide proton-selective (*red*), water-flip-back (*green*) and non-selective (*black*) inversions were used. The four samples and the conditions were: **a** NS5A D2D3 protein (pH 6.5, 278 K); **b** BASP1 (pH 6.5, 278 K); **c** α-synuclein (pH 7.4, 288 K); **d** PV core (pH 7.5, 278 K)

Table 3.2 Ranges of predicted amide proton solvent exchange rates (k_{ex}) (predictions were performed with the SPHERE program (Bai et al. 1993)) and their average over all residues are listed as well as the averages of the apparent longitudinal relaxation time (T_1) constants measured at 800 MHz as shown in Fig. 3.19

	NS5A	BASP-1	α-synuclein	PV core
Conditions				
pH	6.5	2.0	7.4	7.5
T (°C)	5	5	15	5
Predictions				
Range of k_{ex}(s^{-1})	7.6×10^{-2}–$1.2\times10^{+2}$	1.9×10^{-4}–1.3×10^{-3}	1.8×10^{-1}–$2.1\times10^{+2}$	8.7×10^{-2}–$1.6\times10^{+2}$
$\bar{k}_{ex}$ (s^{-1})	1.5	3.7×10^{-4}	$1.7\times10^{+1}$	$1.1\times10^{+1}$
NMR data				
$\bar{T}_{1,hard}$ (s^{-1})	0.92	0.91	2.27	1.46
$\bar{T}_{1,WFB}$ (s^{-1})	0.70	0.89	0.09	0.14
$\bar{T}_{1,sel}$ (s^{-1})	0.21	0.31	0.06	0.09

in a potential sensitivity gain of a factor of 6, according to Eq. 3.11. This clearly motivates the use of amide ^{1}H start pulse schemes that leave aliphatic and water protons unperturbed throughout the NMR experiment.

The simplest and probably most efficient way to achieve longitudinal relaxation enhancement (LRE) is the use of band-selective ^{1}H pulses, especially in the case of amide protons resonating in a frequency range that is well separated from aliphatic and water protons. Pulse sequences exploiting this concept have been termed band-selective excitation short-transient (BEST) experiments (Schanda et al. 2006b;

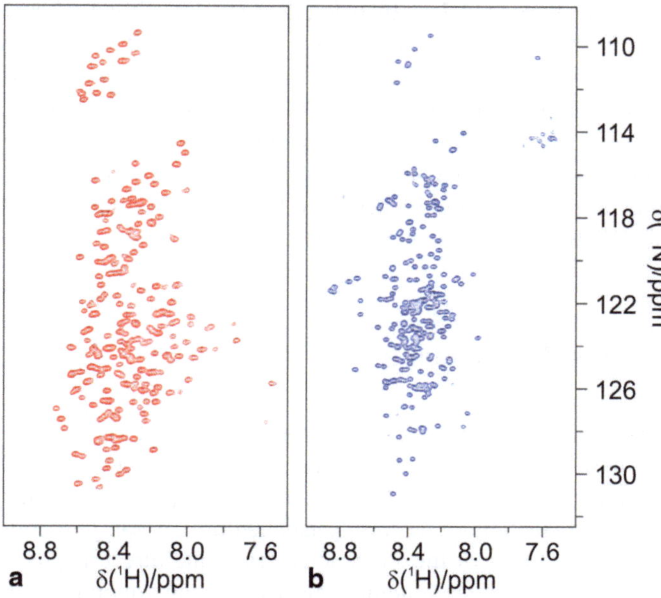

Fig. 3.19 2D ^{1}H-^{15}N BEST-TROSY spectra recorded on two large IDPs (~270 residues), **a** NS5A D2D3 (pH 6.5) and **b** BASP1 (pH 2.0) at 278 K

Lescop et al. 2007; Favier and Brutscher 2011; Solyom et al. 2013). BEST pulse sequence elements are available for all basic pulse sequence elements (building blocks) (Cavanagh et al. 2007), e.g. INEPT, sensitivity-enhanced reverse-INEPT (SE-RINEPT or planar mixing) and TROSY (or double-S_3CT) required for setting up most of the common triple-resonance experiments. In this section we will focus on BEST-TROSY (BT) pulse sequences, which have proven to be particularly useful for the study of IDPs using high magnetic field NMR instruments.

BEST-TROSY combines the advantages of longitudinal relaxation optimization, discussed in the previous section, with those of transverse relaxation-optimized spectroscopy, discussed in Sect. 3.5.4. A special feature of BEST-TROSY is that the proton (H_z) polarization that builds up during the pulse sequence as a consequence of spin relaxation is converted to ^{15}N polarization (N_z) by the final ST2-PT sequence element (Favier and Brutscher 2011). In a conventional TROSY sequence, a large portion of this additional polarization is lost due to ^{15}N T_1 relaxation during the long recovery delay that restores ^{1}H polarization. In BEST-type experiments, however, most of the polarization is conserved because of the short recycle delays used. This provides an additional mechanism for sensitivity and resolution improvement.

The ^{1}H-^{15}N BEST-TROSY sequence and spectra recorded for two large (250 residue) IDPs are shown in Fig. 3.19. Long t_1 acquisition times in the ^{15}N dimension result in highly resolved 2D correlation maps, despite the low chemical shift resolution observed in the ^{1}H dimension. The high resolution obtained in the ^{15}N dimension of a 2D BEST-TROSY spectrum can also be transferred to higher-dimensional BT ^{1}H-^{13}C-^{15}N or BT ^{1}H-^{15}N-^{15}N correlation experiments, as required for sequential resonance assignment (Sect. 3.7.1), by using the semi-CT editing scheme discussed in Sect. 3.5.2. Furthermore, it has been shown that BEST-TROSY implementations of such experiments provide a 20 to 80 % increased SNR compared to BEST-HSQC versions for different IDP samples at 800 MHz ^{1}H frequency (Solyom et al. 2013). BT also results in a more uniform distribution of peak intensities in the spectrum, as the signal enhancement is efficient for residues in transiently structured peptide regions which tend to be the ones characterized by the smallest intensities in optimal conditions for 2D HN correlation experiments.

6 ^{13}C Detected Experiments

The recent progress in instrumental sensitivity (Kovacs et al. 2005) and the development of new NMR experiments made ^{13}C direct detection a useful tool for biomolecular applications (Felli et al. 2013). The first aspect to consider when performing ^{13}C direct detection on uniformly labelled protein samples consists of the presence of large splittings of the ^{13}C resonances in the direct acquisition dimension due to the presence of large homonuclear one-bond couplings, a feature that definitely represents a novelty with respect to ^{1}H direct detection. Indeed, even though ^{13}C-^{13}C couplings are very useful in the design of multidimensional NMR experiments, they are responsible for large signal splitting which, as evident from Fig. 3.20, needs to be suppressed in order to preserve high resolution in the direct acquisition dimension.

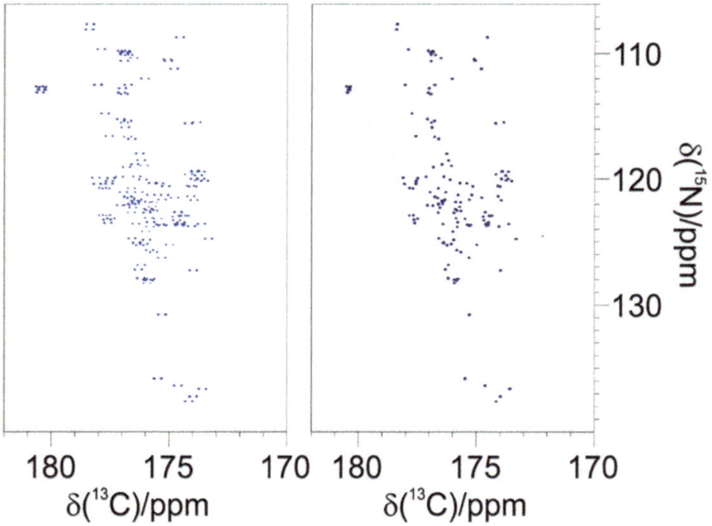

Fig. 3.20 Example of 2D ^{13}C detected CON spectra without and with homonuclear decoupling. The experiments were acquired on α-synuclein (1 mM sample in 20 mM phosphate buffer, pH 6.4, 200 mM NaCl, 0.5 mM EDTA, at 285.5 K) on a 700 MHz Bruker AVANCE equipped with a CPTXO probe

6.1 Homonuclear ^{13}C Decoupling

Achieving homonuclear ^{13}C decoupling in the direct dimension is more complex than in the indirect dimensions because it typically requires the application of radiofrequency pulses at a similar frequency to those of the nuclear spins that are in the process of being detected. A possible solution to this problem consists of sharing the detection period between data acquisition and decoupling pulses (band-selective homonuclear decoupling). This, however, reduces the overall sensitivity of the experiment, and introduces Bloch-Siegert phase shifts (Emsley and Bodenhausen 1990) and decoupling side bands (Waugh 1982) in the spectrum. Recently improved experimental variants exploiting this general idea that however use hard pulses applied at regular intervals during the acquisition period have been proposed (Ying et al. 2014). Alternatively, post-acquisition methods, such as data deconvolution using maximum entropy reconstruction, may be used (Shimba et al. 2003). A last, and arguably the most elegant class of methods for ^{13}C homonuclear decoupling (Felli and Pierattelli 2014b), also known as "virtual decoupling", uses spin-state selection.

Virtual J_{CC} decoupling uses an additional spin evolution delay prior to detection and requires the recording of at least two experiments with different parameter settings. In the first experiment, spin evolution under the J_{CC} coupling is suppressed resulting in an in-phase (IP) line splitting in the detected spectrum, while in the second experiment the J_{CC} coupling evolves for a time $1/(2\ J_{CC})$ resulting in antiphase (AP) line splitting in the final spectrum. A single resonance line (without splitting) is then obtained by calculating the sum and the difference of the two recorded spectra. Finally, the two resulting (sum and difference) spectra are shifted by the amount $J_{CC}/2$ with respect to each other and added up to yield a single line at the correct chemical shift position (Duma et al. 2003a; Duma et al. 2003b; Bertini et al. 2004; Bermel et al. 2006a). This approach, illustrated in Fig. 3.21 for virtual decoupling of $^{13}C^{\alpha}$ in the NMR spectrum of carbonyls, is at the basis of most of the ^{13}C detected multidimensional experiments. In order to work properly, virtual J_{CC} decoupling requires quite uniform J_{CC} couplings in the protein, which is the case for the large one-bond coupling between α carbons ($^{13}C^{\alpha}$) and carbonyls ($^{1}J_{C\alpha C'}$ ≈ 53 Hz). It is worth noting that the two experiments rely on the same number of pulses and lengths of delays, in order to ensure identical signal loss due to pulse imperfections and spin relaxation effects. Another prerequisite of this technique is that the two nuclear spins, e.g. $^{13}C'$ and $^{13}C^{\alpha}$, are sufficiently well separated to allow their selective manipulation through band-selective pulses. There is also a price to pay in terms of sensitivity for the virtual decoupling method, as during the linear combinations necessary to achieve homonuclear decoupling also the thermal noise is increased, thus leading to a reduction of the signal to noise ratio.

Virtual decoupling can be appended to any pulse sequence ending with in-phase $^{13}C'$ transverse coherence (Bermel et al. 2008). Most of the triple resonance experiments based on $^{13}C'$ direct detection end with a coherence transfer step that involves refocusing of either $^{13}C'$-^{15}N or $^{13}C'$-$^{13}C^{\alpha}$ antiphase coherence (or both). Therefore, virtual decoupling can be implemented without appending an additional block by slightly changing the position of the $^{13}C^{\alpha}$ inversion pulses in the last coherence transfer steps, as illustrated in Fig. 3.21.

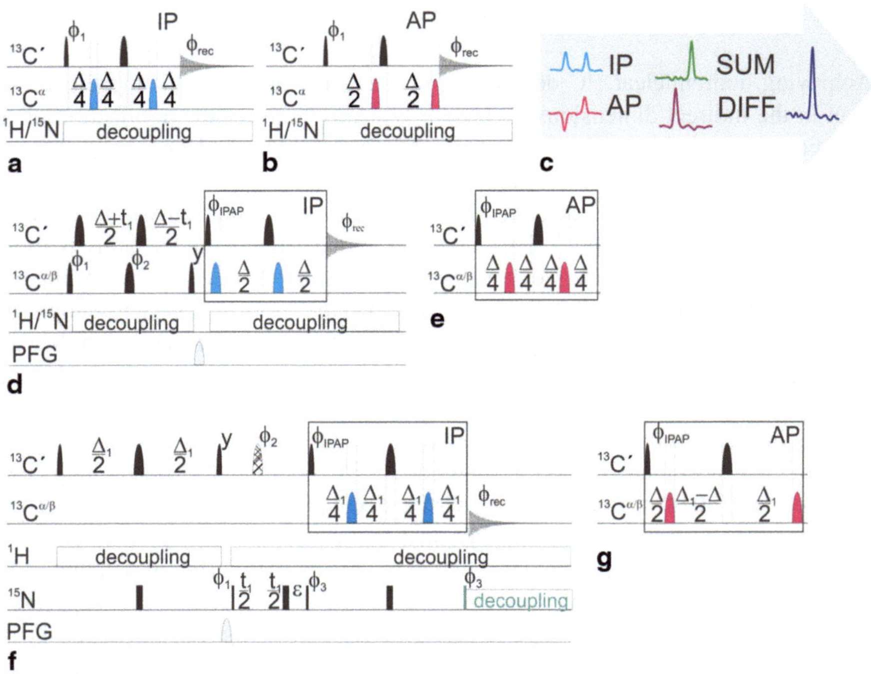

Fig. 3.21 ^{13}C detected 1D and 2D experiments with implemented IPAP decoupling sequence. Implementation of $^{13}C'$-$^{13}C^{\alpha}$ IPAP virtual decoupling building blocks in the 1D mode (**a, b**) and 2D mode: CACO (**d, e**) and CON (**f, g**). Band-selective ^{13}C pulses are denoted by rounded rectangles (*narrow* and wide ones represent 90° and 180° pulses, respectively). The pulses are applied along the x-axis unless otherwise noted. The delays are $\Delta=1/(2J_{C\alpha C'})=9$ ms and $\Delta_1=1/(2J_{NC'})=25$ ms. The phase cycles are: (a) ϕ_1=x, -x and ϕ_{rec}=x, -x; (b) ϕ_1=-y, y and ϕ_{rec}=x, -x; (d) ϕ_1=x, -x ϕ_2=4(x), 4(y) ϕ_{IPAP}=2(x), 2(-x) and ϕ_{rec}=x, 2(-x), x, -x, 2(x), -x; (e) ϕ_1=x, -x ϕ_2=4(x), 4(y) ϕ_{IPAP}=2(-y), 2(y) and ϕ_{rec}=x, 2(-x), x, -x, 2(x), -x; (f) ϕ_1=x, -x ϕ_2=2(x), 2(y) ϕ_3=4(x), 4(y) ϕ_{IPAP}=x and ϕ_{rec}=x, -x, x, 2(-x), x, -x, x; (g) ϕ_1=x, -x ϕ_2=2(x), 2(y) ϕ_3=4(x), 4(y) ϕ_{IPAP}=x and ϕ_{rec}=-y, -x, x, 2(-x), x, -x, x. (**c**) The schematic illustration of the post-acquisition processing for obtaining decoupled spectra.

Finally, as these spin-state selective approaches to achieve homonuclear decoupling perform so well, it is worth mentioning that they can also be implemented for heteronuclear decoupling (Kern et al. 2008; Bermel et al. 2009a). Indeed, a clever variant to achieve heteronuclear ^{15}N decoupling has been proposed both for $^{1}H^{N}$ and $^{13}C'$ direct detection experiments. This can be useful when the $^{13}C'$ coherence lifetimes allow for long acquisition times or to reduce the radiofrequency load/heating on the ^{15}N channel during acquisition, which is particularly useful in the fast pulsing regime (Gil et al. 2013).

6.2 Starting Polarization Source

The other important point to consider in the design/choice of a ^{13}C detected experiment is the starting polarization source. From Sect. 3.1.2, it is immediately clear that a relevant contribution to increase the sensitivity of ^{13}C detected experiments can come from the use of ^{1}H as the starting polarization source (Shimba et al. 2004; Bermel et al. 2009a; O'Hare et al. 2009). In fact, the large one-bond scalar couplings can be easily used to transfer polarization from ^{1}H to the directly bound heteronuclei while still keeping experiments "exclusively heteronuclear". This means that only heteronuclear chemical shifts are frequency labelled in all the spectral dimensions to take advantage of their larger chemical shift dispersion compared to that of ^{1}H, while still exploiting the larger ^{1}H polarization deriving from its higher gyromagnetic ratio. Therefore, several exclusively heteronuclear NMR experiments, such as the (H)CBCACON and (H)CBCANCO experiments (Bermel et al. 2009a), feature ^{1}H as a starting polarization source.

With protons exploited only to increase the experimental sensitivity, it is worth thinking about the possibility of implementing ^{1}H longitudinal relaxation enhancement techniques to reduce the duration of NMR experiments or to increase the sensitivity per unit of time or the resolution. Actually, in these experiments it is very easy to manipulate different sets of ^{1}H spins in a selective way, an essential requirement to promote LRE. Indeed, a simple modification of the initial INEPT block can be introduced to flip-back all the other spins not directly involved in the coherence transfer pathway (H^{flip} method (Bermel et al. 2009b; Bertini et al. 2011b)).

The solution described above for amide protons based on the use of band-selective pulses (BEST approach), can also be implemented in ^{13}C detected experiments (Nováček et al. 2011; Gil et al. 2013). Amide protons indeed resonate in a quite isolated region of the ^{1}H spectra, well separated from the water resonance; they can therefore be selectively manipulated leaving both water and aliphatic resonances unperturbed.

As discussed above, the LRE effect deriving from exchange processes with the solvent is very pronounced in IDPs and adds to that deriving from ^{1}H-^{1}H NOEs, which may contribute at low temperature conditions or in the presence of partially structured elements, even if to a minor extent. Approaching physiological conditions (high temperature and neutral pH), the LRE effect is so large that the recovery of amide polarization to equilibrium becomes extremely fast and recycle delays between transients are therefore no longer necessary. In fact, in this type of experiment the longitudinal recovery already starts before the end of the pulse sequence, as protons get perturbed only in the very initial part of the experiment. Therefore, the optimal set-up consists of an inter-scan delay equal to zero as implemented for the $H^{N\text{-}BEST}$ CON experiment (Gil et al. 2013). Its high sensitivity enables 2D spectra to be collected in a couple of minutes, as shown in Fig. 3.22, making it an ideal technique for *in-cell* applications. However, the use of $^{1}H^{N}$ as the starting polarization source renders the experiment susceptible to the loss of information about prolines and reintroduces a dependence on exchange rates of amide protons with the solvent that impacts the experimental sensitivity.

Alternatively, $^{1}H^{\alpha}$ can be exploited as the starting polarization source. In this way, all backbone sites can be sampled (including prolines) and exchange processes with the solvent are avoided. On the other side of the coin, this also means that only LRE effects deriving from ^{1}H-^{1}H NOEs can be exploited. Therefore, LRE effects become sizeable either at lower temperature or for partially structured proteins. The most convenient way of achieving LRE for $^{1}H^{\alpha}$ consists of exploiting the one-bond

Fig. 3.22 The 2D ^{13}C-^{15}N $H^{N\text{-}BEST}CON$ spectrum of α-synuclein overexpressed in *E. coli* cells acquired in 20 min

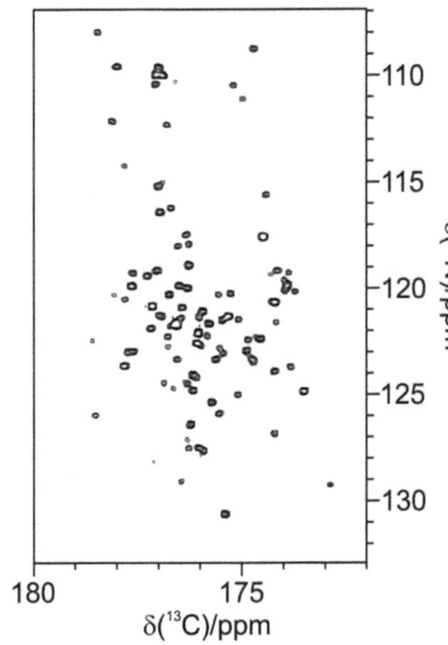

scalar coupling ($^1J_{C\alpha H\alpha}$) with the attached carbon ($^{13}C^{\alpha}$) to selectively manipulate $^1H^{\alpha}$ and flip-back all other protons that are not actively used in the magnetization transfer pathway (H^{flip} method) (Bermel et al. 2009b). Therefore, this trick can be easily implemented in any $^1H^{\alpha}$-start pulse sequence to achieve LRE. Increasing the temperature also provides a simple tool to increase the longitudinal relaxation of $^1H^{\alpha}$ protons that does not require selective manipulation of $^1H^{\alpha}$ protons.

Alternatively, ^{13}C-start versions of such ^{13}C detected experiments can be designed. Indeed ^{13}C spins are not directly involved in chemical exchange processes and are characterized by large chemical shift dispersion. The high flexibility of IDPs results in relatively short ^{13}C T_1 values, which are on the order of seconds, so that recovery delays remain sufficiently short. In addition, the ^1H-^{13}C NOE effect can be exploited in the case of highly flexible IDPs to enhance the signal-to-noise ratio just by irradiating protons during the recovery delay (Bertini et al. 2011b). Therefore, in many cases of practical interest, the sensitivity of ^{13}C-start-^{13}C detected versions of experiments, in particular for the simplest 2D experiments such as CON, CACO and CBCACO, is sufficient to obtain informative spectra (Bermel et al. 2005).

7 From 2D to 3D: From Simple Snapshots to Site-resolved Characterization of IDPs

Multidimensional NMR experiments that encode chemical shift information in several indirect dimensions provide the necessary resolving power to separate correlation peaks from different sites in an IDP. In this section, the general strategy used to achieve site-specific resonance assignment of the NMR signals will be discussed, and a selection of commonly used multidimensional experiments for sequential resonance assignment, as well as complementary techniques for amino acid type identification, will be presented.

7.1 Sequential NMR Assignment: The General Strategy

In order to extract site-specific information from the NMR experiments, each resonance observed in the spectra needs to be associated to a specific nuclear spin of the protein. This task is commonly called sequence-specific (or sequential) resonance assignment. In the early days of biomolecular NMR spectroscopy, resonance assignment was based on the information contained in 2D homonuclear ^{1}H correlation experiments (Wüthrich 1986). The combined analysis of a set of through-bond (COSY, TOCSY) and through-space (NOESY, ROESY) spectra allows sequence-specific assignment of well-folded proteins of moderate size (~100 residues). In the case of highly disordered proteins and IDPs, which are characterized by low chemical shift dispersion in the 2D ^{1}H spectra, this strategy is limited to peptides of less than a few tens of residues. As a consequence, NMR studies of IDPs require uniform ^{13}C and ^{15}N labelled samples that enable the use of 3D (or higher-dimensional) through-bond-only experiments for sequential resonance assignment purposes (Dyson and Wright 2001; Eliezer 2009). These experiments are based on a series of coherence transfer steps, which exploit the large one- and two-bond scalar couplings (Fig. 3.23) (Sattler et al. 1999; Cavanagh et al. 2007).

Starting from a chosen polarization source, typically amide ^{1}H, aliphatic ^{1}H, or ^{13}C, a series of subsequent coherence transfer pathway steps allows the correlation of NMR frequencies of different nuclear spins in the protein with high efficiency. Ideally this strategy could be used to transfer coherence (and thus information) all along the polypeptide chain. In practice, due to spin relaxation effects, coherence transfers are limited only to nuclear spins within the same or neighbouring residues. Therefore, a set of spectra providing complementary connectivity information is required for sequential resonance assignment. The most widely used experiments are introduced in the next sections. They can be divided into three categories: (i) intra-residue, (ii) sequential, and (iii) bi-directional correlation experiments, according to the type of information they contain. A schematic drawing showing

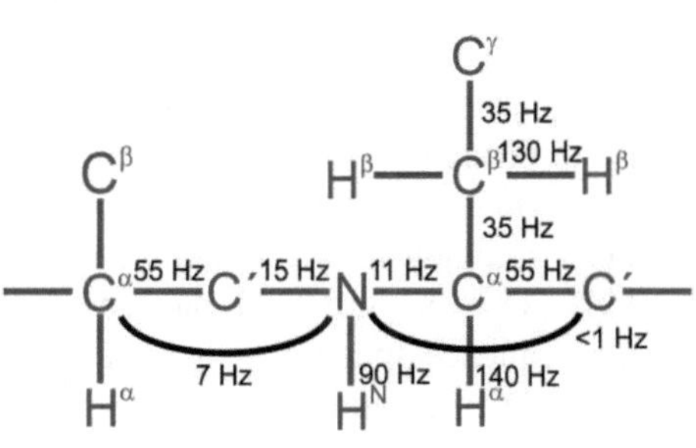

Fig. 3.23 Schematic representation of a protein and the size of the ^1J and ^2J coupling constants that are frequently used for magnetization transfer in ^{13}C, ^{15}N labelled proteins

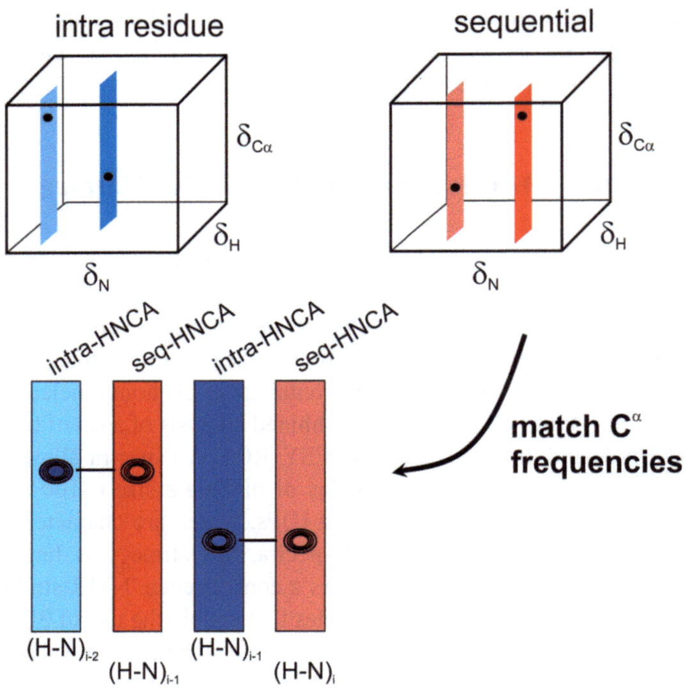

Fig. 3.24 A sequential walk through the NMR spectra for the sequence-specific assignment is illustrated on the example of 3D intra-HNCA and HN(CO)CA (seq-HNCA) experiments

how a pair of such spectra can be used in the sequential resonance assignment procedure is reported in Fig. 3.24.

The basic strategy for sequential assignment is that the different nuclei correlated in a 2D HN or 2D CON fingerprint spectrum, as introduced in Sect. 3.4.2, are connected in a 3D data set with one (or more) additional nuclei, e.g. $^{13}C^{\alpha}$, $^{13}C^{\beta}$, $^{13}C'$, ^{15}N, or $^{1}H^{\alpha}$, that are frequency labelled in the third dimension (indicated as X nuclear spins). ^{1}H-^{15}N or $^{13}C'$-^{15}N pairs are then recognized as belonging to neighbouring residues when the corresponding X frequencies match. In the case of identical or similar resonance frequencies of two or more X nuclei, ambiguities can be solved by combining the information obtained from different pairs of intra- and inter-residue 3D H-N-X or CO-N-X spectra. As a result of this assignment step, chains of sequentially connected residues (^{1}H-^{15}N or $^{13}C'$-^{15}N pairs) are identified. This so-called "sequential assignment walk" might be interrupted by missing correlation peaks, either due to unobservable (line-broadened) NMR signals, e.g. because of pronounced conformational exchange processes or chemical exchange at physiological conditions, missing correlation information, e.g. amide ^{1}H in prolines and/or remaining frequency ambiguities; the information provided by 3D H-N-X and 3D CO-N-X spectra can therefore be combined in order to reduce interruptions in the sequence-specific assignment and ambiguities contributing in this way to the completeness and accuracy of assigned resonances. The final assignment step

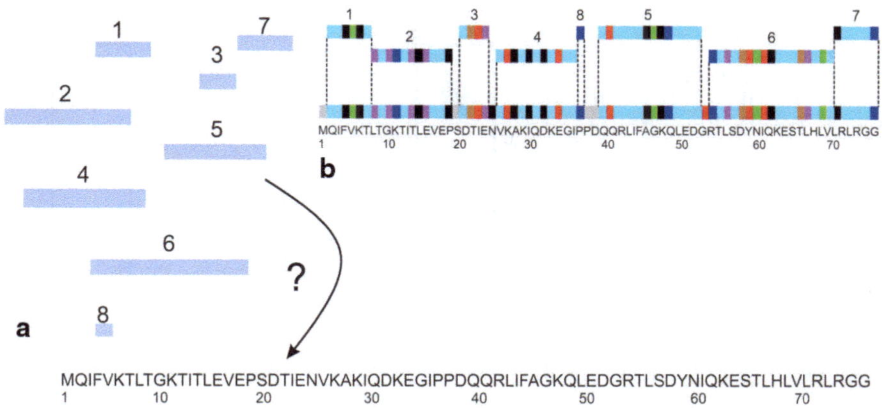

Fig. 3.25 Sequence-specific assignment procedure. The fragments obtained from the sequential walk through the spectra (**a**) are combined with the primary sequence, and sequence-specific assignment is achieved (**b**) by exploiting the amino-acid type information

consists of mapping the identified peptide fragments (correlated NMR nuclei) to their specific positions in the protein sequence. In order to do so, amino acid type information on some (or all) of the residues that form the fragment is required, as illustrated in Fig. 3.25. Such amino acid type information is obtained from side-chain ^{13}C chemical shifts, or from specifically tailored experiments that use spectral editing techniques to differentiate spin-coupling topologies in the amino acid side chains (see Sect. 3.8.4).

8 Sequential NMR Assignment: 3D Experiments

8.1 $^{1}H^{N}$ Detected Experiments

The most common and established resonance assignment approach for proteins uses a set of ^{1}H^{N}-detected 3D correlation experiments (Ikura et al. 1990; Kay et al. 1990). The resulting spectra share as common frequencies those of amide protons in the direct dimension and amide nitrogen atoms in one of the indirect dimensions. The most useful experiments and their underlying coherence transfer pathway(s) are briefly presented in this section (Fig. 3.26). As discussed in Sect. 3.5.4, most amide ^{1}H detected experiments for IDP studies should be implemented as BEST or BEST-TROSY versions in order to enhance experimental sensitivity and spectral resolution (Lescop et al. 2007; Solyom et al. 2013).

The most sensitive coherence transfer pathway correlates the amide ^{1}H^{N} and ^{15}N of amino acid *i* with the ^{13}C′ of the preceding ($i-1$) residue. This experiment, called 3D HNCO (Kay et al. 1990; Grzesiek and Bax 1992b; Schleucher et al. 1993; Solyom et al. 2013), is often used to "count" the number of cross-peaks and therefore to estimate the number of resolved residues that are observed. Additional transfer steps

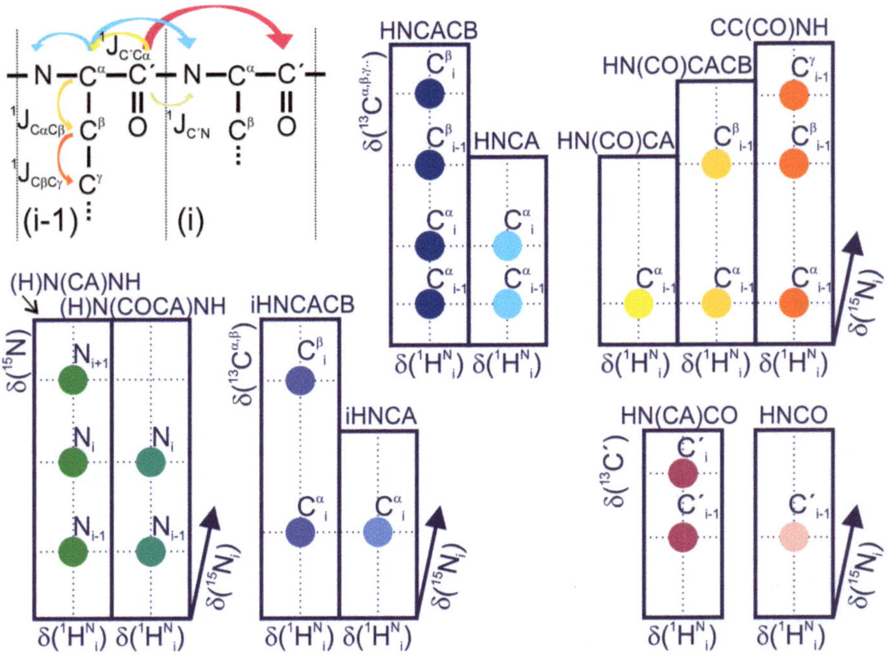

Fig. 3.26 Schematic representation of the $^1H^N$ detected 3D experiments used for the sequence-specific assignment of IDPs

allow correlating the amide $^1H^N$ and ^{15}N with the $^{13}C^\alpha$ and $^{13}C^\beta$ of the preceding ($i-1$) residue, in the so-called 3D HN(CO)CA (Bax and Ikura 1991; Grzesiek and Bax 1992b; Solyom et al. 2013) and 3D HN(CO)CACB (Grzesiek and Bax 1992a; Yamazaki et al. 1994; Solyom et al. 2013) experiments. Note that the intrinsic sensitivity decreases with each additional transfer step, making HN(CO)CACB the least sensitive experiment of the series. For sequential resonance assignment, the information obtained from the experiments correlating nuclei of neighbouring amino acids needs to be complemented with that provided by experiments correlating the amide $^1H^N$ and ^{15}N with the $^{13}C'$, $^{13}C^\alpha$ and $^{13}C^\beta$ of the same (i) residue. This can be achieved, for example, by acquiring 3D HNCA (Kay et al. 1990; Lescop et al. 2007; Grzesiek and Bax 1992b), 3D HNCACB (Wittekind and Mueller 1993; Muhandiram and Kay 1994; Lescop et al. 2007) and 3D HN(CA)CO (Clubb et al. 1992; Kay et al. 1994; Briand et al. 2001; Lescop et al. 2007) experiments belonging to the class of "bi-directional experiments". All of these experiments result in two cross-peaks per residue because of the similar size of the $^1J_{NC\alpha}$ and $^2J_{NC\alpha}$ coupling constants (Fig. 3.23), one corresponding to the desired intra-residue correlation, the other one to the sequential correlation already detected in the previous set of experiments. In order to avoid recording of redundant information and to limit the risk of peak overlaps in the spectra, purely intra-residue correlation experiments, 3D iHNCA (Brutscher 2002; Nietlispach et al. 2002; Solyom et al. 2013), 3D iHNCACB (Brutscher 2002; Nietlispach et al. 2002; Solyom et al. 2013) and 3D iHN(CA)CO (Nietlispach 2004;

Solyom et al. 2013), have been proposed as an alternative to the bi-directional experiments discussed above. The bi-directional and intra-residue experiments are less sensitive than their sequential counterparts and thus, as a rule of thumb, the number of scans should be at least doubled. Recently, improved HNCA+, HNCACB+, and HNCO+ pulse sequences have been introduced to perform bi-directional experiments with improved sensitivity for the sequential correlations (Gil et al. 2014). These experiments are especially useful for the study of fragile IDP samples as they allow the complete information required for sequential assignment to be retrieved from a single set of 3D experiments.

An additional class of experiments, especially useful for IDP samples, is referred to as 3D H-N-N, since their main characteristic is that ^{15}N frequencies are labelled in both indirect dimensions. This allows the sequential assignment walk to be performed by matching ^{15}N frequencies, which, as discussed in Sect. 3.4.2, experience the largest chemical shift dispersion for IDPs lacking a stable structure. A drawback of these pulse schemes is that they require more or longer transfer steps, thus resulting in reduced sensitivity especially for IDP residues involved in transient secondary structure formation. The coherence transfer pathways for the 3D (H)N(COCA)NH and 3D (H)N(CA)NH experiments (Weisemann et al. 1993; Grzesiek et al. 1993b; Bracken et al. 1997; Panchal et al. 2001; Kumar and Hosur 2011) are shown in Fig. 3.27. In the 3D (H)N(COCA)NH, the HNCO sequence is extended passing coherence from ^{13}C′ to ^{13}C$^{\alpha}$ and then to ^{15}N of the neighbouring amino acid, resulting in the correlation of ^{1}H^{N} and ^{15}N of amino acid i with ^{15}N of amino acid i and $i-1$, while the 3D (H)N(CA)NH is an extension of HNCA, yielding correlations with ^{15}N of residues $i-1$, i, and $i+1$. In the case of highly flexible IDPs resulting in large T_2 values, these pulse schemes can be optimized to detect mainly the sequential correlations, while suppressing to a large extent the "diagonal" N_i-N_i correlation peak (Grzesiek et al. 1993b).

A last class of experiments allows the assignment to be extended to the side chain ^{13}C resonances and the provision of valuable information on the side chain length and ^{13}C chemical shifts, important for distinguishing between amino acid types. The most widely used pulse sequence is called 3D (H)C(CO)NH-TOCSY (Montelione

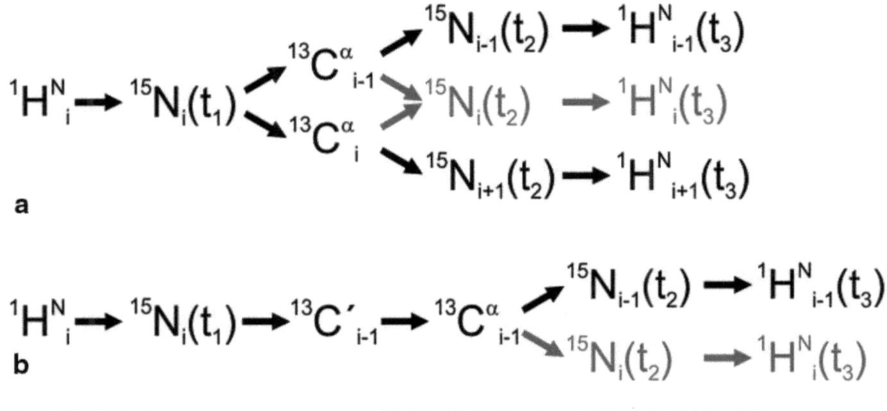

Fig. 3.27 Polarization transfer pathway of (H)N(CA)NH (**a**) and (H)N(COCA)NH (**b**) experiments

et al. 1992; Logan et al. 1992; Logan et al. 1993; Gardner et al. 1996; Grzesiek et al. 1993a). The sequence starts from aliphatic ^{1}H polarization of residue *i*, which is transferred via TOCSY and INEPT steps to the amide group of the following $(i+1)$ residue for final $^{1}H^{N}$ detection.

8.2 ^{13}C Detected Experiments

As discussed in Sect. 3.4, the 2D HN and CON spectra provide complementary tools for the investigation of IDPs. Therefore, similarly to the H^{N}-based experiments discussed above, $^{13}C'$ detection can be extended to higher dimensions, thus providing the required correlation information for sequential resonance assignment of the protein.

A suite of 3D experiments based on $^{13}C'$ detection is shown in Fig. 3.28. The building blocks used for the coherence transfer steps are very similar to those employed in the analogous $^{1}H^{N}$ detected experiments. ^{1}H polarization can be used as a starting source to increase the sensitivity of $^{13}C'$ detected experiments, while still keeping the experiments "exclusively heteronuclear". In all of the spectra recorded with these experiments, the detection dimension is the $^{13}C'$ of residue *i* (C'_i), while the ^{15}N of residue $i+1$ (N_{i+1}) is frequency labelled in one of the indirect dimensions. These C'_i-N_{i+1} correlations are then dispersed in the third dimension by the chemical shift of one (or more) additional nuclear spin(s). In a first set of experiments, CACON and CBCACON (Bermel et al. 2006a; Bermel et al. 2006c; Bermel et al. 2009a), C^{α}_i only or both C^{α}_i and C^{β}_i are edited in the third dimension, resulting in one set of correlation peaks per residue, similarly to the sequential $^{1}H^{N}$ detected

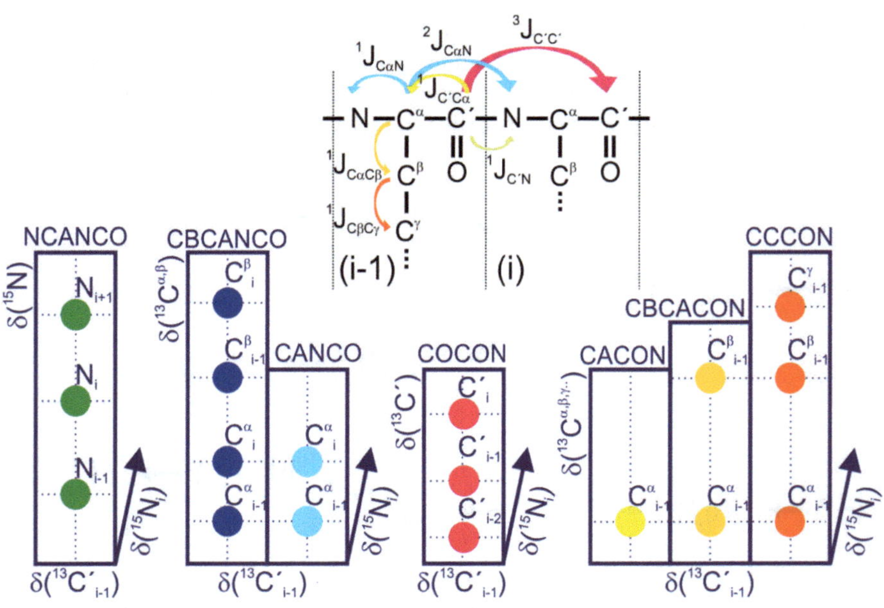

Fig. 3.28 Schematic representation of the $^{13}C'$ detected 3D experiments used for the sequence-specific assignment of IDPs

experiments. In addition, the CCCON-TOCSY experiment (Bermel et al. 2006c; Bermel et al. 2009a) allows the ^{13}C chemical shift information to be extended to all aliphatic side chain carbons of residue *i*. As a result of this set of experiments, each residue of the protein is associated to a set of C^{aliph}_i-C'_i-N_{i+1} correlations. The number of aliphatic carbons detected and their chemical shifts provide information on the amino acid type of the corresponding residue.

The complementary information necessary for sequence-specific assignment is then obtained by acquiring a second set of $^{13}C'$-detected 3D experiments named CANCO and CBCANCO (Bermel et al. 2006c; Bermel et al. 2009a) that correlate the C'_i-N_{i+1} to C^{α}_{i+1} and C^{α}_i (CANCO) or $C^{\alpha}_{i+1}/C^{\beta}_{i+1}$ and C^{α}_i/C^{β}_i (CBCANCO) resulting in two correlation peaks per residue if only C^{α} chemical shifts are detected in the third dimension, or four, if both C^{α} and C^{β} are detected. These experiments thus belong to the class of bi-directional correlation experiments. In addition, the 3D (H)N(CA)NCO (Bermel et al. 2009a) spectrum can be recorded to obtain correlations involving an additional ^{15}N nuclear spin. This experiment thus belongs to the CO-N-N class of spectra (in analogy to the H-N-N ones introduced above for H^N detected experiments). In the 3D (H)N(CA)NCO experiment, the C'_i-N_{i+1} pair is further correlated in the third dimension with the ^{15}N of residues *i*, $i+1$, and $i+2$. Finally, the 3D COCON experiment (Bermel et al. 2006b) was developed to correlate the C'_i-N_{i+1} pair with carbonyls of residues *i*, $i+1$, and $i-1$, providing additional complementary information in this way. The MOCCA mixing scheme significantly improves the sensitivity of this experiment (Bermel et al. 2006b; Felli et al. 2009).

If the whole set of experiments is performed, sequential resonance assignment is obtained by simultaneous matching of C^{α}, C^{β}, C', and N chemical shifts from intra-residue and sequential correlations. The requirement for simultaneous frequency match of four different nuclear spins provides a robust way of solving assignment ambiguities due to resonance overlap in spectra of IDPs.

As also prolines are detected in these spectra, $^{13}C'$ direct detection provides an ideal tool for complete sequence-specific assignment of an IDP, provided that sensitivity is sufficient to enable acquisition of the whole set of experiments in a reasonable overall measurement time (Bermel et al. 2005; Bermel et al. 2006a). Our experience with several IDPs characterized by high flexibility indicates that the complete set of 3D experiments described here can be collected with cryogenic probes optimized for ^{13}C detection and yields the complete sequence-specific assignment without any other additional information. For samples of limited solubility (~100 μM) or using less optimal spectrometer hardware, it is still possible to acquire the most sensitive 2D (CACO, CBCACO, CON) (Bermel et al. 2005) and 3D experiments (CBCACON, CCCON) (Bermel et al. 2006a) in order to complement the spectral information obtained from a series of $^{1}H^{N}$ detected experiments.

8.3 Aliphatic ^{1}H Detected Experiments

An alternative to the two previously discussed approaches based on the two most well resolved 2D spectra to study IDPs (HN and CON), based either on $^{1}H^{N}$ or $^{13}C'$ direct detection, consists in focusing on $^{1}H^{\alpha}$ detection, as proposed for the

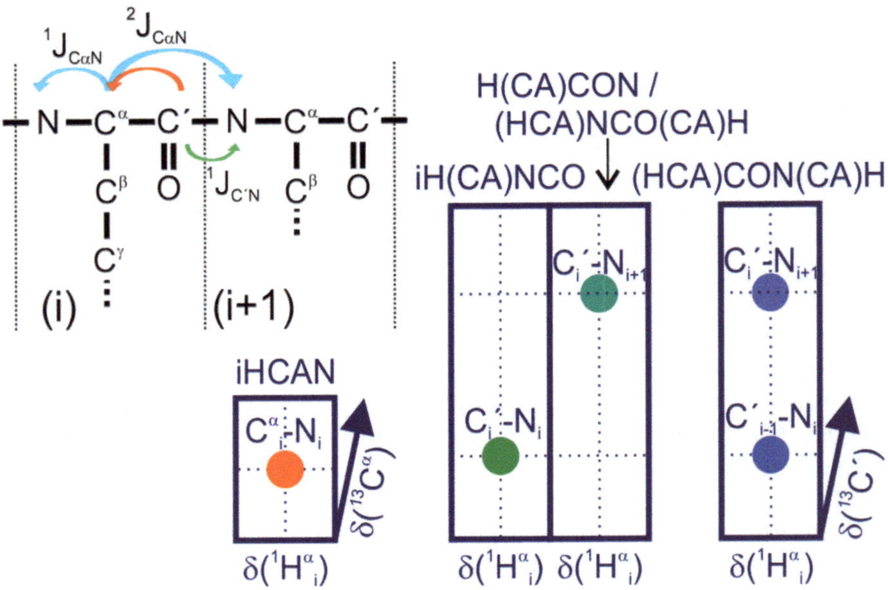

Fig. 3.29 Schematic representation of the $^1H^\alpha$ detected 3D experiments used for the sequence-specific assignment of IDPs

assignment of proline-rich regions (Kanelis et al. 2000). More recently, a set of five triple resonance experiments was developed for sequence-specific resonance assignment of IDPs (Mäntylahti et al. 2010; Mäntylahti et al. 2011). The coherence transfer pathways and correlated frequency information obtained from these experiments are shown in Fig. 3.29. The 3D iH(CA)NCO, iHCAN and (HCA)NCO(CA)H experiments were designed for spin system identification, while the 3D H(CA)CON and (HCA)CON(CA)H provide the complementary connectivity information. The $^1H^\alpha$ signals are clustered in a narrow chemical shift region where also the water proton spins resonate. Resolution in the direct dimension is therefore quite limited, in particular for amino acids of the same kind. Furthermore, excellent water suppression performance is mandatory. Alternatively, the protein can be dissolved in a fully deuterated solvent (99.99 % D_2O), which is characterized by higher viscosity with respect to H_2O and causes an increase in the transverse relaxation rates of the nuclear spins in the protein sample. This approach has been successfully applied to an IDP of about 110 amino acids (Mäntylahti et al. 2011). $^1H^\alpha$-detected experiments might be considered under conditions where H^N detection of the IDP is not feasible, e.g. at near physiological conditions, and when the protein concentration or the experimental setup does not provide the required sensitivity for ^{13}C detection.

8.4 Experiments for Amino Acid Type Editing or Selection

In addition to the experiments discussed in the previous sections, which provide sequential correlation information connecting nuclear spins of neighbouring residues, information on the amino acid type of a given residue is necessary to complete the sequential resonance assignment of a protein. Amino acid type information may be obtained from NMR data by the selective incorporation of labelled (or unlabelled) amino acids at the protein expression level (McIntosh and Dahlquist 1990; Cowburn et al. 2004; Tong et al. 2008), from $^{13}C^{\alpha}$, $^{13}C^{\beta}$ and ^{13}C side chain chemical shift data (Grzesiek and Bax 1993), or from specifically tailored NMR experiments that exploit the particular spin-coupling topologies and chemical shifts in the different amino acid side chains (Wittekind et al. 1993; Olejniczak and Fesik 1994; Yamazaki et al. 1995; Rao et al. 1996; Rios et al. 1996; Dötsch et al. 1996a, 1996b; Pellecchia et al. 1997; Schubert et al. 1999, 2001a, 2001b; Pantoja-Uceda and Santoro 2008). The latter approach has the advantage that unambiguous amino acid type information is obtained from a series of NMR experiments recorded on a single uniformly $^{13}C/^{15}N$ labelled sample. Such experiments, which often rely on band selective pulses to perturb only specific frequency ranges, benefit from the narrow chemical shift ranges of side chain resonances typical of IDPs. A selection of the most common experiments available for amino acid type discrimination is presented in this section.

Conceptually we can distinguish between experiments for *amino acid type selection* and those performing *amino acid type editing*. In the first approach (amino acid type selection), only correlation peaks from specific amino acid types are detected, while in the second approach (amino acid type editing), correlation peaks from different classes of amino acids are separated in different NMR subspectra. The two classes of experiments share, as common features, several ingenious ways to distinguish different amino acids exploiting peculiar spectroscopic features of amino acid side chains such as characteristic chemical shifts, coupling topologies, side chain length, etc., as described in the early NMR literature (Wittekind et al. 1993; Olejniczak and Fesik 1994; Yamazaki et al. 1995; Dötsch et al. 1996a, 1996b; Löhr and Rüterjans 1995; Pellecchia et al. 1997).

The MUSIC (multiplicity selective in-phase coherence transfer) strategy proposed by the Oschkinat group (Schubert et al. 1999; Schubert et al. 2001a; Schubert et al. 2001b), recently extended to ^{13}C direct detection experiments (CAS-NMR) (Bermel et al. 2012a), provides the most complete set of amino acid-selective experiments. Different variants of two basic experiments (HN(CO)CACB and CBCACON) allow the selection of correlations deriving only from specific amino acid types from the basic 2D HN and CON spectra respectively (Fig. 3.30). The resulting spectra are very simple, contributing in this way to reducing the problem of cross-peak overlap. The various pulse schemes mainly differ in the number and lengths of transfer steps required and thus their overall sensitivity. As a rule of thumb, the experiments for shorter side chains are more sensitive than the corresponding pulse sequences for selecting amino acids with longer side chains. The different pulse sequence elements designed to select correlations of specific amino acids can also be implemented in "bidirectional" experiments (HNCACB, CBCAN-CO), providing additional information in this way.

As mentioned before, prolines are abundant in IDPs and can be easily detected in ^{13}C detected experiments; the identification of proline-neighbouring residues is of particular importance for resonance assignment strategies based on $^{1}H^{N}$ detected

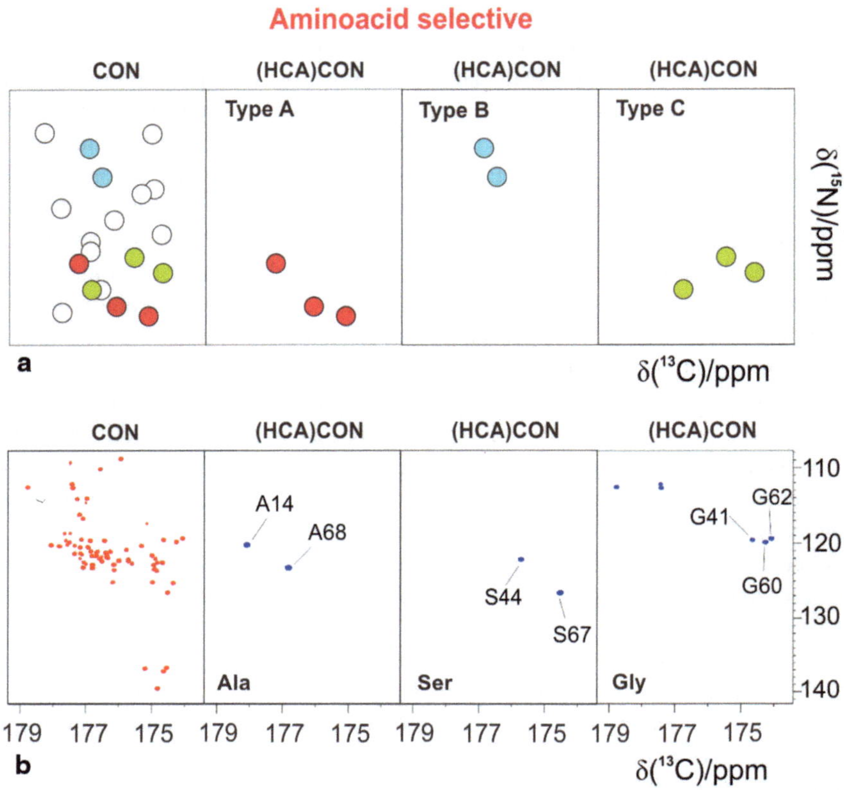

Fig. 3.30 Amino acid-selective experiments to simplify crowded spectra. **a** A schematic representation of a 2D CON spectrum and three 2D (HCA)CON spectra, in which three different amino acids (types A, B and C) are selected. **b** CAS-NMR spectra of Cox17. From *left* to *right*: reference 2D CON spectrum, and (CA)CON spectra from experiments selecting Ala, Ser and Gly of Cox17. The experiments were acquired on Cox17 protein (1.8 mM sample in 20 mM phosphate buffer, pH 7.0, 0.25 mM EDTA, 20 mM DTT, at 307 K) on a 700 MHz Bruker AVANCE spectrometer equipped with a CPTXO probe (Felli et al. 2013)

pulse schemes (Tompa 2002). Proline-selective experiments based on BEST-TROSY HN(COCAN) and iHN(CAN) pulse sequences, exploiting the particular ^{15}N chemical shift of prolines in IDPs, have therefore been proposed that yield increased sensitivity with respect to their corresponding MUSIC counterparts for the identification of proline-neighbouring residues (Solyom et al. 2013). ^{13}C detected variants of the experiments tailored for prolines have also been proposed (Bermel et al 2012a).

Amino acid type editing differs from amino acid-type selection by the fact that correlation peaks for all (observable) residues are detected in the final spectrum, while amino acid type information is obtained by sign encoding. In the simplest type of amino acid type editing, the correlation peaks corresponding to a specific class of amino acids are of opposite sign with respect to all others, thus allowing simple discrimination of this particular class of amino acids. In addition, if a second (reference) experiment is recorded where all NMR signals have the same sign, peaks belonging to different amino acid type classes can be separated in different spectra at the processing level by addition and subtraction of the two recorded data sets. Such a simple sign encoding has been proposed for many basic correlation experiments involving C^{α} and/or C^{β} carbons, requiring only slight modifications of the original pulse sequences (Grzesiek and Bax 1992b; Panchal et al. 2001; Brutscher 2004b). The same concept of sign encoding can also be used to differentiate between more than two classes of amino acids by using HADAMARD spectroscopy (Kupce et al. 2003; Brutscher 2004a). In short, to distinguish between *n* amino acid classes, HADAMARD NMR spectroscopy requires the recording of *n* spectra with different sign encoding according to a particular HADAMARD scheme, also called a HADAMARD matrix. HADAMARD matrices exist for $n=2$, 4, 8, 12, … (multiples of 4). The same HADAMARD matrix employed for encoding is then also used at the processing level to calculate linear combinations of the recorded data set resulting in separate spectra for each amino acid class. This concept (with $n=8$) is used in the HADAMAC (Hadamard-encoded amino acid type editing) experiments (Lescop et al. 2008; Feuerstein et al. 2012; Pantoja-Uceda and Santoro 2012). The sequential HADAMAC (sHADAMAC) experiment, based on the sensitive (H)CBCACONH (Grzesiek and Bax 1992a; Grzesiek and Bax 1993) correlation sequence, provides

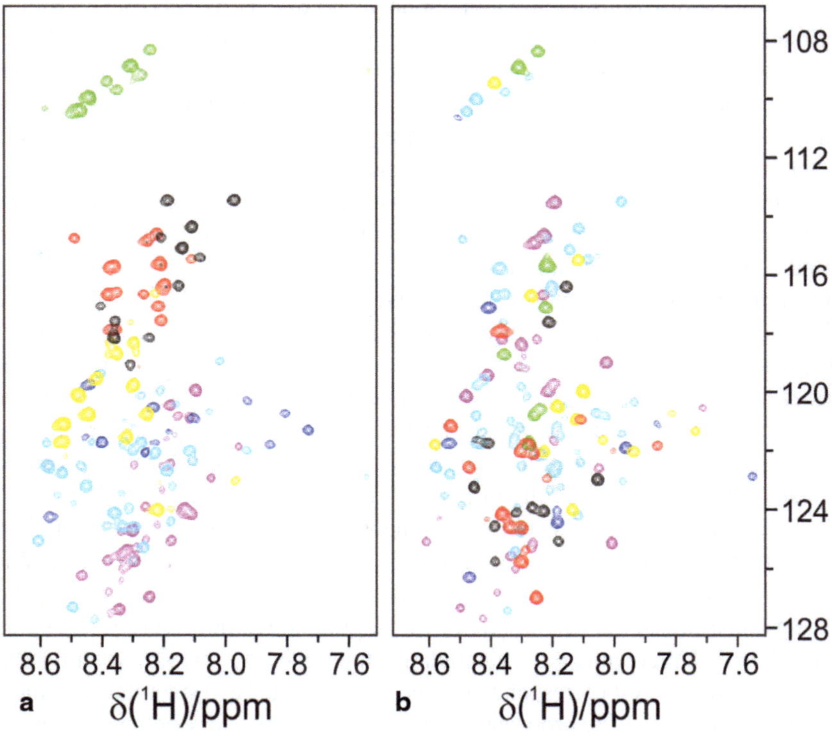

Fig. 3.31 **a** iHADAMAC and **b** HADAMAC spectra acquired on an 800 MHz spectrometer at 298 K on a 90 μM sample of a NS5A fragment. 100 complex points were recorded in the ^{15}N indirect dimension for a spectral width of 2000 Hz using semi-CT editing for both spectra. The total acquisition times were set to 12 h (iHADAMAC) and 2 h (sHADAMAC). The seven subspectra corresponding to the different amino acid type classes are color-coded and superposed on the same graph

amino acid type information for the residue preceding the detected amide group, and allows differentiation between seven classes of amino acids: (1) Val-Ile-Ala, (2) Gly, (3) Ser, (4) Thr, (5) Asn-Asp, (6) His-Phe-Trp-Tyr-Cys and (7) Arg-Glu-Lys-Pro-Gln-Met-Leu (Fig. 3.31b). The same approach has also recently been extended to intra-residue amino acid type editing (iHADAMAC, Fig. 3.31a) (Feuerstein et al. 2012).

8.5 Automated Assignment

The establishment of a vast series of multidimensional NMR experiments has opened the way to the development and improvement of automated assignment programs and to their application to IDPs. The very low chemical shift dispersion and high chemical shift degeneracy typical of disordered proteins have always been the major limitations to the performance of such programs. However, in recent years many efforts have been made to increase the robustness and reliability of automatic assignment procedures and several algorithms are now ready for use. Out of the many, MARS (Jung and Zweckstetter 2004), FLYA (López-Méndez and Güntert 2006) and TSAR (Zawadzka-Kazimierczuk et al. 2012a) are some examples of the most well-established and promising algorithms that can be exploited to assign IDPs.

In general, the programs require as input only a set of experimental peak lists and the primary sequence of the protein to be assigned. The output can be very different depending on the particular algorithm employed. For example, the assignment, usually provided to the user as a text file, can be accompanied by graphical representations and/or additional text files containing, for instance, information about the reliability of the assignment and/or suggestions related to a missing or ambiguous assignment. In addition, to facilitate the manual validation of the assignment, some programs provide assignment results also in suitable formats to be directly loaded and read by tools for spectral analysis.

Most importantly, recent improvements have made the input files more and more flexible, accepting peak lists from any combination of multidimensional experiments provided they are written according to a simple but specific set of rules that describes the magnetization transfer pathways employed. Using all the experimental data simultaneously, also incomplete peak lists in which some peaks are missing can be used; furthermore, peak lists containing redundancy of information can be exploited to increase the reliability of the chemical shift assignment.

Finally, automatic resonance assignment algorithms differ in the calculation time necessary to perform the assignment procedure. For example, programs like TSAR are really fast, since they simply compare the submitted chemical shift values and map them along the amino acid sequence of the protein. On the contrary, programs like MARS and FLYA require more computation time since they employ several iterative cycles to find the best correspondence between predicted and submitted chemical shift values, minimizing the propagation of possible initial errors in the assignment and/or in the experimental peak lists. Therefore, the more peak lists are complete and provide complementary information (for example combining $^{1}H^{N}$ and $^{13}C'$ detection), the less time is required by the assignment algorithms to reach convergence.

9 High-dimensional NMR Experiments (nD, with n > 3)

As discussed in the previous sections, various experimental strategies allow improvement of spectral resolution. However, in the case of severe spectral overlaps, we recommend increasing the dimensionality of the NMR experiments. Indeed, encoding more than three frequencies into a cross-peak observed in multidimensional NMR experiments reduces the chances of accidental signal overlap. In order to take maximum advantage of this approach it is important to maintain the highest resolution in all the detected dimensions. To this end, the combination of fast pulsing techniques with non-uniform sampling strategies and the respective processing approaches, described in detail in Sect. 3.5, is crucial to keep experimental time and spectral resolution in a reasonable range.

By high-dimensional NMR we refer to experiments in which three or more indirect dimensions are used to encode chemical shift information within the same experiment. In many cases, in order to cope with the low dispersion of resonance frequencies typical of IDPs, increasing the number of dimensions provides a unique tool to enhance the resolution in order to accelerate the resonance assignment procedure or to make it possible in difficult cases.

High-dimensional (4-5D) NMR experiments are presented in the next section, focusing on recent developments tailored for IDPs. A very convenient strategy to simplify the inspection and analysis of the resulting spectra is presented, as well as the most useful high-dimensional experiments tailored for IDPs.

9.1 Analysis and Inspection of High-dimensional NMR Spectra

One of the main barriers to the reception and diffusion of high dimensional NMR experiments has always been the idea that more than three dimensions could be conceptually difficult to handle. Indeed, while it is easy to visualize three-dimensional objects, the same does not hold for objects characterized by a higher number of dimensions. A possible solution consists of breaking them down into combinations of objects of lower dimensionality, as also done when inspecting three-dimensional spectra. Indeed most of the available software applications for the analysis of 3D NMR spectra display only two dimensions at a time, extracted at a specific frequency in the third dimension. In a similar way, higher dimensional NMR spectra can be inspected by visualizing only two dimensions at a time, which are associated to a peak in a 2D or 3D reference spectrum in the case of 4D and 5D experiments, respectively. This procedure is possible because, in most cases, higher dimensional experiments are extensions of 2D or 3D spectra in which the new information is encoded in the additional indirect dimensions. This idea is implemented in the sparse multidimensional Fourier transform (SMFT) algorithm (Kazimierczuk et al. 2010b, 2013).

SMFT proposes an innovative strategy to intuitively examine four or five-dimensional spectra in a very simple and straightforward way. Acquired data are processed with the MFT algorithm (mentioned in Sect. 3.5.3), but only at some predefined frequencies (hence the prefix "sparse"): in this way, the inspection of the full spectrum is reduced to that of a limited number of spectral regions that are extremely small with respect to the *n*-dimensional space of the full spectrum. For convenience, these cross-sections are processed as two-dimensional spectra, which are very familiar to almost all NMR users. Therefore, four and five-dimensional spectra appear as a series of 2D spectra to which two or three further frequencies, respectively, are associated. Of course, a prior knowledge of these frequencies is needed: together with the 4D or 5D experiment, respectively a 2D or 3D spectrum, called the "basis spectrum", which shares the same correlations with the higher

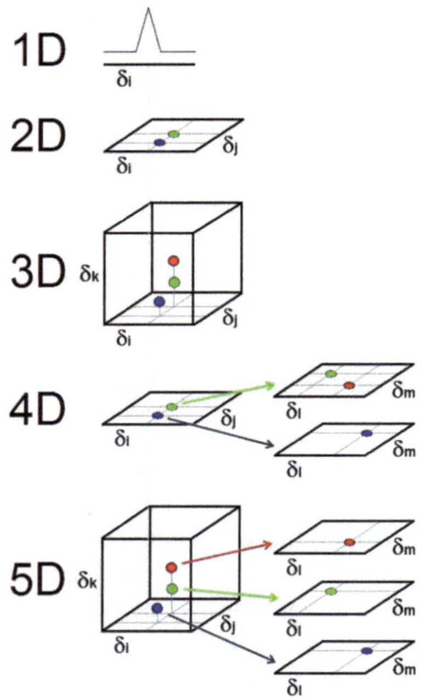

Fig. 3.32 A possible way to visualize high-multidimensional NMR spectra. A schematic illustration of the progressive steps towards experiments of higher dimensionality is reported from the 1D to the 5D case. Spectra up to three dimensions are visualized in a standard way, whereas 4D and 5D spectra are presented by decomposing them into lower dimensionality ones. The algorithm exploits the prior knowledge of the correlations provided by a lower dimensionality spectrum in order to extract only the additional information from a related multidimensional spectrum, thus reducing the number of dimensions that have to be inspected. So, for example, a 4D spectrum can be thought as a 2D basis spectrum, which shares two dimensions with the 4D spectrum (δ_i and δ_j), in which each peak is associated to another 2D spectrum containing the two further dimensions (δ_l and δ_m). In this way, instead of the full spectrum, only a series of cross-sections are effectively computed, in equal number to the peaks detected in the basis spectrum. Similarly, a 5D spectrum can be analysed as a series of two-dimensional spectra (δ_l and δ_m), each one correlated to a given peak of the related 3D basis spectrum (δ_i, δ_j and δ_k)

dimensional one, should be acquired to retrieve them. In this way, the dimensions in which the frequencies of the peaks are already known are not processed, whereas the additional two dimensions containing the new information are fully computed, as the positions of the signals are unknown there. This approach is schematically illustrated in Fig. 3.32 for 4D and 5D experiments.

SMFT facilitates the analysis of high-dimensionality spectra, since this method provides a way to retrieve all the spectral information content without the need to explore the vast *n*-dimensional space of the full spectrum. The visualization of the spectrum through inspection of 2D cross-sections also provides further advantages: for example, it enables the use of automatic peak picking tools, which perform in a very reliable and efficient way, as the signals are well-resolved thanks to the high dimensionality of the experiment. In addition, it allows great amounts of disk space to be saved, since just a small subset of the full spectrum is actually processed; in this way, high digital resolution can be set in all the indirect dimensions.

9.2 Examples of High-dimensional Experiments Tailored for IDPs

The benefits provided by chemical shift labelling of the heteronuclei including ^{13}C direct detection, ^{1}H longitudinal relaxation enhancement and TROSY, all discussed in Sects. 3.5 and 3.6, can be combined with the resolving power of high-dimensional NMR experiments specifically tailored for IDPs. In recent years a large number of high-dimensional experiments has been developed and profitably used to achieve the sequence-specific assignment of several IDPs of medium and large molecular mass. Their number is expected to increase with future methodological advancements and hardware improvements.

Several of the three-dimensional experiments based on coherence transfer via *J*-couplings, discussed in Sects. 3.7 and 3.8, can easily be extended to 4D or 5D versions by explicitly frequency labelling the heteronuclear spins only exploited in 3D experiments for coherence transfer (the nuclear spins generally included in parentheses in the pulse sequence acronyms). Taking the 3D (H)N(COCA)NH as an example, the coherence is transferred through $^{13}C'$ and $^{13}C^{\alpha}$ to N. With the introduction of one or two chemical shift evolution periods, $^{13}C'$ or/and $^{13}C^{\alpha}$, nuclei can also be frequency labelled, extending the experiment to 4D or 5D, respectively. Another interesting example is provided by the extension to 4D of the 3D HN(CA)CO and HN(CO)CA, commonly used for sequence-specific assignment (4D HNCOCA and 4D HNCACO) (Brutscher et al. 1995a; Yang and Kay 1999; Xia et al. 2002). Based on the 2D HN experiment (H^N_i-N_i), they provide sequential correlations via the $^{13}C'$ and $^{13}C^{\alpha}$ nuclei for backbone assignment (C'_{i-1}-C^{α}_{i-1} and C^{α}_i-C'_i/C^{α}_{i-1}-C'_{i-1}, respectively) as shown in Fig. 3.33a. The former can be used to resolve frequency degeneracy in the 3D HNCO spectrum, whereas the latter relies on $^{13}C^{\alpha}$ and $^{13}C'$ chemical shifts to establish sequential correlations. An additional 4D iHNCOCA experiment can also be designed (Konrat et al. 1999). As a general rule, the price to pay for the additional information retrieved in high-dimensional experiments

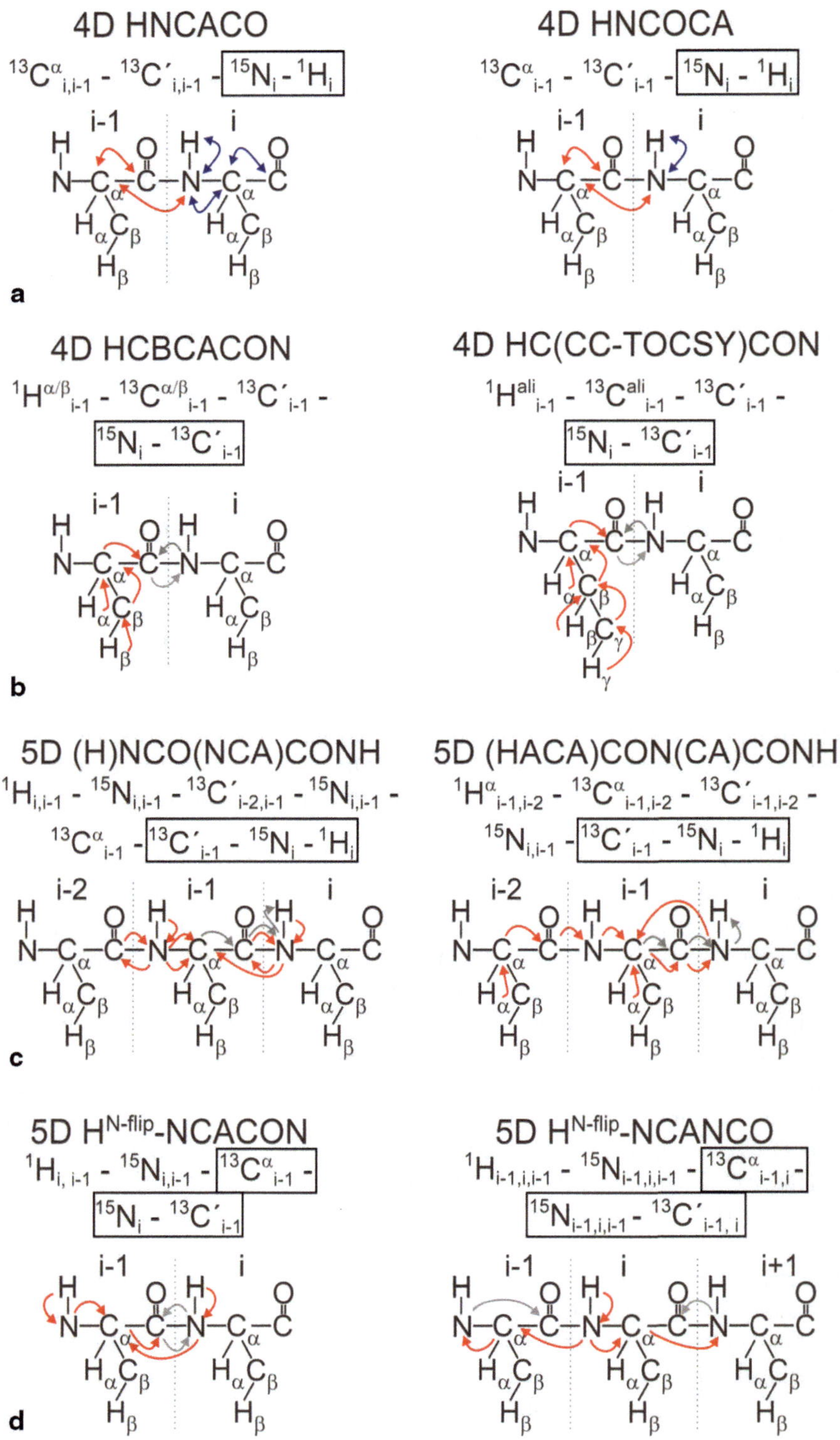

Fig. 3.33 Different examples of the polarization transfer pathways of 4D and 5D experiments

compared to their lower dimensional analogues consists of a loss in experimental sensitivity of at least a factor of $\sqrt{2}$ for each additional dimension introduced, due to the requirement of phase-sensitive quadrature detection and because additional spin relaxation occurs during the frequency editing periods.

An overview of the most useful experiments, exemplifying the many possibilities available to retrieve sufficient information for unambiguous sequence-specific assignment, is presented hereafter. A summary of the experiments discussed, together with the correlations provided, is reported in Tables A.1 and A.2 in the Appendix. The high-dimensional experiments can be grouped into different categories according to their lower dimensional basis 2D (HN or CON) and 3D (HNCO or CACON) spectra as well as to the type of information retrieved in the additional indirect dimensions (amino acid-type information through ^{13}C side chain chemical shifts, sequence-specific information through C$^{\alpha}$, C$^{\beta}$, C′ and N chemical shifts).

Starting from the most simple and intuitive ones, the 5D HCBCACONH (Grzesiek and Bax 1993; Staykova et al. 2008; Kazimierczuk et al. 2010b), 5D HCCCO-NH (Montelione et al. 1992; Clowes et al. 1993; Mobli et al. 2010), 4D HCB-CACON (Bermel et al. 2012b) and 5D HC(CC-TOCSY)CON (Montelione et al. 1992; Logan et al. 1992; Logan et al. 1993; Grzesiek et al. 1993a; Gardner et al. 1996; Hiller et al. 2008) (Fig. 3.33b) provide information on the chemical shifts of the ^{1}H and ^{13}C aliphatic spins of each amino acid (i) by exploiting the HNCO (C'_{i}, N_{i+1}, H^{N}_{i+1}) or the CON (C'_{i}, N_{i+1}) as basis spectra, respectively. Therefore, the related 2D cross-sections that need to be inspected resemble 2D ^{1}H-^{13}C planes which however contain only the correlations observed at the respective frequencies in the basis spectra, in this way providing very clean and straightforward information to identify the residue type and assign side chain ^{1}H and ^{13}C chemical shifts. If necessary, the 5D experiments that exploit the HNCO as the basis 3D spectrum can easily be reduced to their 4D analogues (which exploit the 2D HN as a basis spectrum) by simply omitting the evolution of ^{13}C′ chemical shifts.

Along the same lines, the 2D HN and CON spectra can be used as basis spectra for high-dimensional experiments that provide information to achieve sequence-specific assignment in the additional dimensions. A variety of different experiments can be included in this class. For example, the 4D HNCACB (Zawadzka-Kazimierczuk et al. 2010; Gossert et al. 2011) and 4D HCBCANCO (Bermel et al. 2012b; Nováček et al. 2012) (bidirectional experiments) yield in cross-section the information on the ^{1}H and ^{13}C chemical shifts of amino acids (i) and (i+1) for each cross-peak observed in the 2D HN (N_{i+1}, H^{N}_{i+1}) and 2D CON spectra (C'_{i}, N_{i+1}), respectively.

Another class of experiments provides sequential connectivities through ^{15}N chemical shifts. These are particularly useful because, depending on the experimental variant used, they still exploit the same basis spectra (3D HNCO or 2D HN and 2D CON) and provide information on the ^{15}N chemical shifts of neighbouring residues in the additional dimensions.

This category includes the 4D HN(COCA)NH (Shirakawa et al. 1995; Bracken et al. 1997), 4D HN(CA)NH (Zawadzka-Kazimierczuk et al. 2010), 5D HN(CA)CONH (Kazimierczuk et al. 2010b) and 5D HN(COCAN)CONH (Piai et al. 2014) ($^1H^N$ detected) as well as a variety of different experimental variants based on ^{13}C direct detection described below. The $^1H^N$ detected ones reported above correlate the H^N_i-N_i peak of the 2D ^1H-^{15}N HSQC spectrum (the H^N_i-N_i-C'_{i-1} peaks of the 3D HNCO spectrum, in the 5D case) with the $^1H^N$ and ^{15}N nuclei of neighboring residues. Instead, when amide protons are merely exploited as a starting source of magnetization, ^{15}N and $^{13}C'$ resonances can be used to establish sequential correlations, such as in the 5D (H)NCO(NCA)CONH (Zawadzka-Kazimierczuk et al. 2012b) (N_{i-1}-C'_{i-2} and N_i-C'_{i-1}) and 5D (H)NCO(CAN)CONH (Piai et al. 2014) (N_i-C'_{i-1} and N_{i+1}-C'_i) experiments (Fig. 3.33c).

As regards the ^{13}C detected analogues, they can be acquired in the 4D mode, using the 2D CON as basis experiment, or extended to the 5D mode by exploiting the C^α chemical shift as an additional dimension of the basis experiment. They include 5D NCOCANCO (Nováček et al. 2011), 5D ($H^{N\text{-flip}}$N)CONCACON (Bermel et al. 2013), 5D (HCA)CONCACON (Bermel et al. 2013) and 5D (H)NCO(CAN)CONH (Bermel et al. 2013) experiments. Sequential correlations are established by exploiting ^{15}N and $^{13}C'$ frequencies, both retrieved in the 2D cross-section of the experiments ("CON-CON strategy").

For ^{13}C detection, also a series of experiments that rely on CACO as the basis 2D spectrum have been proposed. The C^α_i-C'_i frequencies of the basis spectrum are correlated to C^α_{i+1}-N_{i+1} and C^α_{i-1}-N_i or to C^α_{i-1}-C'_{i-1} nuclei, respectively through the 4D (H)CANCACO (Bermel et al. 2012b) and the 5D CACONCACO (Nováček et al. 2011) experiments. Finally, the 5D $H^{N\text{-flip}}$NCACON (Bermel et al. 2012b) and 5D $H^{N\text{flip}}$NCANCO (Bermel et al. 2012b) experiments relate the ^{15}N of residue i with those of residue $i+1$ and $i+2$, respectively (Fig. 3.33d).

High-dimensional NMR experiments featuring $^1H^\alpha$-start/$^1H^\alpha$-detection have also recently been designed. These experiments may be useful at temperature and pH conditions in which $^1H^N$ are not detectable and/or in the case of short-lived or low concentration samples that partly limit the use of high-dimensional ^{13}C detected experiments. Extensions to higher dimensionality partly counterbalance the low chemical shift dispersion of $^1H^\alpha$ in the direct dimension (Piai et al. 2014).

In conclusion, due to their high resolving power and increased information content, high-dimensional experiments provide a valuable tool to extend the size and complexity of IDPs that can be characterized by NMR at atomic resolution, provided the sample's relaxation properties allow for multiple (and often long) coherence transfer times. The combination of the different approaches described in this chapter offers a rich toolbox that can be exploited for the investigation of complex IDPs (Fig. 3.34).

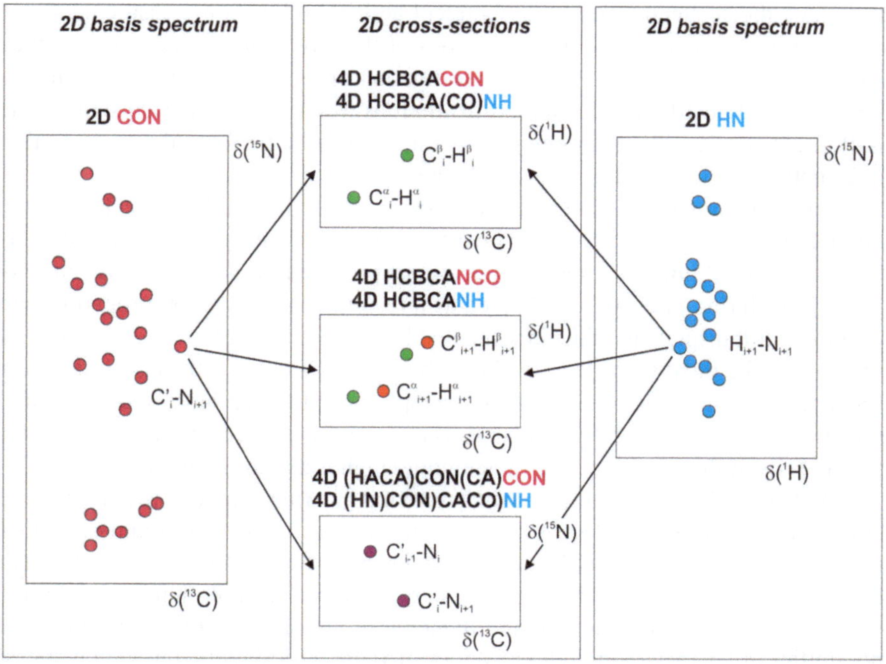

Fig. 3.34 A schematic illustration of a series of ^{13}C and ^{1}H^{N} detected 4D spectra. For each pair of experiments the information provided is the same, except for the direct dimension. Since the 4D experiments share C'_i-N_{i+1} or H_{i+1}-N_{i+1} frequencies, the 2D CON (reported on the *left*) or 2D HN (reported on the *right*) can be respectively used as basis spectra to collect these common frequencies. The new information content of the various 4D experiments can then be easily retrieved inspecting a series of 2D cross-sections, reported in the middle part of the illustration, where the correlation provided by each experiment is shown

10 Conclusions and Perspectives

Thanks to recent improvements in NMR technology, the array of NMR experiments that have been developed, and their optimization for the specific properties of IDPs, a number of tools are available for the atomic resolution structural and dynamic characterization of IDPs.

As of today, IDPs as large as 400 amino acids have been investigated, with two known examples being MAP (Nováček et al. 2013) and Tau (Mukrasch et al. 2009); a number of other examples of high resolution investigations of IDPs in the range between 100 and 300 amino acids have appeared in the literature. Many more high-resolution NMR experimental investigations of IDPs are expected to be accomplished in the near future.

Indeed the sequence-specific assignment and initial structural and dynamic characterization through the analysis of chemical shifts and ^{15}N relaxation rates can readily be achieved through the experiments described in this chapter. Already at this stage the sequence-specific assignment and ^{15}N relaxation measurements are sufficient to describe the overall properties of the IDP in its native state and they can be used as the basis for further investigation of its function by monitoring interactions and post-translational modifications as well as by taking snapshots of the IDP inside whole cells. Many more experiments can be performed to gain further information, as discussed in the next chapters.

Among the many experimental strategies discussed in detail in this chapter, the optimal one for a specific IDP to be investigated can be readily identified by acquiring a small set of initial 1D and 2D spectra that provide information on the sensitivity and resolution that can be achieved as a result of the relaxation properties and chemical shift dispersion of the investigated IDP. In particular, the 2D HN and CON spectra are the most suitable to evaluate the appropriate experimental strategy for sequence-specific assignment. A large panoply of optimized pulse sequences and processing tools are currently available, as are user-friendly computational tools for the analysis of the resulting spectra, including higher dimensional ones.

NMR spectroscopy is continuously improving in terms of hardware performance and experimental approaches. We expect that future progress will aim at further extending the size limit of IDPs that can be investigated at high resolution through NMR, as well as the determination and interpretation of a larger number of observables reporting on the structural and dynamic properties of IDPs, and finally the possibility of characterising the behaviour of IDPs in different aggregation states, ranging from solution to solid state to in-cell, through a combination of different NMR techniques.

Finally, speculating on more long-term perspectives, the development of improved NMR methods to study IDPs is expected to provide a large amount of experimental data on them, contributing to our understanding of the molecular basis responsible for their function and filling a gap of about 50 years with respect to our knowledge on the structural and dynamic behaviour of folded proteins. This is expected to reveal a much larger number of ways in which proteins communicate in the cell. Other expected outcomes of NMR experimental data on IDPs include the improvement of prediction tools, which still suffer from the bias that they are derived from the missing information in the electron density maps in X-ray crystallography data.

Appendix

Table A.1 High-multidimensional ^{1}H detected experiments for backbone and side-chain resonance assignment

^{1}H detected experiments			
Dimensionality	Experiment	Correlations observed	References
Experiments for spin-system identification			
3	HN(CO)CA	H^{N}_{i}-N_{i}-C^{α}_{i-1}	(Bax and Ikura 1991; Grzesiek and Bax 1992b; Solyom et al. 2013)

Table A.1 (continued)

^{1}H detected experiments			
Dimensionality	Experiment	Correlations observed	References
3	HN(CO)CACB	H^N_i-N_i-$C^{\alpha/\beta}_{i-1}$	(Grzesiek and Bax 1992a; Yamazaki et al. 1994; Solyom et al. 2013)
3	(H)C(CC-TOCSY)(CO)NH	C^{ali}_{i-1}-N_i-H^N_i	(Logan et al. 1992; Montelione et al. 1992; Logan et al. 1993; Gardner et al. 1996)
3	iHNCA	H^N_i-N_i-C^{α}_i	(Brutscher 2002; Nietlispach et al. 2002; Solyom et al. 2013)
3	iHNCACB	H^N_i-N_i-$C^{\alpha/\beta}_i$	(Brutscher 2002; Nietlispach et al. 2002; Solyom et al. 2013)
3	iHCAN	H^{α}_i-C^{α}_i-N_i	(Mäntylahti et al. 2010)
4	(H)CBCACONH	$C^{\alpha/\beta}_{i-1}$-C'_{i-1}-N_i-H^N_i	(Grzesiek and Bax 1992a; Grzesiek and Bax 1993)
5	HCBCACONH	$H^{\alpha/\beta}_{i-1}$-$C^{\alpha/\beta}_{i-1}$-C'_{i-1}-N_i-H^N_i	(Grzesiek and Bax 1993; Staykova et al. 2008; Kazimierczuk et al. 2010b)
5	HNCOCACB	H^N_i-N_i- C'_{i-1}- C^{α}_{i-1}-C^{β}_{i-1}	(Hiller et al. 2007; Zawadzka-Kazimierczuk et al. 2012b)
5	HC(CC-TOCSY)CONH	H^{ali}_{i-1}-C^{ali}_{i-1}-C'_{i-1}-N_i-H^N_i	(Logan et al. 1992; Montelione et al. 1992; Logan et al. 1993; Grzesiek et al. 1993a; Grzesiek et al. 1993b; Gardner et al. 1996; Hiller et al. 2008)
Experiments for sequential assignment			
3	HNCO	H^N_i-N_i-C'_{i-1}	(Kay et al. 1990; Grzesiek and Bax 1992b; Schleucher et al. 1993; Solyom et al. 2013)
3	HNCA	H^N_i-N_i-C^{α}_{i-1} & H^N_i-N_i-C^{α}_i	(Kay et al. 1990; Grzesiek and Bax 1992b; Lescop et al. 2007)
3	HNCACB	H^N_i-N_i-$C^{\alpha/\beta}_{i-1}$ & H^N_i-N_i-$C^{\alpha/\beta}_i$	(Wittekind and Mueller 1993; Muhandiram and Kay 1994; Lescop et al. 2007)
3	HN(CA)CO	H^N_i-N_i-C'_{i-1} & H^N_i-N_i-C'_i	(Clubb et al. 1992; Kay et al. 1994; Briand et al. 2001; Lescop et al. 2007)
3	(H)N(COCA)NH	N_{i+1}-N_i-H^N_i	(Grzesiek et al. 1993b; Bracken et al. 1997; Panchal et al. 2001)
3	(H)N(CA)NH	N_{i-1}-N_i-H^N_i & N_{i+1}-N_i-H^N_i	(Grzesiek et al. 1993b; Weisemann et al. 1993)
3	iH(CA)NCO	H^{α}_i-N_i-C'_i	(Mäntylahti et al. 2010)
3	(HCA)NCO(CA)H	H^{α}_i-C'_i-N_{i+1}	(Mäntylahti et al. 2011)
3	H(CA)CON	H^{α}_i-C'_i-N_{i+1}	(Mäntylahti et al. 2010)
3	(HCA)CON(CA)H	H^{α}_i-N_i-C'_{i-1} & H^{α}_i-N_{i+1}-C'_i	(Mäntylahti et al. 2011)
4	HCACON	H^{α}_i-C^{α}_i-C'_i-N_{i+1}	(Kay et al. 1991)
4	HNCOCA	H^N_i-N_i-C'_{i-1}-C^{α}_{i-1}	(Brutscher et al. 1995a; Yang and Kay 1999; Xia et al. 2002)
4	HNCACO	H^N_i-N_i-C^{α}_i-C'_i & H^N_i-N_i-C^{α}_{i-1}-C'_{i-1}	(Yang and Kay 1999; Xia et al. 2002)
4	HNCO, CA	H^N_i-N_i-C'_{i-1}-C^{α}_{i-1} & H^N_i-N_i-C'_{i-1}-C^{α}_i	(Konrat et al. 1999)

Table A.1 (continued)

^{1}H detected experiments

Dimensionality	Experiment	Correlations observed	References
4	HNCACB	$H^N_i-N_i-C^\alpha_{i-1}-C^\beta_{i-1}$ & $H^N_i-N_i-C^\alpha_i-C^\beta_i$	(Gossert et al. 2011)
4	HCBCANH	$H^{\alpha/\beta}_i-C^{\alpha/\beta}_i-N_i-H^N_i$ & $H^{\alpha/\beta}_{i-1}-C^{\alpha/\beta}_{i-1}-N_i-H^N_i$	(Zawadzka-Kazimierczuk et al. 2010)
4	HACANH	$H^\alpha_i-C^\alpha_i-N_i-H^N_i$ & $H^\alpha_{i-1}-C^\alpha_{i-1}-N_i-H^N_i$	(Boucher et al. 1992; Szyperski et al. 1993a)
4	HNCAHA	$H^N_i-N_i-C^\alpha_i-H^\alpha_i$ & $H^N_{i+1}-N_{i+1}-C^\alpha_i-H^\alpha_i$	(Xia et al. 2002)
4	HACA(CO)NH	$H^\alpha_{i-1}-C^\alpha_{i-1}-N_i-H^N_i$	(Boucher et al. 1992)
4	HN(CO)CAHA	$H^N_{i+1}-N_{i+1}-C^\alpha_i-H^\alpha_i$	(Xia et al. 2002)
4	(H)CACO(CA)NH	$C'_i-C^\alpha_i-N_i-H^N_i$ & $C'_{i-1}-C^\alpha_{i-1}-N_i-H^N_i$	(Löhr and Rüterjans 1995)
4	HN(COCA)NH	$H^N_{i+1}-N_{i+1}-N_i-H^N_i$ & $H^N_i-N_i-N_i-H^N_i$	(Shirakawa et al. 1995; Bracken et al. 1997;)
4	HN(CA)NH	$H^N_{i+1}-N_{i+1}-N_i-H^N_i$ & $H^N_{i-1}-N_{i-1}-N_i-H^N_i$	(Zawadzka-Kazimierczuk et al. 2010)
4	HN(CO)CA(CON) CA	$H^N_i-N_i-C^\alpha_{i-1}-C^\alpha_{i-1}$ & $H^N_i-N_i-C^\alpha_{i-1}-C^\alpha_i$	(Bagai et al. 2011)
4	HNCO(N)CA	$H^N_i-N_i-C'_{i-1}-C^\alpha_{i-1}$ & $H^N_i-N_i-C'_{i-1}-C^\alpha_i$	(Bagai et al. 2011)
5	HACACONH	$H^\alpha_{i-1}-C^\alpha_{i-1}-C'_{i-1}-N_i-H^N_i$	(Kim and Szyperski 2003; Hiller et al. 2005; Malmodin and Billeter 2005)
5	HACACONH	$H^\alpha_i-C^\alpha_i-C'_i-N_i-H^N_i$ & $H^\alpha_{i-1}-C^\alpha_{i-1}-C'_{i-1}-N_i-H^N_i$	(Kim and Szyperski 2004)
5	HACA(N)CONH	$H^\alpha_i-C^\alpha_i-C'_{i-1}-N_i-H^N_i$ & $H^\alpha_{i-1}-C^\alpha_{i-1}-C'_{i-1}-N_i-H^N_i$	(Zawadzka-Kazimierczuk et al. 2012b)
5	(HACA)CON(CA) CONH	$C'_{i-2}-N_{i-1}-C'_{i-1}-N_i-H^N_i$ & $C'_{i-1}-N_i-C'_{i-1}-N_i-H^N_i$	(Zawadzka-Kazimierczuk et al. 2012b)
5	HCBCA(CAN) CONH	$H^{\alpha/\beta}_{i-1}-C^{\alpha/\beta}_{i-1}-C'_{i-1}-N_i-H^N_i$ & $H^{\alpha/\beta}_i-C^{\alpha/\beta}_i-C'_{i-1}-N_i-H^N_i$	(Staykova et al. 2008)
5	HN(CA)CONH	$H^N_{i-1}-N_{i-1}-C'_{i-1}-N_i-H^N_i$ & $H^N_i-N_i-C'_{i-1}-N_i-H^N_i$	(Kazimierczuk et al. 2010b)
5	(H)NCO(NCA) CONH	$N_{i-1}-C'_{i-2}-C'_{i-1}-N_i-H^N_i$ & $N_i-C'_{i-1}-C'_{i-1}-N_i-H^N_i$	(Zawadzka-Kazimierczuk et al. 2012b)
5	(H)NCO(CAN) CONH	$N_{i+1}-C'_i-C'_{i-1}-N_i-H^N_i$	(Piai et al. 2014)
5	HN(COCAN) CONH	$H^N_{i+1}-N_{i+1}-C'_{i-1}-N_i-H^N_i$	(Piai et al. 2014)
5	(HACA) CON(CACO) NCO(CA)HA	$C'_{i-1}-N_i-N_{i+1}-C'_i-H^\alpha_i$	(Piai et al. 2014)

Table A.2 High-multidimensional ^{13}C detected experiments for backbone and side-chain resonance assignment

^{13}C detected experiments			
Dimensionality	Experiment	Correlations observed	References
Experiments for spin-system identification			
3	(H)CACON	C^{α}_{i}-N_{i+1}-C'_{i}	(Bermel et al. 2006c; Bermel et al. 2009a)
3	(H)CBCACON	$C^{\alpha/\beta}_{i}$-N_{i+1}-C'_{i}	(Bermel et al. 2006c; Bermel et al. 2009a)
3	(H)C(CC-TOCSY) CON	C^{ali}_{i}-N_{i+1}-C'_{i}	(Bermel et al. 2006c; Bermel et al. 2009a)
4	HCBCACON	$H^{\alpha/\beta}_{i}$-$C^{\alpha/\beta}_{i}$-N_{i+1}-C'_{i}	(Bermel et al. 2012b; Nováček et al. 2012)
4	HC(CC-TOCSY)CON	H^{ali}_{i}-C^{ali}_{i}-N_{i+1}-C'_{i}	(Bermel et al. 2012b)
5	HC(CC-TOCSY) CACON	H^{ali}_{i}-C^{ali}_{i}-C^{α}_{i}-N_{i+1}-C'_{i}	(Nováček et al. 2013)
Experiments for sequential assignment			
3	(H)CANCO	C^{α}_{i}-N_{i+1}-C'_{i} & C^{α}_{i+1}-N_{i+1}-C'_{i}	(Bermel et al. 2006c; Bermel et al. 2009a)
3	(H)CBCANCO	$C^{\alpha/\beta}_{i}$-N_{i+1}-C'_{i} & $C^{\alpha/\beta}_{i+1}$-N_{i+1}-C'_{i}	(Bermel et al. 2006c; Bermel et al. 2009a)
3	($H^{N\text{-}flip}$)N(CA)NCO	N_{i}-N_{i+1}-C'_{i} & N_{i+2}-N_{i+1}-C'_{i}	(Bermel et al. 2009a; Bermel et al. 2012b)
3	(HCA)NCACO	N_{i}-C^{α}_{i}-C'_{i} & N_{i+1}-C^{α}_{i}-C'_{i}	(Bermel et al. 2012b)
3	COCON	C'_{i-1}-N_{i+1}-C'_{i} & C'_{i+1}-N_{i+1}-C'_{i}	(Bermel et al. 2006b)
4	HCBCANCO	$H^{\alpha/\beta}_{i}$-$C^{\alpha/\beta}_{i}$-N_{i+1}-C'_{i} & $H^{\alpha/\beta}_{i+1}$-$C^{\alpha/\beta}_{i+1}$-N_{i+1}-C'_{i}	(Bermel et al. 2012b; Nováček et al. 2012)
4	($H^{N\text{-}flip}$)NCANCO	N_{i}-C^{α}_{i}-N_{i+1}-C'_{i} & N_{i+2}-C^{α}_{i+1}-N_{i+1}-C'_{i}	(Bermel et al. 2012b)
4	($H^{N\text{-}flip}$)NCACON	N_{i}-C^{α}_{i}-N_{i+1}-C'_{i} & N_{i+1}-C^{α}_{i}-N_{i+1}-C'_{i}	(Bermel et al. 2012b)
4	(H)CANCACO	C^{α}_{i-1}-N_{i}-C^{α}_{i}-C'_{i} & C^{α}_{i+1}-N_{i+1}-C^{α}_{i}-C'_{i}	(Bermel et al. 2012b)
5	$H^{N\text{-}flip}$NCANCO	H^{N}_{i}-N_{i}-C^{α}_{i}-N_{i+1}-C'_{i} & H^{N}_{i+2}-N_{i+2}-C^{α}_{i+1}-N_{i+1}-C'_{i}	(Bermel et al. 2012b)
5	$H^{N\text{-}flip}$NCACON	H^{N}_{i}-N_{i}-C^{α}_{i}-N_{i+1}-C'_{i} & H^{N}_{i+1}-N_{i+1}-C^{α}_{i}-N_{i+1}-C'_{i}	(Bermel et al. 2012b)
5	(H)NCOCANCO	N_{i+2}-C'_{i+1}-C^{α}_{i+1}-N_{i+1}-C'_{i}	(Nováček et al. 2011)
5	(H)CACONCACO	C^{α}_{i-1}-C'_{i-1}-N_{i}-C^{α}_{i}-C'_{i}	(Nováček et al. 2011)
5	($H^{N\text{-}flip}$N) CONCACON	C'_{i-1}-N_{i}-C^{α}_{i}-N_{i+1}-C'_{i} & C'_{i}-N_{i+1}-C^{α}_{i}-N_{i+1}-C'_{i}	(Bermel et al. 2013)
5	(HCA)CONCACON	C'_{i-1}-N_{i}-C^{α}_{i}-N_{i+1}-C'_{i} & C'_{i}-N_{i+1}-C^{α}_{i}-N_{i+1}-C'_{i}	(Bermel et al. 2013)
5	(H)CACON(CA)CON	C^{α}_{i}-C'_{i}-N_{i+1}-C'_{i+1}-N_{i+2} & C^{α}_{i}-C'_{i}-N_{i+1}-C'_{i}-N_{i+1}	(Bermel et al. 2013)

References

Arnesano F, Banci L, Bertini I et al (2001) Characterization of the binding interface between the copper chaperone Atx1 and the first cytosolic domain of Ccc2 ATPase. J Biol Chem 276:41365–41376

Arnesano F, Balatri E, Banci L et al (2005) Folding studies of Cox17 reveal an important interplay of cysteine oxidation and copper binding. Structure 13:713–722

Bagai I, Raqsdale SW, Zuiderweg ER (2011) Pseudo-4D triple resonance experiments to resolve HN overlap in the backbone assignment of unfolded proteins. J Biomol NMR 49:69–74

Bai Y, Milne JS, Mayne L et al (1993) Primary structure effects on peptide group hydrogen exchange. Proteins 17:75–86

Banci L, Bertini I, Huber JG et al (1998) Partial orientation of oxidized and reduced cytochrome b_5 at high magnetic fields: magnetic susceptibility anisotropy contributions and consequences for protein solution structure determination. J Am Chem Soc 120:12903–12909

Barbato G, Ikura M, Kay LE et al (1992) Backbone dynamics of calmodulin studied by ^{15}N relaxation using inverse detected two-dimensional NMR spectroscopy; the central helix is flexible. Biochemistry 31:5269–5278

Bax A, Grishaev A (2005) Weak alignment NMR: a hawk-eyed view of biomolecular structure. Curr Opin Struct Biol 15:563–570

Bax A, Ikura M (1991) An efficient 3D NMR technique for correlating the proton and ^{15}N backbone amide resonances with the α-carbon of the preceding residue. J Biomol NMR 1:99–104

Bermel W, Bertini I, Duma L et al (2005) Complete assignment of heteronuclear protein resonances by protonless NMR spectroscopy. Angew Chem Int Ed 44:3089–3092

Bermel W, Bertini I, Felli IC et al (2006a) Novel ^{13}C direct detection experiments, including extension to the third dimension, to perform the complete assignment of proteins. J Magn Reson 178:56–64

Bermel W, Bertini I, Felli IC et al (2006b) Protonless NMR experiments for sequence-specific assignment of backbone nuclei in unfolded proteins. J Am Chem Soc 128:3918–3919

Bermel W, Bertini I, Felli IC et al (2006c) ^{13}C-detected protonless NMR spectroscopy of proteins in solution. Progr NMR Spectrosc 48:25–45

Bermel W, Felli IC, Kümmerle R et al (2008) ^{13}C direct-detection biomolecular NMR. Concepts Magn Reson 32A:183–200

Bermel W, Bertini I, Csizmok V et al (2009a) H-start for exclusively heteronuclear NMR spectroscopy: the case of intrinsically disordered proteins. J Magn Reson 198:275–281

Bermel W, Bertini I, Felli IC et al (2009b) Speeding up ^{13}C direct detection biomolecular NMR experiments. J Am Chem Soc 131:15339–15345

Bermel W, Bertini I, Chill JH et al (2012a) Exclusively heteronuclear ^{13}C-detected amino-acid-selective NMR experiments for the study of instrinsically disordered proteins (IDPs). Chem Bio Chem 13:2425–2432

Bermel W, Bertini I, Gonnelli L et al (2012b) Speeding up sequence specific assignment of IDPs. J Biomol NMR 53:293–301

Bermel W, Felli IC, Gonnelli L et al (2013) High-dimensionality ^{13}C direct-detected NMR experiments for the automatic assignment of intrinsically disordered proteins. J Biomol NMR 57:353–361

Bertini I, Felli IC, Kümmerle R et al (2004) ^{13}C-^{13}C NOESY: a constructive use of ^{13}C-^{13}C spin-diffusion. J Biomol NMR 30:245–251

Bertini I, Felli IC, Gonnelli L et al (2011a) ^{13}C direct-detection biomolecular NMR spectroscopy in living cells. Angew Chem Int Ed 50:2339–2341

Bertini I, Felli IC, Gonnelli L et al (2011b) High-resolution characterization of intrinsic disorder in proteins: expanding the suite of ^{13}C detected NMR experiments to determine key observables. ChemBioChem 12:2347–2352

Bertini I, Luchinat C, Parigi G et al (2011c) Solid-state NMR of proteins sedimented by ultracentrifugation. Proc Natl Acad Sci U S A 108:10396–10399

Billeter M, Neri D, Otting G et al (1992) Precise vicinal coupling constants $^{3}J_{H\alpha N}$ in proteins from nonlinear fits of J-modulated [^{15}N,^{1}H]-COSY experiments. J Biomol NMR 2:257–74

Bloch F (1946) Nuclear induction. Phys Rev 70:460–474

Bloch F (1956) Dynamical theory of nuclear induction. II. Phys Rev 102:104–135

Boucher W, Laue ED, Campbell-Burk SL et al (1992) Improved 4D NMR experiments for the assignment of backbone nuclei in ^{13}C/^{15}N labelled proteins. J Biomol NMR 2:631–637

Bracken C, Palmer AG III, Cavanagh J (1997) (H)N(COCA)NH and HN(COCA)NH experiments for 1H-15N backbone assignments in ^{13}C/^{15}N-labeled proteins. J Biomol NMR 9:94–100

Briand L, Lescop E, Bézirard V et al (2001) Isotopic double-labeling of two honeybee odorant-binding proteins secreted by the methylotrophic yeast Pichia pastoris. Protein Expr Purif 23:167–174

Brutscher B (2002) Intraresidue HNCA and COHNCA experiments for protein backbone resonance assignment. J Magn Reson 156:155–159

Brutscher B (2004a) Combined frequency- and time-domain NMR spectroscopy. Application to fast protein resonance assignment. J Biomol NMR 29:57–64

Brutscher B (2004b) DEPT spectral editing in HCCONH-type experiments. Application to fast protein backbone and side chain assignment. J Magn Reson 167:178–184

Brutscher B, Cordier F, Simorre JP et al (1995a) High-resolution 3D HNCOCA experiment applied to a 28 kDa paramagnetic protein. J Biomol NMR 5:202–206

Brutscher B, Morelle N, Cordier F et al (1995b) Determination of an initial set of NOE-derived distance constraints for the structure determination of ^{15}N/ ^{13}C labeled proteins. J Magn Reson B 109:238–242

Case DA (2000) Interpretation of chemical shifts and coupling constants in macromolecules. Curr Opin Struct Biol 10:197–203

Cavanagh J, Fairbrother WJ, Palmer AG III et al (2007) Protein NMR Spectroscopy. Principles and practice. Academic, San Diego

Chimon S, Shaibat MA, Jones CR et al (2007) Evidence of fibril-like β-sheet structures in a neurotoxic amyloid intermediate of Alzheimer's β-amyloid. Nat Struct Mol Biol 14:1157–1164

Clowes RT, Boucher W, Hardman CH et al (1993) A 4D HCC(CO)NNH experiment for the correlation of aliphatic side-chain and backbone resonances in ^{13}C/^{15}N-labelled proteins. J Biomol NMR 3:349–354

Clubb RT, Thanabal V, Wagner G (1992) A constant-time three dimensional triple-resonance pulse scheme to correlate intraresidue ^{1}HN, ^{15}N, and ^{13}C' chemical shifts in ^{15}N-^{13}C- labeled proteins. J Magn Reson 97:213–217

Coggins BE, Zhou P (2007) Sampling of the NMR time domain along concentric rings. J Magn Reson 184:207–221

Cowburn D, Shekhtman A, Xu R et al (2004) Segmental isotopic labeling for structural biological applications of NMR. Methods Mol Biol 278:47–56

Csizmok V, Felli IC, Tompa P et al (2008) Structural and dynamic characterization of intrinsically disordered human securin by NMR. J Am Chem Soc 130:16873–16879

Dötsch V, Oswald RE, Wagner G (1996a) Amino-acid type-selective triple-resonance experiments. J Magn Reson B 110:107–111

Dötsch V, Oswald RE, Wagner G (1996b) Selective identification of threonine, valine and isoleucine sequential connectivities with a TVI-CBCACONH experiment. J Magn Reson B 110:304–308

Duma L, Hediger S, Brutscher B et al (2003a) Resolution enhancement in multidimensional solid-state NMR spectroscopy of proteins using spin-state selection. J Am Chem Soc 125: 11816–11817

Duma L, Hediger S, Lesage A et al (2003b) Spin-state selection in solid-state NMR. J Magn Reson 164:187–195

Dyson HJ, Wright PE (2001) Nuclear magnetic resonance methods for the elucidation of structure and dynamics in disordered states. Methods Enzymol 339:258–271

Eliezer D (2009) Biophysical characterization of intrinsically disordered proteins. Curr Opin Struct Biol 19:23–30

Emsley L, Bodenhausen G (1990) Phase-shifts induced by transient Bloch-Siegert effect in NMR. Chem Phys Lett 168:297–303

Ernst RR, Bodenhausen G, Wokaun A (1987) Principles of nuclear magnetic resonance in one and two dimensions. Clarendon, Oxford

Farrow NA, Zhang O, Szabo A et al (1995) Spectral density function mapping using ^{15}N relaxation data exclusively. J Biomol NMR 6:153–162

Favier A, Brutscher B (2011) Recovering lost magnetization: polarization enhancement in biomolecular NMR. J Biomol NMR 49:9–15

Felli IC, Pierattelli R (2014a) Novel methods based on ^{13}C detection to study intrinsically disordered proteins. J Magn Reson 241:115–125

Felli IC, Pierattelli R (2014b) Spin-state-selctive methods in solution- and solid-state biomolecular ^{13}C NMR. Prog NMR Spectrosc 84–85:1–13

Felli IC, Pierattelli R, Glaser SJ et al (2009) Relaxation-optimised Hartmann-Hahn transfer for carbonyl-carbonyl correlation spectroscopy using a specifically tailored MOCCA-XY16 mixing sequence for protonless ^{13}C direct detection experiments. J Biomol NMR 43:187–196

Felli IC, Pierattelli R, Tompa P (2012) Intrinsically disordered proteins. In: Bertini I, McGreevy K, Parigi G (eds) NMR of biomolecules: towards mechanistic systems biology. Wiley

Felli IC, Piai A, Pierattelli R (2013) Recent advances in solution NMR studies: ^{13}C direct detection for biomolecular NMR applications. Ann Rep NMR Spectroscop 80:359–418

Felli IC, Gonnelli L, Pierattelli R (2014) In-cell ^{13}C NMR spectroscopy for the study of intrinsically disordered proteins. Nat Protoc 9:2005–2016

Feuerstein S, Plevin MJ, Willbold D et al (2012) iHADAMAC: a complementary tool for sequential resonance assignment of globular and highly disordered proteins. J Magn Reson 214:329–334

Gal M, Edmonds KA, Milbradt AG et al (2011) Speeding up direct ^{15}N detection: hCaN 2D NMR experiment. J Biomol NMR 51:497–504

Gardner KH, Konrat R, Rosen MK et al (1996) An (H)C(CO)NH-TOCSY pulse scheme for sequential assignment of protonated methyl groups in otherwise deuterated ^{15}N, ^{13}C-labeled proteins. J Biomol NMR 8:351–356

Gil S, Hošek T, Solyom Z et al (2013) NMR studies of intrinsically disordered proteins near physiological conditions. Angew Chem Int Ed 52:11808–11812

Gil S, Favier A, Brutscher B (2014) HNCA+, HNCO+, and HNCACB+ experiments: improved performance by simultaneous detection of orthogonal coherence transfer pathways. J Biomol NMR 60:1–9

Gossert AD, Hiller S, Fernández C (2011) Automated NMR resonance assignment of large proteins for protein-ligand interaction studies. J Am Chem Soc 133:210–213

Grzesiek S, Bax A (1992a) Correlating backbone amide and side chain resonances in larger proteins by multiple relayed triple resonance NMR. J Am Chem Soc 114:6291–6293

Grzesiek S, Bax A (1992b) Improved 3D triple-resonance NMR techniques applied to a 31 KDa protein. J Magn Reson 96:432–440

Grzesiek S, Bax A (1993) Amino acid type determination in the sequential assignment procedure of uniformly ^{13}C/^{15}N-enriched proteins. J Biomol NMR 3:185–204

Grzesiek S, Anglister J, Bax A (1993a) Correlation of backbone amide and aliphatic side-chain resonances in ^{13}C/^{15}N-enriched proteins by isotropic mixing of ^{13}C magnetization. J Magn Reson Ser B 101:114–119

Grzesiek S, Anglister J, Ren H et al (1993b) ^{13}C line narrowing by ^{2}H decoupling in ^{2}H/^{13}C/^{15}N-enriched proteins. Application to triple resonance 4D J connectivity of sequential amides. J Am Chem Soc 115:4369–4370

Hiller S, Fiorito F, Wüthrich K et al (2005) Automated projection spectroscopy (APSY). Proc Natl Acad Sci U S A 102:10876–10881

Hiller S, Wasmer C, Wider G et al (2007) Sequence-specific resonance assignment of soluble nonglobular proteins by 7D APSY-NMR spectroscopy. J Am Chem Soc 129:10823–10828

Hiller S, Joss R, Wider G (2008) Automated NMR assignment of protein side chain resonances using automated projection spectroscopy (APSY). J Am Chem Soc 130:12073–12079

Hoch JC, Stern AS (1996) NMR data processing. Wiley-Liss, New York
Holland DJ, Bostock MJ, Gladden LF et al (2011) Fast multidimensional NMR spectroscopy using compressed sensing. Angew Chem Int Ed Engl 50:6548–6551
Hoult DI, Richards RE (1976) The signal-to-noise ratio of the nuclear magnetic resonance experiment. J Magn Reson 24:71–85
Ikura M, Kay LE, Bax A (1990) A novel approach for sequential assignment of ^{1}H, ^{13}C and ^{15}N spectra of larger proteins: heteronuclear triple-resonance three-dimensional NMR spectroscopy. Application to calmodulin. Biochemistry 29:4659–4667
Jung YS, Zweckstetter M (2004) MARS: robust automatic backbone assignment of proteins. J Biomol NMR 30:11–23
Kadeřávek P, Zapletal V, Rabatinová A et al (2014) Spectral density mapping protocols for analysis of molecular motions in disordered proteins. J Biomol NMR 58:193–207
Kanelis V, Donaldson L, Muhandiram DR et al (2000) Sequential assignment of proline-rich regions in proteins: application to modular binding domain complexes. J Biomol NMR 16:253–259
Karplus M (1959) Contact electron-spin coupling of nuclear magnetic moments. J Chem Phys 30:11–15
Kay LE, Torchia DA, Bax A (1989) Backbone dynamics of proteins as studied by ^{15}N inverse detected heteronuclear NMR spectroscopy: application to staphylococcal nuclease. Biochemistry 28:8972–8979
Kay LE, Ikura M, Tschudin R et al (1990) Three-dimensional triple-resonance NMR spectroscopy of isotopically enriched proteins. J Magn Reson 89:496–514
Kay LE, Ikura M, Zhu G et al (1991) Four-dimensional heteronuclear triple resonance NMR of isotopically enriched proteins for sequential assignment of backbone atoms. J Magn Reson 91:422–428
Kay LE, Xu GY, Yamazaki T (1994) Enhanced-sensitivity triple-resonance spectroscopy with minimal H_2O saturation. J Magn Reson Ser A 109:129–133
Kazimierczuk K, Orekhov VY (2011) Accelerated NMR spectroscopy by using compressed sensing. Angew Chem Int Ed Engl 50:5556–5559
Kazimierczuk K, Zawadzka A, Koźmiński W et al (2006) Random sampling of evolution time space and Fourier transform processing. J Biomol NMR 36:157–168
Kazimierczuk K, Zawadzka A, Koźmiński W et al (2007) Lineshapes and artifacts in Multidimensional Fourier Transform of arbitrary sampled NMR data sets. J Magn Reson 188:344–356
Kazimierczuk K, Stanek J, Zawadzka-Kazimierczuk A et al (2010a) Random sampling in multidimensional NMR spectroscopy. Prog NMR Spectrosc 57:420–434
Kazimierczuk K, Zawadzka-Kazimierczuk A, Koźmiński W (2010b) Non-uniform frequency domain for optimal exploitation of non-uniform sampling. J Magn Reson 205:286–292
Kazimierczuk K, Misiak M, Stanek J et al (2012) Generalized Fourier transform for non-uniform sampled data. Top Curr Chem 316:79–124
Kazimierczuk K, Stanek J, Zawadzka-Kazimierczuk A et al (2013) High-dimensional NMR spectra for structural studies of biomolecules. ChemPhysChem 14:3015–3025
Kern T, Schanda P, Brutscher B (2008) Sensitivity-enhanced IPAP-SOFAST-HMQC for fast-pulsing 2D NMR with reduced radiofrequency load. J Magn Reson 190:333–338
Kim S, Szyperski T (2003) GFT NMR, a new approach to rapidly obtain precise high-dimensional NMR spectral information. J Am Chem Soc 125:1385–1393
Kim S, Szyperski T (2004) GFT NMR experiments for polypeptide backbone and $^{13}C_{\beta}$ chemical shift assignment. J Biomol NMR 28:117–130
Kjaergaard M, Poulsen FM (2011) Sequence correction of random coil chemical shifts: correlation between neighbor correction factors and changes in the Ramachandran distribution. J Biomol NMR 50:157–165
Kjaergaard M, Poulsen FM (2012) Disordered proteins studied by chemical shifts. Prog NMR Spectrosc 60:42–51
Kjaergaard M, Brander S, Poulsen FM (2011) Random coil chemical shift for intrinsically disordered proteins: effects of temperature and pH. J Biomol NMR 49:139–149

Konrat R, Yang D, Kay LE (1999) A 4D TROSY-based pulse scheme for correlating $^{1}HN_{i}$, $^{15}N_{i}$, $^{13}C^{a}_{i}$, $^{13}C'_{i-1}$ chemical shifts in high molecular weight, ^{15}N, ^{13}C, ^{2}H labeled proteins. J Biomol NMR 15:309–313
Kovacs H, Moskau D, Spraul M (2005) Cryogenically cooled probes – a leap in NMR technology. Prog NMR Spectrosc 46:131–155
Kumar D, Hosur RV (2011) hNCOcanH pulse sequence and a robust protocol for rapid and unambiguous assignment of backbone ($^{1}H^{N}$, ^{15}N and $^{13}C'$) resonances in $^{15}N/^{13}C$-labeled proteins. Magn Reson Chem 49:575–583
Kupce E, Freeman R (2003) Projection-reconstruction of three-dimensional NMR spectra. J Am Chem Soc 125:13958–13959
Kupce E, Nishida T, Freeman R (2003) Hadamard NMR spectroscopy. Prog NMR Spectr 42: 95–122
Lescop E, Schanda P, Brutscher B (2007) A set of BEST triple resonance experiments for time-optimized protein resonance assignment. J Magn Reson 187:163–169
Lescop E, Rasia R, Brutscher B (2008) Hadamard amino-acid-type edited NMR experiment for fast protein resonance assignment. J Am Chem Soc 130:5014–5015
Levitt MH, Freeman R, Frenkiel T (1982) Broadband heteronuclear decoupling. J Magn Reson 47:328–330
Lipari G, Szabo A (1982) Model-free approach to the interpretation of nuclear magnetic resonance relaxation in macromolecules. 1. Theory and range of validity. J Am Chem Soc 104:4546–4559
Logan TM, Olejniczak ET, Xu RX et al (1992) Side chain and backbone assignments in isotopically labeled proteins from two heteronuclear triple resonance experiments. FEBS Lett 314:413–418
Logan TM, Olejniczak ET, Xu RX et al (1993) A general method for assigning NMR spectra of denatured proteins using 3D HC(CO)NH-TOCSY triple resonance experiments. J Biomol NMR 3:225–231
Löhr F, Rüterjans H (1995) A new triple-resonance experiment for the sequential assignment of backbone resonances in proteins. J Biomol NMR 6:189–197
Löhr F, Rüterjans H (1997) HNCO-E.COSY, a simple method for the stereospecific assignment of side-chain amide protons in proteins. J Magn Reson 124:255–258
López-Méndez B, Güntert P (2006) Automated protein structure determination from NMR spectra. J Am Chem Soc 128:13112–13122
Luan T, Jaravine V, Yee A et al (2005) Optimization of resolution and sensitivity of 4D NOESY using multi-dimensional decomposition. J Biomol NMR 33:1–14
Malmodin D, Billeter M (2005) Multiway decomposition of NMR spectra with coupled evolution periods. J Am Chem Soc 127:13486–13487
Mäntylahti S, Aitio O, Hellman M et al (2010) HA-detected experiments for the backbone assignment of intrinsically disordered proteins. J Biomol NMR 47:171–181
Mäntylahti S, Hellman M, Permi P (2011) Extension of the HA-detection based approach: (HCA) CON(CA)H and (HCA)NCO(CA)H experiments for the main-chain assignment of intrinsically disordered proteins. J Biomol NMR 49:99–109
Matsuki Y, Eddy MT, Herzfeld J (2009) Spectroscopy by integration of frequency and time domain information for fast acquisition of high-resolution dark spectra. J Am Chem Soc 131: 4648–4656
McConnell HM (1958) Reaction rates by nuclear magnetic resonance. J Chem Phys 28:430–431
McIntosh LP, Dahlquist FW (1990) Biosynthetic incorporation of ^{15}N and ^{13}C for assignment and interpretation of nuclear magnetic resonance spectra of proteins. Q Rev Biophys 23:1–38
Mobli M, Stern AS, Hoch JC (2006) Spectral reconstruction methods in fast NMR: reduced dimensionality, random sampling and maximum entropy. J Magn Reson 182:96–105
Mobli M, Stern AS, Bermel W et al (2010) A non-uniformly sampled 4D HCC(CO)NH-TOCSY experiment processed using maximum entropy for rapid protein sidechain assignment. J Magn Reson 204:160–164
Montelione GT, Lyons BA, Emerson SD et al (1992) An efficient triple resonance experiment using carbon-13 isotropic mixing for determining sequence-specific resonance assignments of isotopically-enriched proteins. J Am Chem Soc 114:10974–10975

Morris GA, Freeman R (1979) Enhancement of nuclear magnetic resonance signals by polarization transfer. J Am Chem Soc 101:760–762

Muhandiram DR, Kay LE (1994) Gradient-enhanced triple resonance three-dimensional NMR experiments with improved sensitivity. J Magn Reson Ser B 103:203–216

Mukrasch MD, Bibow S, Korukottu J et al (2009) Structural polimorphism of 441-residue tau at single residue resolution. PLoS Biol 7:e34

Neuhaus D, Williamson M (1989) The nuclear Overhauser effect in structural and conformational analysis. Wiley, New York

Nietlispach D (2004) A selective intra-HN(CA)CO experiment for the backbone assignment of deuterated proteins. J Biomol NMR 28:131–136

Nietlispach D, Ito Y, Laue ED (2002) A novel approach for the sequential backbone assignment of larger proteins: selective intra-HNCA and DQ-HNCA. J Am Chem Soc 124:11199–207

Nováček J, Zawadzka-Kazimierczuk A, Papoušková V et al (2011) 5D ^{13}C-detected experiments for backbone assignment of unstructured proteins with a very low signal dispersion. J Biomol NMR 50:1–11

Nováček J, Haba NY, Chill JH et al (2012) 4D Non-uniformly sampled HCBCACON and ^{1}J(NCα)-selective HCBCANCO experiments for the sequential assignment and chemical shift analysis of intrinsically disordered proteins. J Biomol NMR 53:139–148

Nováček J, Janda L, Dopitová R et al (2013) Efficient protocol for backbone and side-chain assignments of large, intrinsically disordered proteins: transient secondary structure analysis of 49.2 kDa microtubule associated protein 2c. J Biomol NMR 56:291–301

O'Hare B, Benesi AJ, Showalter SA (2009) Incorporating ^{1}H chemical shift determination into 13C-direct detected spectroscopy of intrinsically disordered proteins in solution. J Magn Reson 200:354–358

Olejniczak ET, Fesik SW (1994) Two dimensional nuclear magnetic resonance method for identifying the H^{α}-C^{α} signals of amino acid residues preceding prolines. J Am Chem Soc 116:2215–2216

Otten R, Wood K, Mulder FAA (2009) Comprehensive determination of $^{3}J_{HNH\alpha}$ for unfolded proteins using ^{13}C'-resolved spin-echo difference spectroscopy. J Biomol NMR 45:343–49

Palmer AG III (2004) NMR characterization of the dynamics of biomacromolecules. Chem Rev 104:3623–3640

Palmer AG III, Massi F (2006) Characterization of the dynamics of biomacromolecules using rotating-frame spin relaxation NMR spectroscopy. Chem Rev 106:1700–1719

Palmer AG III, Kroenke CD, Loria JP (2001) Nuclear magnetic resonance methods for quantifying microsecond-to-millisecond motions in biological macromolecules. Methods Enzymol 339:204–238

Panchal SC, Bhavesh NS, Hosur RV (2001) Improved 3D triple resonance experiments, HNN and HN(C)N, for H^{N} and ^{15}N sequential correlations (^{13}C, ^{15}N) labeled proteins: application to unfolded proteins. J Biomol NMR 20:135–147

Pantoja-Uceda D, Santoro J (2008) Amino acid type identification in NMR spectra of proteins via β- and γ-carbon edited experiments. J Magn Reson 195:187–195

Pantoja-Uceda D, Santoro J (2012) New amino acid residue type identification experiments valid for protonated and deuterated proteins. J Biomol NMR 54:145–153

Pellecchia M, Wider G, Iwai H et al (1997) Arginine side chain assignments in uniformly ^{15}N-labeled proteins using the novel 2D HE(NE)HGHH experiment. J Biomol NMR 10: 193–197

Peng JW, Wagner G (1992) Mapping of spectral density function using heteronuclear NMR relaxation measurements. J Magn Reson 98:308–332

Peng JW, Wagner G (1994) Investigation of protein motions via relaxation measurements. Methods Enzymol 239:563–596

Pervushin K, Riek R, Wider G et al (1997) Attenuated T_2 relaxation by mutual cancellation of dipole-dipole coupling and chemical shift anisotropy indicates an avenue to NMR structures of very large biological macromolecules in solution. Proc Natl Acad Sci U S A 94:12366–12371

Piai A, Hošek T, Gonnelli L et al (2014) "CON-CON" assignment strategy for highly flexible intrinsically disordered proteins. J Biomol NMR 60:209–218
Purcell EM, Torrey HC, Pound RV (1946) Resonance absoption by nuclear magnetic moments in solid. Phys Rev 69:37–38
Rao NS, Legault P, Muhandiram DR et al (1996) NMR pulse schemes for the sequential assignment of arginine side-chain H^{ε} protons. J Magn Reson B 113:272–276
Redfield AG (1957) On the theory of relaxation processes. IBM. J Res Develop 1:19–31
Rios C B, Feng W, Tashiro M et al (1996) Phase labeling of C-H and C-C spin-system topologies: application in constant-time PFG-CBCA(CO)NH experiments for discriminating amino acid spin-system types. J Biomol NMR 8:345–350
Sattler M, Schleucher J, Griesinger C (1999) Heteronuclear multidimensional NMR experiments for the structure determination of proteins in solution employing pulsed field gradients. Progr NMR Spectrosc 34:93–158
Schanda P (2009) Fast-pulsing longitudinal relaxation optimized techniques: enriching the toolbox. Prog NMR Spectrosc 55:238–265
Schanda P, Brutscher B (2005) Very fast two-dimensional NMR spectroscopy for real-time investigation of dynamic events in proteins on the time scale of seconds. J Am Chem Soc 127:8014–8015
Schanda P, Forge V, Brutscher B (2006a) HET-SOFAST NMR for fast detection of structural compactness and heterogeneity along polypeptide chains. Magn Reson Chem 44:S177–S184
Schanda P, Van Melckebeke H, Brutscher B (2006b) Speeding up three-dimensional protein NMR experiments to a few minutes. J Am Chem Soc 128:9042–9043
Schleucher J, Sattler M, Griesinger C (1993) Coherence selection by gradients without signal attenuation: application to the three-dimensional HNCO experiment. Angew Chem Int Ed Engl 32:1489–1491
Schubert M, Smalla M, Schmieder P et al (1999) MUSIC in triple-resonance experiments: amino acid type-selective ^{1}H-^{15}N correlations. J Magn Reson 141:34–43
Schubert M, Oschkinat H, Schmieder P (2001a) MUSIC and aromatic residues: amino acid type-selective ^{1}H-^{15}N correlations, III. J Magn Reson 153:186–192
Schubert M, Oschkinat H, Schmieder P (2001b) MUSIC, selective pulses, and tuned delays: amino acid-type selective ^{1}H-^{15}N correlations, II. J Magn Reson 148:61–72
Schwarzinger S, Kroon GJ, Foss TR et al (2001) Sequence-dependent correction of random coil NMR chemical shifts. J Am Chem Soc 123:2970–2978
Selenko P, Wagner G (2007) Looking into live cells with in-cell NMR spectroscopy. J Struct Biol 158:244–253
Selenko P, Frueh DP, Elsaesser SJ et al (2008) In situ observation of protein phosphorylation by high-resolution NMR spectroscopy. Nat Struct Mol Biol 15:321–329
Serber Z, Selenko P, Hänsel R et al (2006) Investigating macromolecules inside cultured and injected cells by in-cell NMR spectroscopy. Nat Protoc 1:2701–2709
Shaka AJ, Keeler J, Freeman R (1983a) Evaluation of a new broadband decoupling sequence: WALTZ-16. J Magn Reson 53:313–340
Shaka AJ, Keeler J, Frenkiel T et al (1983b) An improved sequence for broadband decoupling: WALTZ-16. J Magn Reson 52:335–338
Shaka AJ, Barker PB, Freeman R (1985) Computer-optimized decoupling scheme for wideband applications and low-level operation. J Magn Reson 64:547–552
Shaka AJ, Lee CJ, Pines A (1988) Iterative schemes for bilinear operators; application to spin decoupling. J Magn Reson 77:274–293
Shimba N, Stern AS, Craik CS et al (2003) Elimination of $^{13}C^{\alpha}$ splitting in protein NMR spectra by deconvolution with maximum entropy reconstruction. J Am Chem Soc 125:2382–2383
Shimba N, Kovacs H, Stern AS et al (2004) Optimization of ^{13}C direct detection NMR methods. J Biomol NMR 30:175–179
Shirakawa M, Wälchli M, Shimizu M et al (1995) The use of heteronuclear cross-polarization for backbone assignment of ^{2}H-, ^{15}N- and ^{13}C-labeled proteins: A pulse scheme for triple-resonance 4D correlation of sequential amide protons and ^{15}N. J Biomol NMR 5:323–326
Solomon I (1955) Relaxation processes in a system of two spins. Phys Rev 99:559–565

Solyom Z, Schwarten M, Geist L et al (2013) BEST-TROSY experiments for time-efficient sequential resonance assignment of large disordered proteins. J Biomol NMR 55:311–321

Spera S, Bax A (1991) Empirical correlation between protein backbone conformation and Ca and Cb ^{13}C nuclear magnetic resonance chemical shifts. J Am Chem Soc 113:5490–5492

Staykova DK, Fredriksson J, Bermel W et al (2008) Assignment of protein NMR spectra based on projections, multi-way decomposition and a fast correlation approach. J Biomol NMR 42:87–97

Szyperski T, Wider G, Bushweller JH et al (1993a) 3D ^{13}C-^{15}N-heteronuclear two-spin coherence spectroscopy for polypeptide backbone assignments in ^{13}C-^{15}N-double-labeled proteins. J Biomol NMR 3:127–132

Szyperski T, Wider G, Bushweller JH et al (1993b) Reduced dimensionality in triple resonance experiments. J Am Chem Soc 115:9307–9308

Takeuchi K, Heffron G, Sun ZY et al (2010) Nitrogen-detected CAN and CON experiments as alternative experiments for main chain NMR resonance assignments. J Biomol NMR 47: 271–282

Tamiola K, Acar B, Mulder FAA (2010) Sequence-specific random coil chemical shifts of intrinsically disordered proteins. J Am Chem Soc 132:18000–18003

Theillet FX, Kalmar L, Tompa P et al (2013) The alphabet of intrinsic disorder: I. Act like a Pro: on the abundance and roles of proline residues in intrinsically disordered proteins. Intr Dis Prot 1:e24360

Tjandra N, Grzesiek S, Bax A (1996) Magnetic field dependence of nitrogen-proton J splittings in ^{15}N-enriched human Ubiquitin resulting from relaxation interference and residual dipolar coupling. J Am Chem Soc 118:6264–6272

Tollinger M, Skrynnikov NR, Mulder FAA et al (2001) Slow dynamics in folded and unfolded states of an SH3 domain. J Am Chem Soc 123:11341–11352

Tolman J R, Flanagan JM, Kennedy MA et al (1995) Nuclear magnetic dipole interactions in field-oriented proteins: information for structure determination in solution. Proc Natl Acad Sci U S A 92:9279–9283

Tompa P (2002) Intrinsically unstructured proteins. Trends Biochem Sci 27:527–533

Tong KI, Yamamoto M, Tanaka T (2008) A simple method for amino acid selective isotope labeling of recombinant proteins in *E. coli*. J Biomol NMR 42:59–67

Tugarinov V, Kay LE, Ibraghimov I et al (2005) High-resolution four-dimensional ^{1}H-^{13}C NOE spectroscopy using methyl-TROSY, sparse data acquisition, and multidimensional decomposition. J Am Chem Soc 127:2767–2775

Tycko R (2006) Solid-state NMR as a probe of amyloid structure. Prot Pepr Lett 13:229–34

Vasos PR, Hall JB, Kümmerle R et al (2006) Measurement of ^{15}N relaxation in deuterated amide groups in proteins using direct nitrogen detection. J Biomol NMR 36:27–36

Vuister GW, Bax A (1993) Quantitative J correlation: a new approach for measuring homonuclear three-bond $J(H^{N}H^{a})$ coupling constants in ^{15}N enriched proteins. J Am Chem Soc 115: 7772–7777

Wangsness RK, Bloch F (1953) The dynamical theory of nuclear induction. Phys Rev 89:728–739

Waugh JS (1982) Theory of broadband spin decoupling. J Magn Reson 50:30–49

Weisemann R, Rüterjans H, Bermel W (1993) 3D triple-resonance NMR techniques for the sequential assignment of NH and ^{15}N resonances in ^{15}N- and ^{13}C-labelled proteins. J Biomol NMR 3:113–120

Wishart DS, Sykes BD (1994) The ^{13}C chemical shift index: a simple method for the identification of protein secondary structure using ^{13}C chemical shift data. J Biomol NMR 4:171–180

Wishart DS, Sykes BD, Richards FM (1991) Relationship between nuclear magnetic resonance chemical shift and protein secondary structure. J Mol Biol 222:311–333

Wishart DS, Sykes BD, Richards FM (1992) The chemical shift index: a fast and simple method for the assignment of protein secondary structure through NMR spectroscopy. Biochemistry 31:1647–1651

Wishart DS, Bigam CG, Holm A et al (1995) ^{1}H, ^{13}C and ^{15}N random coil NMR chemical shifts of the common amino acids. I. Investigations of nearest-neighbor effects. J Biomol NMR 5:67–81

Wittekind M, Mueller L (1993) HNCACB, a high-sensitivity 3D NMR experiment to correlate amide-proton and nitrogen resonances with the α- and β-carbon resonances in proteins. J Magn Reson B 101:201–205

Wittekind M, Metzler WJ, Mueller L (1993) Selective correlations of amide groups to glycine alpha protons in proteins. J Magn Reson 101:214–217

Wüthrich K (1986) NMR of proteins and nucleic acids. Wiley, New York

Yao X, Becker S, Zweckstetter M (2014) J Biomol NMR 60(4):231–240

Xia Y, Arrowsmith CH, Szyperski T (2002) Novel projected 4D triple resonance experiments for polypeptide backbone chemical shift assignment. J Biomol NMR 24:41–50

Yamazaki T, Arrowsmith CH, Muhandiram DR et al (1994) A suite of triple resonance NMR experiments for the backbone assignment of ^{15}N, ^{13}C, ^{2}H labeled proteins with high sensitivity. J Am Chem Soc 116:11655–11666

Yamazaki T, Pascal SM, Singer AU et al (1995) NMR pulse schemes for the sequence-specific assignment of arginine guanidino ^{15}N and ^{1}H chemical shifts in proteins. J Am Chem Soc 117:3556–3564

Yang D, Kay LE (1999) TROSY triple-resonance four-dimensional NMR spectroscopy of a 46 ns tumbling protein. J Am Chem Soc 121:2571–2575

Ying J, Li F, Lee JH et al (2014) $^{13}C^{a}$ decoupling during direct observation of carbonyl resonances in solution NMR of isotopically enriched proteins. J Biomol NMR 60:15–21

Zawadzka-Kazimierczuk A, Kazimierczuk K, Koźmiński W (2010) A set of 4D NMR experiments of enhanced resolution for easy resonance assignment in proteins. J Magn Reson 202:109–116

Zawadzka-Kazimierczuk A, Koźmiński W, Billeter M (2012a) TSAR: a program for automatic resonance assignment using 2D cross-sections of high dimensionality, high-resolution spectra. J Biomol NMR 54:81–95

Zawadzka-Kazimierczuk A, Koźmiński W, Sanderová H et al (2012b) High dimensional and high resolution pulse seqeunces for backbone resonance assignment of intrinsically disordered proteins. J Biomol NMR 52:329–337

Zhang YZ (1995) Protein and peptide structure and interactions studied by hydrogen exchange and NMR. University of Pennsylvania, Philadelphia

Zhang H, Neal S, Wishart DS (2003) RefDB: a database of uniformly referenced protein chemical shifts. J Biomol NMR 25:173–195

Chapter 4
Ensemble Calculation for Intrinsically Disordered Proteins Using NMR Parameters

Jaka Kragelj, Martin Blackledge and Malene Ringkjøbing Jensen

Abstract Intrinsically disordered proteins (IDPs) perform their function despite their lack of well-defined tertiary structure. Residual structure has been observed in IDPs, commonly described as transient/dynamic or expressed in terms of fractional populations. In order to understand how the protein primary sequence dictates the dynamic and structural properties of IDPs and in general to understand how IDPs function, atomic-level descriptions are needed. Nuclear magnetic resonance spectroscopy provides information about local and long-range structure in IDPs at amino acid specific resolution and can be used in combination with ensemble descriptions to represent the dynamic nature of IDPs. In this chapter we describe sample-and-select approaches for ensemble modelling of local structural propensities in IDPs with specific emphasis on validation of these ensembles.

Keywords Structure · Dynamics · Conformational ensembles · Experimental validation

1 Introduction

Structural biology is an important branch of the life sciences. The number of protein structures deposited in the Protein Data Bank (PDB)[1] is already exceeding 100000 and underlines the enormous effort that has been invested in solving ever-newer protein structures. The description of protein motion can be seen as the next logical step stemming from this wealth of structural data, strongly supported by the fact that proteins display functional dynamics occurring on a broad range of timescales

[1] www.pdb.org.

M. R. Jensen (✉) · J. Kragelj · M. Blackledge
IBS, University Grenoble Alpes, 38044 Grenoble, France
e-mail: malene.ringkjobing-jensen@ibs.fr

IBS, CNRS, 38044 Grenoble, France

IBS, CEA, 38044 Grenoble, France

I. C. Felli, R. Pierattelli (eds.), *Intrinsically Disordered Proteins Studied by NMR Spectroscopy,* Advances in Experimental Medicine and Biology,
DOI 10.1007/978-3-319-20164-1_4

(Karplus and Kuriyan 2005; Mittermaier and Kay 2006; Henzler-Wildman and Kern 2007; Bernadó and Blackledge 2010). Nuclear magnetic resonance (NMR) spectroscopy is uniquely suited to probing protein dynamics at atomic resolution as a number of experimental parameters report on motions occurring on different time scales ranging from pico- to millisecond (Mittermaier and Kay 2009; Salmon et al. 2011; Göbl et al. 2014).

Protein motion comes in many flavours and can span from local backbone and side chain dynamics in globular, folded proteins (Lindorff-Larsen et al. 2005; Bouvignies et al. 2005; Lange et al. 2008; Salmon et al. 2009; Salmon et al. 2012; Guerry et al. 2013) through the concerted motion of entire domains in multi-domain proteins (Bertini et al. 2007; Yang et al. 2010; Różycki et al. 2011; Francis et al. 2011; Deshmukh et al. 2013; Huang et al. 2014) to intrinsically disordered proteins (IDPs), which represent the most extreme case of protein flexibility (Dyson and Wright 2002; Dunker et al. 2008; Tompa 2012). One way of representing the dynamics of a protein is to capture its characteristics—or more accurately, to explain the experimental NMR data, which depend on the underlying dynamics—with an ensemble of protein structures (Fig. 4.1).

In this chapter we will focus on atomic resolution ensemble descriptions of IDPs on the basis of experimental NMR data with a special emphasis on mapping local conformational propensities. The determination of a single set of three-dimensional atomic coordinates would have little meaning for these conformationally heterogeneous molecules, and ensemble descriptions are therefore necessary in order to build molecular models of IDPs that accurately capture the dynamic behaviour of the polypeptide chains. Special care has to be taken at each step of the ensemble generation protocol to ensure the validity of the obtained ensembles. The way in which ensemble generation protocols are tested and the factors that influence the modelling of the ensembles are therefore important questions that need to be addressed. In this chapter we will discuss these issues with a focus on the application of sample-and-select approaches to mapping local conformational propensities in IDPs.

2 Local Structure in IDPs can be Described by the Dihedral Angle Distributions of Amino Acids

It is expected that a single residue in an IDP will adopt many conformations over the time and ensemble average, and therefore undergoes exchange among many different dihedral angles. The distribution of dihedral angles sampled by a residue may at first seem like a simplistic representation of residual structure but is in reality very practical. The well-known secondary structures, the α-helix and β-sheet, are defined by hydrogen-bonding criteria and also have their own characteristic dihedral angles that are commonly used for annotating secondary structure elements in proteins (Fig. 4.2a, 4.2c) (Kabsch and Sander 1983; Frishman and Argos 1995).

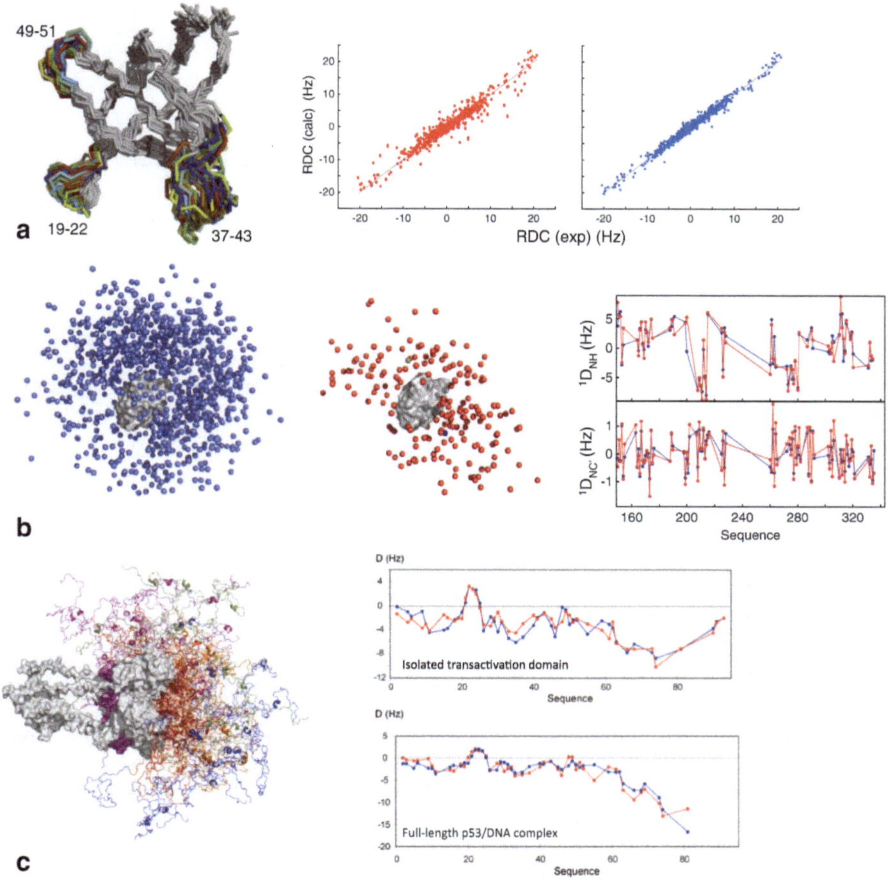

Fig. 4.1 Interpreting NMR data with molecular ensembles to map conformational dynamics in proteins. **a** Dynamics of the SH3 domain from CD2AP derived from NMR residual dipolar couplings (RDCs) measured in multiple, complementary alignment media. An ensemble is shown of the SH3 domain derived from selection of conformational ensembles on the basis of experimental RDCs. The agreement between experimental and back-calculated RDCs is shown for the derived final ensemble (*blue*) and the starting pool of structures from which the ensemble was selected (*red*). Reprinted in part with permission from (Guerry et al. 2013). Copyright 2013 Wiley-VCH. **b** Dynamics of the two-domain splicing factor U2AF65 derived from RDCs and paramagnetic relaxation enhancements induced by *S*-(1-oxyl-2,2,5,5-tetramethyl-2,5-dihydro-1 H-pyrrol-3-yl) methyl methanesulfonothioate (MTSL) spin labels attached at different positions in the two-domain protein. Ensembles of the two-domain protein are shown, where the grey surface represents the location of the domain RRM1, while the location of the second domain RRM2 is shown as spheres positioned at the centre of mass of RRM2. Ensembles are shown representing the initial pool of structures sampling all conformational space (*blue*) and the space occupied by RRM2 after refinement against experimental data (*red*). The agreement between experimental RDCs (*red*) and those back-calculated from the derived ensemble (*blue*) is also shown. Reprinted in part with permission from (Huang et al. 2014). Copyright 2014 American Chemical Society. **c** Conformational ensemble of the intrinsically disordered N-terminal transactivation domain of p53 in the context of the full-length p53-DNA complex (Wells et al. 2008). The ensemble was obtained on

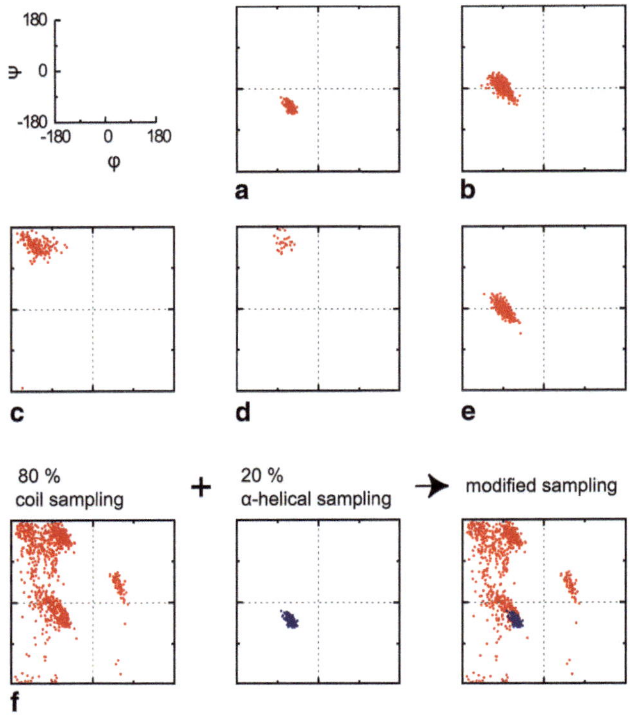

Fig. 4.2 Dihedral angle distributions characteristic of different secondary structure types. **a** Central residues in α-helices. **b** Last residues in α-helices that are C-capped with Schellman loops. **c** Residues in β-strands. **d** Residues in PPII conformations. **e** Residues in type I β-turns. **f** Modification of the dihedral angle sampling of a given residue can be achieved by combining the random coil distribution with an over-sampling of other regions of Ramachandran space (in this case the α-helical region). Dihedral angles for (**a**), (**b**) and (**e**) were extracted from the database embedded in the structure motivator application (Leader and Milner-White 2012). Dihedral angles in (**c**) were extracted from parallel and anti-parallel β-strands from structures with the following PDB codes: 1MLD, 1QCZ, 2CMD, 1XH3, 1OGT and 3GP6. Dihedral angles in (**d**) were extracted from non-proline residues of peptide ligands bound to SH3 domains in a PPII conformation (1BBZ, 1CKA, 1CKB, 1SSH, 1W70, 2DRK, 2DRM, 2O88, 2O9V, 2W0Z, 2W10, 3EG1 and 3I5R)

Other structural motifs can also be identified by their specific ϕ/ψ angles. Apart from the α-helix and β-sheet, poly-L-proline II (PPII) is the only secondary structure that forms linear groups of residues that all adopt the same conformation (Fig. 4.2d) (Hollingsworth et al. 2009). This conformation is particularly interesting as it has been proposed to be significantly populated in IDPs and unfolded states of proteins (Shi et al. 2006; Schweitzer-Stenner 2012). Residues within β-turns also adopt spe-

---

the basis of experimental RDCs (local conformational sampling) and small angle X-ray scattering data (long-range behaviour). The agreement is shown between experimental RDCs (*red*) and those back-calculated from the model ensemble (*blue*) for both the isolated transactivation domain and in the context of the full-length p53/DNA complex. Reprinted in part with permission from (Wells et al. 2008). Copyright 2008 National Academy of Sciences, USA

cific dihedral angles and differ for each β-turn type (I, II, I', II') (Fig. 4.2e). Residues of both N-terminal and C-terminal helix capping motifs have unique dihedral angle distributions (Shen and Bax 2012), and it has been shown that within α-helices of structured proteins, the central residues display different distributions than the C-terminal residues (Fig. 4.2a, 4.2b) (Leader and Milner-White 2011).

Since each structural motif has distinct dihedral angle distributions, we can use them to describe the conformational energy surface of each residue within the disordered protein chain and more importantly also as a metric for the presence of residual secondary structure in IDPs. An increase in sampling of dihedral angles corresponding to the α-helical region will, if sampled at a high enough propensity, give rise to transiently populated α-helices, even in the absence of cooperative effects. IDPs can therefore in general be described as random coils (*i.e.* a peptide chain without specific secondary or tertiary structure), with deviations from this model corresponding to the presence of residual secondary structure (Fig. 4.2f). In order to map the dihedral angle distributions in an IDP we can exploit a number of different NMR parameters as described below.

3 NMR Parameters for Characterizing Local Conformational Propensities in IDPs

NMR is a powerful technique for studying IDPs at atomic resolution and provides many experimental parameters that inform us about local conformational propensities (Jensen et al. 2014). Chemical shifts are the most readily accessible parameters and as a single NMR resonance is usually observed for each nucleus in the spectra of IDPs, the chemical shifts report on the population-weighted average over all conformations sampled in solution up to the millisecond time scale. Chemical shifts are sensitive to the backbone dihedral angle distributions and can, therefore, be interpreted in terms of local conformational propensities. A simple analysis of chemical shifts in IDPs involves the calculation of secondary structure propensities (Marsh et al. 2006; Camilloni et al. 2012; Tamiola and Mulder 2012). This usually relies on characteristic shifts for α-helix, β-sheet and random coil derived from experimental chemical shifts of folded proteins with known three-dimensional structure or from a collection of assigned IDPs (Zhang et al. 2003; De Simone et al. 2009; Tamiola et al. 2010). When deriving conformational propensities it is important to correctly reference the experimental chemical shifts as systematic offsets may lead to erroneous estimates of the amount of secondary structure. It is possible to verify whether the chemical shift is correctly referenced using the secondary structure propensity (SSP) algorithm, which reports the potential reference offset based on the observation that C^{α} and C^{β} secondary chemical shifts are inversely correlated (Marsh et al. 2006).

Scalar couplings measured between nuclei of the protein backbone are also important structural probes in proteins and can be used to map dihedral angle distributions in IDPs. In the same way as chemical shifts, as long as the exchange rate is fast, the scalar couplings represent a population-weighted average over all conformations sampled in solution. The dependence of scalar couplings on the main chain torsion angles can be described using a so-called Karplus relationship (Karplus 1959) that is generally parameterized against experimental scalar couplings measured in proteins of known structure (Smith et al. 1996). One of the commonly measured scalar couplings, the three-bond coupling constant $^3J_{HNH\alpha}$, depends on the backbone dihedral angle ϕ, allowing one to distinguish between α-helical ($^3J_{HNH\alpha}$ <5 Hz) and β-sheet conformations ($^3J_{HNH\alpha}$ >8 Hz) (Vuister and Bax 1993). Other scalar couplings such as $^3J_{C\alpha C\alpha}$, $^3J_{NH\alpha}$, $^3J_{NC\beta}$ and $^3J_{NN}$ report on the ψ angle and in principle provide a more accurate measure of PPII conformations (Graf et al. 2007; Hagarman et al. 2010).

Residual dipolar couplings (RDCs) are obtained by partially aligning the protein molecules in the magnetic field using, for example, a liquid crystal (Rückert and Otting 2000), filamentous phages (Hansen et al. 1998), polyacrylamide gels (Sass et al. 2000), or bicelles (Tjandra and Bax 1997). The inter-nuclear dipolar coupling, which is efficiently averaged to zero by the isotropic rotational tumbling of the molecules in solution, will no longer average to zero and a small part of the dipolar coupling will be measurable (Tolman et al. 1995; Tjandra and Bax 1997). RDCs report on bond vector orientations with respect to a common reference frame and have been used extensively for structure determination of folded proteins as reporters on the relative orientations of secondary structure elements (Prestegard et al. 2004; Blackledge 2005). Since the first measurement of RDCs in an unfolded protein (Shortle and Ackerman 2001) we have significantly advanced in our understanding and interpretation of RDCs in IDPs (Jensen et al. 2009). It is now clear that the RDCs carry contributions from the dihedral angle distribution of the amino acid of interest as well as its nearest neighbours, and the measurement of a single RDC value does therefore not provide a direct "read-out" of residue specific sampling in the same way as chemical shifts and scalar couplings (Huang et al. 2013). In addition, a contribution from the local flexibility of the chain (bulkiness) to the RDCs should be taken into account together with a length-dependent baseline that reflects the polymeric nature of the unfolded chain (Salmon et al. 2010; Huang et al. 2013). In the case of IDPs there is a preference for alignment media that rely on steric interactions between the protein and the medium such that the alignment tensor, and thereby the RDCs, can be predicted directly from the shape of each protein conformation (Zweckstetter and Bax 2000) and averaged over the ensemble.

4 Sample-and-Select Approaches

One way of obtaining representative ensemble descriptions of IDPs on the basis of experimental NMR data is to apply a two-step procedure involving the initial generation of a large pool of structures representing all of the conformational space available to the polypeptide chain (Fig. 4.3). Experimental data are then included in the second step where a set of structures (an ensemble) that agrees with the data is selected, for example using a genetic algorithm.

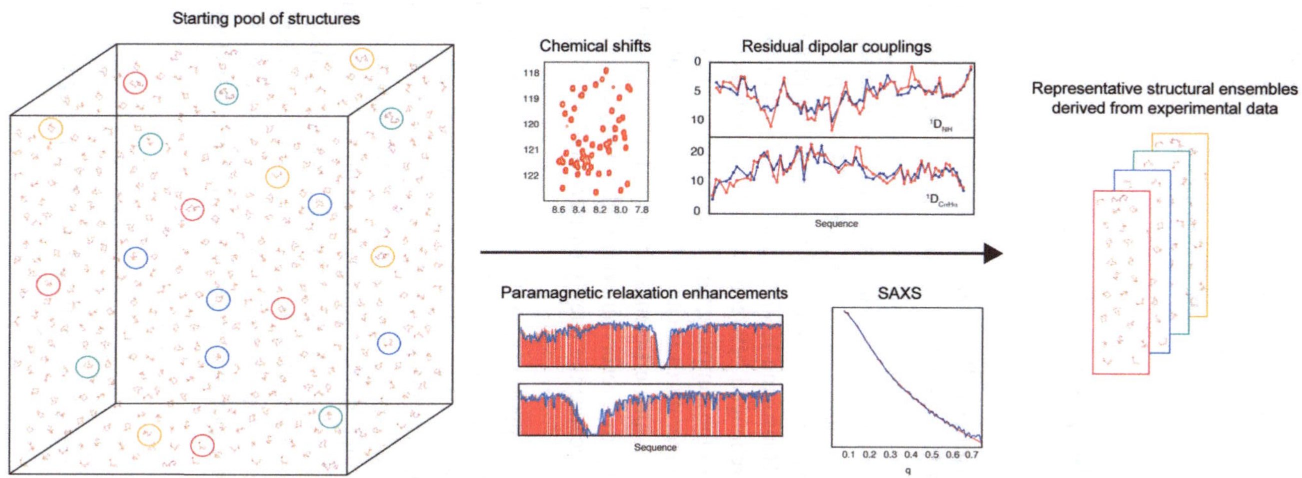

Fig. 4.3 Overview of sample-and-select approaches. Initially a large pool of structures is generated that represents the entire conformational space available to the protein under investigation. Experimental data such as chemical shifts, residual dipolar couplings, paramagnetic relaxation enhancements and small angle X-ray scattering (SAXS) are exploited in a second step to refine this conformational space by selecting sub-ensembles that agree with the experimental data. Reprinted in part with permission from (Jensen et al. 2014). Copyright 2014 American Chemical Society

Different approaches can be used to generate the initial pool of structures, but for a number of reasons it is important that the generated pool covers the entire conformational space of the molecule. Generally, a starting pool can be generated using molecular dynamics approaches or statistical coil generators.

5 Sampling Space Using Molecular Dynamics Simulations

For classical molecular dynamics (MD) simulations, sufficient sampling remains a problem when studying IDPs, even when the simulations are run over long timescales of several hundreds of microseconds (Lindorff-Larsen et al. 2012). Other types of MD simulations address this problem and provide a better sampling; one example is replica exchange molecular dynamics (REMD), which artificially enhances the sampling by exchanging copies of the simulated protein that evolve under different conditions (Hansmann 1997; Sugita and Okamoto 1999). In its simplest form the protein is exchanged between two different temperature reservoirs where at higher temperatures the sampling rate is faster but not physical. When the protein evolves at lower temperatures it can get trapped in a local or global minimum with the end result being an inefficient sampling of the conformational space. Exchanging the protein copy to a reservoir with higher temperature facilitates the sampling of other minima because the energy barriers between them are easier to overcome.

Other approaches include enhanced sampling techniques such as metadynamics, in which a term is added to the force field that penalizes the conformations that have already been explored by the molecule (Leone et al. 2010). The energy penalties accumulate as the protein explores an energy minimum and after some time the protein is forced to explore other minima. Metadynamics can be combined with REMD to enhance the sampling rate even further (Piana and Laio 2007). Accelerated molecular dynamics (AMD) is another approach for enhancing sampling that can be used for IDPs. In this method the free energy surface is modulated by a scaling factor that affects the energy barriers between minima and therefore increases the chance of barrier crossing (Voter 1997; Hamelberg et al. 2004; Pierce et al. 2012).

6 Sampling Space Using Statistical Coil Generators

A starting pool of conformers that is subsequently used in selections can also be produced with a statistical coil generator (Feldman and Hogue 2000; Jha et al. 2005a; Bernadó et al. 2005b; Ozenne et al. 2012a). In this approach a protein molecule is built starting from either end of the chain by adding amino acid after amino acid with a ϕ/ψ angle that is randomly chosen from a database of dihedral angles (the statistical coil library). Each newly added amino acid is checked for steric clashes between the backbone atoms and between simplified representations of side chains. In case of steric clash the newly placed residue is rejected and rebuilt until a suitable conformation is found. Force field bond and angle potentials are not included during the generation of conformers, and the steric clash model is very simple and only defines a certain radius of exclusion for each atom. This approach allows the conformational space to be sampled roughly but efficiently and generates a pool of many different combinations of ϕ/ψ angles for consecutive amino acids.

Statistical coil libraries, which are used for generating the structures, are assembled with the help of databases of high-resolution crystal structures (Serrano 1995; Jha et al. 2005b). The conformational preferences of amino acids in folded proteins differ from those of disordered proteins as most of the residues in folded proteins reside within secondary structure elements, while IDPs are expected to more closely resemble the loop regions. If the α-helices, β-sheets and β-turns are removed from the initial data set of high-resolution crystal structures, only motifs from non-regular loops remain in the database. These loop residues are not restrained by secondary structure hydrogen bonding criteria, unlike for example α-helices, and when a large number of the loop residues are taken into consideration, the potential contributions from the long-range tertiary contacts mostly average out. If we extract the ϕ/ψ angle distributions from the database of loop regions, we obtain a library of amino acid specific distributions of ϕ/ψ angles. These ϕ/ψ angle distributions represent a valid starting point for describing the conformational free energy surface of amino acids within IDPs and can be used in conjunction with statistical coil generators for building ensembles of IDPs. One of these statistical coil generators, Flexible-Meccano, is freely available[2] and is provided with a graphical interface that allows the testing of different sampling regimes by manually modifying the ϕ/ψ sampling of selected amino acids (Ozenne et al. 2012a). Flexible-Meccano calculates NMR observables such as chemical shifts, RDCs, scalar couplings and paramagnetic relaxation enhancements (PREs) from the generated ensembles that allow direct comparison with experimental data (Fig. 4.4).

The statistical coil libraries still have room for improvement in terms of the inclusion of neighbour residue effects, which would be analogous to what has been carried out for random coil chemical shift tabulations (Wishart et al. 1995; Wang and Jardetzky 2002b; Wang and Jardetzky 2002a; De Simone et al. 2009; Tamiola et al. 2010). In fact the neighbour residue correction is often used in the statistical coil libraries for pre-proline residues, because the neighbour effect of prolines on the preceding residue is particularly pronounced. This is due to steric hindrance between the $^{\delta}CH_2$ side chain group of the proline and the NH and $C^{\beta}H_2$ atoms of the preceding residues (MacArthur and Thornton 1991). Again, similarly to random coil chemical shifts, the statistical coil databases could be improved by refining them with the help of experimental data from IDPs themselves (Tamiola et al. 2010).

[2] www.ibs.fr/science-213/scientific-output/software.

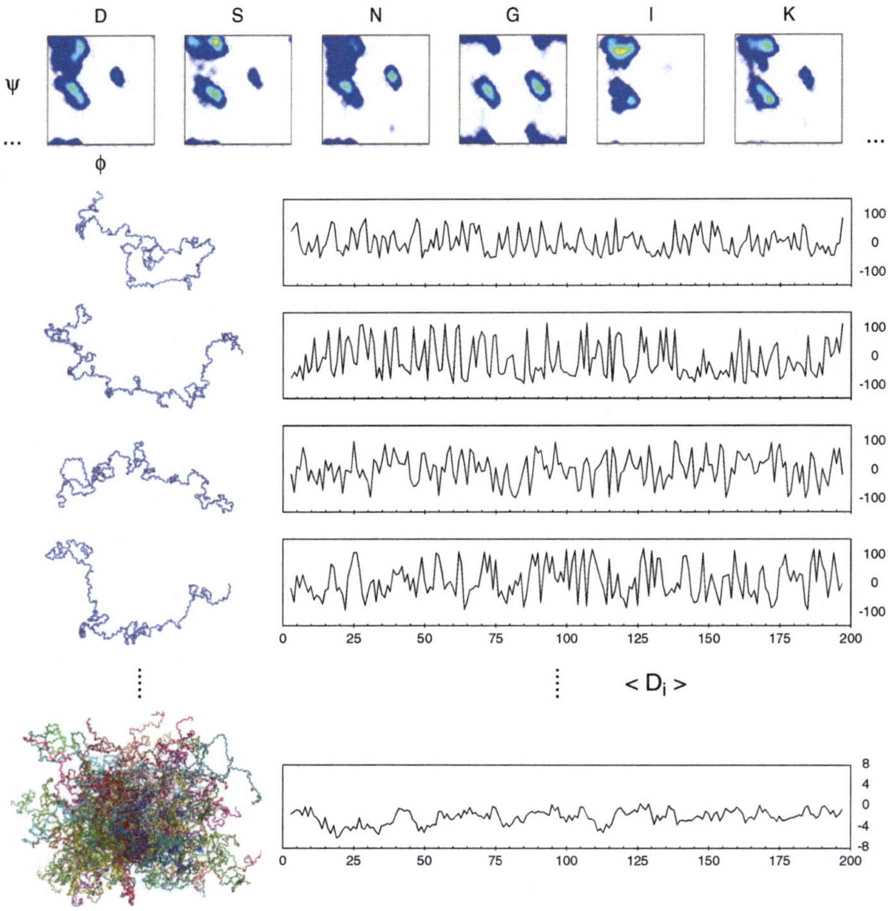

Fig. 4.4 Generation of conformational ensembles of IDPs using a statistical coil generator. Ramachandran plots (ϕ/ψ distributions) are shown for the amino acids D, N, S, G, I and K as derived from loop regions of high-resolution crystal structures. These distributions are used to construct conformations of the protein for a given primary sequence by starting from either the C- or N-terminal end of the protein and building amino acid after amino acid according to randomly chosen ϕ/ψ pairs of the statistical coil library. For each copy of the molecule, experimental NMR data can be calculated and the ensemble-average over multiple copies of the protein can be compared to experimental data. Reprinted in part with permission from (Jensen et al. 2014). Copyright 2014 American Chemical Society

7 Selection of Ensembles on the Basis of Experimental NMR Data

Once the initial pool of structures has been generated, NMR parameters such as scalar couplings, chemical shifts and RDCs can be calculated for each member of the pool. The selection of sub-ensembles proceeds by calculating the averages of the NMR parameters over a given sub-ensemble and comparing them to experimental data. Different approaches have been proposed in the literature for deriving representative ensembles such as ENSEMBLE, which assigns weights to the different conformations of the pool (Marsh and Forman-Kay 2009; Krzeminski et al. 2013), and ASTEROIDS, which relies on a genetic algorithm to select sub-ensembles (Nodet et al. 2009; Jensen et al. 2010), as well as ensemble optimization on the basis of Bayesian weighting (Fisher et al. 2010; Fisher and Stultz 2011). It is important to note that in cases where the IDPs possess transiently populated secondary structures, it is not possible to select an ensemble that matches all the experimental data directly from a pool of statistical coil conformers. The reason for this is that the probability of finding continuous stretches of secondary structure is too low. Therefore, the sample-and-select protocol is often repeated multiple times in an iterative procedure, where the sampling pool is regenerated using the information (local conformational sampling) obtained from ensembles selected in the previous iteration. In this way, the sampling pool is enriched at each step with conformational preferences characteristic of the protein under investigation.

8 Ensemble Representations of the IDP Tau from Chemical Shifts and RDCs

The combination of the statistical coil generator Flexible-Meccano (Bernadó et al. 2005b; Ozenne et al. 2012a) and the ensemble selection algorithm ASTEROIDS has allowed quantitative insight into residue-specific conformational sampling in a number of IDPs involved in neurodegenerative diseases (Bernadó et al. 2005a; Mukrasch et al. 2007; Schwalbe et al. 2014). The protein Tau is a 441 amino acid protein that is intrinsically disordered and undergoes a conformational transition to a pathological form of the same protein. The NMR spectra of Tau have been fully assigned (Narayanan et al. 2010), allowing insight into the conformational preferences of this protein at atomic resolution. A complete set of chemical shifts and $^{1}D_{NH}$ RDCs were obtained for the protein Tau in order to accurately map α-helical, β-strand and PPII populations. Figure 4.5a shows the agreement between experimental data and those back-calculated from selected ASTEROIDS ensembles. The ensemble selections were repeated five times and the conformational sampling of each residue along the sequence of Tau is conveniently represented by their dihedral angle distributions (Fig. 4.5b).

In general, we can learn a lot from these ensembles as they provide quantitative insight into the sampling in different regions of Ramachandran space. Specifically, it is seen that the aggregation nucleation sites in Tau overpopulate the PPII region, suggesting that these conformations represent precursors of aggregation (Fig. 4.5b) (Schwalbe et al. 2014). In addition to these observations, the presence of turn-like motifs can be identified in each of the Tau repeat regions (R1-R4). These turn motifs were also studied in detail previously using AMD simulations of small peptides

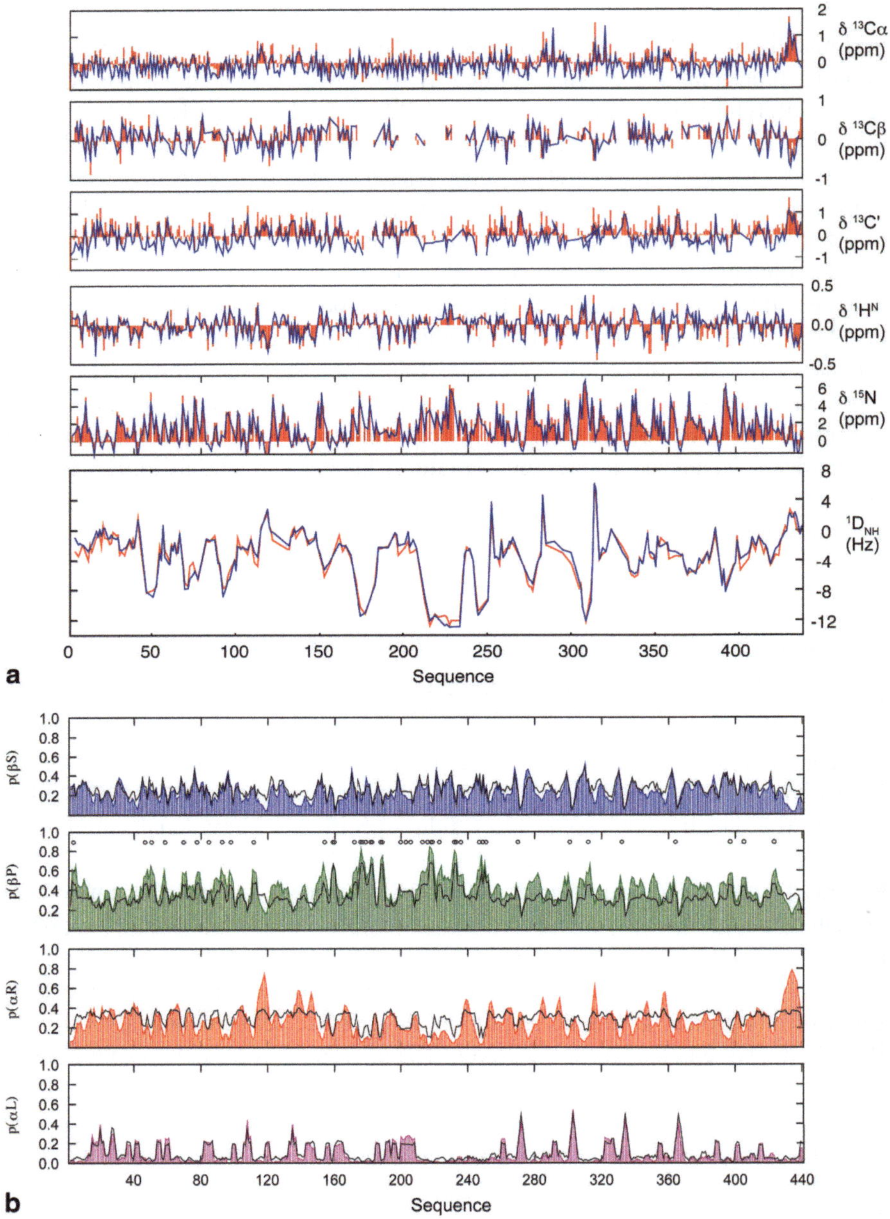

Fig. 4.5 Ensemble representations of the intrinsically disordered Tau protein on the basis of chemical shifts and RDCs. **a** Agreement between experimental (*red*) and back-calculated secondary chemical shifts and RDCs (*blue*) from selected ASTEROIDS ensembles of Tau. **b** Site specific conformational sampling in Tau derived from the selected ensembles (*blue*, *green*, *red*,

of Tau, where it was shown that the AMD derived ϕ/ψ sampling corresponded to type I β-turns (Mukrasch et al. 2007). When this AMD sampling was incorporated into a model ensemble of the smaller K18 construct of Tau, the agreement between experimental and back-calculated $^1D_{NH}$ RDCs improved significantly, proving that these regions indeed adopt type I β-turns as predicted by AMD.

9 The Reference Ensemble Method

The study described above combines different data types to map local conformational sampling. The accuracy with which conformational propensities can be determined depends on the amount of experimental data available for a given system. Assuming that we want to map the population of α-helix, β-strand and PPII conformations for each amino acid of the protein, it would be useful to determine a minimum dataset that would allow this. The α-helical and β-strand propensities can be well characterized with the help of carbon (C^α, C^β, C') chemical shifts, but a residue sampling a statistical coil distribution and a residue sampling exclusively PPII specific dihedral angles have approximately the same carbon chemical shift (Fig. 4.6a). We therefore cannot use carbon chemical shifts to distinguish between the two mentioned sampling regimes. Similarly, most of the RDC types display degeneracy between β-strand and PPII conformations. The $^1D_{NH}$ RDCs are negative for both increased β-strand propensities and increased PPII propensities (Fig. 4.6b).

Selection against synthetic data from a reference ensemble can help reveal such degeneracies and determine the minimum dataset necessary for accurate mapping of the conformational energy landscape. In the reference ensemble approach, an ensemble of structures is generated using either an MD simulation or a statistical coil generator. These structures constitute the target ensemble for which a synthetic dataset is calculated. If our ensemble selection protocol is working without bias and we have sufficient and complementary data types, we should be able to regenerate the local conformational sampling preferences by targeting the synthetic dataset using the sample-and-select approach.

A study by Ozenne et al. demonstrated how useful this approach can be when applied to IDPs (Ozenne et al. 2012b). Initially, an ensemble of a model protein of 60 amino acids of arbitrary sequence was obtained using the statistical coil generator Flexible-Meccano, where three distinct regions of the protein over-sampled the α-helical, β-strand and PPII region of Ramachandran space (50 % additional sampling in each region compared to the statistical coil). Different types of ensemble-averaged chemical shifts and RDCs were calculated for this ensemble

---

magenta) compared to standard statistical coil distributions (*black*). Populations are reported for four different regions of Ramachandran space corresponding to right- (αR, *red*) and left-handed α-helix (αL, *magenta*), β-strand (βS, *blue*) and PPII conformations (βP, *green*). Circles indicate the presence of proline residues. Reprinted in part with permission from (Schwalbe et al. 2014). Copyright 2014 Elsevier

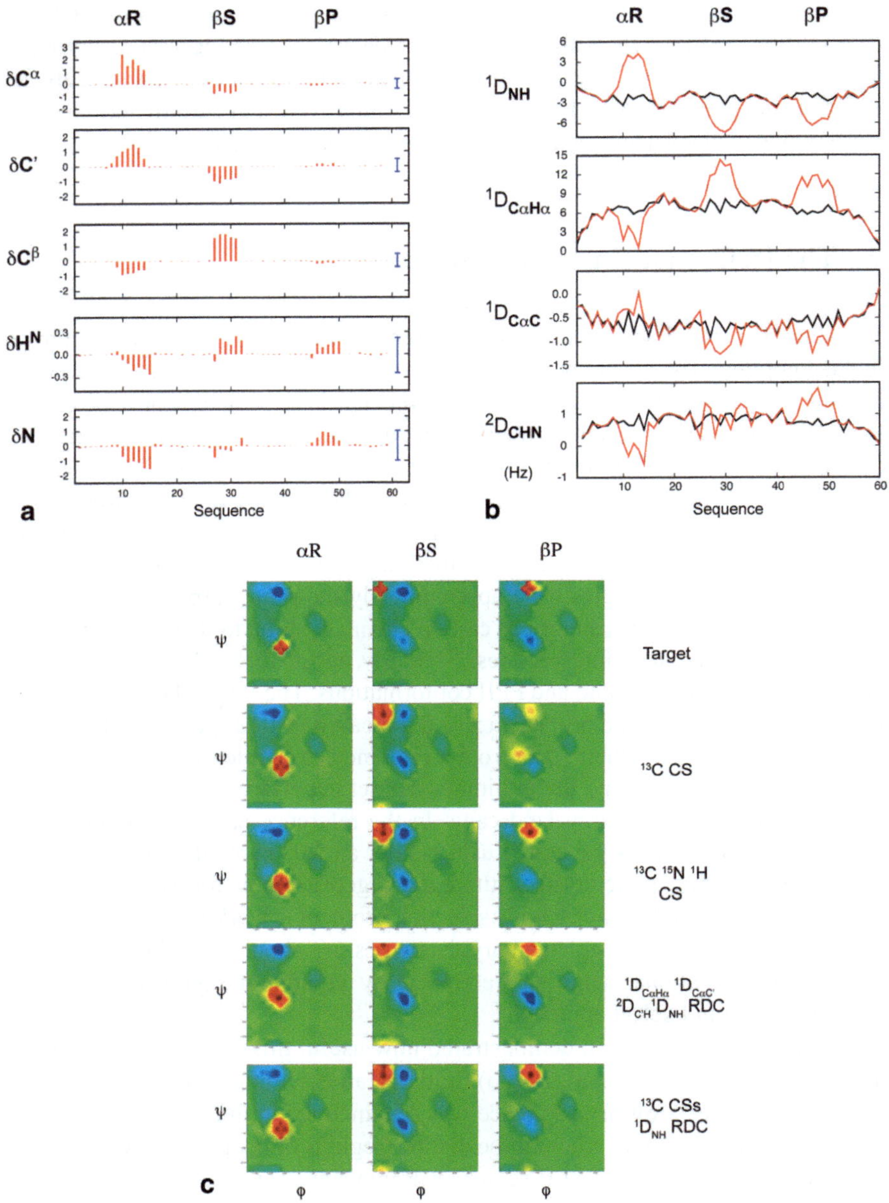

Fig. 4.6 Testing the accuracy with which experimental data can map local conformational propensities in IDPs using the reference ensemble method. **a** Synthetic chemical shift dataset calculated for an ensemble of a model protein of 60 amino acids of arbitrary sequence. Three different regions of the protein over-sample the α-helical, β-strand and PPII region, while the remaining regions sample statistical coil conformations. The difference is shown between the predicted chemical shifts for this ensemble and the ensemble-averaged chemical shifts for a statistical coil ensemble. **b** Synthetic RDC dataset calculated for the model protein over-sampling the three different regions of Ramachandran space (*red*) compared to RDCs predicted for a statistical coil ensemble of the same protein (*black*). **c** Selection of sub-ensembles using ASTEROIDS on the basis of different

and used as targets in a selection protocol using the genetic algorithm ASTEROIDS by starting from a statistical coil pool, i.e. a pool without any particular secondary structure preferences. After the ensemble selection using only data for carbon chemical shifts (C^{α}, C^{β}, C'), the ϕ/ψ sampling was reproduced in the regions with enhanced α-helical and β-strand sampling, but not in the region with enhanced PPII sampling (Fig. 4.6c). The study further showed that inclusion of either backbone ^{15}N and $^{1}H^{N}$ chemical shifts or $^{1}D_{NH}$ RDCs in the selection procedure allowed a reproduction of the PPII sampling in the third biased region of the target ensemble, while inclusion of both backbone chemical shifts and RDCs represents a robust and accurate way to map the local conformational sampling of IDPs (Fig. 4.6c). Calibration of ensemble generation protocols against a synthetic target can therefore tell us if we are able to reproduce the sampling of a synthetic ensemble, and consequently also a real ensemble with the same characteristics. The reference ensemble method can also be used in a quantitative way by adding Gaussian noise to the synthetic dataset to determine the accuracy with which the conformational space of IDPs can be mapped using different data types (Ozenne et al. 2012b).

10 Taking into Account Cooperatively Formed Secondary Structures in IDPs

When an IDP contains a longer stretch of a cooperatively formed structure, such as an α-helix, the sample-and-select approach does not work as efficiently. An α-helix can be stabilized by many cooperative interactions (Muñoz and Serrano 1995; Doig 2002). For example, the effect of helix capping can span several amino acids further down the protein sequence and can affect the stability of the helix as a whole. Apart from capping interactions and the regular backbone-backbone i to $i+3$ hydrogen bonding pattern, many other stabilizing interactions are present between i and $i+3$ residues and between residues even further away. As helices in IDPs can span more than ten residues, we expect that amino acids that are far apart in the primary sequence should contribute together to the formation of the helix.

Statistical coil generators take into account amino acid type conformational preferences that are mainly local. As a consequence the sampling in the statistical coil library can correctly sample α-helical conformations in selected regions of the protein; however, the chance of building a long helix without an interruption is relatively small. For example, with a statistical coil library with 80 % helical sampling, the probability of forming a helical element consisting of six consecutive amino acids is 0.8^6, which is around 25 %. The probability of forming a longer α-helix with

combinations of the synthetic chemical shift and RDC datasets. Ramachandran plots of the target (*top line*) and the results of the selections employing different data types are shown. Reprinted in part with permission from (Ozenne et al. 2012b). Copyright 2012 American Chemical Society

a high enough population to fit the data is therefore very low. Approaches using MD simulations experience a similar problem when it comes to long cooperatively folded helices (or other secondary structures), and breaks in helices are often observed throughout the simulations.

A solution to this problem is to generate many different starting ensembles where each ensemble incorporates an α-helix with a different start and end point, calculate the ensemble-averaged NMR data for each of these ensembles, and subsequently find the best combination of ensembles with corresponding populations that agree with the experimental data. Essentially this corresponds to enriching the initial starting pool with cooperatively formed α-helices in specific regions of the protein that are known to over-sample the α-helical region of Ramachandran space.

This approach was developed and applied to the C-terminal intrinsically disordered domain, N_{TAIL}, of Sendai virus nucleoprotein, which undergoes induced α-helical folding of its molecular recognition element upon binding to its partner protein PX (Jensen et al. 2008). The $^1D_{NH}$ RDCs measured in N_{TAIL} were positive within the molecular recognition element and showed a characteristic dipolar wave pattern consistent with the formation of cooperatively formed α-helices (Fig. 4.7a) (Jensen and Blackledge 2008). The experimental RDCs were fitted with models of increasing complexity, i.e. starting from a statistical coil model and increasing the number of helical ensembles until a satisfactory fit was obtained (Fig. 4.7a). For each model the populations of the helical elements were optimized to best agree with the experimental data. Data reproduction evidently improves as the number of helical ensembles increases, and a standard F-test was therefore used to test for the statistical significance of this improvement. It was found that three helical ensembles with different populations in exchange with a disordered form of the protein are needed to describe the experimental RDCs (Fig. 4.7b). Interestingly, all the selected helical ensembles are preceded by aspartic acids or serines, which are the most common N-capping residues in helices of folded proteins (Fig. 4.7c, 4.7d). An N-capping residue stabilizes a helix by forming a hydrogen bond between its side chain and the backbone amides at position 2 or 3 in the helix (Fig. 4.7c). Importantly, this indicates that the helices preferentially being populated in solution in N_{TAIL} are stabilized by N-capping interactions, and that the helical formation is being promoted by strategically placed aspartic acids and serines in the primary sequence. The partial pre-structuration of N_{TAIL} in its free state suggests that the interaction with PX occurs through conformational selection, where one of the helices is selected by the partner protein in order to form the complex (Hammes et al. 2009).

11 Choosing an Appropriate Ensemble Size

A scoring function that measures the agreement between the experimental and simulated data for the model ensemble is applied during the ensemble selection procedure. A measure commonly used is chi square ($\chi^2 = \sum (s_i - m_i)^2 / \sigma_i$) where s_i represents the back-calculated data from the ensemble, m_i represents the

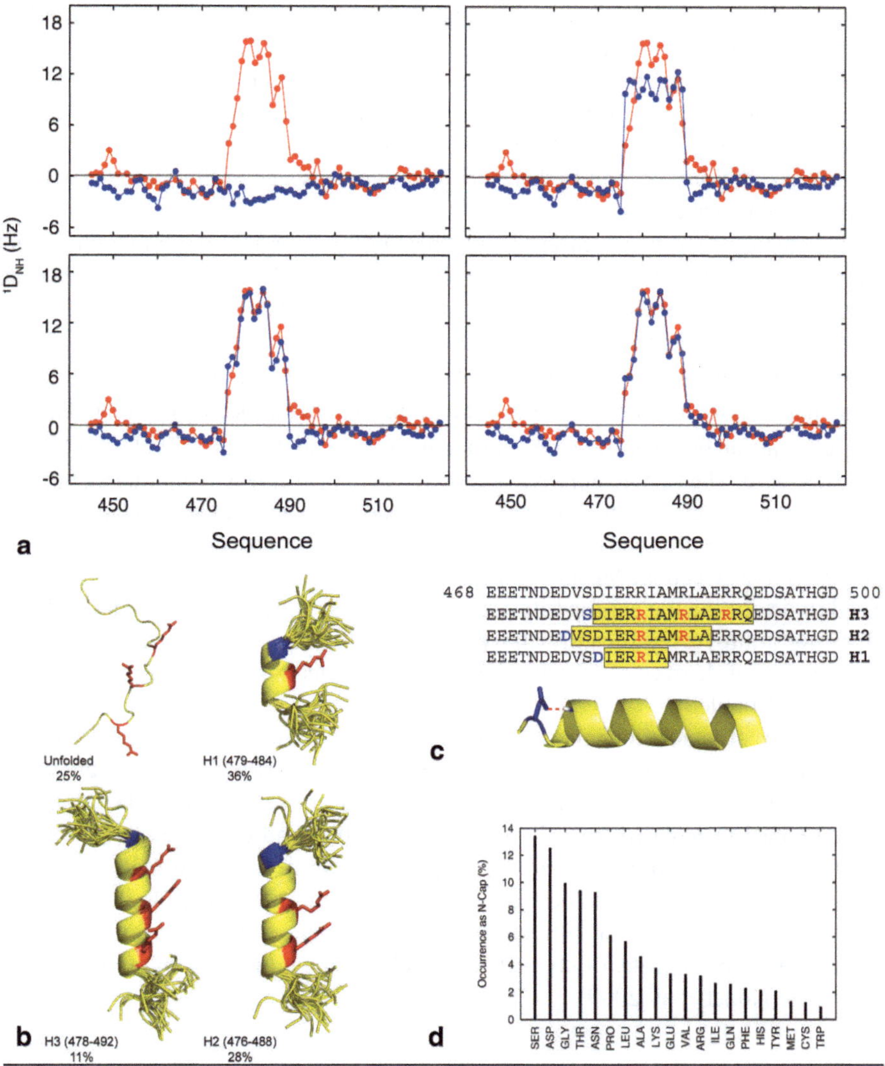

Fig. 4.7 Analysis of cooperatively formed α-helices within the molecular recognition element of the C-terminal domain, N_{TAIL}, of the Sendai virus nucleoprotein. **a** Reproduction of experimental $^1D_{NH}$ RDCs in N_{TAIL} for models with an increasing number, N, of helical ensembles: $N=0$ (*top, left*), $N=1$ (*top, right*), $N=2$ (*bottom, left*), $N=3$ (*bottom, right*). Experimental RDCs are shown in red, while back-calculated RDCs from the different models are shown in *blue*. **b** Molecular representation of the equilibrium of the molecular recognition element of N_{TAIL} in solution. The four different helical states are presented as a single structure for the completely disordered form and as twenty randomly selected conformers for the three helical states. The molecular recognition arginines are displayed in *red*, while N-capping residues are shown in *blue*. **c** The amino acid sequence of the molecular recognition element of N_{TAIL} showing that the selected helical elements are all preceded by aspartic acids or serine residues. The cartoon representation illustrates an N-capping aspartic acid side chain-backbone interaction. **d** The occurrence of different amino acid types as N-capping residues in helices of folded proteins. Reprinted in part with permission from (Jensen et al. 2008). Copyright 2008 American Chemical Society

measured data, and σ_i is the experimental error associated with the different NMR parameters. One has to take care in order not to over-fit the experimental data. Over-fitting happens when the difference between the simulated and experimental data is minimized during the fitting process not in order to improve the physical model describing the system, but because the model is modified to fit the random error and noise contributions. A good fit therefore always means a good fit within the defined experimental error.

Ensemble size also influences the goodness of the fit and the ensemble should not be too small or too large. The ensemble obtained in the selection procedure is not accurate if it is composed of too few structures and therefore does not represent the conformational heterogeneity present in solution. In this case, with too few conformers in the ensemble, we say that we are over-restraining or also under-fitting. On the other hand, as we increase the ensemble size, the number of parameters (e.g. dihedral angles) that can be independently adjusted increases and the total χ^2 value will therefore decrease. The fit may improve because of an improvement in our model, but also because inaccuracies in the model are compensated by newly added structures.

There are tests that can help us decide on the ensemble size that we should choose for ensemble selection. Most commonly a plot of final χ^2 against ensemble size is used to determine the appropriate size for a given set of experimental data. The fit does not improve significantly above a certain ensemble size, and the increase in the number of degrees of freedom introduced by selecting a larger ensemble is no longer justifiable.

An alternative method, and in principle a more correct one, is to use cross-validation procedures where a part of the experimental data is left out of the ensemble selection procedure. Ensembles of different sizes are selected and the "passive" data are back-calculated from the selected ensembles and compared to the experimental data. The optimal reproduction of the passive data will normally occur for the most appropriate ensemble size. This procedure has for example been used to obtain the most appropriate ensemble size (200 structures) for describing the local conformational sampling of urea-denatured ubiquitin on the basis of multiple types of RDCs (Nodet et al. 2009).

12 Ensemble Size in Relation to Convergence Properties of NMR Parameters

When optimizing the size of the selected ensembles, one also needs to consider the convergence characteristics of the different NMR parameters when averaged over the sub-ensembles. We say that convergence of a parameter has been reached when the addition of one more conformer to the ensemble does not perturb the calculated average parameter within a predefined limit. The convergence of parameters is particularly important as the use of too few structures in the selected ensembles will force the fitting procedure to accommodate fluctuations in the averaged NMR parameters that do not necessarily correspond to specific conformational propensities, thereby potentially leading to incorrect residue-specific conformational sampling.

The number of conformers needed for a certain simulated parameter to converge depends on its variance. This is the reason for the different convergence properties of RDCs and chemical shifts. Chemical shifts are sensitive to the local chemical environment and are affected by main and side chain dihedral angles, amino acid identity, ring current effects and hydrogen bonding. When chemical shifts are predicted in IDPs the most important factor is the dihedral angle distribution. Carbon (C^α, C^β, C') and proton H^α chemical shifts depend mostly on the ϕ/ψ angles of the residue of interest, while the chemical shifts of the nitrogen (N) atom and the amide proton (H^N) depend mostly on the ψ angle of the preceding residue. The fact that chemical shifts can be predicted from local structure only makes them a well-behaved parameter when it comes to convergence. Sufficient sampling of the ϕ/ψ space of a single amino acid can even be achieved with only a few hundred structures. As a consequence, when selecting ensembles against experimental chemical shifts, 100–200 structures are sufficient for achieving convergence of the predicted chemical shifts (Fig. 4.8a, 4.8b, 4.8c). Scalar couplings also report on local conformational features of the polypeptide chain, and similarly to chemical shifts, a hundred conformers in the model ensemble suffice for achieving convergence.

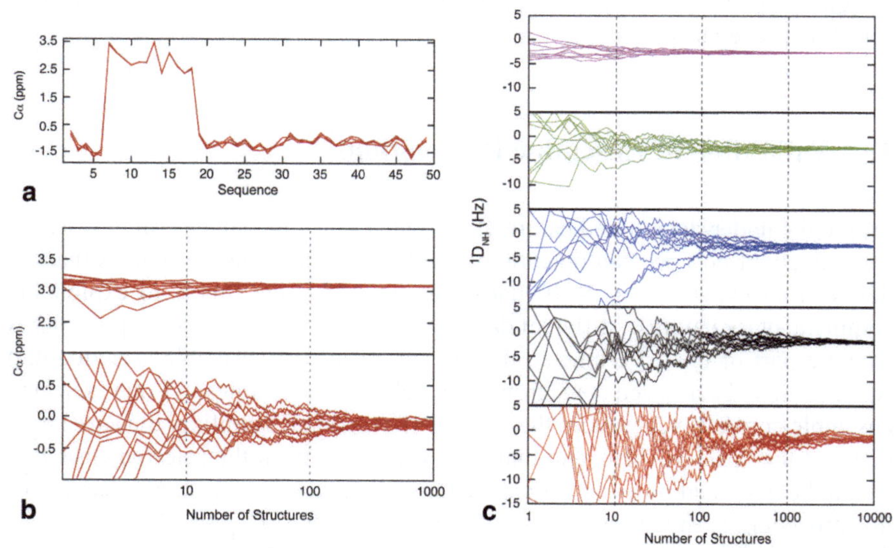

Fig. 4.8 Convergence of experimental NMR parameters over structural ensembles. **a** Secondary Cα chemical shifts averaged over 250 conformers generated using Flexible-Meccano for a model protein of 50 amino acids of arbitrary sequence. The results for five different ensemble averages are shown. Residues 7–18 populate the α-helical region of Ramachandran space, while the remaining residues adopt random coil conformations. **b** Ensemble-averaged secondary Cα chemical shifts for increasing ensemble size for residue 15 of the model protein. Ensemble-averaged secondary Cα chemical shifts for increasing ensemble size for residue 32 of the model protein. **c** Convergence of $^1D_{NH}$ RDCs over a structural ensemble with an increasing number of conformers for a model protein of 76 amino acids. Results are shown for the calculation using a global alignment tensor (*red*) and employing different sizes of short segments for calculating the alignment tensor: 25 (*black*), 15 (*blue*), 9 (*green*) and 3 (*pink*) amino acids. Reprinted in part with permission from (Nodet et al. 2009). Copyright 2009 American Chemical Society

As mentioned above, RDCs depend on both local and long-range structure, i.e. on the conformational sampling of the residue itself and immediate neighbours as well as intra-peptide long-range contacts. The large number of combinations of dihedral angle pairs that potentially all give different RDC values combined with the large range of RDCs calculated from a single structure make the convergence of the RDC average much slower. In addition, RDCs converge more slowly for longer polypeptide chains and for an IDP of 100 amino acids, more than 10,000 structures are needed in order to achieve convergence of the RDC average (Fig. 4.8d). In order to overcome this problem, we can divide the protein chain into shorter, uncoupled segments and predict the RDCs for the central amino acid of each segment (Marsh et al. 2008), thereby achieving sufficient convergence of the RDC average with only a few hundred structures (Fig. 4.8d). The disadvantage of this approach is that we remove any information about long-range structure from the predicted RDCs; however, this information can be reintroduced by multiplying the predicted RDCs by a baseline that takes into account the chain-like nature of the IDP (Nodet et al. 2009; Salmon et al. 2010). Our ability to separate the contribution to the RDCs from local conformational sampling and long-range interactions allows convergence of the RDC average with an ensemble of only a few hundred structures. The use of short segments for the calculation of RDCs therefore appears essential when using RDCs in ensemble selection procedures.

13 Validation of Ensemble Descriptions

Due to the under-determined nature of ensemble selections in general, it is useful to think about how we can potentially validate the structural ensembles that we derive from experimental NMR data. One way of doing this is to exploit the complementary nature of different data types and use cross-validation procedures where a part of the experimental data is left out of the ensemble selection and subsequently back-calculated from the selected ensembles. If the selected ensemble correctly reproduces the local conformational sampling, the agreement between the "passive" data and that back-calculated from the selected ensemble should be good and no systematic deviations should be observed. An example of this procedure is shown in Fig. 4.9 where experimental $^{1}D_{NH}$ RDCs measured in Tau protein are compared to the RDCs extracted from ASTEROIDS ensembles of Tau selected on the basis of chemical shift data alone. The agreement between the two sets of data is excellent and even the turn motifs in the repeat regions of Tau—where positive $^{1}D_{NH}$ RDCs are observed experimentally—are reproduced by the chemical shift ensemble. This type of procedure therefore validates the local conformational sampling of Tau derived from chemical shifts only.

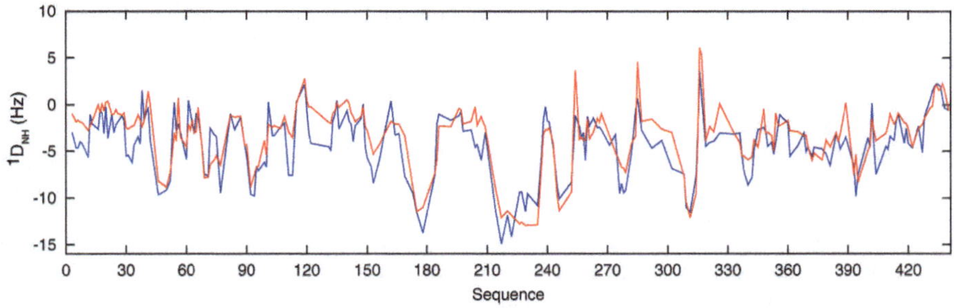

Fig. 4.9 Validation of structural ensembles of IDPs derived from experimental NMR data. An example is shown of cross-validation of experimental $^1D_{NH}$ RDCs of the IDP Tau. Experimental data are shown in red, while back-calculated RDCs from an ensemble of Tau selected by the genetic algorithm ASTEROIDS on the basis of experimental chemical shifts only are shown in *blue*

14 Conclusions and Outlook

Ensemble descriptions have in recent years emerged as the preferred tool for representing the structural and dynamic properties of IDPs and their functional complexes. Within such descriptions it is assumed that the protein adopts a continuum of rapidly interconverting structures, and the determination of these representative ensembles is one of the major challenges in the studies of IDPs. In this chapter we have described how different NMR data types can be combined with sample-and-select approaches to map local conformational propensities in IDPs. In particular, we have emphasized some of the pitfalls associated with these approaches such as under- and over-restraining, and we have discussed ways to validate the derived structural ensembles. Validating structural ensembles is particularly important if we are to use these ensembles in the future for the prediction of other, independent experimental observables or for the development of small molecules that can interfere with the biological function of IDPs.

References

Bernadó P, Blackledge M (2010) Structural biology: proteins in dynamic equilibrium. Nature 468:1046–1048

Bernadó P, Bertoncini CW, Griesinger C et al (2005a) Defining long-range order and local disorder in native α-synuclein using residual dipolar couplings. J Am Chem Soc 127:17968–17969

Bernadó P, Blanchard L, Timmins P et al (2005b) A structural model for unfolded proteins from residual dipolar couplings and small-angle X-ray scattering. Proc Natl Acad Sci U S A 102:17002–17007

Bertini I, Gupta YK, Luchinat C et al (2007) Paramagnetism-based NMR restraints provide maximum allowed probabilities for the different conformations of partially independent protein domains. J Am Chem Soc 129:12786–12794

Blackledge M (2005) Recent progress in the study of biomolecular structure and dynamics in solution from residual dipolar couplings. Prog Nucl Magn Reson Spectrosc 46:23–61
Bouvignies G, Bernadó P, Meier S et al (2005) Identification of slow correlated motions in proteins using residual dipolar and hydrogen-bond scalar couplings. Proc Natl Acad Sci U S A 102:13885–13890
Camilloni C, De Simone A, Vranken WF et al (2012) Determination of secondary structure populations in disordered states of proteins using nuclear magnetic resonance chemical shifts. Biochemistry 51:2224–2231
De Simone A, Cavalli A, Hsu S-TD et al (2009) Accurate random coil chemical shifts from an analysis of loop regions in native states of proteins. J Am Chem Soc 131:16332–16333
Deshmukh L, Schwieters CD, Grishaev A et al (2013) Structure and dynamics of full-length HIV-1 capsid protein in solution. J Am Chem Soc 135:16133–16147
Doig AJ (2002) Recent advances in helix-coil theory. Biophys Chem 101–102:281–293
Dunker AK, Silman I, Uversky VN et al (2008) Function and structure of inherently disordered proteins. Curr Opin Struct Biol 18:756–764
Dyson HJ, Wright PE (2002) Coupling of folding and binding for unstructured proteins. Curr Opin Struct Biol 12:54–60
Feldman HJ, Hogue CW (2000) A fast method to sample real protein conformational space. Proteins 39:112–131
Fisher CK, Stultz CM (2011) Constructing ensembles for intrinsically disordered proteins. Curr Opin Struct Biol 21:426–431
Fisher CK, Huang A, Stultz CM (2010) Modeling intrinsically disordered proteins with bayesian statistics. J Am Chem Soc 132:14919–14927
Francis DM, Różycki B, Koveal D et al (2011) Structural basis of p38a regulation by hematopoietic tyrosine phosphatase. Nat Chem Biol 7:916–924
Frishman D, Argos P (1995) Knowledge-based protein secondary structure assignment. Proteins 23:566–579
Göbl C, Madl T, Simon B et al (2014) NMR approaches for structural analysis of multidomain proteins and complexes in solution. Prog Nucl Magn Reson Spectrosc 80:26–63
Graf J, Nguyen PH, Stock G, Schwalbe H (2007) Structure and dynamics of the homologous series of alanine peptides: a joint molecular dynamics/NMR study. J Am Chem Soc 129:1179–1189
Guerry P, Salmon L, Mollica L et al (2013) Mapping the population of protein conformational energy sub-states from NMR dipolar couplings. Angew Chem Int Ed Engl 52:3181–3185
Hagarman A, Measey TJ, Mathieu D et al (2010) Intrinsic propensities of amino acid residues in GxG peptides inferred from amide I' band profiles and NMR scalar coupling constants. J Am Chem Soc 132:540–551
Hamelberg D, Mongan J, McCammon JA (2004) Accelerated molecular dynamics: a promising and efficient simulation method for biomolecules. J Chem Phys 120:11919–11929
Hammes GG, Chang Y-C, Oas TG (2009) Conformational selection or induced fit: a flux description of reaction mechanism. Proc Natl Acad Sci U S A 106:13737–13741
Hansen MR, Mueller L, Pardi A (1998) Tunable alignment of macromolecules by filamentous phage yields dipolar coupling interactions. Nat Struct Biol 5:1065–1074
Hansmann UHE (1997) Parallel tempering algorithm for conformational studies of biological molecules. Chem Phys Lett 281:140–150
Henzler-Wildman K, Kern D (2007) Dynamic personalities of proteins. Nature 450:964–972
Hollingsworth SA, Berkholz DS, Karplus PA (2009) On the occurrence of linear groups in proteins. Protein Sci 18:1321–1325
Huang J, Ozenne V, Jensen MR et al (2013) Direct prediction of NMR residual dipolar couplings from the primary sequence of unfolded proteins. Angew Chem Int Ed Engl 52:687–690
Huang J, Warner LR, Sanchez C et al (2014) Transient electrostatic interactions dominate the conformational equilibrium sampled by multidomain splicing factor U2AF65: a combined NMR and SAXS study. J Am Chem Soc 136:7068–7076

Jensen MR, Blackledge M (2008) On the origin of NMR dipolar waves in transient helical elements of partially folded proteins. J Am Chem Soc 130:11266–11267

Jensen MR, Houben K, Lescop E et al (2008) Quantitative conformational analysis of partially folded proteins from residual dipolar couplings: application to the molecular recognition element of Sendai virus nucleoprotein. J Am Chem Soc 130:8055–8061

Jensen MR, Markwick PRL, Meier S et al (2009) Quantitative determination of the conformational properties of partially folded and intrinsically disordered proteins using NMR dipolar couplings. Structure 17:1169–1185

Jensen MR, Salmon L, Nodet G et al (2010) Defining conformational ensembles of intrinsically disordered and partially folded proteins directly from chemical shifts. J Am Chem Soc 132:1270–1272

Jensen MR, Zweckstetter M, Huang J-R et al (2014) Exploring free-energy landscapes of intrinsically disordered proteins at atomic resolution using NMR spectroscopy. Chem Rev 114:6632–6660

Jha AK, Colubri A, Freed KF et al (2005a) Statistical coil model of the unfolded state: resolving the reconciliation problem. Proc Natl Acad Sci U S A 102:13099–13104

Jha AK, Colubri A, Zaman MH et al (2005b) Helix, sheet, and polyproline II frequencies and strong nearest neighbor effects in a restricted coil library. Biochemistry 44:9691–9702

Kabsch W, Sander C (1983) Dictionary of protein secondary structure: pattern recognition of hydrogen-bonded and geometrical features. Biopolymers 22:2577–2637

Karplus M (1959) Contact electron-Spin coupling of nuclear magnetic moments. J Chem Phys 30:11–15. doi:10.1063/1.1729860

Karplus M, Kuriyan J (2005) Molecular dynamics and protein function. Proc Natl Acad Sci U S A 102:6679–6685

Krzeminski M, Marsh JA, Neale C et al (2013) Characterization of disordered proteins with ENSEMBLE. Bioinformatics 29:398–399

Lange OF, Lakomek N-A, Farès C et al (2008) Recognition dynamics up to microseconds revealed from an RDC-derived ubiquitin ensemble in solution. Science 320:1471–1475

Leader DP, Milner-White EJ (2011) The structure of the ends of α-helices in globular proteins: effect of additional hydrogen bonds and implications for helix formation. Proteins 79:1010–1019

Leader DP, Milner-White EJ (2012) Structure Motivator: a tool for exploring small three-dimensional elements in proteins. BMC Struct Biol 12:26. doi:10.1186/1472-6807-12-26

Leone V, Marinelli F, Carloni P et al (2010) Targeting biomolecular flexibility with metadynamics. Curr Opin Struct Biol 20:148–154

Lindorff-Larsen K, Best RB, Depristo MA et al (2005) Simultaneous determination of protein structure and dynamics. Nature 433:128–132

Lindorff-Larsen K, Trbovic N, Maragakis P et al (2012) Structure and dynamics of an unfolded protein examined by molecular dynamics simulation. J Am Chem Soc 134:3787–3791

MacArthur MW, Thornton JM (1991) Influence of proline residues on protein conformation. J Mol Biol 218:397–412

Marsh JA, Forman-Kay JD (2009) Structure and disorder in an unfolded state under nondenaturing conditions from ensemble models consistent with a large number of experimental restraints. J Mol Biol 391:359–374

Marsh JA, Singh VK, Jia Z et al (2006) Sensitivity of secondary structure propensities to sequence differences between α- and γ-synuclein: implications for fibrillation. Protein Sci 15:2795–2804

Marsh JA, Baker JMR, Tollinger M et al (2008) Calculation of residual dipolar couplings from disordered state ensembles using local alignment. J Am Chem Soc 130:7804–7805

Mittermaier A, Kay LE (2006) New tools provide new insights in NMR studies of protein dynamics. Science 312:224–228

Mittermaier AK, Kay LE (2009) Observing biological dynamics at atomic resolution using NMR. Trends Biochem Sci 34:601–611

Mukrasch MD, Markwick PRL, Biernat J et al (2007) Highly populated turn conformations in natively unfolded tau protein identified from residual dipolar couplings and molecular simulation. J Am Chem Soc 129:5235–5243

Muñoz V, Serrano L (1995) Helix design, prediction and stability. Curr Opin Biotechnol 6:382–386
Narayanan RL, Dürr UHN, Bibow S et al (2010) Automatic assignment of the intrinsically disordered protein Tau with 441-residues. J Am Chem Soc 132:11906–11907
Nodet G, Salmon L, Ozenne V et al (2009) Quantitative description of backbone conformational sampling of unfolded proteins at amino acid resolution from NMR residual dipolar couplings. J Am Chem Soc 131:17908–17918
Ozenne V, Bauer F, Salmon L et al (2012a) Flexible-meccano: a tool for the generation of explicit ensemble descriptions of intrinsically disordered proteins and their associated experimental observables. Bioinformatics 28:1463–1470
Ozenne V, Schneider R, Yao M et al (2012b) Mapping the potential energy landscape of intrinsically disordered proteins at amino acid resolution. J Am Chem Soc 134:15138–15148
Piana S, Laio A (2007) A bias-exchange approach to protein folding. J Phys Chem B 111:4553–4559
Pierce LCT, Salomon-Ferrer R, Augusto F et al (2012) Routine access to millisecond time scale events with accelerated molecular dynamics. J Chem Theory Comput 8:2997–3002
Prestegard JH, Bougault CM, Kishore AI (2004) Residual dipolar couplings in structure determination of biomolecules. Chem Rev 104:3519–3540
Różycki B, Kim YC, Hummer G (2011) SAXS ensemble refinement of ESCRT-III CHMP3 conformational transitions. Structure 19:109–116
Rückert M, Otting G (2000) Alignment of biological macromolecules in novel nonionic liquid crystalline media for NMR Experiments. J Am Chem Soc 122:7793–7797
Salmon L, Bouvignies G, Markwick PRL et al (2009) Protein conformational flexibility from structure-free analysis of NMR dipolar couplings: quantitative and absolute determination of backbone motion in ubiquitin. Angew Chem Int Ed Engl 48:4154–4157
Salmon L, Nodet G, Ozenne V et al (2010) NMR characterization of long-range order in intrinsically disordered proteins. J Am Chem Soc 132:8407–8418
Salmon L, Bouvignies G, Markwick P et al (2011) Nuclear magnetic resonance provides a quantitative description of protein conformational flexibility on physiologically important time scales. Biochemistry 50:2735–2747
Salmon L, Pierce L, Grimm A et al (2012) Multi-timescale conformational dynamics of the SH3 domain of CD2-associated protein using NMR spectroscopy and accelerated molecular dynamics. Angew Chem Int Ed Engl 51:6103–6106
Sass HJ, Musco G, Stahl SJ et al (2000) Solution NMR of proteins within polyacrylamide gels: diffusional properties and residual alignment by mechanical stress or embedding of oriented purple membranes. J Biomol NMR 18:303–309
Schwalbe M, Ozenne V, Bibow S et al (2014) Predictive atomic resolution descriptions of intrinsically disordered hTau40 and α-synuclein in solution from NMR and small angle scattering. Structure 22:238–249
Schweitzer-Stenner R (2012) Conformational propensities and residual structures in unfolded peptides and proteins. Mol Biosyst 8:122–133
Serrano L (1995) Comparison between the phi distribution of the amino acids in the protein database and NMR data indicates that amino acids have various phi propensities in the random coil conformation. J Mol Biol 254:322–333
Shen Y, Bax A (2012) Identification of helix capping and β-turn motifs from NMR chemical shifts. J Biomol NMR 52:211–232
Shi Z, Chen K, Liu Z et al (2006) Conformation of the backbone in unfolded proteins. Chem Rev 106:1877–1897
Shortle D, Ackerman MS (2001) Persistence of native-like topology in a denatured protein in 8 M urea. Science 293:487–489
Smith LJ, Bolin KA, Schwalbe H et al (1996) Analysis of main chain torsion angles in proteins: prediction of NMR coupling constants for native and random coil conformations. J Mol Biol 255:494–506
Sugita Y, Okamoto Y (1999) Replica-exchange molecular dynamics method for protein folding. Chem Phys Lett 314:141–151

Tamiola K, Mulder FAA (2012) Using NMR chemical shifts to calculate the propensity for structural order and disorder in proteins. Biochem Soc Trans 40:1014–1020

Tamiola K, Acar B, Mulder FAA (2010) Sequence-specific random coil chemical shifts of intrinsically disordered proteins. J Am Chem Soc 132:18000–18003

Tjandra N, Bax A (1997) Direct measurement of distances and angles in biomolecules by NMR in a dilute liquid crystalline medium. Science 278:1111–1114

Tolman JR, Flanagan JM, Kennedy MA et al (1995) Nuclear magnetic dipole interactions in field-oriented proteins: information for structure determination in solution. Proc Natl Acad Sci U S A 92:9279–9283

Tompa P (2012) Intrinsically disordered proteins: a 10-year recap. Trends Biochem Sci 37:509–516

Voter AF (1997) Hyperdynamics: accelerated molecular dynamics of infrequent events. Phys Rev Lett 78:3908–3911

Vuister GW, Bax A (1993) Quantitative J correlation: a new approach for measuring homonuclear three-bond J(HNH.α.) coupling constants in ^{15}N-enriched proteins. J Am Chem Soc 115:7772–7777

Wang Y, Jardetzky O (2002a) Investigation of the neighboring residue effects on protein chemical shifts. J Am Chem Soc 124:14075–14084

Wang Y, Jardetzky O (2002b) Probability-based protein secondary structure identification using combined NMR chemical-shift data. Protein Sci 11:852–861

Wells M, Tidow H, Rutherford TJ et al (2008) Structure of tumor suppressor p53 and its intrinsically disordered N-terminal transactivation domain. Proc Natl Acad Sci U S A 105:5762–5767

Wishart DS, Bigam CG, Holm A et al (1995) ^{1}H, ^{13}C and ^{15}N random coil NMR chemical shifts of the common amino acids. I. Investigations of nearest-neighbor effects. J Biomol NMR 5:67–81

Yang S, Blachowicz L, Makowski L et al (2010) Multidomain assembled states of Hck tyrosine kinase in solution. Proc Natl Acad Sci U S A 107:15757–15762

Zhang H, Neal S, Wishart DS (2003) RefDB: a database of uniformly referenced protein chemical shifts. J Biomol NMR 25:173–195

Zweckstetter M, Bax A (2000) Prediction of sterically induced alignment in a dilute liquid crystalline phase: aid to protein structure determination by NMR. J Am Chem Soc 122:3791–3792

Chapter 5
NMR Spectroscopic Studies of the Conformational Ensembles of Intrinsically Disordered Proteins

Dennis Kurzbach, Georg Kontaxis, Nicolas Coudevylle and Robert Konrat

Abstract Intrinsically disordered proteins (IDPs) are characterized by substantial conformational flexibility and thus not amenable to conventional structural biology techniques. Given their inherent structural flexibility NMR spectroscopy offers unique opportunities for structural and dynamic studies of IDPs. The past two decades have witnessed significant development of NMR spectroscopy that couples advances in spin physics and chemistry with a broad range of applications. This chapter will summarize key advances in NMR methodology. Despite the availability of efficient (multi-dimensional) NMR experiments for signal assignment of IDPs it is discussed that NMR of larger and more complex IDPs demands spectral simplification strategies capitalizing on specific isotope-labeling strategies. Prototypical applications of isotope labeling-strategies are described. Since IDP-ligand association and dissociation processes frequently occur on time scales that are amenable to NMR spectroscopy we describe in detail the application of CPMG relaxation dispersion techniques to studies of IDP protein binding. Finally, we demonstrate that the complementary usage of NMR and EPR data provide a more comprehensive picture about the conformational states of IDPs and can be employed to analyze the conformational ensembles of IDPs.

Keywords Intrinsically disordered proteins · Biomolecular NMR · Protein metastructure · EPR spectroscopy · Paramagnetic relaxation · NMR spin relaxation

1 Introduction

Intrinsically disordered proteins (IDPs) have attracted great attention in recent years based on their importance in eukaryotic life and their important roles in protein interaction networks. Their sampling of a vast and heterogeneous conformational

R. Konrat (✉) · D. Kurzbach · G. Kontaxis · N. Coudevylle
Department of Computational and Structural Biology, Max F. Perutz Laboratories, University of Vienna, Campus Vienna Biocenter 5, 1030 Vienna, Austria
e-mail: robert.konrat@univie.ac.at

I. C. Felli, R. Pierattelli (eds.), *Intrinsically Disordered Proteins Studied by NMR Spectroscopy,* Advances in Experimental Medicine and Biology,
DOI 10.1007/978-3-319-20164-1_5

space allows for the interaction with multiple binding partners at the same time. This structural plasticity and adaptability is considered to be the reason for IDPs to engage in weak regulatory networks. The inherent structural flexibility of IDPs, however, requires the application of appropriate experimental methods since X-Ray crystallography cannot access the distribution of conformational states of these proteins. In contrast, NMR spectroscopy has been developed into a powerful structural biology technique that offers unique opportunities for structural and dynamic studies of IDPs. Here we summarize recent experimental methodologies that have been developed to analyze the complex NMR spectra typically found for IDPs, spectral simplification strategies employing bio-organic chemistry based isotope-labeling approaches, cross-correlated NMR spin relaxation approaches to probe dihedral angle averaging, and finally paramagnetic relaxation enhancements strategies.

2 Experimental Techniques

NMR based methodology has emerged to characterize the structural dynamics of IDPs. Among those are: Hydrogen exchange rates, NMR chemical shifts and residual dipolar couplings (RDC) that can be used to evaluate local transient secondary structure elements with atomic resolution, whereas paramagnetic relaxation enhancements (PRE) report on transient long-range contacts (Dyson and Wright 2004).

2.1 NMR Spectral Assignment of IDPs

NMR signal assignment is well established for globular proteins. Typically, a suite of triple-resonance experiments is used to find sequential connectivities between neighboring residues. These experimental strategies rely on coherence transfer steps involving backbone ^{13}C, ^{15}N and ^{1}H nuclei. Application of these efficient techniques to IDPs is hampered because of severe spectral overlap (Fig. 5.1 shows a comparison of prototypical ^{1}H-^{15}N HSQC spectra obtained for folded proteins and IDPs) and due to significant chemical exchange with bulk water that reduces ^{1}H^{N} signal intensities leading to low signal-to-noise (S/N) ratios. While the latter can be partly overcome by measurements at low temperature and/or low pH, signal overlap problems required the development of novel NMR techniques. Exploiting improved instrumental sensitivities substantial improvements were made (Chap. 3) due to: (i) non-uniform sampling technologies enabling high-dimensionality (>4D) experiments (Kazimierczuk et al. 2009), (ii) faster acquisition of NMR experiments making use of longitudinal relaxation enhancements (Schanda et al. 2006) and (iii) direct heteronuclei (^{13}C) detection using cryoprobe technology (avoiding exchange problems) (Bermel et al. 2012; Bertini et al. 2011; Novacek et al. 2011). The problems of poor signal dispersion and extensive signal overlap found for IDPs

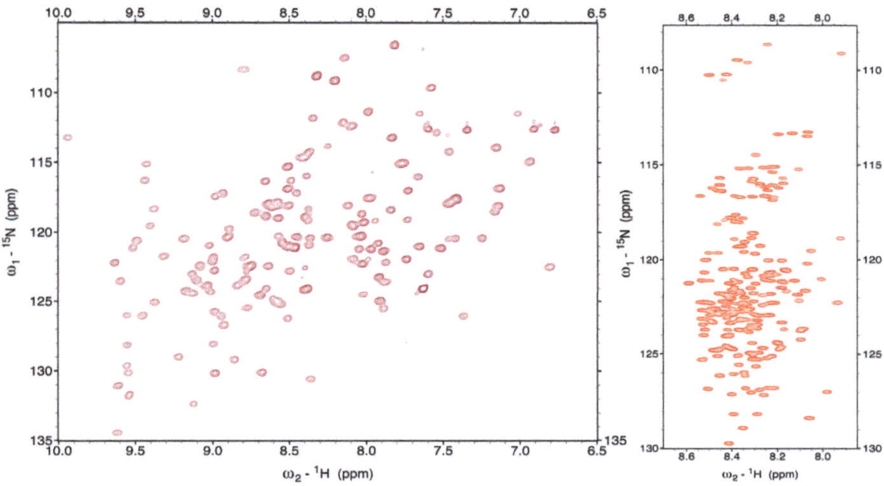

Fig. 5.1 Comparison between spectra of stably folded (globular) and intrinsically disordered proteins. For comparison, $^{1}H^{N}$-^{15}N HSQC spectra for prototypical folded **a** lipocalin Q83 and unfolded **b** BASP1 are shown

are overcome by high-dimensionality spectra (>4D) using non-uniform sampling (NUS) of indirect dimensions together with appropriate processing schemes, e.g., Sparse Multidimensional Fourier Transform (SMFT) processing (Kazimierczuk et al. 2010a; Kazimierczuk et al. 2010b; Motackova et al. 2010; Zawadzka-Kazimierczuk et al. 2012). The analysis of high-dimensionality spectra is straightforward as the relevant frequency regions can easily be identified based on some a priori knowledge of peak locations (resonance frequencies) known from lower dimensionality spectra (2D, 3D, HNCO, HNCA) acquired before. Representative strip plots illustrating experimentally observed connectivities used for sequential signal assignment in IDPs are shown in Fig. 5.2.

Given that NMR spectroscopy of IDPs (due to their favorable spin relaxation properties) is typically not limited by sensitivity but rather spectral resolution, relaxation-optimized detection schemes were shown to lead to further improvements. Recently, a 3D BEST–TROSY-HNCO experiment has been described capitalizing on relaxation-optimized excitation schemes (Solyom et al. 2013). Additionally, given the fact that proline residues are highly abundant in IDPs, BEST-TROSY-optimized Pro-edited 2D ^{1}H-^{15}N experiments have been developed, that either detect ^{1}H-^{15}N correlations of residues following a proline (Pro-HNcocan) or preceding a proline (Pro-iHNcan) (Solyom et al. 2013).

2.1.1 Specific Isotope-Labelling Strategies

Despite the availability of efficient NMR methodology for signal assignment of IDPs it is foreseeable that larger and more complex IDPs will require further

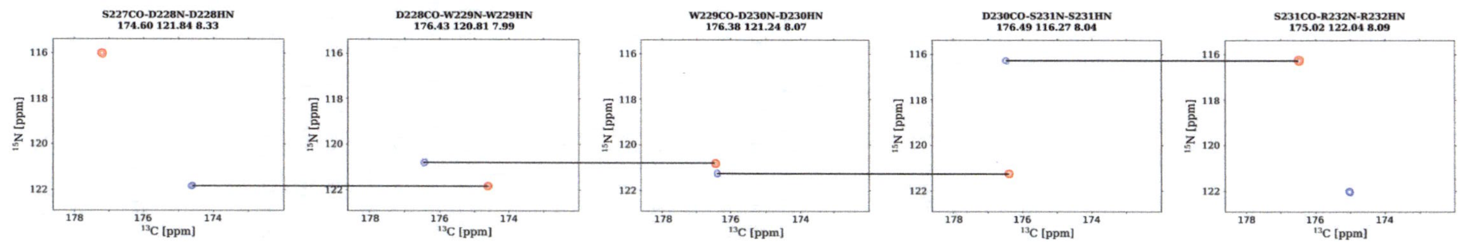

Fig. 5.2 2D spectral planes for consecutive amino acids in hOPN obtained by SMFT processing of the 5D randomly sampled signal. 2D cross-sections of 5D hNCOncaCONH (N_i—CO_{i-1} & N_{i-1}—CO_{i-2}). Spins that were used for coherence transfer are depicted in lower case, while spins that are recorded during the indirect dimensions are given in upper case

spectral simplification strategies. Specific isotope-labeling strategies have been demonstrated to simplify the spectral complexity of large globular proteins and thereby extend the realm of biomolecular NMR spectroscopy. Highly advanced NMR-experiments, together with diverse stable isotope labeling techniques, have therefore been developed in order to maximize the number of attainable structural and dynamic parameters while reducing the spectral complexity. The different isotope labeling strategies exploit the biosynthetic machinery by supplementing the growth medium with suitably labeled late stage amino acid precursor compounds. Metabolic precursor compounds, such as pyruvate- (Guo et al. 2009), α-ketoacid-(Goto et al. 1999) or acetolactate-derivatives (Gans et al. 2010) have been efficiently applied in cell-based protein expression systems, leading to incorporation of stable isotopes at well-defined positions in the target macromolecules. Recently, it was demonstrated that α-ketoacids are very versatile precursor compounds for specific labeling (Lichtenecker et al. 2004; Lichtenecker et al. 2013a, 2013b) since problems of Cα stereochemistry can be avoided via this synthetic route. Starting from α-ketobutyrate and α-ketoisovalerate suitable compounds for selective Ile- as well as Val- and Leu-labeling are available. Another important step was the development of a novel approach for independent leucine labelling using α-ketoisocaproate as a direct metabolic precursor without interfering with the valine metabolic pathway (Lichtenecker et al. 2013a, 2013b). Most recently, this approach was extended to the aromatic amino acids phenylalanine and tyrosine (further extensions to tryptophan are underway) (Lichtenecker et al. 2013c). Although originally designed for the generation of unique isotope-labeling patterns of methyl groups and aromatic ring systems the approach also allows for specific backbone labelling (Cα and/or C′ positions). As has been shown already, specific backbone labelling can be efficiently used to edit, for example, 2D ^{15}N-^{1}H HSQC (by employing HNCO or HNCA-type pulse schemes and only recording the 2D HN plane) (Lichtenecker et al. 2013a, 2013b, 2013c). The availability of specifically labeled amino acids offers exciting possibilities for spectral simplification of large and complex IDPs. Here we demonstrate the approach with an application to the large intrinsically disordered linker domain of BRCA1. Figure 5.3 shows a comparison of 2D ^{15}N-^{1}H HSQC spectra obtained on uniformly ^{15}N-labelled ^{15}N(u)-BRCA1 (a) and 2D HN HNCO planes obtained with either (b) ^{15}N(u),^{13}C(Phe-$^{13}C'$) or (c) ^{15}N(u),^{13}C(Val-$^{13}C'$)-labelled BRCA1. The resulting significant spectral simplification allows for straightforward detection of spectral changes upon, for example, ligand binding. Figure 5.3d shows intensity changes observed for BRCA1 upon binding to the oncogenic transcription factor complex Myc-Max (Kurzbach et al. manuscript in preparation). Given the obtainable significant spectral simplification it can be anticipated that these labeling strategies together with existing high-dimensionality NMR experiments will broaden the scope and applicability of the concept *"IDPbyNMR"*.

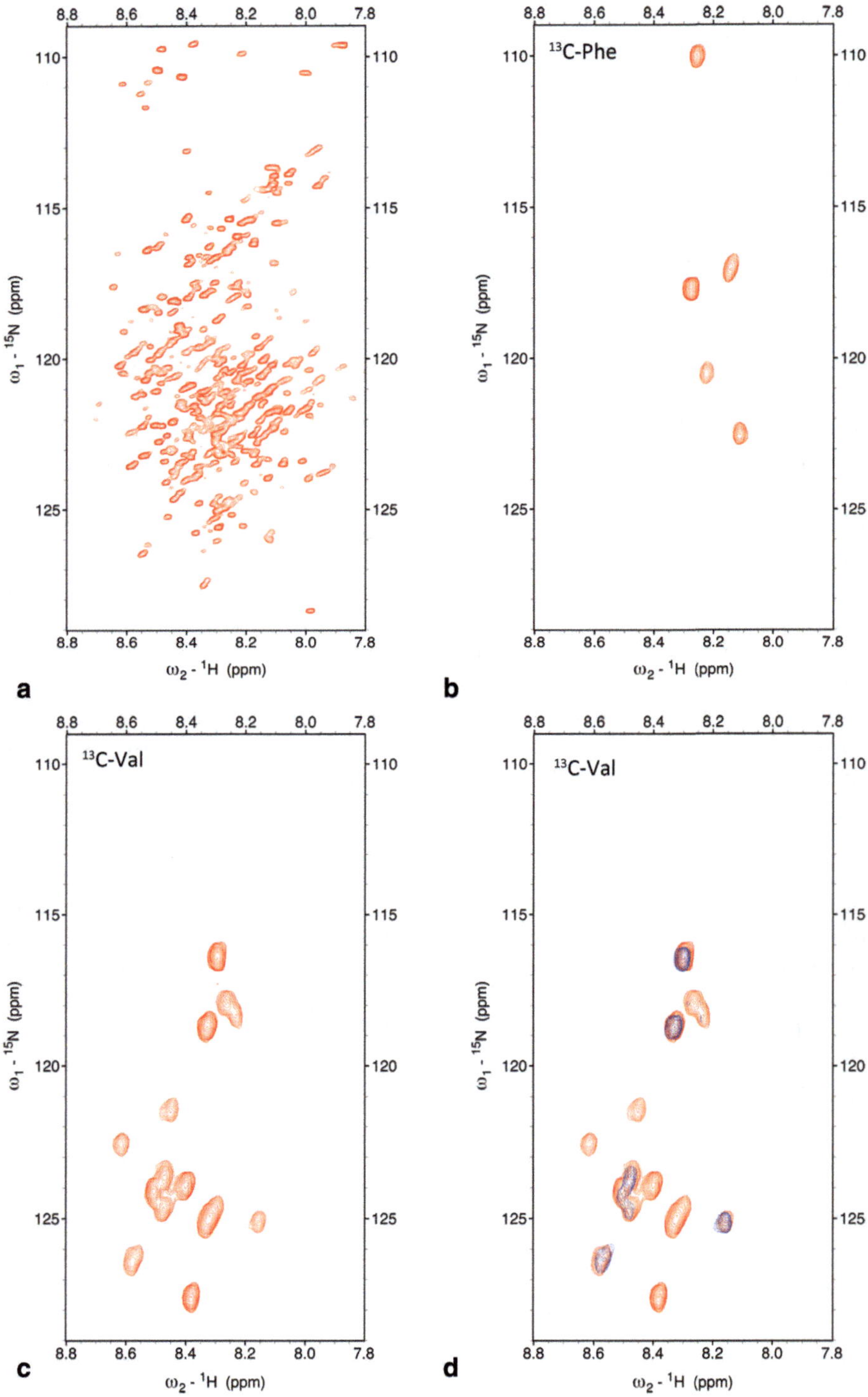

Fig. 5.3 Illustration of spectral simplification in highly crowded IDP NMR spectra using selective amino acid precursor labelling **a** 2D ^{15}N-^{1}H HSQC of truncated BRCA1 (219–504), 2D ^{15}N-^{1}H taken form an HNCO experiment using respective **b** C′-labelled Phe and **c** C′-labelled Val keto-

2.2 NMR Chemical Shifts

NMR Chemical shifts have been demonstrated to be sensitive reporters of backbone conformation and thus provide valuable information about local structural propensities of IDPs (reviewed in Kragelj et al. 2013). Typically, deviations from random coil values are employed to probe local geometries in IDPs and quantitatively assess secondary structure elements (secondary structure propensities, SSP) in IDPs (Chap. 3) (Marsh et al. 2006; Camilloni et al. 2012; Tamiola and Mulder 2012). More sophisticated analysis scheme of NMR chemical shift data involve ensemble approaches (Chap. 4) (Choy and Forman-Kay 2001; Fisher et al. 2010 Fisher and Stultz 2011; Ozenne et al. 2012). Applications of this methodology have been summarized in comprehensive reviews (Kragelj et al. 2013; Mittag and Forman-Kay 2007). Interestingly, NMR chemical shifts have also been shown to provide information about protein dynamics (Berjanskii and Wishart 2006). The inverse weighted sum of backbone secondary chemical shifts for Cα, CO, Cβ, N and Hα nuclei were used to define a Random Coil Index (RCI). Although originally defined for the analysis of globular proteins, applications to IDPs will be feasible given the growing number of experimental studies.

2.3 Residual Dipolar Couplings (RDCs)

Residual dipolar couplings (RDCs) have been developed into a powerful experimental probe for structural dynamics of proteins in solution. Dissolving proteins in anisotropic media with restricted overall motional averaging lead to non-zero RDCs that are experimentally observable in NMR spectra (Tjandra and Bax 1997). Applications to IDPs are straightforward and have already been reported. For example, negative $^{1}D_{NH}$ RDCs are found for segments in which the NH vector is largely oriented perpendicular to the polypeptide chain (extended conformations). Conversely, positive $^{1}D_{NH}$ values are found for α-helical segments (Mohana-Borges et al. 2004). Again, a more sophisticated ensemble approach provides information about specific structural properties such as transient secondary and tertiary structures (Chap. 4) (Bernadó et al. 2005; Wells et al. 2008). Despite the tremendous success of these applications of RDCs in the past care has to be taken in the case of IDPs and careful control experiments have to be employed to ensure that the conformational ensemble is not significantly perturbed by the anisotropic alignment media. A more comprehensive review of the field can be found elsewhere (Salmon et al. 2010).

---

acid precursor compounds (as described by Lichtenecker et al.) **d** Location of the interaction site in BRCA1 to the proto-oncogenic transcription factor complex Myc-Max. HNCO peaks of free (*orange*) and bound to Myc-Max (*blue*) are overlaid

2.4 Paramagnetic Relaxation Enhancements (PREs)

Undoubtedly the most relevant experimental approach to probe transient long-range contacts in IDPs employs the measurement of paramagnetic relaxation enhancements (PREs) (Kosen 1989). As ^{1}H-^{1}H nuclear Overhauser effects (NOEs) are characterized by pronounced distance dependence, conventional NOESY experiments are not sensitive enough to probe distances beyond approximately 6 Å, particularly, as the effective populations of compact sub-states are generally rather small in IDPs. To study paramagnetic relaxation enhancements the protein under investigation is chemically modified by attaching paramagnetic spin labels at defined positions. Typically, the thiol groups of Cys residues (introduced via site-directed mutagenesis) are used to covalently attach the spin label. It has to be noted that the introduction of paramagnetic spin labels into the protein affects both chemical shifts (pseudo contact shifts, PCS) and/or signal intensities via dipolar relaxation between the unpaired electron and the $^{1}H^{N}$ and ^{15}N nuclei (Otting 2010). Depending on the specific spin label used these effects will be different. Applications of PCS to IDPs are now also feasible due to the availability of novel ligands for lanthanide ions and will be a promising additional tool as PCSs report both on distances and orientations relative to the principal axes frame of the paramagnetic center. So far, however, paramagnetic relaxation enhancement (PREs) was the most common experimental parameter used for the analysis of IDPs' tertiary structures in solution. The presence of the paramagnetic spin label (e.g. nitroxide moiety, TEMPO or MTSL) leads to an enhancement in transverse relaxation rates, R_2, depending on the inverse sixth power of the average distance ($1/r^6$) between the unpaired electron and the observed nucleus. For the quantitative analysis of PRE data two approaches have been proposed. In the first approach PRE data are converted into distances or distance ranges using well-established methodology (Battiste and Wagner 2000) that can subsequently be used in, for example, MD simulations to calculate conformational ensembles (Lindorff-Larsen et al. 2004). A second approach involves a more sophisticated extended model-free model for the time–dependency of PRE effects (Salmon et al. 2010). Several applications to IDPs have been reported demonstrating the validity of the approach (Allison et al. 2009, Bibow et al. 2011; Marsh and Forman-Kay 2011; Pinheiro et al. 2011).

Despite the popularity and the robustness of the PRE approach applications to IDPs are still far from trivial. Firstly, the identification of suitable spin label attachment sites without prior knowledge of potentially more compact substructures is not a trivial task as the introduction of the spacious spin label at positions that are relevant for the compact tertiary structure will inevitably perturb the structures. In the worst case, as observed for globular proteins, single point mutations can have detrimental effects on the structural stability of proteins. Thus, additional, entirely primary sequence-based analysis tools are needed for the reliable definition of attachment sites. Secondly, it has been shown that the pronounced distance dependence of PREs can lead to significant bias in the derived ensemble, although this can be partly improved by invoking independent, complementary experimental data (e.g. SAXS) (Allison et al. 2009).

3 Novel Concepts and Experimental Tools for IDP Research

Although NMR is a well-established technique for solution studies of IDPs recent experimental developments highlight the need for further methodological and theoretical developments to provide a comprehensive description of the conformational ensembles of IDPs. Here we summarize novel physico-chemical concepts for the description of IDPs (protein meta-structure) and how these theoretical concepts can be fruitfully combined with new NMR experimental techniques (e.g., cross-correlated NMR spin relaxation, NMR observation of sparsely populated excited states and the complementary usage of EPR and NMR spectroscopy) for structural dynamics studies of IDPs.

3.1 A Physico-Chemical Concept for Protein Disorder

The observation of dynamic conformational ensembles populated by IDPs in solution mandates a reassessment of the *order-disorder dichotomy* (Konrat 2009). Despite the fact that "ordered" and "disordered" proteins have significantly different (tertiary) structural stabilities they share similar residue-residue interaction patterns leading to analogous protein folding funnels governed by the primary sequence. Thus the major differences between ordered and disordered proteins are merely the heights of energy barriers separating the various thermally accessible conformational substates. Specifically, as globular proteins can partly populate different(ly) unfolded states, the conformational ensemble of disordered proteins can also comprise a significant number of compact structures stabilized by favourable long-range interactions. In order to overcome the limitations of the popular dichotomic partitioning of proteins, the meta-structure was introduced as a novel concept for protein sequence analysis (Konrat 2009). In this conceptual view a protein is described as a network of interacting residues. The nodes in the network are the individual amino acids whereas edges connecting two nodes indicate spatial proximity in the 3D structure. It should be noted that in this conceptual view the intricate mutual couplings between amino acids and the resulting cooperative character of the protein are retained. Details of the methodology and applications have been described (Konrat 2009). In brief, the meta-structure of the protein is encoded by two sequence-derived parameters, compactness and local secondary structure. Residue-specific compactness values quantify the spatial neighborhood of individual residues within the 3D protein structures. Residues deeply buried in the interior of a 3D structure display large compactness values whereas surface exposed and conformationally flexible exhibit small (even negative) values. The meta-structure derived local secondary structure parameter is defined in analogy to the NMR $^{13}C\alpha$ secondary chemical shift, with positive values for α-helices and negative values for extended conformations. It has already been shown that this novel approach is very useful for the analysis of IDPs (Konrat 2009; Mayer et al. 2012) since a large

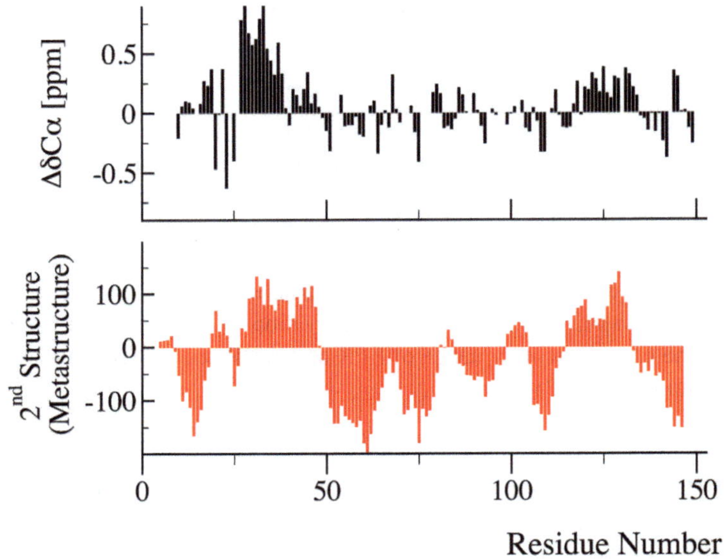

Fig. 5.4 Comparison between experimental NMR data and primary sequence-derived local secondary structure elements in the microtubule-binding domain of the IDP MAP1B light chain. (**a**, *top*) ^{13}C secondary chemical shifts (ΔδCα), (**b**, *bottom*) meta-structure derived local 2nd structure. The meta-structure derived values are defined according to NMR convention (positive values: α-helical elements; negative: extended conformations or β-strands)

scale comparison of calculated compactness values of IDPs (taken from the DisProt database) with well-folded proteins deposited in the PDB database showed that IDPs display significantly smaller compactness values (~230) compared to their well-folded counter parts (~330) and thus suggesting that compactness values are valuable quantitative probes for structural compaction of proteins (Konrat 2009). Furthermore, calculated local secondary structure parameters are reliable parameters for the identification of (transient) α-helices and β-strands or polyproline II helices (Geist et al. 2013). A meta-structure analysis together with a comparison to NMR data for a prototypical IDP is given in Fig. 5.4. Overall, the meta-structure values convincingly compare with experimental NMR secondary chemical shifts or NMR-derived secondary structure propensities. Novel applications to large-scale, sequence-based protein analysis and selection (e.g., identification of IDPs displaying significant local α-helical preformation) are feasible and have already been suggested (Geist et al. 2013).

In addition to large-scale (bioinformatics) sequence analysis of primary sequence information meta-structure data can be used to improve NMR applications to IDP structural studies. Here we outline how meta-structure data (e.g., compactness) can be used to optimize PRE-measurements. Despite their ease of implementation, PRE applications to IDPs (actually to proteins in general) are limited due to uncertainties in the identification of appropriate spin label attachment sites without prior structural information. In order to overcome these limitations, it was suggested to use meta-structure derived compactness data to identify suitable sites of spin label attachment

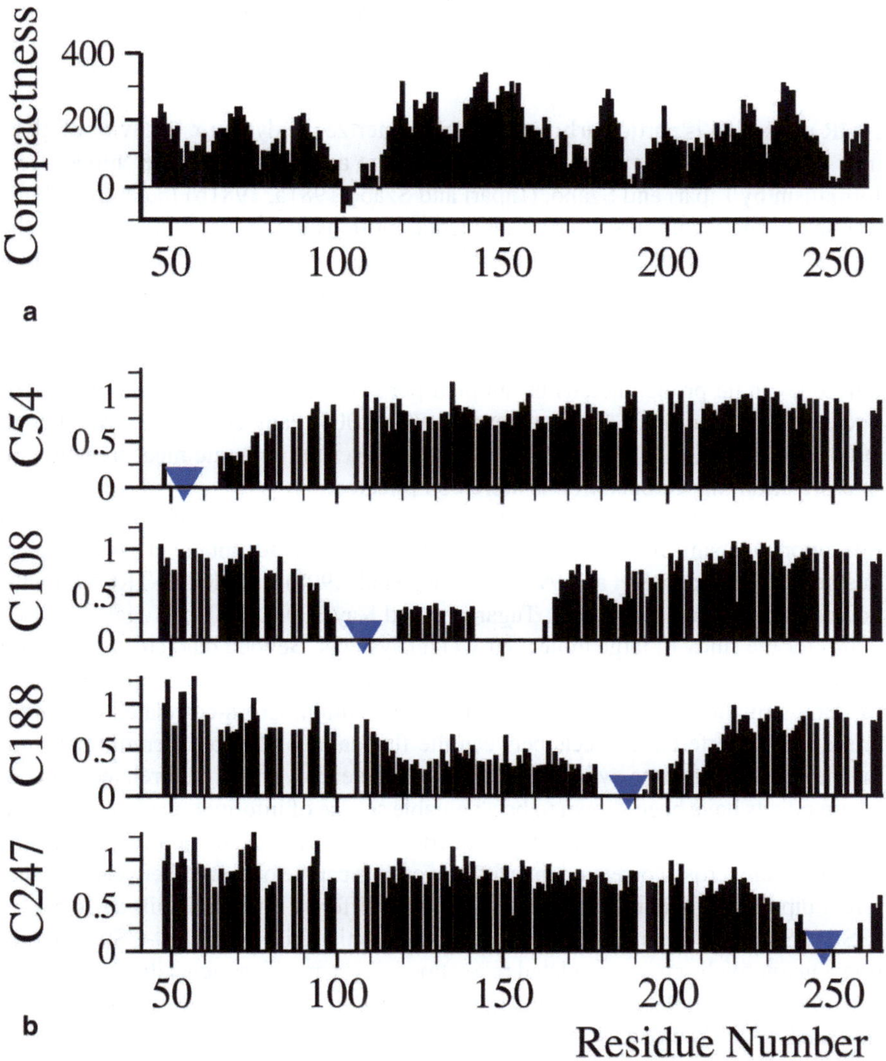

Fig. 5.5 Rational identification of appropriate spin labeling sites based on meta-structure derived compactness data (see text for details). Large compactness values are found for compact regions of the proteins, whereas small compactness values indicate conformationally flexible residue positions (and thus suitable for attaching the spin label) **a** Meta-structure derived compactness values and **b** experimental PREs obtained on the IDP Osteopontin. Spin label attachment sites were selected using small compactness values. (Platzer et al.)

(Platzer et al. 2011). The rationale behind the approach is that small compactness values indicate surface exposed residues in the protein structure and that the sites of spin label attachment should therefore be selected based on small compactness values as for these regions tight side chain interactions or packing can safely be neglected. Figure 5.5 shows compactness and PRE data for a prototypical IDP.

3.2 Cross-Correlated Relaxation (CCR) to Define Backbone Conformation

NMR relaxation is particularly suited to characterize the dynamic behavior of globular proteins especially in the [ps-to-ns] regime as described in the 'model-free' formalism by Lipari and Szabo. (Lipari and Szabo 1981a, 1981b) In contrast, IDPs frequently show complex internal segmental motions in the [ns] range and thus the separation of internal and global correlation times, a basic underlying assumption of the Lipari-Szabo and extended Lipari-Szabo (Clore et al. 1990) formalism is likely to be inadequate. In general ^{15}N relaxation is employed to characterize protein backbone dynamics, although it may only yield an incomplete picture, because only one single probe nucleus is sampled per residue. Other less commonly used nuclei are $^{13}C'$ and/or $^{13}C^{\alpha}$. (Chang and Tjandra 2005; Pang et al. 2002; Wang et al. 2005; Wang et al. 2006; Zeng et al. 1996) However the accurate measurement and interpretation of ^{13}C relaxation is more complicated.

A less commonly known approach uses the interference effects of cross-correlated relaxation mechanisms, which have been utilized to great advantage in TROSY (Pervushin et al. 1997; Riek et al. 1999; Salzmann et al. 1998) and Methyl-TROSY methodology (Tugarinov et al. 2003; Tugarinov and Kay 2004a, 2004b; Tugarinov et al. 2004) for the study of large molecular weight systems. Beyond that, cross-correlated NMR relaxation (CCR) has attracted substantial interest in the past as a powerful tool to study structure and dynamics of proteins in solution. Cross-correlated relaxation arises from interference effects between the fluctuations of two different relaxation mechanisms of rank two, which are active simultaneously and in a correlated manner. These effects have been shown to be a valuable source of information about structure and dynamics of proteins, since their concerted effect is related to their relative geometry. Typically cross-correlated interference effects can be observed between two different dipolar (DD) interactions (DD-DD), two different chemical shift anisotropies (CSA-CSA) or between a dipolar and a chemical shift anisotropy (DD-CSA) interaction. Qualitatively, cross-correlated relaxation effects manifest themselves as unequal relaxation of multiplet components in the absence of decoupling. Best known is the so-called TROSY effect, which arises from the correlated action of the ^{15}N ($^{1}H^{N}$) chemical shift anisotropy (CSA) and the ^{15}N-$^{1}H^{N}$ dipolar (DD) interaction. This results in an asymmetric broadening of the $^{15}N\{^{1}H^{N}\}$ doublet of a backbone amide moiety and thus the downfield component of the ^{15}N doublet and the upfield component of the ^{1}H doublet relaxes much more slowly than the other half of the multiplet by an amount 2η. In the simple case of an amide N-H^{N} moiewty:

$$\eta = {}^{1}/_{15} * (\mu_0 / 4\pi)\gamma_N \gamma_H \hbar / r_{NH}{}^{3} * \gamma_N B_0 (\sigma_{\|} - \sigma_{\perp})\{4J(0) + 3J(\omega_N)\}{}^{1}/_{2}\,(3\cos^2\theta - 1)$$

where ($\sigma_{\|}$-$\sigma_{\perp}$) is the chemical shift anisotropy of the nitrogen (assumed to be axially symmetric in this case) and θ is the angle between the orientation of $\sigma_{\|}$ and the NH bond vector (typically <20°) all other symbols have their usual meaning. Normally, $J(0) >> J(\omega_N)$ and $J(0) = S^2\tau_c$ (assuming isotropic re-orientation). Thus either geometric (θ) or, in the case of known geometry, dynamic information comprising S^2 and/or τ_c can be extracted from the analysis of cross-correlated relaxation. (Tjandra et al. 1996) In general, cross-correlated effects manifest themselves through relaxation-mediated interconversion (Pelupessy et al. 1999; Pelupessy et al. 2003a, 2003b; Schwalbe et al. 2001) between in- and anti-phase coherences depending on the initial conditions and the active relaxation mechanisms with their respective Hamiltonians. Of particular interest for structural studies are cross-correlated relaxation effects, which are centered on different atoms, with one or more rotatable bonds in between, as their dependence on relative orientation allows determination of intervening torsion angles (Reif et al. 1997). These can be either cross-correlation rates between two dipolar interactions, which are typically the N-H^N and C^α-H^α bond vectors of a protein backbone (Kloiber et al. 2002; Reif et al. 1997), or cross-correlations between a chemical shift anisotropy tensor, typically the carbonyl atom, and a dipolar interaction, e.g., the C^α-H^α interaction (Yang et al. 1998; Yang et al. 1997). For a compilation of common cross-correlated relaxation experiments and their applications see (Schwalbe et al. 2001).

Dipole-Dipole cross-correlation rates between adjacent N-H^N and C^α-H^α bond vectors can be calculated as (Pelupessy et al. 1999; Reif et al. 1997)

$$\Gamma_{C\alpha H\alpha,NH} = {}^2/_5 (\mu_0 / 4\pi)^2 \gamma_C \gamma_N \gamma_H{}^2 \hbar^2 / (r_{CH}{}^3 r_{NH}{}^3) S^2 \tau_c{}^1/_2 (3\cos^2\theta - 1)$$

where all symbols have their usual meaning. The projection angle θ between the two bond vectors can be related to the backbone dihedral angles φ or ψ (assuming standard trans-peptide geometry) according to

$$\cos(\theta) = 0.163 + 0.819\cos(\psi - 120)$$

for inter-residue, sequential N-H^N(i), C^α-H^α(i-1) and

$$\cos(\theta) = 0.163 - 0.819\cos(\varphi - 60)$$

for intra-residue N-H^N(i), C^α-H^α(i) pairs (Reif et al. 1997).

Dipole-CSA cross-correlated rates between the CSA tensor of the carbonyl and the C^α-H^α dipolar interactions can be calculated as: (Goldman 1984; Yang et al. 1997)

$$\Gamma_{C',C_\alpha H_\alpha} = {}^4/_{15} (\mu_0 / 4\pi)\hbar \gamma_C \gamma_H / r_{CH}{}^3 * \gamma_C B_0 S^2 \tau_c{}^1 (f_x + f_y + f_z)$$

where the $f_{x,y,z}$ are the projections of the dipolar vector onto the carbonyl CSA tensor

$$f_{x,y,z} = {}^1/_2 * \sigma_{xx,yy,zz} (3\cos^2\theta_{x,y,z} - 1).$$

These projection angles are again related to the Ramachandran angle φ (in case of sequential, inter-residue C′(i-1)- C^α-H^α(i)) or ψ (intraresidue C′ (i-1)- C^α-H^α(i-1) cross-correlation effects) as

$$\cos\theta_x = -0.3095 + 0.3531\cos(\psi - 120)/(\phi + 120)$$

$$\cos\theta_y = -0.1250 = 0.8740\cos(\psi - 120)/(\phi + 120)$$

$$\cos\theta_z = -0.9426\sin(\psi - 120)/(\phi + 120)$$

assuming standard, trans peptide geometry and standard parameters of the carbonyl CSA tensor (Kloiber and Konrat 2000). Thus cross-correlated rates can be related to torsion angles φ and ψ through 'pseudo Karplus' relations. It should be noted that all cross-correlated rates scale with their effective order parameter. Experimentally, cross-correlated rates can be obtained through the relaxation of multiple-spin coherences (ZQ/DQ) involving the nuclei of interest (e.g. $^{15}N/^{13}C^{\alpha}$ or $^{13}C'/^{13}C^{\alpha}$). It is possible to obtain the extent to which different interactions involving these spins are correlated to each other, and thus geometric information regarding the intervening dihedral angle(s) can be extracted.

CCRs have therefore become a powerful tool in solution-state NMR for obtaining torsion angle restraints. Since their introduction a number of experiments have been developed to obtain CCRs and thus φ and ψ angles in proteins. In contrast to methods based on ^3J scalar couplings, in theory, at least, no calibration is required for structural interpretation. For folded, globular proteins their order parameter S^2 is well-defined and fairly uniform within structured regions. In contrast, this assumption may not be justified for IDPs and thus complications arise and the interpretation of cross-correlated relaxation should be done in a qualitative or semi-quantitative way. A number of experiments have been described to measure cross-correlated relaxation rates in proteins. (Schwalbe et al. 2001). Among these, two basic types of experiments can be distinguished:

I. The cross-correlated rates are active during a coupled evolution period so that a multiplet lineshape is observed in an indirect spectral dimension. If it is of the constant time (CT) type, the cross-correlation rates are extracted from the amplitudes of the multiplet structure of the cross peak as ratios of individual multiplet components. This is also referred to as J-resolved CT-Γ spectroscopy. If the evolution period is not a constant time (CT) delay, then it is rather the linewidths of the multiplet components that is individually affected.
II. The cross-correlated relaxation rates are active during a constant non-evolution delay. In that case couplings are refocused and the cross-correlated rates are determined from the intensity ratios of two different data sets: one, in which the decay of the original state is observed (the so-called 'reference' experiment) and one, in which the resulting state arising from relaxation-mediated conversion is observed (the so-called cross experiment). This is usually referred to as quantitative Γspectroscopy. A typical experiment for the measurement of $\Gamma_{(HN(i)-H\alpha C\alpha(i))}$ (Kloiber et al. 2002) is shown in Fig. 5.6.

To select, which CCR is active during a specified delay, their evolution can be controlled and refocused by suitably applied inversion pulses on one or more of the nuclei involved, in a manner similar to refocusing shifts and J-couplings in correlation spectra, except that one must be aware that the product of two Hamiltonians is effective and thus must consider the overall behavior of the CCR. (Chiarparin et al. 1999; Schwalbe et al. 2001) Due to the favorable (slow) relaxation properties of IDPs even small cross-correlated relaxation rates can normally be measured with sufficient precision, because long mixing delays do not result in severe S/N penalties. Problems with signal overlap due to their small shift dispersion are usually resolved through the use of multi-dimensional techniques, sometimes in combination with nun uniform

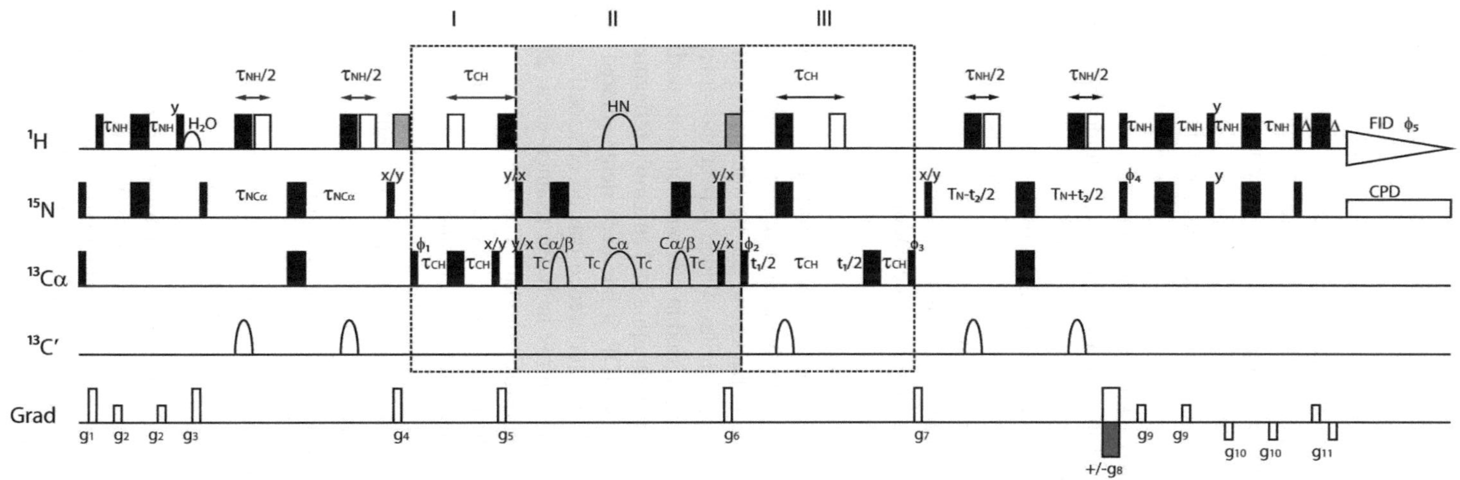

Fig. 5.6 Pulse sequence for the measurement of intraresidue N-H^N and C$^\alpha$-H$^\alpha$ dipole-dipole cross-correlated relaxation in doubly labeled proteins. Narrow and wide rectangles indicate 90° and 180° pulses, round elements indicate selective and shaped pulses. The ^{1}H/^{15}N/^{13}C carriers were centered at H_2O (4.70 ppm), center of the amide region (120.0 ppm) and center of $^{13}C^\alpha$ (56.0 ppm). The critical, simultaneous, selective inversion pulses on $^1H^N$ and $^{13}C^\alpha$ in the center of the relaxation period were of the RE-BURP type. The ^{1}H 180° pulses indicated by grey rectangles serve to restore ^{1}H magnetization to its equilibrium. This quantitative Γ experiment is of the HNCA type. The $^{13}C^\alpha$/^{15}N DQ/ZQ period (4T$_C$, II), during which interference between the N-H^N and C$^\alpha$-H$^\alpha$ dipoles is active, is shown in grey. Typically two experiments are recorded which select either $2C^\alpha_x N_x$ or $8C^\alpha_y N_y H^\alpha_z H^N_z$ at the end of the relaxation period 4T$_C$. When the double anti-phase term is selected by shifting the phases of the ^{13}C and ^{15}N 90° pulses at the end of the relaxation period (II), the C$^\alpha$-H$^\alpha$ and N-H^N couplings are rephased for detection during ^{13}C shift evolution (period III) and during ^{15}N shift evolution by shifting the position if the ^{1}H 180° pulses to the positions indicated by the open rectangles. Alternatively, double antiphase can be prepared before the relaxation period (II) by shifting the position of the indicated ^{1}H 180° pulses during the ^{15}N to $^{13}C^\alpha$ INEPT and during the period (I) together with the phases of the ^{13}C and ^{15}N 90° pulses preceding period II). Delays: τ_{NH}, τ_{CH}, $\tau_{NC\alpha}$, T_C, T_N are 2.25, 1.6, 12.5, 7.0 and 12.5 ms. Phase cycling φ_1=4(x),4(−x), φ_2=x, −x, φ_3=2(x),2(−x), φ_4=x, φ_5=x,−x,−x,x,−x,x,x,−x. Quadrature detection in F1is achieved by States-TPPI (Marion et al. 1989) on φ_2, whereas quadrature detection in F2 is achieved by inverting the sign of gradient g8 together with the phase φ_4 according to Rance-Kay (Kay et al. 1992)

sparse sampling (NUS) of the indirect spectral dimensions. (Konrat 2014) Normally, a single experimental cross-correlation rate does not provide unambiguous geometrical information but due to the inherent symmetry of the DD and CSA interactions is consistent with several dihedral angles. To resolve these ambiguities a method has been suggested exploiting the simultaneous analysis of different complementary $^{1}H^{N}$-^{15}N and $^{1}H^{\alpha}$-$^{13}C^{\alpha}$ DD and $^{13}C'$ D-CSA cross-correlation rates for the extraction of unambiguous and reliable dihedral angles along the protein backbone. A typical implementation of this protocol simultaneously utilizes $\Gamma_{C\alpha H\alpha(i),NH(i)}$, $\Gamma_{C\alpha H\alpha(i-1),NH(i)}$, $\Gamma_{C'(i-1),\,C\alpha H\alpha(i-1)}$, $\Gamma_{C'(i-1),\,C\alpha H\alpha(i)}$, $\Gamma_{C\alpha H\alpha(i-1),\,C\alpha H\alpha(i)}$ rates together with some limited qualitative ^{3}J-scalar coupling information (either $^{3}J_{C'C'}$ or $^{3}J_{HNH\alpha}$). This is sufficient to determine torsion angles along a protein backbone. Furthermore, this approach has been shown to be reasonably robust and insensitive to small amplitude dynamics (Kloiber et al. 2002). It has been demonstrated that in favorable cases determination of protein backbone folds from cross-correlation spectroscopy, supplemented with limited J-coupling information is feasible. In the case of structured, globular proteins application of cross-correlation derived dihedral restraints is quite straightforward in structure determination protocols by applying them as dihedral restraints. The situation is more complicated when applying such methodology to IDPs due to their dynamic behavior. In IDPs small S^2 order parameters flatten out the pseudo-Karplus curves, which relate them to dihedral angles, and complicate quantitative interpretation of cross-correlation rates. Due to large-scale conformational dynamics only information regarding time- and ensemble-averaged backbone conformation and conformational propensities of the protein backbone can be obtained. Still this may be useful in defining conformational ensembles, which reflect the structural propensities obtained by NMR, especially, when used in combination with modified structure calculation protocols including time-averaging and ensemble refinement.

As a first example of an application to an IDP, it has been demonstrated that intra-residue $^{1}H(i)$-$^{15}N(i)$-$^{13}C'(i)$ dipolar-CSA interference can be efficiently used to discriminate between type-I and type-II β-turns in IDPs. (Stanek et al. 2013) The experiment is based on a relaxation pathway originally designed for measurements of dihedral angles in globular proteins. To improve spectral resolution the experiment was run as a 4-dimensional experiment and combined with non-uniform sampling techniques required in order to overcome the spectral overlap problem encountered in IDPs. Since IDPs populate Ramachandran space in a rather unique way and substantially sample β-turn (I, II) and polyproline II helical conformations, this novel experimental approach can be efficiently used to assess these (non α-helical, non β-strand) conformations in IDPs.

In this first application the experiment was also used to detect subtle local structural changes in IDPs upon pH-induced structural compaction. (Stanek et al. 2013) In another application a set of five cross-correlated rates ($\Gamma_{C\alpha H\alpha(i),NH(i)}$, $\Gamma_{C\alpha H\alpha(i-1),NH(i)}$, $\Gamma_{C'(i-1),C\alpha H\alpha(i-1)}$, $\Gamma_{C'(i-1),C\alpha H\alpha(i)}$, $\Gamma_{C\alpha H\alpha(i-1),C\alpha H\alpha(i)}$) was measured for the partially unstructured protein Myc, which acts as an proto-oncogenic transcription factor and belongs to the basic-helix-loop-helix-zipper (bhlhzip) protein family, to explore the feasibility of CCR based structure determination protocols and to characterize its conformational propensities utilizing CCRs.

Figure 5.7 shows CRR data for residue Gln 125 of Myc, which displays a characteristic α-helical signature in its NOE pattern and $^{13}C^{\alpha}$ secondary chemical shift

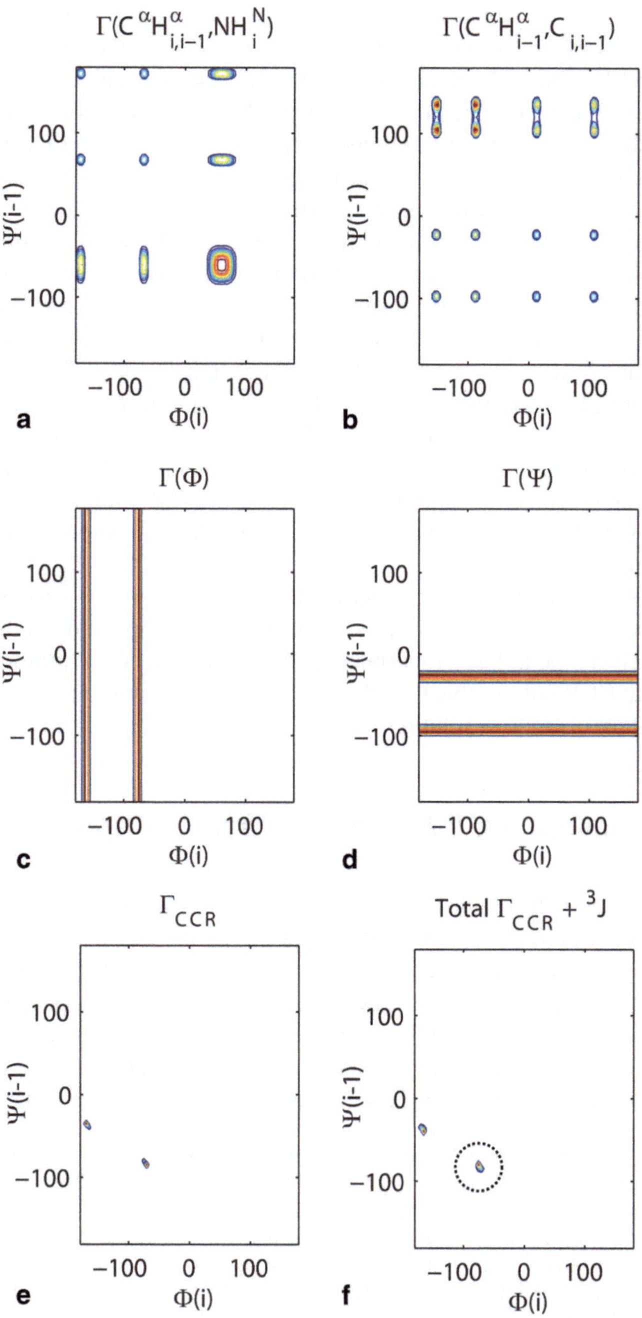

Fig. 5.7 Automated backbone dihedral angle determination using cross-correlated relaxation (CCR). The figure shows a schematic description how the combined usage of different, complementary cross-correlation rates can be used to unambiguously identify the backbone dihedral angles φ and ψ. The figure demonstrates the generation of the dihedral angles probability surface or Z-surface for residue Gln 125, which shows a clearly α-helical signature, of the partially unstructured protein Myc. **a** combined Z-surface for D-D cross-correlated relaxation:

in agreement with previous structural investigations (Fieber et al. 2001, Nair and Burley 2003). Beyond structural information, cross-correlated relaxation provides a versatile tool to study the dynamics of protein backbone and anisotropy, because they report on the generalized order parameter S^2. Problems arise though in terms of the possible complexity of a full description of peptide plane dynamics. Recent studies utilizing multi-nuclear relaxation (including $^{13}C'$ and $^{13}C^{\alpha}$) have hinted at substantial motional anisotropy of the peptide moiety, which makes it very difficult to consistently integrate ^{15}N and ^{13}C relaxation data into a unified model of protein dynamics. As cross-correlated relaxation usually involves interactions, which point into different orientations in three-dimensional space, it could help to better define the anisotropic dynamics of peptide re-orientation and define amplitudes and timescales of dynamic modes in three-dimensional space. The magnitude and precise nature of anisotropy of peptide dynamics is, yet, a matter of controversy (Bytchenkoff et al. 2005; Carlomagno et al. 2000; Chang and Tjandra 2005; Vogeli and Yao 2009). Furthermore, cross-correlated relaxation could be ideally suited to study long-range collective motions, typical of segmental motions in IDPs with the only limitation of efficient excitation of long-range MQCs in sequentially adjacent peptide fragments

Given the sensitivity of CCR experiments to subtle structural changes, together with the diversity of CCR experiments, which are at the disposition of the investigator and which can be tailored to a specific structural or dynamic problem, it can be expected that CCR will be able to make a substantial contribution to the study of the dynamic nature of IDPs in solution.

3.3 *A Combined NMR/EPR Approach for IDP Research*

In addition to only NMR-based approaches and due to the fact that experimental parameters measured by electron paramagnetic resonance (EPR) and NMR spectroscopy depend differently on motional averaging, a novel approach to look at IDPs was recently proposed (Kurzbach et al. 2013). While solution NMR provides ensemble averages, pulsed EPR spectroscopy is performed at low temperature where transitions between different states are quenched and individual states can be probed. In a first proof of principle the methodology was applied to the IDP

$Z^{D-D}\{\Gamma^{DD}(H^{\alpha}C^{\alpha}_{i},H^{N}N_{i})(\varphi_{i}), \Gamma^{DD}(H^{\alpha}C^{\alpha}_{i-1},H^{N}N_{i})(\psi_{i-1})\}$, **b** combined Z-surface for D-CSA cross-correlated relaxation $Z^{D-CSA}\{\Gamma^{DCSA}(H^{\alpha}C^{\alpha}_{i-1},C'_{i-1})(\psi_{i-1}), \Gamma^{D}(H^{\alpha}C^{\alpha},C')(\varphi_{i})\}$, **c** combined Z-surface for cross-correlated relaxation rates reporting on the backbone angle φ_i $Z^{\varphi i}\{\Gamma^{DD}(H^{\alpha}C^{\alpha}_{i},H^{N}N_{i})(\varphi_{i}), \Gamma^{DD}(H^{\alpha}C^{\alpha},C')(\varphi_{i})\}$, **d** combined Z-surface for cross-correlated relaxation rates reporting on the backbone angle ψ_{i-1} $Z^{\psi i-1}\{ \Gamma^{DD}(H^{\alpha}C^{\alpha}_{i-1},H^{N}N_{i})(\psi_{i-1}), \Gamma^{DCSA}(H^{\alpha}C^{\alpha}_{i-1},C'_{i-1}))(\psi_{i-1})\}$, **e** combined Z-surface for all cross-correlated relaxation rates $Z^{CCR}\{\Gamma^{DD}(H^{\alpha}C^{\alpha}_{i},H^{N}N_{i})(\varphi_{i}), \Gamma^{DD}(H^{\alpha}C^{\alpha}_{i-1},H^{N}N_{i})(\psi_{i-1}), \Gamma^{DCSA}(H^{\alpha}C^{\alpha}_{i-1},C'_{i-1})(\psi_{i-1}), \Gamma^{DD}(H^{\alpha}C^{\alpha},C')(\varphi_{i}), \Gamma^{DD}(H^{\alpha}C^{\alpha}_{i-1}, H^{\alpha}C^{\alpha}_{i})(\varphi,\psi_{i-1})\}$, **f** Combined total Z-surface including ^{3}J coupling information $Z^{Total}=Z^{CCR}*Z^{J}$. Of the remaining last two possible backbone conformations at position Gln 125 the correct, α-helical conformation, indicated in (f) can be easily identified, because the alternative solution falls far outside the allowed region of the Ramachandran diagram. Details can be found elsewhere (Kloiber et al. 2002)

Osteopontin (OPN), a cytokine involved in metastasis of several kinds of cancer. Conformational substates of OPN were probed by applying the EPR-based method double electron-electron resonance (DEER) spectroscopy to six spin-labeled Cys-double mutants of OPN. It is important to note that DEER experiments yield non-averaged data and reveal intra-molecular dipole-dipole coupling between the two spins of the labels of a double mutant. The detected signal modulation is related to the dipolar coupling frequency that in turn depends on the inter-spin distance as r^{-3}. It is important to note that the analysis tools established for stably folded proteins fail in the case of IDPs as a consequence of the rather broad pair-distribution functions between the two spin labels of a double mutant. To avoid over-interpretation (over-fitting) of data, the observed DEER data were quantitatively analyzed via an effective modulation depth, Δ_{eff}, that is an approximate measure of the average inter-spin distance for broad P(R) distributions. To probe structural stability of the IDP the Δ_{eff} values were measured as a function of urea concentration (Fig. 5.8). Several double mutants (C54-C108, C108-C188, C188-C247, C54-C188, C108-C247 and C54-C247) were prepared using low compactness values as selection criteria (see above). Most importantly, while most of the double mutants showed a smooth decrease upon urea denaturation, for the double mutant C54-C247 an unexpected sigmoidal Δ_{eff}-derived denaturation profile with urea concentration was observed (Fig. 5.8b). This unexpected and unprecedented sigmoidal urea dependence of an IDP clearly indicates significant populations of stably and cooperatively folded tertiary structures in the structural ensemble of OPN. The conformational ensemble of OPN thus contains both, cooperatively folded and unfolded, extended conformations. It is also important to note, that EPR and NMR (PRE) experiments under high NaCl concentrations showed that not only hydrophobic interactions contribute to the OPN's structural stability, but

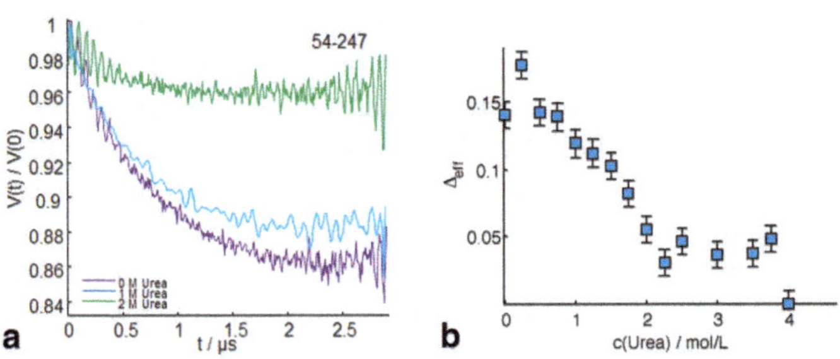

Fig. 5.8 Solution structural probing of IDPs using EPR-based double electron-electron resonance (DEER) spectroscopy (Kurzbach et al. 2013). **a** DEER time traces of the double Cys-mutant C54-C247 of the IDP Osteopontin (OPN) at different urea concentrations. The modulation depth, $\Delta_{eff}=1.0-V(t=3\mu s)/V(0)$ is a direct measure of structural compaction ((Kurzbach et al.)). Decreased Δ_{eff} at higher urea concentration is due to global unfolding of the protein. **b** Identification of cooperatively folded substates in the ensemble of the OPN by measuring Δ_{eff} for the OPN double Cys-mutant C54-C247 as a function of urea concentration compaction. The sigmoidal dependence of Δ_{eff} *vs* urea concentration clearly shows the existence of a cooperatively folded substate (Kurzbach et al. 2013). Error bars stem from signal noise

also electrostatics play a crucial role in the stabilization of compact structures of OPN in solution (Kurzbach et al. 2013). Taken together the data clearly indicated that the term *"intrinsically disordered"* does not apply for the *"IDP"* OPN but rather points to the subtle interplay of electrostatic and hydrophobic interactions for the realization of diverse structural ensembles of "*rheomorphic*" proteins (Holt and Sawyer 1993). The surprisingly detailed picture of the conformational ensemble of OPN obtained by this novel approach indicates valuable applications to studies of structural dynamics of IDPs.

3.4 NMR Studies of Excited States of Proteins

IDPs are characterized by rugged energy landscapes devoid of distinct energy barriers and therefore display significant structural plasticity and undergo large structural rearrangements. A comprehensive characterization of the solution structures of IDPs thus requires studies of conformational dynamics. NMR spectroscopy is destined for these studies and a plethora of different experiments are available providing detailed information about motional dynamics on different time scales. Fast (ps-ns) time scale motions are probed by ^{15}N spin relaxation experiments (^{15}N-T_1,T_2 and ^{15}N-$^{1}H^{N}$ NOEs) and analyzed using well-established theoretical frameworks (e.g., the 'model-free' formalism (Lipari and Szabo 1981a)). Slower motions occurring on μs-ms time scales are investigated by CPMG-type experiments introduced decades ago that turned into a powerful experimental methodology applicable even to very large molecular weight systems as demonstrated by Kay and co-workers (Baldwin and Kay 2009). The particular uniqueness of NMR spin relaxation measurements is the fact that detailed information about internal motions can be discerned. In case of globular, stably folded proteins the analysis relies on distinctly different correlation times describing overall tumbling and internal motions.

Over the past decades a multitude of Carr-Purcell-Meiboom-Gill (CPMG) relaxation dispersion (RD) techniques have been developed for the investigation of μs-ms exchange processes between different states in protein dynamics.(Kloiber et al. 2011; Korzhnev and Kay 2008; Korzhnev et al. 2004b; Schanda et al. 2008) These methods were developed primarily for folded substates. Yet they are applicable for intrinsically disordered proteins, too. The principle appearance of CPMG dispersion profiles (in the case of two-site exchange between states A and B) in the presence and absence of exchange phenomena is shown in Fig. 5.9a. The observed effective relaxation rate, $R_2^{eff}(\nu_{cp})$, is affected by the frequency of pulses in the CPMG train, ν_{cp}. A fit of the dispersion profile R_2^{eff} vs ν_{cp} yields the populations of the two states, P_A and P_B and the chemical shift difference, $\delta\omega$, and the rate or time constant of exchange k_{ex} or τ_{ex} between them. In the absence of (detectable) exchange a population-weighted average is detected of the individual $R_{2,i}$ values of the different interconverting species, i, present in the system. Typically, CPMG RD is used to detect sparsely populated states in two-site exchange systems (for which $P_A >> P_B$ holds). Thus $R_{2,A}$ is effectively detected in the absence of exchange as base line value $R_{2,0}$. The principle pulse scheme for the detection of RD profiles of protein

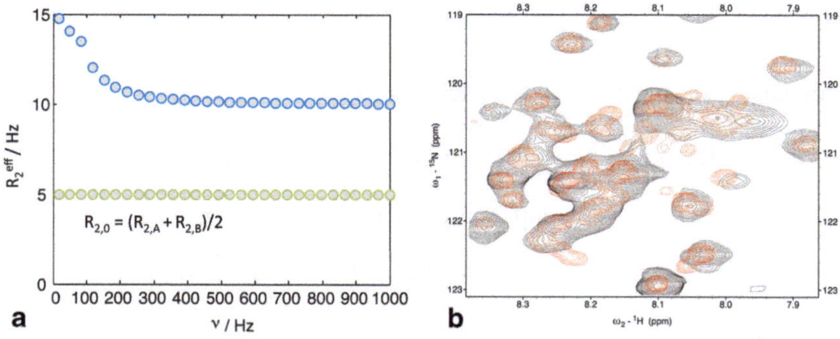

Fig. 5.9 **a** Calculated CPMG RD data in the presence (*blue*) and absence (*green*) of exchange **b** 2D ^1H-^{15}N HSQC like correlation obtained from the application of a two-dimensional pulse-sequence for measurement of CPMG relaxation dispersion data for the oncogenic protein MAX are shown at 500 and 800 MHz for comparison

backbone or side chain (Tollinger et al. 2001) X-^1H pairs (X = ^{15}N, ^{13}C)(Kloiber et al. 2011, Zintsmaster et al. 2008) are well-known. In general, a CPMG pulse train can be incorporated into a two-dimensional X-^1H correlation detection scheme yielding relaxation dispersion data for one of the two nuclei, depending on to which the CPMG train is applied. The time, τ_{cp}, between pulses in the CPMG pulse train is related to the observed (normalized) intensity by $I(\nu_{cp})/I_0 = \exp(-R_2^{eff}(\nu_{cp})T_{cp})$ with the pulse frequency $\nu_{cp} = (4\tau_{cp})^{-1}$ (Tollinger et al. 2001). I_0 denotes the reference cross peak intensity in the absence of a CPMG element in the pulse sequence and T_{cp} the constant duration of the CPMG block.

For intrinsically disordered proteins a certain technical problem is that CPMG RD data should be recorded at two different field strength to avoid underdetermination of data fits and to yield reliable sets of the parameters P_A, P_B and $\delta\omega\tau_{ex}$. At lower field strength (500–600 MHz) severe signal overlap poses an experimental problem, since the recorded data typically appear as 2D correlation spectra and IDP spectra (monomeric state at 35 °C) are confined to a narrow ppm range. In Fig. 5.9b a typical region of a 2D correlation spectrum of an intrinsically disordered protein, MAX (Myc associated factor X), extracted from a CPMG RD data set is shown both at 500 and 800 MHz for comparison. Since the constant length relaxation interval in CPMG pulse sequences has to be adjusted to the $R_{2,0}$ values of the dominant state of a protein of interest, it should further be taken into account that $R_{2,0}$ of IDPs is typically much smaller than that of folded proteins. Thus, T_{cp} might become unfamiliarly long.

Typically, three time regimes are specified in the realm of NMR: fast exchange, if $(\delta\omega\tau_{ex})^2 << 1$, intermediate exchange $(\delta\omega\tau_{ex})^2 \sim 1$ and slow exchange $(\delta\omega\tau_{ex})^2 >> 1$. τ_{ex} denotes the effective time constant of interconversion between the two states. For these regimes different models, comprising varying simplifications are appropriate for fitting CPMG RD data. From density matrix type of considerations one yields:

Fast exchange(Allerhand and Gutowsky 1964, Luz and Meiboom 1963):

$$R_{2,A}\mathrm{eff}(\nu_{cp}) = R_{2,A} + P_A P_B (\delta\omega)^2 \tau_{cp}$$

Intermediate exchange(Carver and Richards 1972):

$$\begin{aligned} R_{2,A}\mathrm{eff}(\nu_{cp}) = & -\tau_{cp}^{-1}\{-\tau_{cp}(R_{2,A} + R_{2,B} + k_A + k_B)/2 \\ & + \ln[(D_+ + \cos h^2\eta_+ - D_- + \cos^2\eta_-)^{1/2} \\ & + D_+ + \sinh^2\eta_+ - D_- + \sin^2\eta_-)^{1/2}]\} \end{aligned}$$

$$D_\pm = \pm 1 + 2(\psi + 2(\delta\omega)^2)/([\psi + 2\delta\omega(R_{2,A} - R_{2,B} + k_A - k_B)]^{1/2}$$

$$\eta_\pm = \tau_{cp}/4\sqrt{2}\{\pm\psi + [\psi^2 + (2\delta\omega(R_{2,A} - R_{2,B} + k_A - k_B))^2]^{1/2}\}^{1}/_{2}$$

$$\psi = (2\delta\omega(R_{2,A} - R_{2,B} + k_A - k_B))^2 - (\delta\omega)^2 + 4k_A^{-1}k_B^{-1}$$

Slow exchange(Tollinger et al. 2001):

$$R_{2,A}^{\mathrm{eff}}(\nu_{cp}) = R_{2,A} + k_A - k_A \sin(\delta\omega\tau_{cp})/(\delta\omega\tau_{cp})$$

k_A and k_B are the forward and backward exchange rate constants and $P_{A/B}=k_{A/B}^{-1}/(k_A^{-1}+k_B^{-1})$.

There are several useful software packages for the analysis of CPMG data in different exchange regimes (CATIA; http://pound.med.utoronto.ca/~flemming/catia/ or RD NMR http://reldispnmr.spinrelax.at). These programs provide simple graphical user interfaces and parameter setups facilitating data analysis. Typically, these programs assume that $R_{2,A} = R_{2,B}$. throughout the data fitting procedure. Yet, for IDPs $R_{2,A}$ might differ significantly from $R_{2,B}$, if the observed exchange comprises, e.g., the transition into a sparsely populated folded state in which local backbone flexibility drastically differs from the unfolded state. In such a case the respective relaxation rates of ^{15}N amides might differ by an order of magnitude. This might be due to a ligand induced folding or mere sampling of a quite heterogeneous conformational space. The case of different $R_{2,A/B}$ has been thoroughly investigated by Ishima and Torchia (Ishima and Torchia 2006; Ishima and Torchia 1999). They provide evaluation methods for the induced errors of fitted parameters, but also mention that these errors are generally not too large, even if the difference between $R_{2,A}$ and $R_{2,B}$ is significant. Thus, analysis of IDP transitions into folded states by means of CPMG RD can be performed too with the aid of the above-mentioned programs despite of the oversimplifying assumption that $R_{2,A} = R_{2,B}$.

IDPs are often involved into complicated substrate recognition and coupled folding and binding events. Thus, it might be that the nature of the exchange process is not clear a-priori, e.g., whether a two- or multi-site exchange is observed, e.g., due to ligand association and dissociation in combination with major conformational modifications (Kurzbach et al. 2014). In such a case double- and zero quantum CPMG analysis might be employed to screen the nature of the exchange process as suggested by Kay and coworkers (Orekhov et al. 2004). This technique is also applicable to improve data quality if amide proton spins are negatively influenced by contributions from neighboring protons (Korzhnev et al. 2004b).

Protein-ligand association and dissociation processes of IDPs are frequently too fast to be observed by means of relaxation dispersion, such that the exchange frequency is not covered by one of the above mentioned exchange regimes. Nevertheless, conformational transitions of IDPs might yield effects that effectively lead to exchange on all timescales. An example is MAX that partially exists as a monomeric IDP under physiological conditions and body temperatures that might, hence, be of physiological importance. Yet, MAX also populates a stable (well-known) homodimer comprising a coiled-coil motif with a leucine zipper subunit (Fieber et al. 2001; Sauve et al. 2004).

Figure 5.10a shows RD traces for two of MAX residues, S22 and E49 at 35 and 40 °C. Distinct changes in relaxation dispersion profiles as a function of temperature can be observed hinting towards temperature dependent conformational sampling of the folded state. Figure 5.10b shows the correlation between $\delta\omega(^{15}N)$ gained from single-quantum CPMG RD data fitting and from the frequency difference of 2D correlation spectra of both the pure disordered MAX monomer and the pure folded homodimer. Agreement of $\delta\omega$ indicates a reasonable data fit and thus validates the choice of the model. Each data point corresponds to one single residue. The appearance of $\delta\omega(^{15}N)$ in a 2D correlation plane is exemplarily illustrated in Fig. 5.9b. Error bars arise either from uncertainties of the numerical data fit (cf. Fig. 5.10a) or from uncertainties in resonance frequency of the low-populated state, since corresponding 2D correlations spectra for both species can only be detected at drastically different experimental conditions that favor this (otherwise, under CPMG conditions low-populated) state.

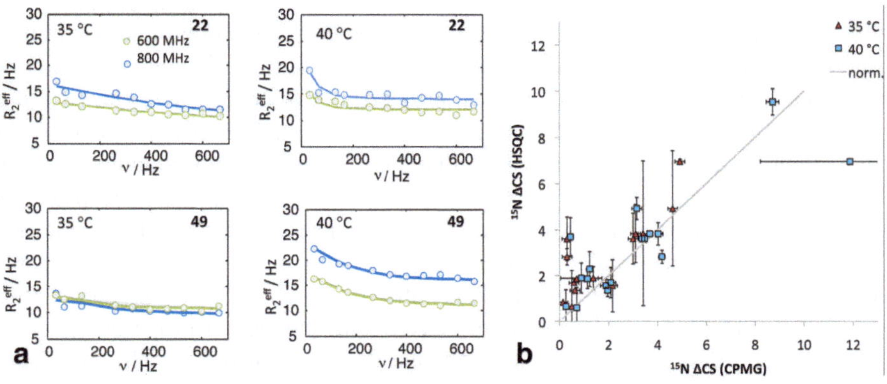

Fig. 5.10 **a** Relaxation dispersion profiles of Max residues S22 and E49 at 35 and 40 °C **b** correlation of ^{15}N-Chemical shift differences from HSQC data and fits of CPMG traces. A reliable fit will yield the shifts obtained from 2D correlation spectra

Further, note that IDPs frequently show complex internal motions and the classification of internal and global correlation times by means of the Lipari-Szabo and extended Lipari-Szabo formalism is likely to be insufficient (Rule and Hitchens 2006). Thus, in order to gain correct exchange rates between folded and unfolded state and their populations not only single residue fits should be taken into account but a global data fit of all available RD profiles by only one set of P_a and P_b, while only $\delta\omega$ varies residue-dependent, is reasonable to perform (Sára et al. 2014). Global fitting of the data shown in Fig. 5.3 yield a population of the MAX dimer of ~0.9% at 35 °C. In general populations as low as 0.5% can be assessed and studied by means of CPMG RD (Korzhnev and Kay 2008; Mittag and Forman-Kay 2007; Sugase et al. 2007). However, it should be noted that up to today conformational transitions of IDPs, i.e., conformational sampling, taking place on the detectable time regime by CPMG at ambient temperatures were only rarely reported (Neudecker et al. 2006; Tollinger et al. 2006). Although CPMG is a promising alternative for the characterization of transient conformational sub-states, IDPs frequently sample their accessible conformational spaces too fast in order to be detected by means of RD. Other solution-state, NMR-based techniques, yet, still yield ensemble-averaged data sets in most cases and low-populated sub-states become inaccessible by these means. The observation of conformational sampling with MAX, however, shows that in some cases, e.g., of sampling of rigidly folded structures, conformational sampling of IDPs is well within the time scale of CPMG experiments. IDP-ligand association and dissociation processes, yet, frequently take place on time scales that are ideally suited for CPMG RD detection of the low-populated (bound/free) state of the IDP (Sugase et al. 2007), since it is within the very nature of IDPs to form transient complexes—often associated with regulatory cell function—with their natural targets.

3.5 Post-Translational Modifications in IDPs

Post-translational modifications (PTMs) are an essential step in protein maturation and biosynthesis that allows a tight regulation of the activity, interactions, cellular location, lifetime and physico-chemical properties of proteins. Thus, knowing and understanding the impact of PTMs on proteins is essential in order to properly address their function, and becomes particularly critical for the understanding of IDPs physiological properties. Indeed, IDP and disordered regions are more prone to PTMs than folded proteins (Khoury et al. 2011), and due to their ability to interact with multiple partners, IDPs are often acting as hubs in intricate regulation pathways (Liu et al. 2009). Thus, PTMs will contribute to this versatility by modulating and regulating their cellular location, activity and interaction properties (Deribe et al. 2010). Additionally, due to the inherent flexibility of IDPs, PTMs will have large impacts on their conformational ensemble and alter their structural dynamics (Errington and Doig 2005; Maltsev et al. 2012; Mao et al. 2010; Meyer and Möller 2007; Theillet et al. 2014). Unfortunately, so far, an overwhelming majority of NMR studies completely neglect the impact of PTMs on the properties and structural dynamics of IDPs and are consequently devoid of biological significance.

The traditional argument for neglecting PTMs is that biomolecular NMR requires large amount of isotopically labeled and homogeneously modified protein, which (i) cannot be obtained by the use of recombinant technologies that lack the enzymatic machinery involved in mammalian PTMs (ii) rules out the usage of mammalian expression systems where the costs will be too high, the expression yield too low and the PTMs probably heterogeneous. However, recently developed chemical and biochemical tools allow the homogeneous modification of large amounts of isotopically labeled recombinant protein either during or subsequently to recombinant expression.

Phosphorylation is one of the most common PTMs, and the easiest to achieve *in vitro*. IDP phosphorylation results in a modified net charge and strongly affects the structural dynamics of the IDP (Errington and Doig 2005; Mao et al. 2010). Phosphorylation will often regulate interaction properties. For example, site-specific phosphorylation of GAP-43 and BASP-1 hampers their interactions with calmodulin and phosphorylation of MARCKS will regulate its interaction with calmodulin and F-actin (Liu et al. 2009; Maekawa et al. 1994; Tejero-Diez et al. 2000). Additionally, many IDPs are hyperphosphorylated. For example, hyperphosphorylation of the neuronal protein *tau* abolishes its interaction with microtubules and promotes the formation of pathogenic paired helical fragments (Beharry et al. 2014); hyperphosphorylation of the secreted osteopontin is required for its interaction with Ca^{2+} and CD44 (Hunter 2013; Kazanecki et al. 2007) (Fig. 5.11). Finally, the most striking example of IDP hyperphosphorylation comes from the work of Mittag et al. on the protein Sic1 (Borg et al. 2007; Mittag et al. 2010). They could show that the intrinsic dynamics of Sic1 combined with multiple phosphorylation generates a mean electric field mediating the interaction with Cdc4 in an ultrasensitive manner, in order to form a so-called "fuzzy complex" (Mittag et al. 2008). This last example clearly illustrates the futility of studying unmodified polypeptide and the fascinating intrinsic versatility of IDP. *In vitro* phosphorylation of recombinant protein can be relatively easily achieved using either purified enzyme (recombinant or endogenous) or even cell extracts and a plethora of protocols have already been published (Landrieu et al. 2006; Selenko et al. 2008). Additionally, a noticeable recent application of fast acquisition NMR is to monitor *in vitro* phosphorylation in a site specific and time resolved manner (Theillet et al. 2013; Liokatis et al. 2010).

Secreted IDPs, or disordered part of transmembrane proteins, are very likely to undergo glycosylation and/or tyrosine sulfation. Both modifications are expected to have significant effects on the structural dynamics and interaction properties of IDPs (Meyer and Möller 2007). Tyrosine sulfation in particular has been shown to regulate many extracellular interactions (Kehoe and Bertozzi 2000). Constitutively uniformly N-glycosylated (in a mammalian-like fashion) and isotopically labeled (^{15}N and/or ^{13}C) proteins can be obtained by expression or fermentation in *Pichia pastoris* (Guo et al. 2001; Wood and Komives 1999), an expression system that is also well suited for proteins rich in disulfide bridges. Expression of recombinant

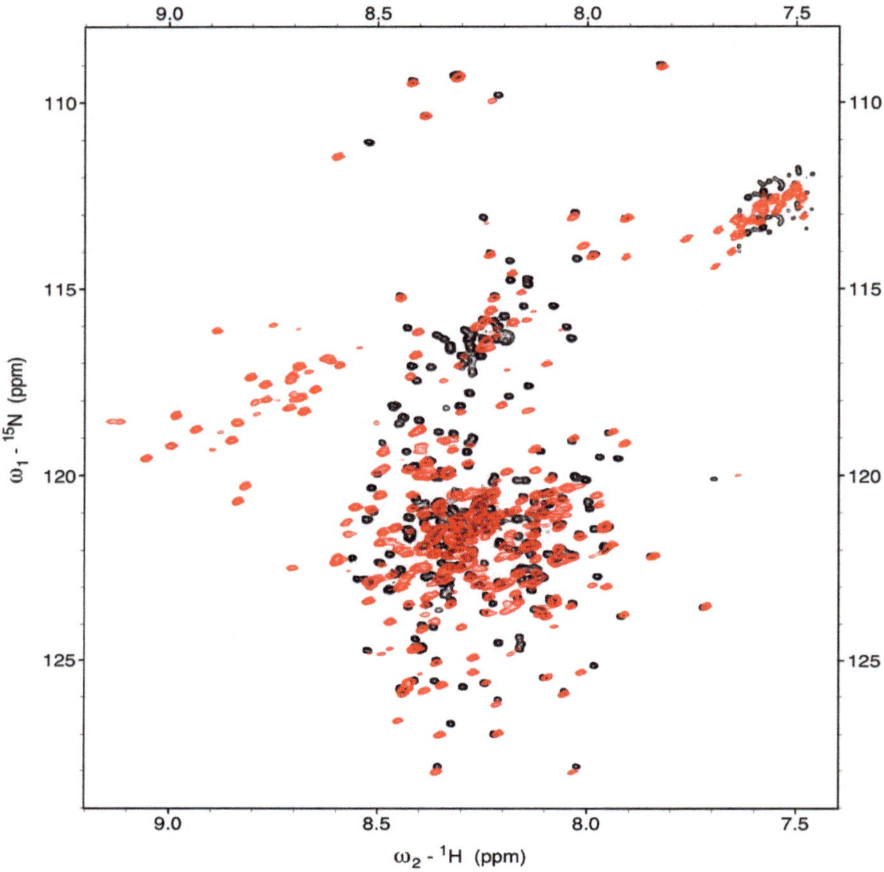

Fig. 5.11 Effect of Ser/Thr phosphorylation on IDPs. Overlay of the ^{1}H-^{15}N HSQC spectra of unmodified osteopontin (*black resonances*) and hyperphosphorylated osteopontin (*red resonances*)

isotopically labeled protein containing sulfotyrosine on specific positions can also fairly easily be achieved in *E. coli* by exploiting an expanded genetic code that inserts sulfotyrosine in response to the amber nonsense codon (Liu et al. 2009). The elegant usage of the amber codon has been perfected by Liu and co-workers who designed a plasmid that encodes for both the $tRNA_{CUA}$ and a modified aminoacyl-tRNA syntethase specific for sulfotyrosine. Interestingly, this general strategy can be used to co-translationally insert non-natural amino acids in proteins expressed in *E. coli* and *Pichia pastoris* (Young and Schultz 2010). Of particular interest for the biomolecular magnetic resonance community is the usage of this strategy to co-translationally incorporate spin labels and allows the straightforward measurement of NMR-PRE and EPR data (Fleissner et al. 2009).

Acylation is a very common PTM where the length of the fatty acid attached to the protein can vary from C2 (acetyl) to C14 or C16 (mirystoyl or palmitoyl). Many neuronal IDPs appear to be acylated; α-synuclein experiences an N-terminal acetylation (Bartels et al. 2011), MARCKS and BASP-1 are myristoylated (Zakharov et al. 2003) (Fig. 5.12), GAP-43 is palmitoylated (Arnaudon et al. 1993). N-terminal acylation, mirystoylation and palmitoylation are generally involved in sub cellular trafficking and membrane association, the longer the acyl chain the more likely the protein is to interact with membranes, but it has also been shown that N-terminal acylation of α-synuclein enhances the helical propensity of the N-terminal part of the protein (Maltsev et al. 2012) and increases the affinity of α-synuclein towards calmodulin by a factor of 10 (Gruschus et al. 2013). Similarly, BASP-1 can only interact with calmodulin in its mirystoylated form (Matsubara et al. 2004). Palmitoylation of recombinant protein can be achieved *in vitro* using the relatively expansive palmitoyl-CoA (Veit 2000). The condensation to the reduced or activated thiol group of a cysteine residue is spontaneous and quantitative at slightly basic pH. Kim et al. recently proposed an alternative cheaper method in order to mimic palmitoylation by attaching a thioalkyl chain (Kim et al. 2014). N-terminal acetylation and mirystoylation, on the other hand, can be achieved directly in eukaryotic cells. In both cases, this is achieved by co-expressing in the cell the enzyme responsible for the PTM, the fission yeast NatB acetylation complex (Johnson et al. 2010) or the human N-myristoyl-transferase (Gluck et al. 2010).

Acetylation can occur on lysine residues, which are also prone to methylation. These modifications of the lysine side chains modulate protein interactions and are highly relevant in the well-known cases of the disordered N-terminal "tail" domains (NTDs) of the core histones and the C-terminal tail domain (CTD) of linker histones, where lysine acetylation and methylation will modulate the interaction between the histone and its many partners (Latham and Dent 2007). Both modifications can easily be performed unspecifically *in vitro*, by condensation to Acetyl-CoA or by reductive methylation (Means and Feeney 1968). Considering the relative abundance of lysine residues it might be desirable to use a site-specific method. The production of recombinant protein acetylated or methylated on specific lysines is possible by using the same strategy as for the incorporation of non-natural amino acids. In the case of acetylation, the expression system is co-transformed with a vector containing the $tRNA_{CUA}$ and a N-acetyllysyl-tRNA syntethase and the incorporation of the acetylated lysine occurs co-translationally (Neumann et al. 2008). In the case of lysine methylation, a protected lysine is incorporated co-translationally at the desired position by co-transforming the expression system with a vector carrying the $tRNA_{CUA}$ and a modified tRNA syntethase. After purification of the protein, the other amino groups are protected using an orthogonal protecting group, the lysine side chain of interest can then be specifically deprotected, methylated *in vitro* and finally remove the orthogonal protection from the rest of the amino groups (Nguyen et al. 2010). Moreover, it is interesting to note that reductive methylation can be done with ^{13}C labeled formaldehyde, providing a very sensitive probe which can be exploited to study structure, dynamics or interactions by NMR (Geist et al. 2013). Finally, as for phosphorylation, fast acquisition NMR can be used to monitor *in vitro* acetylation and methylation in a site specific and time resolved manner (Liokatis et al. 2010; Theillet et al. 2012).

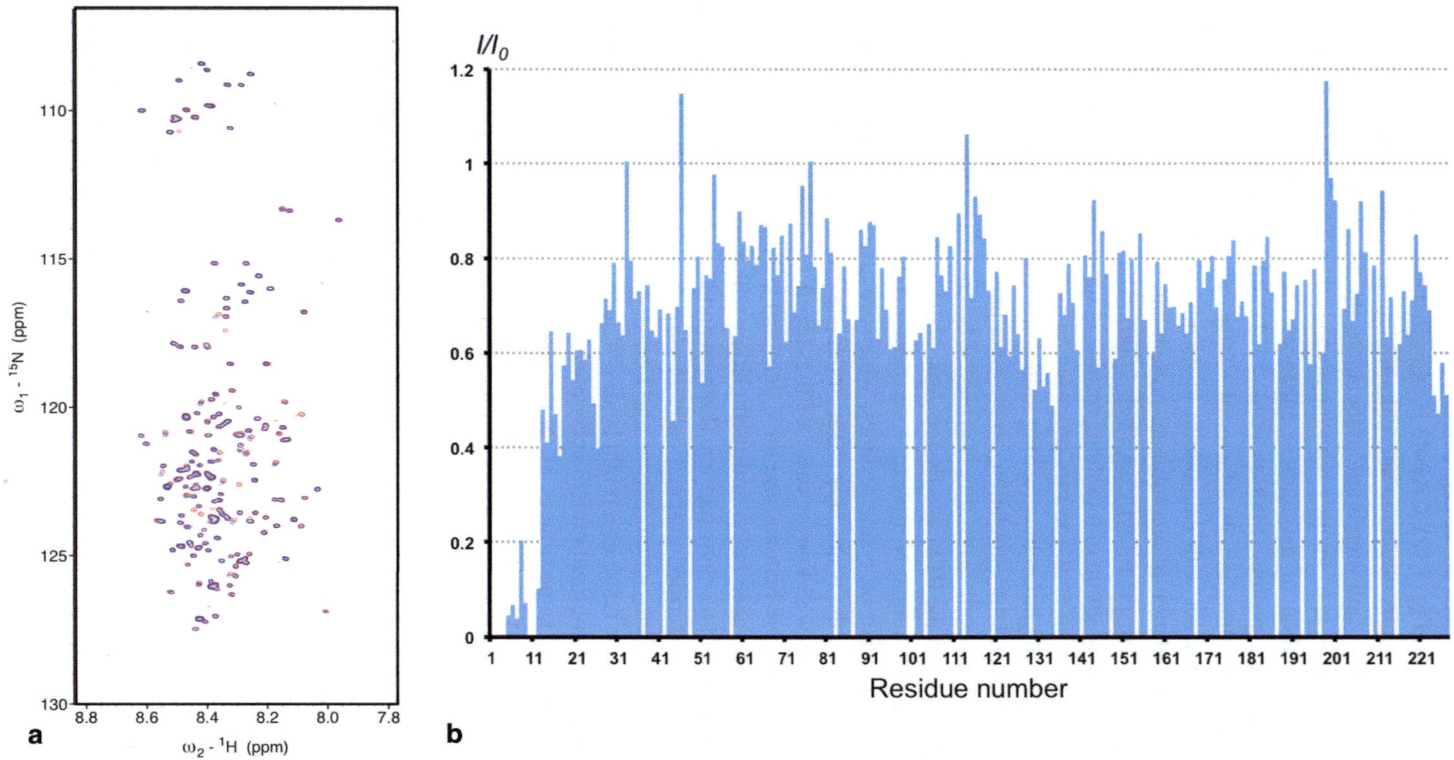

Fig. 5.12 Effect of acylation on IDPs. **a** Overlay of the ^{1}H-^{15}N HSQC spectra of unmodified BASP-1 (*blue resonances*) and myristoylated BASP-1 (*red resonances*). **b** Intensity ratio of the spectra between unmodified and myristoylated BASP-1. The 2 spectra are virtually identical but the intensity comparison reveals that N-myristoylation seems to promote the formation of micelles and also affect parts of the protein distant from the N-myristoylation site

To conclude, the last 10 years have witnessed the development of many chemical, biochemical and molecular biological tools that can be used to generate recombinant proteins with precisely controlled PTMs. Most of these tools will certainly benefit from further refinements and sophistications in the near future. Additionally, they are all suitable for biomolecular NMR as they allow for isotopic labeling, which opens exciting possibilities for biochemical studies and NMR–aided biochemistry of intrinsically disordered as well as natively folded proteins.

3.6 The IDP Conformational Ensemble

Although IDPs are generally annotated as unstructured/disordered there is ample NMR experimental evidence that the conformational space of IDPs is very heterogeneous and comprises both extended, marginally stable as well as stably, even cooperatively folded compact states with distinct side-chain interaction patterns (Kurzbach et al. 2013). The challenging problem in the structural characterization of IDPs is therefore the definition of a representative conformational ensemble sampled by the polypeptide chain in solution. To date the commonly used conceptual approaches are: (1) ensemble averaging using restrained MD simulations or Monte Carlo sampling incorporating experimental constraints as driving force and (2) the assignment of populations to a large pool of structures that have been *pre-generated* by employing different experimental constraints (e.g., PREs, chemical shifts, RDCs, SAXS) (Choy and Forman-Kay 2001; Bernadó et al. 2005; Fisher et al. 2010; Fisher and Stultz 2011; Ozenne et al. 2012). Due to limited sampling of the enormously large conformational space accessible to IDPs there are still doubts remaining about how representative the resulting ensembles are. It should be noted that a similar conclusion was made for the unfolded state of proteins (Rose et al. 2006), as experimental findings and theoretical considerations have provided evidence that the unfolded state is not a featureless structural ensemble but rather comprises distinct conformations retaining a surprisingly high degree of structural preformation. This structural preformation is a direct consequence of the existence of autonomously folded structural (sub)domains comprising basic structural elements (e.g., super-secondary structure elements, closed loops) (Levitt and Chothia 1976) (Berezovsky and Trifonov 2002; Trifonov and Frenkel 2009) (Rose et al. 2006). Detailed analysis of protein structures revealed that the fundamental building blocks of proteins typically consist of residue stretches of 20–25 amino acids length (Berezovsky and Trifonov 2002). A recent bioinformatics study revealed that protein structures can be regarded as tessellations of basic units (Parra et al. 2013). These data suggest a building principle capitalizing on the existence of pre-defined fundamental structural motifs that are combined in a combinatorial and (pseudo)-repetitive fashion. The inherent symmetry or higher order correlations in protein structures are also relevant in the context of energy landscape theory, as it was predicted that funnelled landscapes and low energy structures are more easily realized when symmetry prevails (Wolynes 1996). Since both IDPs and their folded counterparts share the same composing amino acids with similar physico-chemical properties it can thus be concluded that fundamental building principles of protein structures can be exploited for the generation of reliable and meaningful structural ensembles of IDPs by finding and using adequate sequence alignment techniques to identify structural homologues and existing basic motifs. The future strategy will rely on a pre-generated large pool of structures from which most suitable conformations are selected using experimental (e.g., PRE, chemical shifts, RDC, SAXS) constraints. To this end appropriate sequence alignments techniques will be necessary. Preliminary experiments suggest that meta-structure based sequence alignments of IDPs to sequences taken from the PDB structural database can indeed reveal hidden similarities and structural building blocks in IDPs that can be subsequently used to generate meaningful conformational ensembles.

4 Conclusion

IDPs seriously challenge classical structural biology that, historically, has emphasized only structural aspects of proteins, the spatial arrangements and mutual interactions of atoms in unique conformations. However, proteins do not exist in single conformations and are thus characterized by a funnel-like energy landscape and exchange between many different conformational isomers (substates). Fundamental biological processes involve protein interactions (protein domain exchanges/swapping, conformational adaptations or switches, induced-fit vs. conformational selection). Most importantly, the protein-funnel conceptual view provides a unified physical framework for globular proteins and IDPs. While stably folded, globular proteins display a smooth bottom with only few, very narrow, (structurally similar) minima, IDPs sample broad, but rugged energy surfaces with low barriers and a large number of accessible and energetically comparable minima. It is interesting to note, that the problem of characterizing IDPs has a parallel in the history of polymer science where the application of quantitative statistical mechanics allowed for the successful explanation of the dependence of physical properties of polymeric materials on molecular weight distributions (Dill 1999). For the characterization of the properties of IDPs, too, it seems that statistics will be required to fully grasp their physico-chemical properties that endow them with unique molecular properties and subsequent functionalities. Clearly, the comprehensive description of intrinsically disordered proteins will involve information about structure, dynamics and thermodynamics. As outlined in this chapter, NMR (in conjunction with EPR) can provide valuable information about cooperative effects in IDPs. An intriguing question to be addressed will be: "What is the relationship between the geometry of the complex IDP energy landscape and the nature of conformational transitions between different states?" While in stably folded proteins transitions between different conformational states often occur as (discontinuous) first-order phase transitions, IDPs will experience more complex phase transitions and conformational averaging might also proceed in a continuous manner. NMR has already been developed into a uniquely powerful technique to study conformational exchange processes (folding-unfolding processes, phase transitions) and provided unique insight into the structures and dynamics of low-populated (excited) protein states in solution. Although new computational tools and theoretical concepts will still be required to properly address the phase transition behavior of proteins, NMR spectroscopy is undoubtedly destined to play an important role in this fascinating area of structural biology research.

Acknowledgments The work of the author was supported in part by the FWF (P20549-N19 and W-1221-B03). The authors are grateful to all members of the group for providing experimental data, figures, valuable discussions and comments to the manuscript. The fruitful cooperations with Wiktor Kozminski (University of Warsaw), Bernhard Brutscher (ISB Grenoble) and Phil Selenko (FMP Berlin) and colleagues are also gratefully acknowledged.

References

Allerhand A, Gutowsky HS (1964) Spin-Echo NMR studies of chemical exchange.1. Some general aspects. J Chem Phys 41:2115

Allison JR, Varnai P, Dobson CM et al (2009) Determination of the free energy landscape of α-synuclein using spin label nuclear magnetic resonance measurements. J Am Chem Soc 131:18314–18326

Arnaudon L, Assmann R, Billan J et al (1993) Measurement of the mass of the Z-Boson and the energy calibration of lep. Phys Lett B 307:187–193

Baldwin AJ, Kay LE (2009) NMR spectroscopy brings invisible protein states into focus. Nat Chem Biol 5:808–814

Bartels T, Choi JG, Selkoe DJ (2011) α-Synuclein occurs physiologically as a helically folded tetramer that resists aggregation. Nature 477:107–110

Battiste JL, Wagner G (2000) Utilization of site-directed spin labeling and high-resolution heteronuclear nuclear magnetic resonance for global fold determination of large proteins with limited nuclear overhauser effect data. Biochemistry 39:5355–5365

Beharry C, Cohen LS, Di J et al (2014) Tau-induced neurodegeneration: mechanisms and targets. Neurosci Bull 30:346–358

Berezovsky IN, Trifonov EN (2002) Back to units of protein folding. J Biomol Struct Dyn 20: 315–316

Berjanskii M, Wishart DS (2006) NMR: prediction of protein flexibility. Nat Protoc 1:683–688

Bermel W, Bertini I, Chill J et al (2012) Exclusively heteronuclear ^{13}C-detected amino-acid-selective NMR experiments for the study of intrinsically disordered proteins (IDPs). Chembiochem 13:2425–2432

Bernadó P, Blanchard L, Timmins P et al (2005) A structural model for unfolded proteins from residual dipolar couplings and small-angle X-ray scattering. Proc Natl Acad Sci U S A 102:17002–17007

Bertini I, Felli IC, Gonnelli L et al (2011) High-resolution characterization of intrinsic disorder in proteins: expanding the suite of ^{13}C-detected NMR spectroscopy experiments to determine key observables. Chembiochem 12:2347–2352

Bibow S, Ozenne V, Biernat J et al (2011) Structural impact of proline-directed pseudophosphorylation at AT8, AT100, and PHF1 epitopes on 441-residue tau. J Am Chem Soc 133: 15842–15845

Borg M, Mittag T, Pawson T et al (2007) Polyelectrostatic interactions of disordered ligands suggest a physical basis for ultrasensitivity. Proc Natl Acad Sci U S A 104:9650–9655

Bytchenkoff D, Pelupessy P, Bodenhausen G (2005) Anisotropic local motions and location of amide protons in proteins. J Am Chem Soc 127:5180–5185
Camilloni C, De Simone A, Vranken WF et al (2012) Determination of secondary structure populations in disordered states of proteins using nuclear magnetic resonance chemical shifts. BioChemistry 51:2224–2231
Carlomagno T, Maurer M, Hennig M et al (2000) Ubiquitin backbone motion studied via NH^N–C`Cα Dipolar–Dipolar and C`–C`Cα/NH^N CSA-dipolar cross-correlated relaxation. J Am Chem Soc 122:5105–5113
Carver JP, Richards RE (1972) General 2-site solution for chemical exchange produced dependence of T^2 upon Carr-Purcell pulse separation. J Magn Reson 6:89
Chang SL, Tjandra N (2005) Temperature dependence of protein backbone motion from carbonyl ^{15}C and amide ^{15}N NMR relaxation. J Magn Reson 174:43–53
Chiarparin E, Pelupessy P, Ghose R et al (1999) Relaxation of two-spin coherence due to cross-correlated fluctuations of dipole-dipole couplings and anisotropic shifts in NMR of N-15,C-13-labeled biomolecules. J Am Chem Soc 121:6876–6883
Choy WY, Forman-Kay JD (2001) Calculation of ensembles of structures representing the unfolded state of an SH3 domain. J Mol Biol 308:1011–1032
Clore GM, Szabo A, Bax A et al (1990) Deviations from the simple 2-parameter model-free approach to the interpretation of N-15 nuclear magnetic-relaxation of proteins. J Am Chem Soc 112:4989–4991
Deribe YL, Pawson T, Dikic I (2010) Post-translational modifications in signal integration. Nat Struct Mol Biol 17:666–672
Dill KA (1999) Polymer principles and protein folding. Protein Sci 8:1166–1180
Dyson HJ, Wright PE (2004) Unfolded proteins and protein folding studied by NMR. Chem Rev 104:3607–3622
Errington N, Doig AJ (2005) A phosphoserine-lysine salt bridge within an α-helical peptide, the strongest α-helix side-chain interaction measured to date. Biochemistry 44:7553–7558
Fieber W, Schneider ML, Matt T et al (2001) Structure, function, and dynamics of the dimerization and DNA-binding domain of oncogenic transcription factor v-Myc. J Mol Biol 307:1395–1410
Fisher CK, Stultz CM (2011) Constructing ensembles for intrinsically disordered proteins. Curr Opin Struct Biol 21:426–431
Fisher CK, Huang A, Stultz CM (2010) Modeling intrinsically disordered proteins with bayesian statistics. J Am Chem Soc 132:14919–14927
Fleissner MR, Brustad EM, Kalai T et al (2009) Site-directed spin labeling of a genetically encoded unnatural amino acid. Proc Natl Acad Sci U S A 106:21637–21642
Gans P, Hamelin O, Sounier R et al (2010) Stereospecific isotopic labeling of Methyl groups for NMR spectroscopic studies of high-molecular-weight proteins. Angew Chem Int Edit 49:1958–1962
Geist L, Henen MA, Haiderer S et al (2013) Protonation-dependent conformational variability of intrinsically disordered proteins. Protein Sci 22:1196–1205
Gluck JM, Hoffmann S, Koenig BW et al (2010) Single vector system for efficient N-myristoylation of recombinant proteins in E. coli. PloS ONE 5:e10081
Goldman M (1984) Interference effects in the relaxation of a pair of unlike spin-1/2 nuclei. J Magn Reson 60:437–452
Goto NK, Gardner KH, Mueller GA et al (1999) A robust and cost-effective method for the production of Val, Leu, Ile (delta 1) methyl-protonated N-15-, C-13-, H-2-labeled proteins. J Biomol NMR 13:369–374
Gruschus JM, Yap TL, Pistolesi S et al. (2013) NMR structure of calmodulin complexed to an N-terminally acetylated α-synuclein peptide. Biochemistry 52(20):3436–3445
Guo RT, Chou LJ, Chen YC et al (2001) Expression in Pichia pastoris and characterization by circular dichroism and NMR of rhodostomin. Proteins 43:499–508
Guo CY, Geng C, Tugarinov V (2009) Selective backbone labeling of proteins using {1,2-C-13(2)}-pyruvate as carbon source. J Biomol NMR 44:167–173

Holt C, Sawyer L (1993) Caseins as rheomorphic proteins—interpretation of primary and secondary structures of the α-S1-Caseins, β-Caseins and κ-Caseins. J Chem Soc Faraday T 89:2683–2692

Hunter GK (2013) Role of osteopontin in modulation of hydroxyapatite formation. Calcif Tissue Int 93:348–354

Ishima R, Torchia DA (1999) Estimating the time scale of chemical exchange of proteins from measurements of transverse relaxation rates in solution. J Biomol NMR 14:369–372

Ishima R, Torchia DA (2006) Accuracy of optimized chemical-exchange parameters derived by fitting CPMG R_2 dispersion profiles when $R_2^{0a} \neq R_2^{0b}$. J Biomol NMR 34:209–219

Johnson M, Coulton AT, Geeves MA et al (2010) Targeted amino-terminal acetylation of recombinant proteins in *E. coli*. PloS ONE 5:e15801

Kay LE, Keifer P, Saarinen T (1992) Pure absorption gradient enhanced heteronuclear single quantum correlation spectroscopy with improved sensitivity. J Am Chem Soc 114:10663–10665

Kazanecki CC, Uzwiak DJ, Denhardt DT (2007) Control of osteopontin signaling and function by post-translational phosphorylation and protein folding. J Cell Biochem 102:912–924

Kazimierczuk K, Zawadzka A, Kozminski W (2009) Narrow peaks and high dimensionalities: exploiting the advantages of random sampling. J Magn Reson 197:219–228

Kazimierczuk K, Stanek J, Zawadzka-Kazimierczuk A et al (2010a) Random sampling in multidimensional NMR spectroscopy. Prog Nucl Magn Reson Spectrosc 57:420–434

Kazimierczuk K, Zawadzka-Kazimierczuk A, Kozminski W (2010b) Non-uniform frequency domain for optimal exploitation of non-uniform sampling. J Magn Reson 205:286–292

Kehoe JW, Bertozzi CR (2000) Tyrosine sulfation: a modulator of extracellular protein-protein interactions. Chem Biol 7:R57–R61

Khoury GA, Baliban RC, Floudas CA (2011) Proteome-wide post-translational modification statistics: frequency analysis and curation of the swiss-prot database. Scientific reports 1

Kim JH, Peng D, Schlebach JP et al. (2014) Modest effects of lipid modifications on the Structure of Caveolin-3. Biochemistry 53(27):4320–4322

Kloiber K, Konrat R (2000) Measurement of the protein backbone dihedral angle phi based on quantification of remote CSA/DD interference in inter-residue $^{13}C'(i-1)$-$^{13}C\alpha(i)$ multiple-quantum coherences. J Biomol NMR 17:265–268

Kloiber K, Schuler W, Konrat R (2002) Automated NMR determination of protein backbone dihedral angles from cross-correlated spin relaxation. J Biomol NMR 22:349–363

Kloiber K, Spitzer R, Tollinger M et al (2011) Probing RNA dynamics via longitudinal exchange and CPMG relaxation dispersion NMR spectroscopy using a sensitive C-13-methyl label. Nucleic Acids Res 39:4340–4351

Konrat R (2009) The protein meta-structure: a novel concept for chemical and molecular biology. Cell Mol Life Sci 66:3625–3639

Konrat R (2014) NMR contributions to structural dynamics studies of intrinsically disordered proteins. J Magn Reson 241:74–85

Korzhnev DM, Kay LE (2008) Probing invisible, low-populated states of protein molecules by relaxation dispersion NMR spectroscopy: An application to protein folding. Acc Chem Res 41:442–451

Korzhnev DM, Salvatella X, Vendruscolo M et al (2004a) Low-populated folding intermediates of Fyn SH3 characterized by relaxation dispersion NMR. Nature 430:586–590

Korzhnev DM, Kloiber K, Kay LE (2004b) Multiple-quantum relaxation dispersion NMR spectroscopy probing millisecond time-scale dynamics in proteins: theory and application. J Am Chem Soc 126:7320–7329

Kosen PA (1989) Spin labeling of proteins. Methods Enzymol 177:86–121

Kragelj J, Ozenne V, Blackledge M et al (2013) Conformational propensities of intrinsically disordered proteins from NMR chemical shifts. ChemPhysChem 14:3034–3045

Kurzbach D, Platzer G, Schwarz TC et al (2013) Cooperative unfolding of compact conformations of the intrinsically disordered protein osteopontin. Biochemistry 52:5167–5175

Kurzbach D, Schwarz TC, Platzer G et al (2014) Compensatory adaptations of structural dynamics in an intrinsically disordered protein complex. Angew Chem Int Ed 53:3840–3843

Landrieu I, Lacosse L, Leroy A et al (2006) NMR analysis of a Tau phosphorylation pattern. J Am Chem Soc 128:3575–3583
Latham JA, Dent SY (2007) Cross-regulation of histone modifications. Nat Struct Mol Biol 14:1017–1024
Levitt M, Chothia C (1976) Structural patterns in globular proteins. Nature 261:552–558
Lichtenecker R, Ludwiczek ML, Schmid W et al (2004) Simplification of protein NOESY spectra using bioorganic precursor synthesis and NMR spectral editing. J Am Chem Soc 126:5348–5349
Lichtenecker RJ, Coudevylle N, Konrat R et al (2013a) Selective isotope labelling of leucine residues by using α-ketoacid precursor compounds. ChemBioChem 14:818–821
Lichtenecker RJ, Weinhaupl K, Reuther L et al (2013b) Independent valine and leucine isotope labeling in *Escherichia coli* protein overexpression systems. J Biomol NMR 57:205–209
Lichtenecker RJ, Weinhaupl K, Schmid W et al (2013c) α-Ketoacids as precursors for phenylalanine and tyrosine labelling in cell-based protein overexpression. J Biomol NMR 57:327–331
Lindorff-Larsen K, Kristjansdottir S, Teilum K et al (2004) Determination of an ensemble of structures representing the denatured state of the bovine acyl-coenzyme A binding protein. J Am Chem Soc 126:3291–3299
Liokatis S, Dose A, Schwarzer D et al (2010) Simultaneous detection of protein phosphorylation and acetylation by high-resolution NMR spectroscopy. J Am Chem Soc 132:14704–14705
Lipari G, Szabo A (1981a) A model-free approach to the interpretation of NMR relaxation in macromolecules. Biophys J 33:A307–A307
Lipari G, Szabo A (1981b) Pade approximants to correlation-functions for restricted rotational diffusion. J Chem Phys 75:2971–2976
Liu CC, Cellitti SE, Geierstanger BH et al (2009) Efficient expression of tyrosine-sulfated proteins in *E. coli* using an expanded genetic code. Nat Protoc 4:1784–1789
Luz Z, Meiboom S (1963) Nuclear magnetic resonance study of the protolysis of trimethylammonium ion in aqueous solution-order of the reaction with respect to solvent. J Chem Phys 39:366–370
Maekawa S, Murofushi H, Nakamura S (1994) Inhibitory effect of calmodulin on phosphorylation of NAP-22 with protein kinase C. J Biol Chem 269:19462–19465
Maltsev AS, Ying J, Bax A (2012) Impact of N-terminal acetylation of α-synuclein on its random coil and lipid binding properties. Biochemistry 51:5004–5013
Mao AH, Crick SL, Vitalis A et al (2010) Net charge per residue modulates conformational ensembles of intrinsically disordered proteins. Proc Natl Acad Sci U S A 107:8183–8188
Marion D, Ikura M, Tschudin R et al (1989) Rapid recording of 2D NMR-spectra without phase cycling—application to the study of hydrogen-exchange in proteins. J Magn Reson 85:393–399
Marsh JA, Forman-Kay JD (2011) Ensemble modeling of protein disordered states: Experimental restraint contributions and validation. Proteins 80(20):556–572
Marsh JA, Singh VK, Jia Z et al (2006) Sensitivity of secondary structure propensities to sequence differences between α- and γ-synuclein: implications for fibrillation. Protein Sci 15:2795–2804
Matsubara M, Nakatsu T, Kato H et al (2004) Crystal structure of a myristoylated CAP-23/NAP-22 N-terminal domain complexed with Ca^{2+}/calmodulin. EMBO J 23:712–718
Mayer C, Slater L, Erat MC et al (2012) Structural analysis of the plasmodium falciparum erythrocyte membrane protein 1 (PfEMP1) intracellular domain reveals a conserved interaction epitope. J Biol Chem 287:7182–7189
Means GE, Feeney RE (1968) Reductive alkylation of amino groups in proteins. Biochemistry 7:2192–2201
Meyer B, Möller H (2007) Conformation of glycopeptides and glycoproteins. Top Curr Chem 267:187–251
Mittag T, Forman-Kay JD (2007) Atomic-level characterization of disordered protein ensembles. Curr Opin Struct Biol 17:3–14
Mittag T, Orlicky S, Choy WY et al (2008) Dynamic equilibrium engagement of a polyvalent ligand with a single-site receptor. Proc Natl Acad Sci U S A 105:17772–17777

Mittag T, Marsh J, Grishaev A et al (2010) Structure/function implications in a dynamic complex of the intrinsically disordered Sic1 with the Cdc4 subunit of an SCF ubiquitin ligase. Structure 18:494–506

Mohana-Borges R, Goto NK, Kroon GJ et al (2004) Structural characterization of unfolded states of apomyoglobin using residual dipolar couplings. J Mol Biol 340:1131–1142

Motackova V, Novacek J, Zawadzka-Kazimierczuk A et al (2010) Strategy for complete NMR assignment of disordered proteins with highly repetitive sequences based on resolution-enhanced 5D experiments. J Biomol NMR 48:169–177

Nair SK, Burley SK (2003) X-ray structures of Myc-Max and Mad-Max recognizing DNA. Molecular bases of regulation by proto-oncogenic transcription factors. Cell 112:193–205

Neudecker P, Zarrine-Afsar A, Choy WY et al (2006) Identification of a collapsed intermediate with non-native long-range interactions on the folding pathway of a pair of Fyn SH3 domain mutants by NMR relaxation dispersion spectroscopy. J Mol Biol 363:958–976

Neumann H, Peak-Chew SY, Chin JW (2008) Genetically encoding N(epsilon)-acetyllysine in recombinant proteins. Nat Chem Biol 4:232–234

Nguyen DP, Garcia Alai MM, Virdee S et al (2010) Genetically directing varepsilon-N, N-dimethyl-L-lysine in recombinant histones. Chem Biol 17:1072–1076

Novacek J, Zawadzka-Kazimierczuk A, Papouskova V et al (2011) 5D ^{13}C-detected experiments for backbone assignment of unstructured proteins with a very low signal dispersion. J Biomol NMR 50:1–11

Orekhov VY, Korzhnev DM, Kay LE (2004) Double- and zero-quantum NMR relaxation dispersion experiments sampling millisecond time scale dynamics in proteins. J Am Chem Soc 126:1886–1891

Otting G (2010) Protein NMR using paramagnetic ions. Annu Rev Biophys 39:387–405

Ozenne V, Bauer F, Salmon L et al (2012) Flexible-meccano: a tool for the generation of explicit ensemble descriptions of intrinsically disordered proteins and their associated experimental observables. Bioinformatics 28:1463–1470

Pang Y, Buck M, Zuiderweg ER (2002) Backbone dynamics of the ribonuclease binase active site area using multinuclear (^{15}N and ^{13}CO) NMR relaxation and computational molecular dynamics. BioChemistry 41:2655–2666

Parra RG, Espada R, Sanchez IE et al (2013) Detecting repetitions and periodicities in proteins by tiling the structural space. J Phys Chem B 117:12887–12897

Pelupessy P, Chiarparin E, Ghose R et al (1999) Efficient determination of angles subtended by Cα-Hα and N-H(N) vectors in proteins via dipole-dipole cross-correlation. J Biomol NMR 13:375–380

Pelupessy P, Espallargas GM, Bodenhausen G (2003a) Symmetrical reconversion: measuring cross-correlation rates with enhanced accuracy. J Magn Reson 161:258–264

Pelupessy P, Ravindranathan S, Bodenhausen G (2003b) Correlated motions of successive amide N-H bonds in proteins. J Biomol NMR 25:265–280

Pervushin K, Riek R, Wider G et al (1997) Attenuated T_2 relaxation by mutual cancellation of dipole-dipole coupling and chemical shift anisotropy indicates an avenue to NMR structures of very large biological macromolecules in solution. Proc Natl Acad Sci U S A 94:12366–12371

Pinheiro AS, Marsh JA, Forman-Kay JD et al (2011) Structural signature of the MYPT1-PP1 interaction. J Am Chem Soc 133:73–80

Platzer G, Schedlbauer A, Chemelli A et al (2011) The metastasis-associated extracellular matrix protein osteopontin forms transient structure in ligand interaction sites. Biochemistry 50: 6113–6124

Reif B, Hennig M, Griesinger C (1997) Direct measurement of angles between bond vectors in high-resolution NMR. Science 276:1230–1233

Riek R, Wider G, Pervushin K et al (1999) Polarization transfer by cross-correlated relaxation in solution NMR with very large molecules. Proc Natl Acad Sci U S A 96:4918–4923

Rose GD, Fleming PJ, Banavar JR et al (2006) A backbone-based theory of protein folding. Proc Natl Acad Sci U S A 103:16623–16633

Rule GS, Hitchens TK (2006) Fundamentals of protein NMR spectroscopy. Springer, Dordrecht

Salmon L, Nodet G, Ozenne V et al (2010) NMR characterization of long-range order in intrinsically disordered proteins. J Am Chem Soc 132:8407–8418
Salzmann M, Pervushin K, Wider G et al (1998) TROSY in triple-resonance experiments: new perspectives for sequential NMR assignment of large proteins. Proc Natl Acad Sci U S A 95:13585–13590
Sára T, Schwarz TC, Kurzbach D et al. (2014) Magnetic resonance access to transiently formed protein complexes. ChemistryOpen 3(3):115–123
Sauve S, Tremblay L, Lavigne P (2004) The NMR solution structure of a mutant of the max b/HLH/LZ free of DNA: insights into the specific and reversible DNA binding mechanism of dimeric transcription factors. J Mol Biol 342:813–832
Schanda P, Van Melckebeke H, Brutscher B (2006) Speeding up three-dimensional protein NMR experiments to a few minutes. J Am Chem Soc 128:9042–9043
Schanda P, Brutscher B, Konrat R et al (2008) Folding of the KIX domain: characterization of the equilibrium analog of a folding intermediate using $^{15}N/^{13}C$ relaxation dispersion and fast $^{1}H/^{2}H$ amide exchange NMR spectroscopy. J Mol Biol 380:726–741
Schwalbe H, Carlomagno T, Hennig M et al (2001) Cross-correlated relaxation for measurement of angles between tensorial interactions. Methods Enzymol 338:35–81
Selenko P, Frueh DP, Elsaesser SJ et al. (2008) In situ observation of protein phosphorylation by high-resolution NMR spectroscopy. Nat Struct Mol Biol 15:321–329
Solyom Z, Schwarten M, Geist L et al (2013) BEST-TROSY experiments for time-efficient sequential resonance assignment of large disordered proteins. J Biomol NMR 55:311–321
Stanek J, Saxena S, Geist L et al (2013) Probing local backbone geometries in intrinsically disordered proteins by cross-correlated NMR relaxation. Angew Chem Int Ed Engl 52:4604–4606
Sugase K, Lansing JC, Dyson HJ et al. (2007) Tailoring relaxation dispersion experiments for fast-associating protein complexes. J Am Chem Soc 129:13406
Tamiola K, Mulder FA (2012) Using NMR chemical shifts to calculate the propensity for structural order and disorder in proteins. Biochem Soc Trans 40:1014–1020
Tejero-Diez P, Rodriguez-Sanchez P, Martin-Cofreces NB et al (2000) bFGF stimulates GAP-43 phosphorylation at ser41 and modifies its intracellular localization in cultured hippocampal neurons. Mol Cell Neurosci 16:766–780
Theillet FX, Liokatis S, Jost JO et al (2012) Site-specific mapping and time-resolved monitoring of lysine methylation by high-resolution NMR spectroscopy. J Am Chem Soc 134:7616–7619
Theillet FX, Rose HM, Liokatis S et al (2013) Site-specific NMR mapping and time-resolved monitoring of serine and threonine phosphorylation in reconstituted kinase reactions and mammalian cell extracts. Nat Protoc 8:1416–1432
Theillet FX, Binolfi A, Frembgen-Kesner T et al. (2014) Physicochemical properties of cells and their effects on intrinsically disordered proteins (IDPs). Chem Rev 114(13):6661–6714
Tjandra N, Bax A (1997) Direct measurement of distances and angles in biomolecules by NMR in a dilute liquid crystalline medium. Science 278:1111–1114
Tjandra N, Szabo A, Bax A (1996) Protein backbone dynamics and N-15 chemical shift anisotropy from quantitative measurement of relaxation interference effects. J Am Chem Soc 118:6986–6991
Tollinger M, Skrynnikov NR, Mulder FAA et al (2001) Slow dynamics in folded and unfolded states of an SH3 domain. J Am Chem Soc 123:11341–11352
Tollinger M, Kloiber K, Agoston B et al (2006) An isolated helix persists in a sparsely populated form of KIX under native conditions. Biochemistry 45:8885–8893
Trifonov EN, Frenkel ZM (2009) Evolution of protein modularity. Curr Opin Struct Biol 19: 335–340
Tugarinov V, Kay LE (2004a) ^{1}H, ^{13}C-^{1}H, ^{1}H dipolar cross-correlated spin relaxation in methyl groups. J Biomol NMR 29:369–376
Tugarinov V, Kay LE (2004b) An isotope labeling strategy for methyl TROSY spectroscopy. J Biomol NMR 28:165–172

Tugarinov V, Hwang PM, Ollerenshaw JE et al (2003) Cross-correlated relaxation enhanced 1^{H}(bond)^{13}C NMR spectroscopy of methyl groups in very high molecular weight proteins and protein complexes. J Am Chem Soc 125:10420–10428
Tugarinov V, Sprangers R, Kay LE (2004) Line narrowing in methyl-TROSY using zero-quantum ^{1}H-^{13}C NMR spectroscopy. J Am Chem Soc 126:4921–4925
Veit M (2000) Palmitoylation of the 25-kDa synaptosomal protein (SNAP-25) in vitro occurs in the absence of an enzyme, but is stimulated by binding to syntaxin. Biochem J 345(Pt 1):145–151
Vogeli B, Yao L (2009) Correlated dynamics between protein HN and HC bonds observed by NMR cross relaxation. J Am Chem Soc 131:3668–3678
Wang T, Frederick KK, Igumenova TI et al (2005) Changes in calmodulin main-chain dynamics upon ligand binding revealed by cross-correlated NMR relaxation measurements. J Am Chem Soc 127:828–829
Wang T, Weaver DS, Cai S et al (2006) Quantifying Lipari-Szabo modelfree parameters from ^{13}CO NMR relaxation experiments. J Biomol NMR 36:79–102
Wells M, Tidow H, Rutherford TJ et al (2008) Structure of tumor suppressor p53 and its intrinsically disordered N-terminal transactivation domain. Proc Natl Acad Sci U S A 105:5762–5767
Wolynes PG (1996) Symmetry and the energy landscapes of biomolecules. Proc Natl Acad Sci U S A 93:14249–14255
Wood MJ, Komives EA (1999) Production of large quantities of isotopically labeled protein in Pichia pastoris by fermentation. J Biomol NMR 13:149–159
Yang DW, Konrat R, Kay LE (1997) A multidimensional NMR experiment for measurement of the protein dihedral angle psi based on cross-correlated relaxation between (H α-^{13}C α)-H-1 dipolar and ^{13}C' (carbonyl) chemical shift anisotropy mechanisms. J Am Chem Soc 119:11938–11940
Yang D, Gardner KH, Kay LE (1998) A sensitive pulse scheme for measuring the backbone dihedral angle psi based on cross-correlation between 13C α- 1H α dipolar and carbonyl chemical shift anisotropy relaxation interactions. J Biomol NMR 11:213–220
Young TS, Schultz PG (2010) Beyond the canonical 20 amino acids: expanding the genetic lexicon. J Biol Chem 285:11039–11044
Zakharov VV, Capony JP, Derancourt J et al (2003) Natural N-terminal fragments of brain abundant myristoylated protein BASP1. Biochim Biophys Acta 1622:14–19
Zawadzka-Kazimierczuk A, Kozminski W, Sanderova H et al (2012) High dimensional and high resolution pulse sequences for backbone resonance assignment of intrinsically disordered proteins. J Biomol NMR 52:329–337
Zeng L, Fischer MW, Zuiderweg ER (1996) Study of protein dynamics in solution by measurement of ^{13}C α- ^{13}CO NOE and ^{13}CO longitudinal relaxation. J Biomol NMR 7:157–162
Zintsmaster JS, Wilson BD, Peng JW (2008) Dynamics of ligand binding from C-13 NMR relaxation dispersion at natural abundance. J Am Chem Soc 130(43):14060–14061

Chapter 6
Recombinant Intrinsically Disordered Proteins for NMR: Tips and Tricks

Eduardo O. Calçada, Magdalena Korsak and Tatiana Kozyreva

Abstract The growing recognition of the several roles that intrinsically disordered proteins play in biology places an increasing importance on protein sample availability to allow the characterization of their structural and dynamic properties. The sample preparation is therefore the limiting step to allow any biophysical method being able to characterize the properties of an intrinsically disordered protein and to clarify the links between these properties and the associated biological functions.

An increasing array of tools has been recruited to help prepare and characterize the structural and dynamic properties of disordered proteins. This chapter describes their sample preparation, covering the most common drawbacks/barriers usually found working in the laboratory bench. We want this chapter to be the bedside book of any scientist interested in preparing intrinsically disordered protein samples for further biophysical analysis.

Keywords Protein expression · Protein purification · Isotopic enrichment · Protein tags · Recombinant expression · Isotope labeling · Protein solubility · Purification · Heat-stable proteins · Hydrodynamic volume

---

E. O. Calçada (✉)
Magnetic Resonance Center (CERM), University of Florence, Via Luigi Sacconi 6, 50019 Sesto Fiorentino, Italy
e-mail: eocalcada@gmail.com

M. Korsak · T. Kozyreva
Giotto Biotech, Via Madonna del Piano 6, 50019
Sesto Fiorentino, Italy
e-mail: korsak@giottobiotech.com

T. Kozyreva
e-mail: kozyreva@giottobiotech.com

I. C. Felli, R. Pierattelli (eds.), *Intrinsically Disordered Proteins Studied by NMR Spectroscopy,* Advances in Experimental Medicine and Biology,
DOI 10.1007/978-3-319-20164-1_6

1 Introduction

It is becoming evident that intrinsically disordered proteins (IDPs) are not fully disordered, but have all sorts of transient, short and long-range structural organisations that are function-related.

The structural and functional study of biomolecules is a highly interdisciplinary field that requires a correlation between different biophysical techniques. Although at first glance the study of IDPs seems very similar to the traditional analysis of structured proteins, specific skills and different approaches are needed to deal with IDP different properties (Uversky 2011). Structural biology requires a large number of steps to convert DNA sequence information into protein samples, including the selection of the proper expression constructs and vectors, the setting of the right growth conditions, and efficient purification strategies. The following is intended to assist in the sample preparation of IDPs from genome browsing to sample preparation and analysis. Different methodologies will be addressed with special tips and tricks highlighted that are helpful for overcoming common drawbacks/barriers usually found when working with IDPs.

2 Genome Browsing and Bioinformatics Analysis

Intrinsic structural disorder is a widespread phenomenon, especially in eukaryotes, where conservative bioinformatics predictions suggest that 5–15 % of proteins are IDPs, and about 35–50 % of proteins have intrinsically disordered regions (IDRs) longer than 30 residues (Ward et al. 2004). It has been accepted that disorder is needed for signalling among various living systems, and increases with organism complexity (Dunker and Obradovic 2001). Indeed, 75 % of the signalling proteins in mammals are predicted to contain long disordered regions (Dunker et al. 2008) (Fig. 6.1).

The web tools currently available are incredibly easy to use and, in most cases, very accurate information can be obtained quickly. Advanced genome tools for studying biology made an incredible amount of biological data available with a

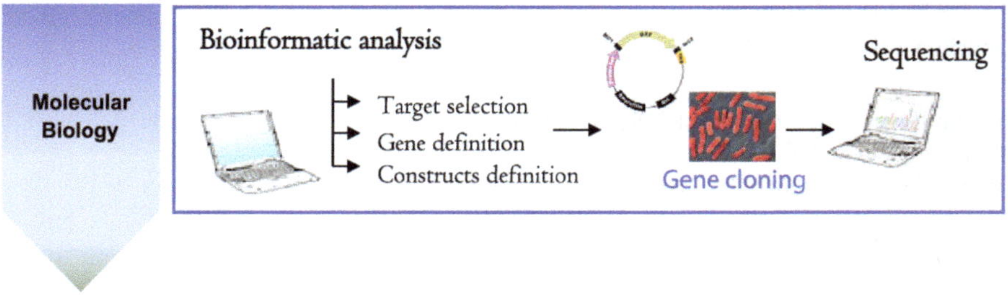

Fig. 6.1 Molecular biology - bioinformatic analysis

concomitant proliferation of biological databases and web software tools, i.e. data banks containing information on DNA, protein sequences, expression profiles, protein ensembles and structures. One example is the recently created protein ensemble database[1] for IDPs (Chap. 11) (Varadi et al. 2013).

The first step when beginning to work with a new protein, whether you are a protein hunter looking for an interesting target to study or if you've just joined on-going scientific work, is the sequence analysis. DNA sequence analysis by bioinformatic tools is useful either to amplify the gene from a natural source or for the synthetic gene construction. In the latter case, the sequence of the target protein should be adapted to the expression system used. Make sure to analyse the domain composition, presence of signal peptides and inert-membrane helices, as well as disordered regions in order to create suitable expression constructs. Even if the bioinformatic analysis is just a prediction, it could be useful in any case in order to obtain preliminary information on the target protein. In the case of IDPs, the prediction of disordered domains is essential (Chap. 9).

A number of approaches have been developed to predict regions of protein disorder. These methods can be broadly classified into several different categories: *ab initio*, clustering and meta or consensus.

The majority of these predictors are available through public servers, and links to many of them can be found in the "Disordered Protein Database" (Disprot)[2], (Sickmeier et al. 2007), the "Database of Disordered Protein Prediction" (D^2P^2)[3] (Oates et al. 2013), and in the recently created IDPbyNMR website[4].

The large majority of available prediction methods depends almost exclusively on sequence information. In other words, nothing more than the primary sequence is needed in order to make a prediction. Disordered regions in proteins are predicted using features extracted from the primary sequence in conjunction with statistical models. In clustering methods, tertiary structure models are predicted for the target protein, and then these models are superimposed by carrying out structural alignments.

The generation of disorder prediction tools started from a comparison between the amino acid sequences of IDPs and those of structured globular proteins, which resulted in a number of significant differences including amino acid composition, sequence complexity, hydrophobicity, aromaticity, charge, and flexibility (Dunker et al. 2001). For example, IDPs are significantly depleted in hydrophobic (Ile, Leu, and Val) and aromatic (Trp, Tyr, and Phe) amino acid residues, which form and stabilize the hydrophobic core of folded globular proteins. These residues are called order-promoting amino acids. On the other hand, IDPs/IDRs are substantially rich in polar (Arg, Gln, Glu, Lys, and Ser) and structure-breaking (Gly and Pro) disorder-promoting amino acid residues (Dunker et al. 2001; Radivojac et al. 2007). Among the twenty common amino acid residues, proline is the most disordered-promoting

[1] http://pedb.vib.be.

[2] http://www.dabi.temple.edu/disprot/index.php.

[3] http://d2p2.pro.

[4] http://www.idpbynmr.eu/home.

(Theillet et al. 2013). The accuracy of predictors is regularly assessed as part of the Critical Assessment of Structure Prediction (CASP) experiment[5] (Monastyrskyy et al. 2011). It is therefore now possible to predict the tendency of a polypeptide chain to be disordered based on its primary sequence with an approximate accuracy of 80 % (He et al. 2009).

To obtain more accurate disorder predictions, a good option is the use of meta predictors such as PONDR-FIT (Xue et al. 2010), which combine the output of several individual predictors.

After obtaining the preliminary view of your sequence with information about folded and unfolded regions, you can think about creating a set of constructs by omitting folded regions, creating shorter segments containing overlays on the terminals, or simply substituting some amino acid.

Consider the possibility of creating different domain constructs as well as the full-length construct.

IDPs are often able to bind to different partners or to act as hub proteins, and in this way play an important role in a variety of different processes. IDPs are particularly relevant for viruses, which need to exploit simple amino acid sequences (short linear motifs, SLiMs), which are well-exposed and ready to function, in order to interact with crucial key proteins from the host organism (Chap. 9). They are also known as molecular recognition features (MoRFs) and different tools for their prediction are available such as SLiMfinder[6] (Davey et al. 2010), MoRFpred[7] (Disfani et al. 2012) and Anchor[8] (Mészáros et al. 2009).

Using the BLAST alignment it is possible to establish the homology of the target protein with some proteins with known three-dimensional structure. In this case one can consider the use of the MODELLER software to generate a homology comparative structural model of the target protein. The development of comparative software allows the study of the homology modelling of the protein's three-dimensional structures (Eswar et al. 2006). With this tool, the user provides a sequence alignment of the protein of interest and obtains a model based on known related Protein Data Bank[9] (PDB) structures. For example, a model containing all of the non-hydrogen atoms calculated by comparative protein structure modelling based on the satisfaction of spatial restraints can be obtained using MODELLER (Šali and Blundell 1993). MODELLER also performs *de novo* modelling of loops in protein structures, multiple alignments of protein sequences and structures, optimization of

[5] http://predictioncenter.org.

[6] http://bioware.ucd.ie/~compass/biowareweb/Server_pages/slimfinder.php.

[7] http://biomine-ws.ece.ualberta.ca/MoRFpred/index.html.

[8] http://anchor.enzim.hu.

[9] http://www.pdb.org/pdb/home/home.do.

various models of protein structure with respect to a defined objective, searching of sequence databases, clustering, comparison of protein structures, and so on.

Once the protein sequence has been defined, the easiest way to obtain the corresponding gene is by ordering the synthesized gene of interest suitable for your expression system. With the development of new instruments that facilitate the production of biological material, the synthesis of genes containing the DNA sequence of a target protein is now feasible. Many companies such as Invitrogen's GenArt®, OriGene, Eurofins MWG Operon, GenScrip, and DNA2.0 among others provide web tools to order these genes, with several possibilities and strategies, including the optimization of the codons for the specific expression organism. It is important to highlight that the over-expression of human proteins in *E. coli* systems could be compromised, resulting in low expression yields, if the open reading frame (ORF) of the protein contains codons infrequently used by *E. coli*, the so-called "rare codons". In particular, codons for arginine (AGG, AGA, CGA), isoleucine (ATA), leucine (CTA) and proline (CCC) should be avoided (Schenk et al. 1995). Different web tools, such as those offered by Genscript[10] and the United States National Institutes of Health (NIH)[11], are available to check the presence of rare codons related to the desired expression system. Nevertheless, the gene of interest can be directly cloned from the organism cDNA, if available. The cloning strategy should be designed carefully, as it could be the basis of a successful work.

> IDPs usually have a high proline content. Avoid rare codons in the DNA sequence, as they could lead to a low expression yields.

3 Expression Plasmid Generation

The standard procedure to express a recombinant protein is to carry out a screening of different constructs to identify the most efficient conditions for downstream purposes. The first step of the cloning process consists of the amplification of the target gene from a DNA template or plasmid through a polymerase chain reaction (PCR) using specific primers. After purification, the amplified product is inserted into a specific expression vector. Different vectors may be selected in order to obtain native protein or protein fused with different tags. The tags vary in size starting from 6-His to fusion proteins of 10–20 kDa or even 40 kDa proteins such as maltose-binding protein (MBP). They can enhance the expression level, increase solubility, and be very useful for the subsequent purification procedure due to their chemical properties (Esposito and Chatterjee 2006). Later on these tags may be removed by

---

[10] http://www.genscript.com/cgi-bin/tools/rare_codon_analysis.

[11] http://nihserver.mbi.ucla.edu/RACC/.

proteolytic cleavage with specific enzymes such as factor Xa, enterokinase (EK) or tobacco etch virus (TEV) protease (Arnau et al. 2006; Malhotra 2009).

Even if the tags are considered a good strategy, always consider expressing the native protein sequence.

The classic method to insert an amplified PCR product into a vector is to use restriction enzymes that cleave DNA at specific recognition sites. Both the DNA and the cloning vector have to be treated with two restriction enzymes that create compatible ends. Later on these ends are joined together by a ligation reaction performed by the bacteriophage T4 DNA ligase. Finally an aliquot of the product of the reaction is transformed in suitable *E. coli* strains such as DH5α, TOP10, JM109 etc. *E. coli* competent cells and positive clones are screened by PCR screening followed by DNA sequencing. However, the classic cloning strategy is sometimes not feasible for the preparation of different constructs in parallel due to the lack of the suitable restriction sites common for the target gene and available vectors. Together with low efficiency and false positive clones, this technique is not the best for high-throughput cloning. Therefore, other cloning strategies have been developed that exploit ligation-independent cloning such as Gateway® (Invitrogen), TOPO cloning, and most recently Electra™ (DNA2.0) (Katzen 2007).

Many companies offer the possibility to clone the synthetic gene directly into a desired expression plasmid.

Gateway® cloning technology has been one of the most used strategies and enables rapid and highly efficient simultaneous transfer of DNA sequences into multiple vector systems for protein expression and functional analysis while maintaining orientation and reading frame. It basically consists of the generation of an expression silent entry clone that can be further recombined in the several expression vectors without the use of any restriction enzyme, taking advantage of the site-specific recombination properties of bacteriophage lambda. There are many ways to create an entry clone but the most straightforward method is directional TOPO® cloning. The ligation reaction of the PCR product to the pENTR vector is accomplished by topoisomerase I. After isolation of the entry clone, the second step is to generate an expression vector. This is done by recombination of the gene on the entry clone with the final expression vector, performed by LR Clonase®. The different antibiotic resistances of the vectors allow fast clone selection. A large selection of Gateway® expression vectors is available for the expression of native proteins as well as proteins fused with tags. One of the versions of the Gateway® Cloning System is pENTR/TEV/D-TOPO. This version of the Gateway® Cloning System includes the TEV recognition site on the N-terminus of the protein. The following expression destination vectors can be used in order to create the expression clones: pDEST-17 (conferring 6x histidine N-terminus), pETG-30A (conferring GST plus 6x histidine N-terminus), and pDEST-His-MBP (conferring MBP plus 6x histidine N-terminus), pETG-20A (conferring Trx plus 6x histidine N-terminus), and pTH34 (conferring GB1 plus 6x histidine N-terminus), among others. It is important to highlight that the use of this methodology will result in the expression protein including extra residues on the N-terminus. For example, using the expression vector pDEST-17 that contains the 6xHis tag, the expressed and purified target protein will have 44 extra amino acids. After tag removal using TEV protease, the final construct will contain 4 extra residues on the N-terminus, GSFT.

> Added fusion tags can be removed upon protease cleavage, but many of them can result in the addition of an extra amino acids sequence.

Site-directed mutagenesis is a standard technique used to make point mutations, replace amino acids, and delete or insert single or multiple adjacent amino acids. Point mutations in which a single nucleotide is exchanged but the new codon specifies the same amino acid are called silent mutations. They basically code for the same amino acid, and are an easy way to avoid the "rare codons" that may decrease expression yield. If just a few silent mutations are sufficient, this approach is valuable; otherwise synthetic genes optimized for *E. coli* expression, or specific *E. coli* strains (such as CodonPLUS or pRARE containing strains), should be used for improving yields.

Briefly, site directed mutagenesis uses the double-stranded DNA vector template containing the target gene and two complementary synthetic oligonucleotide primers, both containing the desired mutation. The primers are mixed with the DNA vector template and extended during PCR cycles performed by a high fidelity DNA polymerase. The PCR product is then treated with DpnI, an endonuclease that will digest the DNA template due to its high specificity to Dam methylated and hemimethylated DNA isolated from *E. coli* strains. The new copies of the DNA PCR product were never methylated and are thus not digested. The digested solution is then transformed into XL1-Blue super-competent cells and subsequently subjected to sequencing analysis.

> Use site-directed mutagenesis to create silent mutations of rare codons, increasing the expression yield.

4 Protein Expression

Several host systems are available for protein production including fungi, plant, bacteria, insect, yeast and mammalian cells (Shatzman 1995). The choice of the expression system for the high-level production of recombinant proteins depends on

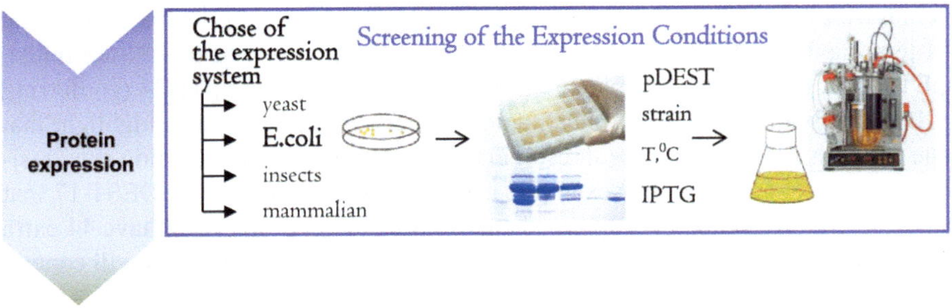

Fig. 6.2 Protein expression - screening of the expression conditions

many factors, including biological activity of the target protein, post-translational modifications, cell growth features, intracellular and extracellular characteristics, and expression levels. The many advantages of the use of *Escherichia coli (E. coli)* have ensured that it has remained a valuable organism for the high-level production of recombinant proteins; it is the easiest, quickest and cheapest expression system (Tong et al. 2008). Protein expression using the yeast expression system is also a good solution; however, for ^{13}C labelling, which requires a correct promoter for the carbon source (Weinhandl et al. 2014), it can be an extremely expensive solution compared with the *E. coli* expression system for isotopically enriched proteins. A wealth of biochemical and genetic knowledge of *E. coli* has driven the development of a variety of strategies for achieving high-level protein expression. The major challenges for obtaining high protein yields at low cost involve several aspects such as expression vector design, transcriptional regulation (promoter), mRNA stability, translational regulation (initiation and termination), host design considerations, codon usage, and the culture conditions to increase the expression of the protein of interest (Jana and Deb 2005) (Fig. 6.2).

Expression of isotopically labelled proteins in *E. coli* is much cheaper as compared to yeast due to carbon-source costs.

The expression condition should be tested once the expression system has been selected. Therefore, the best approach is to use parallel test expression strategies (Lesley 2009). A preliminary expression test can be performed using a small volume in order to find the best conditions to be reproduced in large volumes to obtain soluble recombinant protein. It is important to take the following factors into consideration for protein expression: culture medium, temperature, optical density, inducer concentration and induction time. A library of different *E. coli* strains possessing various properties that could be advantageous in the expression of a certain protein is available. Common examples are the *E. coli* strains BL21(DE3) (the standard version) or some variants such as BL21(DE3)pLysS, which encode the T7 lysozyme to decrease the background expression level of target genes under the control of the T7 promoter, but do not interfere with expression levels following induction by isopropyl β-D-1-thiogalactopyranoside (IPTG), Rosetta(DE3) and Codon Plus for genes containing rare codons, Origami(DE3) for proteins containing disulfide bridges, and Gold(DE3) for increasing expression yields.

Depending on the research purpose the cells can be grown in different types of media: rich media, lysogeny broth (LB), yeast extract and tryptone broth (YT), Terrific broth, NZY and the minimal medium M9. When an isotopically labelled protein is necessary, the M9 medium is a good choice as [^{13}C] enriched glucose and [^{15}N] enriched ammonium sulphate/chloride are used as the sole ^{13}C and ^{15}N sources. In order to increase the expression yield of isotopically labelled protein, one good solution could be the use of the Marley method (Marley et al. 2001). The cells transformed with the expression plasmid are initially grown in rich medium until high optical densities, and are then centrifuged and exchanged into an isotopically defined minimal media enriched with $(^{15}NH_4)_2SO_4$ (1 g/L) and (^{13}C) glucose (4 g/L). Labelled protein can also be produced in commercially available and isotopically enriched rich media, e.g. Silantes. This could be particularly advantageous for the expression of deuterated labelled protein as the cell adaptation process is generally easier. Small-scale test expression is the first step in the expression of the target protein; it is important to reproduce the expression conditions required for large-scale production as much as possible.

> In some cases the isotopically labelled carbon and nitrogen sources may give different expression yields and protein solubility. Consider a small test expression using isotopically labelled nutrients.

Care should be taken in controlling the concentrations of metal ions in the culture medium in general and when working with metallo-proteins in particular. For example, zinc(II) is essential for many cellular processes, including DNA synthesis, transcription, and translation, but its excess can be toxic (Babich and Stotzky 1978; Kindermann et al. 2005). In order to find the optimal quantity of zinc additive, different trials should be performed using controlled minimal media instead of rich LB (Outten and O'Halloran 2001). The amount of zinc can be tested by comparing zinc-depleted cultures with those containing zinc(II) in a concentration range of 10–400 μM (obtained adding $ZnCl_2$ or $ZnSO_4$).

> When working with a metalloprotein, the expression tests should be performed taking into account the concentration of the required metal ion.

In order to avoid hundreds of different screening conditions, one can start with a couple of sets and alter specific conditions if needed, according to the preliminary results obtained. Three different *E. coli* strains induced at a single optical density

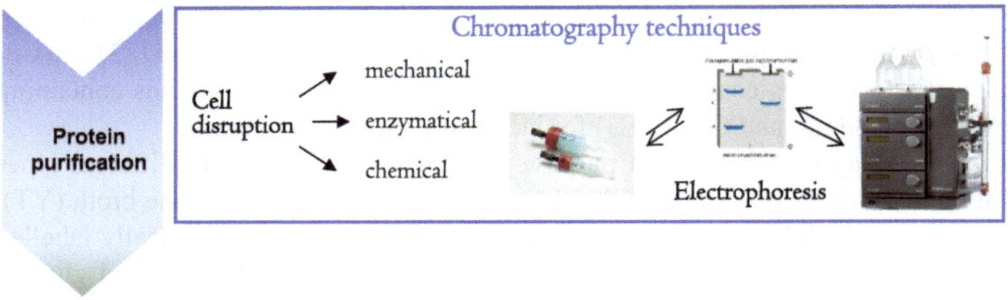

Fig. 6.3 Protein purification - Chromatography techniques

of 0.6, using a single IPTG concentration of 0.5 mM, three different expression temperatures (17, 30 and 37 °C) and two expression times (4 and 16 h) will result in 18 different conditions. In case of non-satisfactory results, one can try different *E. coli* strains, optical densities or IPTG concentrations. Cells should be harvested, disrupted and normalized to the final optical density. The presence of soluble protein is checked with sodium dodecyl sulfate polyacrylamide gel electrophoresis (SDS-PAGE).

One important aspect of IDPs when running an SDS-PAGE is abnormal protein size. IDPs typically run on SDS-PAGE as though they exhibit a higher molecular mass. This aberrant migration occurs due to the different amino acid composition caused by the high acidic residue content of some IDPs (Graceffa et al. 1992; Armstrong and Roman 1993).

If all the trials to obtain soluble protein fail, the protein in the insoluble fraction can be recovered through a refolding process. In some cases the latter approach may even become the most efficient method to obtain the protein in good yield, of course after proving that the protein obtained through the refolding process has the same properties of the native one.

5 Protein Purification

The protein purification strategies rely on the biophysical and biochemical properties of each specific protein. The purification strategy could be summarized in three main steps (Fig. 6.4), which may or may not be performed depending on the required purity of the final sample. In the first step the target protein is isolated from the cells and protected from possible degradation. The main bulk of impurities could then be removed by heating, passing the lysate through ion exchange, hydrophobic interaction or affinity columns. In the final step the samples could undergo size exclusion or high-resolution chromatography for the removal of trace impurities. Of course the order in which the various purification steps are performed depends on the specific case.

The peculiar amino acid composition of IDPs will contribute to specific biophysical properties that are different from those of folded proteins. Several characteristics must

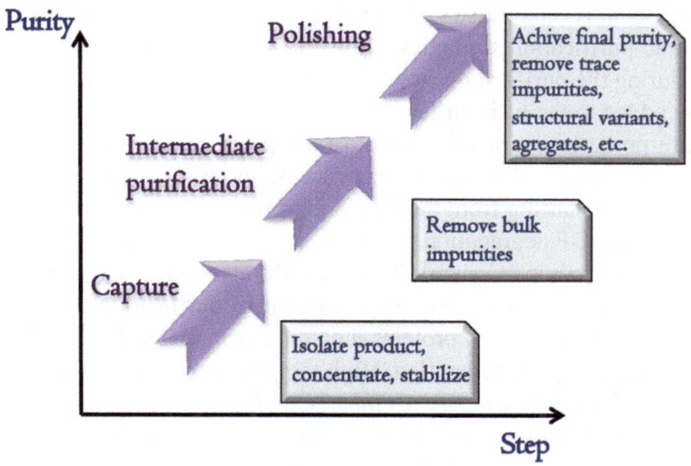

Fig. 6.4 Protein purification-three step purification strategy

be considered, for example whether or not the protein is soluble and, if not, which agents might help solubilizing it. Is the protein sensitive to variations in salt concentration, pH, temperature, or oxidation, particularly by oxygen? Is the protein labile at high or low concentrations? A set of preliminary experimental trials should be performed to learn which reagents might be present during the purification, which one must be avoided, and under which conditions the protein has to be stored. Designing the protein purification strategy for a new protein from scratch requires a preliminary bioinformatics study to predict several biochemical and biophysical characteristics as well as the study of previously published purification strategies (Hunt 2005). Several factors should be taken into account for both soluble and insoluble proteins, including: ionic strength, pH, temperature, oxygen concentration, and protein concentration. For example, if the protein contains several free cysteine residues, the use of reducing agents and of anaerobic conditions during the purification steps may be required to avoid the formation of non-physiological multimers and oligomers.

The localization of the expressed protein within the cell (soluble in cytoplasm or periplasm, or present in inclusion bodies (IBs)) makes a strong contribution to the choice of the purification strategy (Linn 2009). In each case the isolation is performed in different ways. Assuming the protein was not excreted into the growth medium, cell lysis is the first step for protein purification. The disruption of the cells can be performed through several techniques, ranging from mechanical to detergent-based methods. For instance, the French press is an efficient method, though it heats the final lysate, while freeze-thaw or enzymatic and detergent lysis are considered mild methods but less efficient. A good choice can be the cell disruption performed by sonication, which comprises pulsed, high frequency sound waves to mechanically agitate and lyse the cells. The isolation of proteins using sonication should be done carefully because the mechanical wave energy will heat the sample. To avoid sample warming, sonication should be performed using an ice bath and with short pulses, with intervals to allow the temperature to decrease.

Sonication methods can be also performed inside an anaerobic chamber (glove box) for anaerobic purification strategies. If degradation of the protein target has been observed during the purification steps, different cocktails of protease inhibitors can be directly added to the lysis buffer.

IDPs are prone to degradation due to their properties, which make the amino acid chain fairly exposed and allow easy access to proteases. The use of protease inhibitor cocktails starting from the first step of protein purification may be a good solution to avoid IDP degradation. Working at low temperatures may also help decreasing protease activity. Conversely, some IDPs are thermo-stable at high temperatures, property which can be exploited as an initial purification step. By warming the cell extract, several folded proteins including proteases will precipitate and can be easily separated by centrifugation. High temperature can cause some conformational changes in protein structure. Comparison of the pure protein sample obtained with and without heating should be performed by using circular dichroism (CD), dynamic light scattering (DLS) and nuclear magnetic resonance (NMR). Different temperatures and durations of sample exposure to the heat can be checked in order to find the optimal conditions where the IDP can be isolated without being damaged.

Check how thermo-stable your IDP is. It may be helpful for the purification steps.

Once the protein fraction is isolated from the cells, the following purification step can be performed in a multitude of chromatography runs. The methodology should always be optimized to reach an efficient protocol in terms of yield, speed and costs. Among all the different chromatography techniques, we will describe the three most used ones: immobilised metal ion affinity chromatography (IMAC), ion exchange chromatography (IEX) and size exclusion chromatography (SEC).

Immobilized metal ion affinity chromatography (IMAC) is currently the most used affinity technique exploiting the interaction between chelated transition metal ions (such as Zn^{2+}, Co^{2+} or Ni^{2+}) and the side chains of specific amino acids (such as histidines) on the protein. For the purpose, a so-called "histidine-tag" can be introduced either at the beginning or at the end of the protein primary sequence to enhance the interaction with the IMAC column.

Take advantage of the specific metal affinity of your protein in the choice of metal ions for IMAC.

In basic terms, by using IMAC the target protein is tightly bound to the resin matrix and the impurities are washed out with increasing concentrations of imidazole, which acts as a competitive agent respect to histidine side chains. At higher imidazole concentrations the target protein is eluted at an almost pure concentration (Block et al. 2009).

Fused proteins can be separated from the histidine-tag by enzymatic digestion to cleave the tags. After tag cleavage the separation of tag and target protein can be accomplished by the second IMAC chromatography step.

To avoid high quantities of imidazole during the IMAC protein elution process, which can interfere with the metal binding properties of the protein itself, an elution buffer with low pH could be used to favour the protonation of histidines.

IEX separates proteins on the basis of a reversible interaction between the polypeptide chain and a specific charge ligand attached to a chromatographic matrix. The isoelectric point (pI) of the target protein must be known to guide the choice of chromatographic conditions. The sample is loaded in conditions that favour specific binding such as specific pH and low ionic strength in order to enhance the interaction between the target protein and column matrix. The unbound impurities are washed out and the bound protein is eluted by varying the pH or the ionic strength of the elution buffer. If the overall net charge of the protein is positive, a cationic IEX resin must be used, while if it is negative, an anionic IEX resin must be used. The buffer pH should be at least ± 1 unit different from the protein pI.

SEC is a separation technique based on the hydrodynamic radius of the proteins. The column matrix is composed of precisely sized beads containing pores of given sizes. Larger proteins whose hydrodynamic dimensions are too big to fit inside any pore will only have access to the mobile phase between the beads, and will be excluded as they will just follow the solvent flow and reach the end of the column before molecules of a smaller size. Proteins with smaller hydrodynamic dimensions will be drawn into the pores by diffusion, and have access to the mobile phase inside and between the beads. Therefore, smaller molecules will have a long distance to cross with several small retention times between the diffusion movements through the bead pores. Due to larger retention, smaller hydrodynamic molecules will elute last during the size exclusion separation. SEC can be used to separate proteins by size and shape, to exchange the buffer, and also to isolate protein mixtures and separate monomers, multimers or oligomers. In the case of folded proteins, it can also be used to have an estimate of the molecular mass by performing a molecular weight distribution analysis using available standards. Salts are necessary to avoid ionic interaction with the resin.

Hydrodynamic volume is one of the most important IDP biophysical parameters to be taken into consideration.

SEC profiles are therefore dependent on the hydrodynamic volume of a protein, which is one of the most important and fundamental structural parameters of a protein molecule. Hydrodynamic volume is a prerequisite for an accurate classification of a protein conformation. It changes dramatically depending on whether the protein hydrodynamic dimension is compact like a folded protein or extended or partially extended like an IDP.

> Comparison between the protein in native conditions and in the presence of a chaotropic agent allows a better understanding of how compact or extended the IDP hydrodynamic volume is. A comparative analysis can easily be addressed by SEC, CD, DLS and SAXS.

Comparing two proteins with the same molecular weight, a well-folded protein will have a smaller hydrodynamic radius while an IDP will have a bigger one, behaving like a large folded protein with SEC. The SEC retention times for a folded protein and an IDP of the same molecular mass will therefore be very different and the IDP will elute first. SEC has been used for over three decades for the separation of unfolded and folded proteins (Gupta 1983). However, due to the particular characteristics of IDPs, SEC can be used for the analytical study of the conformational IDP properties in solution where the size and shape of molecules are the prime separation parameters (Uversky 2013). SEC is usually performed at 4 °C as the last purification step for the preparation of high purity samples. For example, a column of at least one meter high connected to a water thermostatic cooling system at 4 °C can be one of the best solutions for separating proteins with large hydrodynamic volumes.

> When using analytical SEC to have information on the properties of IDPs, do not rely only on globular proteins as standard samples but consider other IDPs for comparison of the results.

Purifying and refolding a protein from the insoluble fraction could be a challenging task and should be planned carefully. The strategy can be based on a previous bioinformatics analysis of the target protein combined with knowledge of the state-of-the-art of similar protein systems. Although many protocols have already been published and summarized in many reviews and book chapters (Vincentelli et al. 2004; Singh et al. 2005; Cowieson et al. 2006; Qoronfleh et al. 2007; Burgess et al. 2009), each protein is unique and requires a specific approach to the refolding process.

The first step in refolding is solubilisation of the inclusion bodies. The solubilising agent/denaturant could be a chaotropic agent such as GdHCl and urea, or a detergent, and should be prepared in a controlled pH buffer. As for the other techniques already described in this work, the key concept for refolding is the systematic, parallel screening of multiple refolding conditions. Many additives may prove useful in refolding, but preventing aggregation and precipitation upon refolding is crucial for refolding at low protein concentrations. However, many variables in refolding should be controlled such as pH, temperature, salt concentration, redox environment, and the presence of divalent metal ions.

Redox Agents

Various redox pairs can be used including reduced and oxidized cysteine or glutathione as well as a reducing agent such as β-mercaptoethanol (BME), dithiothreitol (DTT) or tris (2-carboxyethyl) phosphine hydrochloride (TCEP) to control the oxidation state of the protein. Since the cytoplasm of *E. coli* is highly reducing, most internal proteins are in the reduced state. If the protein contains both native disulfide bonds and free cysteines, the redox couple should be introduced into the refolding system in order to achieve optimal native disulfide bond formation, keeping native free cysteine residues.

> Consider the presence of disulfide bridges; the addition of a high concentration of the reductant can break them.

Salt Concentration

To prevent undesirable hydrophobic interactions, the ionic strength of the solution could be increased by the addition of salt starting from 150 mM.

pH

In general the pH of the buffer should be at least 1 pH unit different from the pI in order to avoid a zero net charge of the protein, making it prone to precipitation. Some protocols rely only on a single pH refolding procedure (Coutard et al. 2012).

Temperature

Most refolding procedures are carried out at room temperature, which is low enough to prevent thermal damage to the protein and high enough to increase the thermal motion of the molecules, an important aspect that allows them to reach their native state. Screening of different temperature conditions might be useful to optimize sample conditions.

Proline and Arginine

Proline is considered an osmoprotectant and has been found to be effective in increasing solubility both *in vitro* and *in vivo* (Ignatova and Gierasch 2006). Arginine can decrease aggregation by slowing the rate of protein–protein interactions by supramolecular assembly formation in solution. However, effective concentrations are reported to be in the range of 0.5–1.0 M (Das et al. 2007).

Glycerol

Glycerol has been found to be an excellent refolding additive in many cases, and is usually used in the 5–30 % range.

Detergents

Detergents can increase solubility, preventing aggregation during refolding. At low concentrations they bind weakly to exposed hydrophobic regions, preventing aggregation.

As their concentration decreases they dissociate and allow reformation of the native structure. At high concentrations detergents are denaturants, but at low concentrations they can act as an artificial chaperone, promoting refolding without aggregation. One important aspect of detergents is the critical micelle concentration (CMC), the concentration at which micelles begin to form. The CMC value can be subject to buffer conditions such as pH and ionic strength. The CMC should be checked once for each buffer system as the manufacturer only provides a few values/conditions. Tables exist that report several values for each specific buffer condition (Brito and Vaz 1986; Jumpertz et al. 2011).

Once the protein of interest has been solubilised and all the refolding buffer conditions have been defined, refolding can be attempted. The refolding procedure is the removal of the denaturant agent, allowing the protein to reach its native state. Refolding can therefore be performed by dilution, multi-step dialysis, single dialysis, or with on-column refolding. Dialysis is one of the most used methods but it is time- and reagent-consuming. Affinity tagging of the recombinant protein can give the possibility of efficient and rapid on-column refolding and purification. If it doesn't precipitate inside the column, the refolded protein can be purified performing a few washing steps followed by elution. Alternatively, dilution refolding can be performed by reverse dilution (the addition of refolding buffer to denatured protein with mixing between each addition), by flash dilution (the addition of denatured protein to refolding buffer quickly), and by drip dilution (the addition of denatured protein to refolding buffer very slowly, drop-by-drop, allowing refolding at low concentration).

Several commercial products have been developed to help identify suitable refolding conditions such as EMD/Novagen's iFOLD kits, Pierce Biotechnology's ProMatrix™ and AthenaES's QuickFold™.

6 Sample Handling of Intrinsically Disordered Proteins

The preparation of samples for different biophysical and biochemical characterisations should be prepared according to the technical limits of each technique/method (Fig. 6.5).

Stability

Stability is one of the first conditions to check. One can try to perform the degradation test at various temperatures in order to understand if the protein is susceptible to degradation.

> For the sake of protein long-term stability, work in the presence of protease inhibitor cocktails and at low temperatures during the purification steps.

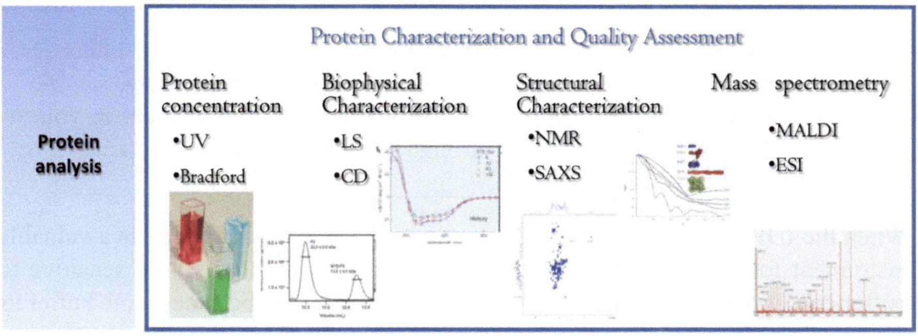

Fig. 6.5 Protein analysis - protein characterization and quality assessment

Anaerobic Purification
Proteins containing cysteine residues must be handled inside an anaerobic chamber under nitrogen atmosphere to prevent the oxidation of cysteine residues. All buffers used for the purification and sample preparation steps should be extensively degassed with nitrogen or argon. Reducing agents may be added to keep the cysteine residues reduced.

Reducing Agents
Reducing agents can be exogenous sulfhydryl containing reducing agents such as DTT, or non-sulfhydryl reducing agents such as TCEP. The control of the redox state of the IDP target is crucial for a successful purification. The optimal reducing activity of DTT is in the 6.5–9.0 pH range, while TCEP has a wide optimal reducing activity pH range spanning from 1.5 to 9.0. In some cases TCEP was reported to be more useful than DTT for protein sample preparation (Getz et al. 1999; Krezel et al. 2003).

6.1 *Measuring Protein Concentration*

Once the protein sample is pure, stable and its conditions are known, it is important to determine the protein concentration accurately.

Electronic spectroscopy can be generally used when tryptophan (Trp), tyrosine (Tyr), and phenylalanine (Phe) are present in the protein sequence, as aromatic residues absorb light in the UV-range with absorption maxima around 280 nm. However, Trp, Tyr, and Phe absorption spectra of IDPs can be compromised as typically IDPs are depleted of these residues (Dunker et al. 2001).

The concentrations can be calculated using the Beer-Lambert law, $A=\varepsilon cl$, where A is the absorbance value at the chosen wavelength, c is the sample concentration (M) and l is the length of the light path through the sample (cm). The theoretical molar extinction coefficient ε at a specific wavelength can be calculated, for example using the ExPASyProtparam[12] tool (Gasteiger et al. 2005).

[12] http://web.expasy.org/ protparam/.

Some reagents used during the purification procedures such as imidazole or DTT exhibit absorbance at 280 nm. Remember to use the blank sample containing the same concentration of these reagents that is present in your sample. The best blank is the flow-through of a centrifugation.

When the IDP doesn't contain any aromatic residues, a refractometer is a valuable instrument for measuring protein concentration. This method is very sensitive to buffer conditions, so a standard curve should be created in the exact same buffer as the protein of interest. The preparation of a *similar* buffer will not work. The last step dialysis buffer, or even better the flow-through buffer, should be used when concentrating a pure protein sample. The same buffer should be used for the preparation of a standard-protein solution of known concentration to create a standard curve measured on the refractometer instrument. The target protein concentration can then be measured with high accuracy by interpolation.

A refractometer is very useful for IDPs that lack aromatic residues, but a standard curve should be created with exactly the same buffer as the one of the target IDP.

6.2 Handling IDPs by NMR

The use of NMR to study IDPs requires protein sample enrichment by the use of isotopic labelled sources of carbon and nitrogen and sometimes also deuterium. This will allow the use of advanced techniques such as the recently developed multidimensional NMR experiments (Bermel et al. 2012).

The optimization of NMR sample conditions is the most important task before starting NMR data acquisition. The quality of the sample can be easily checked by 1D ^{1}H spectra or even better by 2D ^{1}H ^{15}N high resolution heteronuclear single quantum coherence spectroscopy (HSQC), in particular for IDPs due to their typical low chemical shift dispersion (Chap. 3).

The most suitable NMR experiments to evaluate the feasibility of a complete NMR characterisation of the protein are the 2D experiments that correlate the amide nitrogen with the directly bound amide proton (HSQC) or carbonyl (CON). The low chemical shift dispersion provides the first indication that the protein is not characterized by a stable 3D structure. The heteronuclear correlations involving backbone nitrogen nuclei, both for H^{N}-N and C′-N, can be detected with several different variations of the experiments that may help to increase sensitivity, increase resolution, decrease experimental time, increase the number of detected cross peaks and so on (Chap. 3). The spectral quality can often also be improved through minor changes in the experimental conditions (temperature, pH, buffer, salt, etc.).

The analysis of individual peak line widths and signal to noise intensities should be addressed, in particular when playing with different buffers, pH, salt concentration and most important the protein sample concentration. A highly concentrated protein sample is always required for NMR spectroscopy but special care should be taken to avoid protein aggregation.

IDPs are often considered particularly prone to aggregation, but this represents only a small fraction of the disordered proteome (Theillet et al. 2014). As IDP protein concentration is an important factor in the induction of protein aggregation, a concentration dependence study should be performed. The temperature and pH should also be monitored as well as the use of different NMR fields for relaxation studies.

The suite of experiments generally used for the sequence-specific assignment of folded proteins can also be applied to IDPs, taking care to optimize the experimental set-up for resolution using a high number of acquired data points. The long magnetization transfer pathways and small scalar couplings, which generally drastically reduce the sensitivity of these experiments when applied to the study of folded proteins, have less of an impact when used with IDPs since the high protein flexibility causes an increase in coherence lifetime. Experiments with long coherence transfer delays as well as with multiple coherence transfer steps can therefore be planned. Better descriptions of suitable NMR experiments to study IDPs are reported in Chap. 3.

The highly flexible nature of IDPs induces extensive conformational averaging, reducing the nuclear chemical shift dispersion (Bertini et al. 2012). Taking flexibility to its extreme, chemical shifts progressively collapse to those of random coil polypeptides, causing extensive resonance overlap. The intrinsic chemical shift dispersion increases from protons to heteronuclei (^{13}C, ^{15}N). Therefore, exclusively heteronuclear NMR experiments based on ^{13}C direct detection have crucial relevance for IDP studies (Bermel et al. 2006a, 2006b; Braun et al. 1994; Zhang et al. 1997), of course in combination with ^{1}H detected experiments, as all the information available is welcome when studying complex systems. In addition, the determination of ^{15}N relaxation rates provides accurate information on the motional properties of the backbone for each amino acid, as well as a general estimation of the expected transverse relaxation rates, which of course have a large impact on the overall sensitivity of multidimensional NMR experiments.

6.3 Complementary Biophysical Techniques

NMR is known as the best technique to study IDPs at atomic level, providing detailed residue-specific information. However, it is useful to complement the information obtained from NMR with other biophysical methodologies to better understand all kinds of function-related transient, short, and long-range structural organisations (Chap. 7, Chap. 8).

Different biophysical methods can be used according to the information needed to complement NMR data. Mass spectrometry (MS), circular dichroism (CD), light scattering (LS), including dynamic light scattering (DLS) and small angle X-ray scattering (SAXS) are excellent complementary techniques of NMR data (Chap. 7, Chap. 8). In this section we will focus our attention on IDP sample preparation for different biophysical techniques.

The potential of MS for analysing proteins is due to the advances gained through the development of soft ionization techniques such as electrospray ionisation (ESI) and matrix-assisted laser desorption ionisation (MALDI), which can transform biomolecules into ions. ESI can be efficiently interfaced with separation techniques, expanding the range of applications in the Life Sciences (Di Marco and Bombi 2006). MALDI has the advantage of producing singly charged ions of peptides and proteins, minimizing spectral complexity.

These strategies can be used to retrieve accurate molecular weight measurements, determine the purity of a sample, verify amino acid substitutions, detect post-translational modifications, calculate the number of disulphide bridges, and analyse intermolecular interactions (e.g. protein-protein binding). Application of hydrogen-deuterium exchange mass spectroscopy can provide valuable information about the localisation of flexible regions and protein folding (Kaltashov et al. 2013).

Non-denaturing ESI-MS has recently been used to study IDPs. Although ESI can be used to study the conformational states of a protein, recent advances have been made to implement ion mobility electron spray ionisation (IM-ESI) to better understand additional IDP charge and shape characteristics. This technology has the potential to separate and discern different conformational families even for IDPs, so that the conformational properties of different forms of the same protein can be compared and structural events taking place during the transition from co-populated conformations can be monitored, giving interesting insights into their respective conformational behaviour patterns (Knapman et al. 2013).

NMR requires higher sample concentrations for the implementation of some specific experiments, but the higher concentration samples can promote intermolecular interactions that may lead IDPs to aggregate in solution. This question can be easily addressed by a concentration dependence studied by MS using a large spectral window to look for higher molecular mass species present.

Use MS as one of the complementary techniques to characterize the different conformational states of IDPs.

CD is an excellent tool that allows the quick evaluation of the presence of secondary structural elements (e.g. arrangement of peptide bonds in secondary structure elements such as helices and strands) of proteins in solution and is based on differential absorption of left- and right-handed circularly polarized light, which allows for the assessment of the secondary structural properties of a protein or protein regions (Chap. 7).

IDPs present particular CD characteristics different from those of folded proteins and also different from random coil polypeptides, presenting specific conformational preferences and thus revealing partially populated secondary structure content. These dynamic secondary structure elements can be stabilized or perturbed by temperature, by different chemical agents such as solvents, pH, ionic strength, and reducing agents, by post-translational modifications such as phosphorylation, and by the presence of metal ligands. The secondary structural properties of IDPs can therefore be studied by CD measurements, changing the chemical conditions to analyse the nature of the intrinsically disordered protein samples.

Proteins present CD bands in the far ultraviolet (UV) or amide region (175–250 nm), providing information about the secondary structure content, mostly based on the asymmetric conformation attained by the main polypeptide backbone. CD bands in the far-UV region are characteristic for different types of secondary structure. The α helix structure displays the most invariable band pattern: a characteristic spectrum with a positive band at 190 nm and two negative bands at 208 and 222 nm. Beta sheet elements, however, are more variable, with a positive band at around 198 nm and a single negative band ranging from 214 to 218 nm, depending on the type of structure. The random coil conformation is characterized by a negative band below 200 nm (Kelly et al. 2005).

Special care should be devoted to sample preparation. All samples prepared for CD measurements should be of the highest possible purity due to the fact that contaminations lead to deceptive results. The high concentrations of chloride and nitrate, as well as common chelators (ethylene glycol tetraacetic acid (EGTA)/ethylenediaminetetraacetic acid (EDTA)) and reducing agents (dithiothreitol and 2-mercaptoethanol) should be avoided in the buffer. It is also not advisable to use certain organic solvents (dioxane, dimethyl sulfoxide) and some biological buffers such as HEPES, PIPES, and MES. If the IDP target is oxygen sensitive, anaerobic conditions should be established.

The CD data can be fitted using the secondary structure estimation programs such as K2D3[13] (Louis-Jeune et al. 2012).

> CD allows the determination of whether or not a purified protein is folded and how sample conditions affect its conformation or stability.

DLS, also known as photon correlation spectroscopy or quasi-elastic light scattering, analyses the temporal fluctuations of the light scattering intensity caused by hydrodynamic motions in solution. DLS is therefore an appropriate technique to determine the hydrodynamic radius (R_H) of a protein, also known as the Stokes radius. The R_H value reflects the apparent size adopted by the solvated tumbling molecule, making it possible to monitor the expansion or compaction of protein molecules.

[13] http://www.ogic.ca/projects/k2d3/.

Considering a spherical particle such as a folded protein, DLS can also be useful for determining molar mass. For IDPs, R_H is the most informative parameter DLS provides, allowing a characterization considering the buffer, concentration and temperature conditions. A comparison of the measured R_H radii with those of particular reference states, such as the compactly folded or fully unfolded states, is very informative. Comparing the R_H of native and denaturant conditions (such as 8 M urea or 6 M GdHCl) will provide information on how far the IDP under study is from the completely random coil state.

To characterize IDPs, it might be interesting to compare DLS data with proteins of the same molecular mass and also with well studied standard IDPs such as α-synuclein, ensuring the measurements are made under the same conditions for a proper R_H comparison.

A useful procedure to compare the measured Stokes radius for an IDP through DLS is to use well-known proteins with similar number of residues but with different disordered states, measured in the same conditions.

Coupled with chromatography, DLS also provides information about the aggregation state of the proteins. An essential point in sample preparation for DLS measurements is removal of interfering dust particles, for example by filtration of the protein solution. It is important to remove unwanted scattering events that will exclude coherence and destroy determinations of the diffusion coefficient and therefore the size of the sample of interest.

The use of DLS attached to a fast protein liquid chromatography (FPLC) with SEC system might be applied for the investigation of the aggregation state of the IDPs, as well as efficient separation of their different assemblies such as oligomers.

Although DLS allows an easy and quick way to determine the value of the hydrodynamic Stokes radius (R_H), obtainment of the geometric radius of gyration (R_g) through this technique is usually hampered by an excessively small size of the protein. This limitation can be overcome by utilization of SAXS. The ratio of R_g and R_H (R_g/R_H) provides useful shape information about a protein molecule, being the most informative parameter comparing SAXS and DLS data for the same IDP measured at the same buffer, temperature and concentration conditions.

Among many parameters that describe a molecule's size and shape, R_H is the most important information that can be obtained from DLS measurements of IDPs. A comparison with R_g from SAXS should be made.

SAXS complemented with NMR structural calculations is the major technique for describing IDP structural ensembles. SAXS provides low-resolution structural characterisations of biological macromolecules in solution, contributing with information on the IDP's hydrodynamic behaviour and the topology of the polypeptide chain. In a SAXS experiment, samples containing soluble protein are exposed to an X-ray beam. The different scattered beam intensities are further recorded by a detector as function of the scattering angles of the soluble protein, giving rise to an isotropic scattering intensity (I). The solvent scattering is subtracted and the background corrected intensity is presented as a radially averaged one-dimensional curve I(s) (Petoukhov and Svergun 2013). A monodisperse solution is required to take advantage of all the analysis potential of SAXS. SAXS can be very challenging to use in the case of polydisperse IDP sample solutions. For monodisperse solutions of non-interacting identical and randomly oriented proteins, the SAXS curve is proportional to the scattering of a single particle averaged over all orientations. The scattering profile therefore carries information about the major geometrical parameters of the particle. In particular, the molecular mass (M) of the solute and its radius of gyration (Rg) are derived from the slope of the Guinier plot. Moreover, the values of the hydrated particle volume (V) and its specific surface (S) can be obtained using the so-called Porod invariant (Petoukhov and Svergun 2013).

Many IDPs are prone to aggregate, making SAXS measurements a challenging task. The sample preparation conditions may need to be redesigned according to the SAXS instrumentation capabilities.

SAXS has been actively used to characterize the conformational flexibility of IDPs (Bernadó et al. 2007; Bernadó and Svergun 2012a). A complete description of SAXS techniques and data acquisition and analysis for IDPs is presented in Chap. 8.

SAXS not only provides shapes, oligomeric states, and quaternary structures of folded protein complexes, but also allows for the quantitative analysis of flexible systems. This fact can be successfully used to study the conformational flexibility of IDPs (Bernadó et al. 2007; Bernadó and Svergun 2012b) and allows the exploration of different protein conformations in response to variations in external conditions such as buffer composition, ionic strength, pH, and temperature. Temperature measurements are particularly useful for studies of the thermodynamic characteristics of IDPs. However, in order to avoid radiation damage, SAXS measurements are usually made below room temperature.

A set of the different conditions to be studied should be designed in advance in order to estimate the number of conditions/samples to measure. Conditions such as buffer composition, ionic strength, pH, and temperature, among others, can be used to analyse the conformational properties of the IDP under investigation.

Sample preparation for SAXS experiments should follow the standard guidelines (Jacques et al. 2012). SAXS experiments typically require a highly pure, monodisperse protein solution in a concentration range from 1 to 10 mg/ml in order to fulfil the condition of a “dilute” solution. The concentrations must be determined accurately as it is necessary to appropriately normalize the scattering data and thus to estimate the effective molecular mass of the solute. If the sample is aggregated, the scattering data will be difficult or even impossible to interpret. A typical sample volume required for each single measurement depends on the SAXS station and a volume of about 50 µl is typically required for each single measurement. Each SAXS experiment at a given condition should be prepared in at least three different concentrations. Each condition such as buffer, ionic strength, pH, and temperature therefore requires 1–2 mg of purified sample. The concentration range can be prepared by dilution of a concentrated stock, if it is known that the protein is not affected by aggregation. In case the protein tends to aggregate it is better to prepare a stock of diluted protein sample and then prepare the concentrated samples immediately before the SAXS experiments. Is important to note that for each experiment at a given condition, the scattering of the buffer is also measured by SAXS for further subtraction; the background can therefore be corrected and the intensity can be presented as a radially averaged one-dimensional curve I(s).

All scattering-based techniques such as DLS, SAXS and small angle neutron scattering (SANS) require very homogeneous samples. The presence of even a small fraction of the aggregated material is known to dramatically affect the scattering profile, making data interpretation difficult. The buffer composition must precisely match the composition of the sample. Even a small mismatch in the chemical composition of the solvent between the buffer and the sample may lead to difficulties during background subtraction. The best approach is to use the dialysis buffer in which the protein was prepared.

Different strategies and software packages are available for the analysis of the SAXS data. One example is the protocol based on the use of the program ATSAS, which is in use at the German Electron Synchrotron DESY (Konarev et al. 2006; Petoukhov et al. 2012). The ATSAS package also includes the Ensemble Optimization Method (EOM), which has been developed as a strategy for the structural characterization of IDPs (Bernadó et al. 2007). This program takes into consideration a large amount of conformations that are in equilibrium and is especially useful for flexible systems.

Although SAXS provides a low-resolution structural characterization of biological macromolecules, it is an excellent complementary method to solution NMR to study IDPs.

Acknowledgments Kathleen McGreevy and Leonardo Gonnelli are gratefully acknowledged for their comments to the manuscript.

References

Armstrong DJ, Roman A (1993) The anomalous electrophoretic behavior of the human papillomavirus type 16 E7 protein is due to the high content of acidic amino acid residues. Biochem Biophys Res Commun 192:1380–1387. doi:10.1006/bbrc.1993.1569

Arnau J, Arnau J, Lauritzen C et al (2006) Current strategies for the use of affinity tags and tag removal for the purification of recombinant proteins. Protein Expr Purif 48:1–13. doi:10.1016/j.pep.2005.12.002

Babich H, Stotzky G (1978) Toxicity of zinc to fungi, bacteria, and coliphages: influence of chloride ions. Appl Environ Microbiol 36:906–914

Bermel W, Bertini I, Felli IC et al (2012) Speeding up sequence specific assignment of IDPs. J Biomol NMR 53:293–301. doi:10.1007/s10858-012-9639-0

Bernadó P, Svergun DI (2012a) Structural analysis of intrinsically disordered proteins by small-angle X-ray scattering. Mol BioSyst 8:151–167. doi:10.1039/c1mb05275f

Bernadó P, Svergun DI (2012b) Analysis of intrinsically disordered proteins by small-angle X-ray scattering. Methods Mol Biol 896:107–122. doi:10.1007/978-1-4614-3704-87

Bernadó P, Mylonas E, Petoukhov MV et al (2007) Structural characterization of flexible proteins using small-angle X-ray scattering. J Am Chem Soc 129:5656–5664. doi:10.1021/ja069124n

Block H, Maertens B, Spriestersbach A et al (2009) Immobilized-metal affinity chromatography (IMAC): a review. Meth Enzymol 463:439–473. doi:10.1016/S0076-6879(09)63027-5

Brito RMM, Vaz WLC (1986) Determination of the critical micelle concentration of surfactants using the Fluorescent-Probe N-Phenyl-1-Naphthylamine. Anal Biochem 152:250–255. doi:10.1016/0003-2697(86)90406-9

Burgess R, Richard R, Murray P (2009) Refolding solubilized inclusion body proteins. Methods Enzymol 463:259–282

Coutard B, Danchin EGJ, Oubelaid R et al (2012) Single pH buffer refolding screen for protein from inclusion bodies. Protein Expr Purif 82:352–359. doi:10.1016/j.pep.2012.01.014

Cowieson NP, Wensley B, Listwan P et al (2006) An automatable screen for the rapid identification of proteins amenable to refolding. Proteomics 6:1750–1757. doi:10.1002/pmic.200500056

Das U, Hariprasad G, Ethayathulla AS et al (2007) Inhibition of protein aggregation: supramolecular assemblies of arginine hold the key. PLoS ONE 2:e1176. doi:10.1371/journal.pone.0001176

Davey NE, Haslam NJ, Shields DC et al (2010) SLiMFinder: a web server to find novel, significantly over-represented, short protein motifs. Nucleic Acids Res 38:W534–W539. doi:10.1093/nar/gkq440

Di Marco VB, Bombi GG (2006) Electrospray mass spectrometry (ESI-MS) in the study of metal–ligand solution equilibria. Mass Spectrom Rev 25:347–379. doi:10.1002/mas.20070

Disfani FM, Hsu W-L, Mizianty MJ et al (2012) MoRFpred, a computational tool for sequence-based prediction and characterization of short disorder-to-order transitioning binding regions in proteins. Bioinformatics 28:i75–i83. doi:10.1093/bioinformatics/bts209

Dunker AK, Obradovic Z (2001) The protein trinity—linking function and disorder. Nat Biotechnol 19:805–806. doi:10.1038/nbt0901-805

Dunker AK, Lawson JD, Brown CJ et al (2001) Intrinsically disordered protein. J Mol Graph Model 19:26–59

Dunker AK, Oldfield CJ, Meng J et al (2008) The unfoldomics decade: an update on intrinsically disordered proteins. BMC Genomics 9(Suppl 2):S1. doi:10.1186/1471-2164-9-S2-S1

Esposito D, Chatterjee DK (2006) Enhancement of soluble protein expression through the use of fusion tags. Curr Opin Biotechnol 17:353–358. doi:10.1016/j.copbio.2006.06.003

Eswar N, Webb B, Marti-Renom MA et al (2006) Comparative protein structure modeling using MODELLER. Curr Protoc Bioinformat UNIT 5.6. doi:10.1002/0471250953.bi0506s15

Gasteiger E, Hoogland C, Gattiker A et al (2005) Protein identification and analysis tools on the ExPASy server. 571–607. doi:10.1385/1-59259-890-0:571

Getz EB, Xiao M, Chakrabarty T et al (1999) A comparison between the sulfhydryl reductants tris(2-carboxyethyl)phosphine and dithiothreitol for use in protein biochemistry. Anal Biochem 273:73–80. doi:10.1006/abio.1999.4203

Graceffa P, Jancsó A, Mabuchi K (1992) Modification of acidic residues normalizes sodium dodecyl sulfate-polyacrylamide gel electrophoresis of caldesmon and other proteins that migrate anomalously. Arch Biochem Biophys 297:46–51. doi:10.1016/0003-9861(92)90639-E

Gupta BB (1983) Determination of native and denatured milk proteins by high-performance size exclusion chromatography. J Chromatogr A 282:463–475. doi:10.1016/S0021-9673(00)91623-6

Hunt I (2005) From gene to protein: a review of new and enabling technologies for multi-parallel protein expression. Protein Expr Purif 40:1–22. doi:10.1016/j.pep.2004.10.018
Ignatova Z, Gierasch LM (2006) Inhibition of protein aggregation in vitro and in vivo by a natural osmoprotectant. Proc Natl Acad Sci U S A 103:13357–13361. doi:10.1073/pnas.0603772103
Jacques DA, Guss JM, Svergun DI, Trewhella J (2012) Publication guidelines for structural modelling of small-angle scattering data from biomolecules in solution. Acta Crystallogr D Biol Crystallogr 68:620–626. doi:10.1107/S0907444912012073
Jana S, Deb JK (2005) Strategies for efficient production of heterologous proteins in *Escherichia coli*. Appl Microbiol Biotechnol 67:289–298. doi:10.1007/s00253-004-1814-0
Jumpertz T, Tschapek B, Infed N et al (2011) High-throughput evaluation of the critical micelle concentration of detergents. Anal Biochem 408:64–70. doi:10.1016/j.ab.2010.09.011
Kaltashov IA, Bobst CE, Abzalimov RR (2013) Mass spectrometry-based methods to study protein architecture and dynamics. Protein Sci 22:530–544. doi:10.1002/pro.2238
Katzen F (2007) Gateway ®recombinational cloning: a biological operating system. Expert Opin Drug Discov 2:571–589. doi:10.1517/17460441.2.4.571
Kelly SM, Jess TJ, Price NC (2005) How to study proteins by circular dichroism. Biochem. Biophys. Acta (BBA)—Proteins Proteomics 1751:119–139. doi:10.1016/j.bbapap.2005.06.005
Kindermann B, Döring F, Fuchs D et al (2005) Effects of increased cellular zinc levels on gene and protein expression in HT-29 cells. Biometals 18:243–253. doi:10.1007/s10534-005-1247-y
Knapman TW, Valette NM, Warriner SL et al (2013) Ion mobility spectrometry-mass spectrometry of intrinsically unfolded proteins: trying to put order into disorder. Curr Anal Chem 9:181–191. doi:10.2174/1573411011309020004
Konarev PV, Petoukhov MV, Volkov VV et al (2006) ATSAS 2.1, a program package for small-angle scattering data analysis. J Appl Crystallogr 39:277–286. doi:10.1107/S0021889806004699
Krezel A, Latajka R, Bujacz GD et al (2003) Coordination properties of tris(2-carboxyethyl) phosphine, a newly introduced thiol reductant, and its oxide. Inorg Chem 42:1994–2003. doi:10.1021/ic025969y
Lesley SA (2009) Parallel methods for expression and purification. Methods. Enzymol. 463:767–785
Linn S (2009) Strategies and considerations for protein purifications. Methods. Enzymol. 463:9–19
Louis-Jeune C, Andrade-Navarro MA, Perez-Iratxeta C (2012) Prediction of protein secondary structure from circular dichroism using theoretically derived spectra. Proteins 80:374–381. doi:10.1002/prot.23188
Malhotra A (2009) Tagging for protein expression. In: Methods in Enzymology. Elsevier, pp 239–258
Mészáros B, Simon I, Dosztányi Z (2009) Prediction of protein binding regions in disordered proteins. PLoS Comput Biol 5:e1000376. doi:10.1371/journal.pcbi.1000376
Oates ME, Romero P, Ishida T et al (2013) D^2P^2: database of disordered protein predictions. Nucleic Acids Res 41:D508–D516. doi:10.1093/nar/gks1226
Outten CE, O'Halloran ATV (2001) Femtomolar sensitivity of metalloregulatory proteins controlling zinc homeostasis. Science 292:2488–2492. doi:10.1126/science.1060331
Petoukhov MV, Svergun DI (2013) Applications of small-angle X-ray scattering to biomacromolecular solutions. Int J Biochem Cell Biol 45:429–437. doi:10.1016/j.biocel.2012.10.017
Petoukhov MV, Franke D, Shkumatov AV et al (2012) New developments in the ATSAS program package for small-angle scattering data analysis. J Appl Crystallogr 45:342–350. doi:10.1107/S0021889812007662
Qoronfleh MW, Hesterberg LK, Seefeldt MB (2007) Confronting high-throughput protein refolding using high pressure and solution screens. Protein Expr Purif 55:209–224. doi:10.1016/j.pep.2007.05.014
Radivojac P, Iakoucheva LM, Oldfield CJ et al (2007) Intrinsic disorder and functional proteomics. Biophys J 92:1439–1456. doi:10.1529/biophysj.106.094045
Šali A, Blundell TL (1993) Comparative protein modelling by satisfaction of spatial restraints. J Mol Biol 234:779–815. doi:10.1006/jmbi.1993.1626

Schenk PM, Baumann S, Mattes R et al (1995) Improved high-level expression system for eukaryotic genes in Escherichia coli using T7 RNA polymerase and rare ArgtRNAs. Biotechniques 19:196–200

Shatzman AR (1995) Expression systems. Curr Opin Biotechnol 6:491–493

Sickmeier M, Hamilton JA, LeGall T et al (2007) DisProt: the database of disordered proteins. Nucleic Acids Res 35:D786–D793. doi:10.1093/nar/gkl893

Singh SM, Singh SM, Panda AK et al (2005) Solubilization and refolding of bacterial inclusion body proteins. J Biosci Bioeng 99:303–310. doi:10.1263/jbb.99.303

Theillet F-X, Kalmar L, Tompa P et al (2013) The alphabet of intrinsic disorder I. Act like a Pro: on the abundance and roles of proline residues in intrinsically disordered proteins. Intrinsically Disord Protein 1:0–12

Theillet F-X, Binolfi A, Frembgen-Kesner T et al (2014) Physicochemical properties of cells and their effects on intrinsically disordered proteins (IDPs). Chem Rev 140627063652000. doi:10.1021/cr400695p

Tong KI, Yamamoto M, Tanaka T (2008) A simple method for amino acid selective isotope labeling of recombinant proteins in *E. coli.* J Biomol NMR 42:59–67. doi:10.1007/s10858-008-9264-0

Uversky VN (2011) Intrinsically disordered proteins from A to Z. Int J Biochem Cell Biol 43:1090–1103. doi:10.1016/j.biocel.2011.04.001

Uversky VN (2013) A decade and a half of protein intrinsic disorder: biology still waits for physics. Protein Sci 22:693–724. doi:10.1002/pro.2261

Varadi M, Kosol S, Lebrun P et al (2013) pE-DB: a database of structural ensembles of intrinsically disordered and of unfolded proteins. Nucleic Acids Res. doi:10.1093/nar/gkt960

Vincentelli R, Canaan S, Campanacci V et al (2004) High-throughput automated refolding screening of inclusion bodies. Protein Sci 13:2782–2792. doi:10.1110/ps.04806004

Ward JJ, Sodhi JS, McGuffin LJ et al (2004) Prediction and functional analysis of native disorder in proteins from the three kingdoms of life. J Mol Biol 337:635–645. doi:10.1016/j.jmb.2004.02.002

Weinhandl K, Winkler M, Glieder A et al (2014) Carbon source dependent promoters in yeasts. Microb Cell Fact 13:5. doi:10.1186/1475-2859-13-5

Xue B, Dunbrack RL, Williams RW et al (2010) PONDR-FIT: a meta-predictor of intrinsically disordered amino acids. Biochim Biophys Acta 1804:996–1010. doi:10.1016/j.bbapap.2010.01.011

Chapter 7
Biophysical Methods to Investigate Intrinsically Disordered Proteins: Avoiding an "Elephant and Blind Men" Situation

Vladimir N. Uversky

Abstract Intrinsically disordered proteins (IDPs) and hybrid proteins possessing ordered domains and intrinsically disordered protein regions (IDPRs) are highly abundant in various proteomes. They are different from ordered proteins at many levels, and an unambiguous representation of an IDP structure is a difficult task. In fact, IDPs show an extremely wide diversity in their structural properties, being able to attain extended conformations (random coil-like) or to remain globally collapsed (molten globule-like). Disorder can differently affect different parts of a protein, with some regions being more ordered than others. IDPs and IDPRs exist as dynamic ensembles, resembling "protein-clouds". IDP structures are best presented as conformational ensembles that contain highly dynamic structures interconverting on a number of timescales. The determination of a unique high-resolution structure is not possible for an isolated IDP, and a detailed structural and dynamic characterization of IDPs cannot typically be provided by a single tool. Therefore, accurate descriptions of IDPs/IDPRs rely on a multiparametric approach that includes a host of biophysical methods that can provide information on the overall compactness of IDPs and their conformational stability, shape, residual secondary structure, transient long-range contacts, regions of restricted or enhanced mobility, etc. The goal of this chapter is to provide a brief overview of some of the components of this multiparametric approach.

Keywords Historical background · Biphysical techniques · CD · FTIR · RR · Single molecule techniques

---

V. N. Uversky (✉)
Department of Molecular Medicine and USF Health Byrd Alzheimer's Research Institute, Morsani College of Medicine, University of South Florida, Tampa, FL 33612, USA
e-mail: vuversky@health.usf.edu

Biology Department, Faculty of Science, King Abdulaziz University, P.O. Box 80203, Jeddah 21589, Kingdom of Saudi Arabia

Institute for Biological Instrumentation, Russian Academy of Sciences, Pushchino 142290, Moscow Region, Russia

I. C. Felli, R. Pierattelli (eds.), *Intrinsically Disordered Proteins Studied by NMR Spectroscopy,* Advances in Experimental Medicine and Biology,
DOI 10.1007/978-3-319-20164-1_7

1 Intrinsic Disorder in a Few Bullet Points

As part of a book on the study of intrinsically disordered proteins (IDPs) by nuclear magnetic resonance (NMR), this chapter does not require a lengthy introduction of the protein intrinsic disorder phenomenon. Therefore, there is an exceptional possibility to introduce IDPs and hybrid proteins possessing both ordered domains and intrinsically disordered protein regions (IDPRs) in the form of a few short bullet points.

- IDPs and IDPRs are biologically active in the absence of unique structures (Dunker et al. 1998; Wright and Dyson 1999; Uversky et al. 2000a; Dunker et al. 2001; Tompa 2002; Daughdrill et al. 2005; Uversky and Dunker 2010);
- They are found in all of the proteomes characterized so far (Dunker et al. 2000; Ward et al. 2004; Dunker et al. 2001; Uversky and Dunker 2010; Uversky 2010);
- Their amino acid sequences/compositions are very different from the sequences and amino acid compositions of ordered proteins and domains (Dunker et al. 1998; Dunker et al. 2001; Uversky 2002b; Uversky et al. 2000a; Uversky and Dunker 2010; Williams et al. 2001; Romero et al. 2001; Radivojac et al. 2007; Vacic et al. 2007; Garner et al. 1998);
- IDPs and IDPRs are predictable from just the amino acid sequence (He et al. 2009);
- Structurally, IDPs and IDPRs are highly heterogeneous and can be more or less compact, possess smaller or larger amounts of flexible secondary structure, and contain smaller or larger numbers of tertiary contacts (Dunker and Obradovic 2001; Uversky and Dunker 2010; Uversky 2002b; Daughdrill et al. 2005);
- Although IDPs/IDPRs exist as highly dynamic ensembles (Dunker and Uversky 2010), their structures can be described reasonably well by a rather limited number of lower-energy conformations (Choy and Forman-Kay 2001; Huang and Stultz 2008);
- The functions of IDPs/IDPRs complement the functions of ordered proteins (Iakoucheva et al. 2002; Dunker et al. 2005; Uversky et al. 2005);
- Functionally, IDPs/IDPRs can be grouped into several broad classes such as assemblers, chaperones, display sites, effectors, entropic chains, and scavengers (Tompa 2002);
- Some IDP functions are described as “entropic chain activities” since they rely entirely on the constant motion of extended random-coil-like polypeptides (Dunker et al. 2001);
- IDPs are infrequently responsible for enzymatic catalysis, except for rare occasions in which collapsed IDPs exhibit catalytic activity (Uversky et al. 1996; Pervushin et al. 2007; Vamvaca et al. 2008; Woycechowsky et al. 2008);
- IDPRs commonly contain sites of various posttranslational modifications (Dunker et al. 2002; Iakoucheva et al. 2004);
- IDPs are promiscuous binders and are able to interact with nucleic acids, metal ions, heme groups, other small molecules, proteins, polysaccharides, and membrane bilayers (Uversky et al. 2000a; Dunker et al. 2002);

- Some IDPRs are able to interact specifically with multiple structurally unrelated partners (Oldfield et al. 2008);
- Many IDPs/IDPRs can participate in one-to-many and many-to-one binding (Dunker et al. 2005; Uversky et al. 2005);
- They frequently serve as hubs in protein interaction networks (Dunker et al. 2005; Uversky et al. 2005);
- IDPs/IDPRs can undergo disorder-to-order transitions caused by interaction with specific binding partner(s) (Dyson and Wright 2002; Oldfield et al. 2005);
- Some IDPRs can adopt different conformations upon binding to different partners (Oldfield et al. 2008);
- IDPs/IDPRs display a wide range of binding modes that produce a multitude of rather unusual complexes (Uversky 2011);
- Some IDPs/IDPRs do not fold (partially or completely) in their bound state, forming so-called disordered, dynamic, or fuzzy complexes (Nash et al. 2001; Mittag et al. 2008; Mittag et al. 2010; Uversky 2011);
- IDPs are commonly associated with the pathogenesis of various human diseases (Uversky et al. 2008).

2 Multiparametric Approach in the Structural Analysis of IDPs/IDPRs

Due to the highly dynamic nature of IDPs/IDPRs, their structures do not converge to a single conformation during the experimental time frame, giving rise to cloud-like conformational ensembles and suggesting that structural analysis of these proteins/regions is not an easy task. Since the determination of a unique high-resolution structure is not possible, a set of rather complex methods must be used to obtain experimental constraints on the ensemble of states that is sampled by the intrinsically disordered polypeptide chain (Uversky and Dunker 2012c).

It should be considered that all of the techniques used for the structural characterization of a protein have restrictions. For example, even though nuclear magnetic resonance (NMR) spectroscopy is considered to be the technique of choice for providing high-resolution structural information on IDPs in solution (Chaps. 2–5) (Daughdrill et al. 2005; Eliezer 2009; Jensen et al. 2010), this technique has some logical limitations such as protein size (increasing size leads to slower tumbling and shorter spin-spin relaxation times as well as increasingly complex spectra), increased redundancy due to the presence of tandem repeats, a lack of spectral dispersion due to similar environments for the various residues, and extreme line broadening due to the motional dynamics in the millisecond to microsecond time scale (which typically precludes direct NMR studies of molten globules). Furthermore, NMR information arises from nuclei and their local environments, so NMR approaches provide very little information regarding the overall size and shape of the IDPs (though overall size information is provided by diffusion estimates from line broadening) (Uversky and Dunker 2012c).

Due to their highly heterogeneous nature and conformational fluctuations occurring at multiple time scales, the full spectrum of structural and dynamic characteristics of IDPs cannot be gained by a single tool and clearly requires a multiparametric approach (Uversky and Dunker 2012c). The use of such a multiparametric approach for the structural and dynamic characterization of IDPs gives a number of important advantages. In essence, multiparametric analysis resembles the compound eyes of insects (Uversky and Dunker 2012c) which, compared with simple eyes, possess a very large view angle, and can detect fast movement (Völkel et al. 2003). Many of the tools in the modern arsenal of biophysical techniques that can be used to characterize the dynamic structure of IDPs have been the subject of focused reviews and books (Receveur-Brechot et al. 2006; Daughdrill et al. 2005; Eliezer 2009; Jensen et al. 2010; Longhi and Uversky 2010; Uversky and Dunker 2012a, b, c).

The biophysical tools included into this set of "compound eyes" of the multiparametric approach were elaborated to provide information on the different structural levels of a protein such as the overall compactness, residual secondary structure, transient long-range contacts, shape, conformational stability, and presence of regions of restricted or enhanced mobility. Obviously, a complete picture can be obtained only through the simultaneous application of various techniques sensitive to the different structural levels of a protein molecule. In other words, the use of the multiparametric approach allows one to avoid the "elephant and blind men" situation described below (see the Box 7.1).

Box 7.1. A multiparametric approach to the structural and conformational analysis of intrinsically disordered proteins allows researchers to avoid the "blind men and an elephant" situation by generating a complete picture of an IDP (elephant) based on the techniques sensitive to its different structural levels (elephant's leg, tail, trunk, ear, belly, and tusk).

The "elephant and blind men" story

The story below is reproduced from http://www.jainworld.com/literature/story25.htm.

Once upon a time, there lived six blind men in a village. One day the villagers told them, "Hey, there is an elephant in the village today."

They had no idea what an elephant is. They decided, "Even though we would not be able to see it, let us go and feel it anyway." All of them went where the elephant was. Everyone of them touched the elephant.

"Hey, the elephant is a pillar," said the first man who touched his leg.

"Oh, no! It is like a rope," said the second man who touched the tail.

"Oh, no! It is like a thick branch of a tree," said the third man who touched the trunk of the elephant.

"It is like a big hand fan" said the fourth man who touched the ear of the elephant.

"It is like a huge wall," said the fifth man who touched the belly of the elephant.

"It is like a solid pipe," Said the sixth man who touched the tusk of the elephant.

They began to argue about the elephant and everyone of them insisted that he was right. It looked like they were getting agitated. A wise man was passing by and he saw this. He stopped and asked them, "What is the matter?" They said, "We cannot agree to what the elephant is like." Each one of them told what he thought the elephant was like. The wise man calmly explained to them, "All of you are right. The reason every one of you is telling it differently because each one of you touched the different part of the elephant. So, actually the elephant has all those features what you all said."

"Oh!" everyone said. There was no more fight. They felt happy that they were all right."

3 Warning: You are Dealing With an IDP!

The vast majority of techniques used for the structural characterization of IDPs/IDPRs were initially elaborated for the analysis of the structural properties and conformational behaviour of ordered proteins. Therefore, these techniques were not originally intended to provide information on IDPs/IDPRs without unique structure. To avoid potential misinterpretations, extreme caution should be used while interpreting data generated by these structure-centred approaches, since information on the presence of intrinsic disorder often results from the absence of a signal characteristic for the ordered protein. Again, a simultaneous analysis of a given protein by several techniques sensitive to the different structural levels provides the most unambiguous characterization of its disordered ensemble (Uversky and Dunker 2012c).

4 Techniques for the Analysis of Intrinsically Disordered Proteins

4.1 X-Ray Crystallography: Regions with Missing Electron Density and High B-Factor

In X-ray crystallographic experiments, the increased flexibility of atoms in disordered regions leads to non-coherent X-ray scattering and makes them "invisible", therefore giving rise to regions with missing electron density. Many proteins in the Protein Data Bank (PDB)[1] have portions of their sequences missing from the determined structures (so-called missing electron density) (Bloomer et al. 1978; Bode et al. 1978), with these unobserved atoms being assumed to be disordered. Curiously, the use of the term "disorder" by crystallographers when describing missing regions of electron density has a long history (Arnone et al. 1971). Many studies have demonstrated the biased nature of the PDB toward ordered proteins (Peng et al. 2004; Gerstein 1998). In fact, intrinsically disordered proteins are under-represented in the PDB, with roughly 6 % of the proteins in the dataset of PDB proteins that share less than 25 % sequence identity presenting long (≥30 consecutive amino acids) regions of missing residues (Le Gall et al. 2007). Although some missing density corresponds to wobbly, structured domains rather than to intrinsically disordered ensembles, such wobbly domains are evidently not very common among the long regions of missing electron density (Radivojac et al. 2004).

Traditionally, searching the PDB for regions with missing electron density (e.g. by using the "REMARK 465 MISSING RESIDUE" option of the PDB file) represents one of the classical approaches for finding experimentally validated IDPRs. This approach is used, for example, in MobiDB[2] to show PDB-based experimental disorder in a query protein (Di Domenico et al. 2012).

Besides the regions of missing electron density, crystallized proteins often contain regions with a high B-factor (the B-factor of the α-carbon and the B-factor averaged over the four backbone atoms are the commonly used measures of the residue flexibility of folded proteins (Karplus and Schulz 1985; Vihinen et al. 1994; Kundu et al. 2002)), which in crystal structures of macromolecules reflects the uncertainty in atom positions in the model and often represents the combined effects of thermal vibrations and static disorder (Rhodes 1993). In order to differentiate between flexible but ordered regions and IDPRs, comparisons were made among four categories of protein flexibility: low-B-factor ordered regions, high-B-factor ordered regions, short disordered regions, and long disordered regions (with the last two categories being selected as the short and long regions of missing electron density, respectively) (Radivojac et al. 2004). The high-B-factor regions were shown to be more similar to IDPRs than to ordered regions with low-B-factor. Furthermore, the observed distinctive amino acid biases of high-B-factor ordered regions, short IDPRs, and long IDPRs clearly indicated that the sequence determinants for these flexibility categories differ from one another, suggesting that the amino acid attributes

[1] http://www.rcsb.org.

[2] http://mobidb.bio.unipd.it/.

that specify flexibility and intrinsic disorder are distinct and not merely quantitative differences on a continuum (Radivojac et al. 2004).

4.2 NMR: Another High-Resolution Technique for the Characterization of Protein Disorder

Since the detailed consideration of the peculiarities of application of NMR-based techniques for the structural characterization of IDPs/IDPRs is given in several preceding chapters (Chaps. 3–5), only a few very basic points are given below. Heteronuclear multidimensional NMR can be used for gaining precise structural information on IDPs/IDPRs and can also provide direct measurement of the mobility of IDPRs. Recent advances in this technology have allowed the complete assignment of resonances for several unfolded and partially folded proteins (Chap. 3), as well as the disordered fragments of folded proteins (Wright and Dyson 1999; Dyson and Wright 2002; Eliezer 2007, 2009; Mittag and Forman-Kay 2007; Jensen et al. 2009).

Although long-range contacts in IDPs are transient and difficult to detect by traditional NMR approaches (such as chemical shifts or long-range nuclear Overhauser effects (NOEs) typically used for obtaining topological distance constraints in well-structured proteins), paramagnetic relaxation enhancement (PRE) has been shown to be a highly successful tool for the unambiguous detection of the long-range contacts in disordered protein ensembles (Eliezer 2009).

4.3 Techniques for the Analysis of Intrinsic Disorder in Cells

In-cell NMR spectroscopy represents a very promising approach for the structural characterization of IDPs in their natural environments, i.e. within cells. In fact, the acquisition of heteronuclear multidimensional NMR spectra of biomacromolecules inside living cells is the only currently available technique for investigating the 3D structure and dynamics of proteins at atomic detail in the intracellular environment. Successful in-cell characterizations of IDPs have been reported for both bacterial and eukaryotic cells (Dedmon et al. 2002; Li et al. 2008; McNulty et al. 2006; Bodart et al. 2008; Binolfi et al. 2012; Freedberg and Selenko 2014; Takaoka et al. 2013; Selenko and Wagner 2007; Theillet et al. 2014). Peculiarities of the in-cell NMR analysis of IDPs are described elsewhere in this book (Chap. 10) and therefore will not be discussed in this chapter.

Another spectroscopic technique, Fourier transform infrared (FTIR) microspectroscopy (Orsini et al. 2000), can also be used for the characterization of complex biological systems (Shaw et al. 1999; Shaw and Mantsch 1999; Heraud and Tobin 2009) such as intact cells, tissues, and even whole model organisms (Wood et al. 2008; Ami et al. 2008; Walsh et al. 2009; Kretlow et al. 2006; Ami et al. 2004; Ami et al. 2012). This non-invasive and label free approach provides a unique molecular fingerprint that contains information on the composition and structure of the main cellular biomolecules such as proteins, nucleic acids, lipids, and carbohydrates.

Therefore, FTIR microspectroscopy provides an invaluable tool for the *in situ* analysis of various biological processes that take place within the biological system. This spectroscopic approach was successfully used to monitor *in situ* protein folding and aggregation (Diomede et al. 2010; Gonzalez-Montalban et al. 2008; Doglia et al. 2008; Kneipp et al. 2003; Choo et al. 1996), cell differentiation (Ami et al. 2008; Tanthanuch et al. 2010), and cancerogenesis (Kelly et al. 2009; Schultz et al. 1997).

Utilization of the susceptibility of IDPs to the 20 S proteasome is an alternative approach for the operational determination of these proteins inside the cell (Tsvetkov et al. 2008; Tsvetkov and Shaul 2012). In fact, both IDPs and ordered proteins are subjected to proteasomal degradation. However, unlike ordered proteins that have to unfold prior degradation within the 20 S catalytic subunit of the proteasome, there is no crucial need of protein unfolding step for the IDPs to be degraded by the proteasome. As a result, IDPs can be degraded by the 20 S proteasome subunit by default, without being polyubiquitinated or undergoing any other modifications. However, they can escape degradation by default by a number of mechanisms, with a more general one being interaction with a partner, a "nanny" protein. Therefore, one can define IDP by conducting a set of specially designed cell free and cell culture experiments (Tsvetkov et al. 2008; Tsvetkov and Shaul 2012). IDPs and IDPRs are also characterized by an increased susceptibility to proteolytic degradation *in vitro* and therefore limited proteolysis can be used to indirectly confirm their increased flexibility (Dunker et al. 2001; Iakoucheva et al. 2001a; Fontana et al. 1986, 1993, 1997a, b).

Fast relaxation imaging (FReI) is a promising method based on making movies of fast protein dynamics inside living cells (Ebbinghaus et al. 2010; Dhar et al. 2012; Dhar and Gruebele 2011). Here, a combination of temperature jumps and Förster resonance energy transfer (FRET) imaging is used to probe biomolecular dynamics and stability inside a single living cell with high spatiotemporal resolution (Ebbinghaus et al. 2010). Protein dynamics are measured by jumping the temperature of the cell up by a few degrees with an infrared laser, and then monitoring FRET by imaging fluorescence in the donor and acceptor (Dhar et al. 2012). The corresponding FReI sample consists of live cells expressing a recombinant protein of interest sandwiched between two fluorescent proteins that comprise a FRET pair. The protein stability and folding/aggregation/binding kinetics *in vivo* are evaluated based on the assumption that the average distance between FRET pairs will change upon (un)folding or aggregation (Ebbinghaus et al. 2010; Dhar et al. 2012; Dhar and Gruebele 2011).

4.4 Low-Resolution Spectroscopic Techniques

4.4.1 Tools for Protein Secondary Structure Analysis

Low levels of ordered secondary structure may be detected by several spectroscopic techniques including far ultra violet (UV) CD (Adler et al. 1973; Provencher and

Glockner 1981; Johnson 1988; Woody 1995; Fasman 1996; Kelly and Price 1997; Vassilenko and Uversky 2002), optical rotatory dispersion (ORD) (Chen et al. 2003a), Fourier transform infrared spectroscopy (FTIR) (Uversky et al. 2000a), Raman optical activity (Smyth et al. 2001), and deep-UV resonance Raman spectroscopy (Xu et al. 2005; Xu et al. 2008).

4.4.1.1 Some Basic Principles of Circular Dicroism (CD) and Optical Rotation Dispersion (ORD)

Circular dicroism (CD) and optical rotation (OR) phenomena are defined by the unequal absorption and refraction of left- and right-handed circularly polarized light by matter. These phenomena occur when asymmetric/chiral molecules (enantiomers) or molecules/chromophores in the asymmetric environment interact with polarized light. The electric field, E, of the linearly polarized light oscillates sinusoidally in a single plane. When viewed from the front, the sinusoidal wave can be visualized as the resultant of two vectors of equal length, which trace out circles, one which rotates clockwise (E_R) and the other which rotates counterclockwise (E_L). Therefore, a linearly polarized wave can be considered a superposition of a right- and a left-handed circular polarized wave of the same frequency and phase.

The asymmetric or optically active molecules may absorb right- and left-handed circularly polarized light to different extents and also have different indices of refraction for the two waves. As a result of this unequal refraction and absorption of left- and right-handed circularly polarized light, the plane of the light wave is rotated and the addition of the differently absorbed E_R and E_L components generates the elliptically polarized light. The phenomenon of different refraction is associated with optical rotatory dispersion (ORD), whereas different absorption of left- and right-handed circularly polarized light gives rise to circular dichroism (CD).

OR originates in the optically active media, where left- and right-circularly polarized waves propagate with different speeds due to the fact that these media have different refractive indices for left- and right-handed circularly polarized light: $n_L \neq n_R$. Due to the different refractive index for the two waves, their electric field vectors rotate at different speeds. Superposition of the two waves after a distance d yields a linearly polarized wave with the polarization plane rotated by an angle of rotation α that depends on the wavelength and the difference of the two refractive indices $\Delta n = n_L - n_R$. In other words, because of the inequality of n_L and n_R, there is the inclination of the plane of polarization of the superposition of left- and right handed light after passing through the optically active medium. One should keep in mind that the typical difference of the two refractive indices Δn is in the order of 10^{-6}, being therefore very small in comparison with the n_L and n_R values, which are usually between 1.0 and 1.8. However, one can measure Δn by measuring the rotation of the linearly polarized light. OR is typically reported as specific rotation [α] defined by the following equation: $[\alpha] = 100 \times \alpha / lC$, where l is the pathlength in dm and C is the concentration in g/100 ml. A plot of [α] versus wavelength represents the ORD spectrum.

The absorption of light by a chromophore is characterized by an absorption band that has a maximum at a particular wavelength λ_{max}. In the case of an optically active substance, a corresponding ORD spectrum is characterized by specific anomalous behavior in the neighborhood of the absorption band. Here, the OR absolute magnitude at first varies rapidly with wavelength, crosses zero at absorption maximum and then again varies rapidly with wavelength but in opposite direction. This phenomenon is known as Cotton effect, which is said to be positive if the optical rotation first increases as the wavelength decreases, and negative if the rotation first decreases (Djerassi 1960).

CD spectroscopy is a simple and powerful technique for the assessment of the secondary structure of a protein or protein domain. In fact, the alignment of the chromophores of the amides of the polypeptide backbone of proteins in ordered arrays of different secondary structure elements results in the shifting of their optical transitions or leads to the splitting of their optical transitions into multiple transitions due to "exciton" interactions (Greenfield 2006).

CD is reported either in units of $\Delta\varepsilon$, which is the difference in absorbance of E_R and E_L by an asymmetric molecule (i.e., the difference in the extinction coefficients for the right- and left-circularly polarized light), or in degrees (ellipticity). Here, ellipticity is defined as the angle whose tangent is the ratio of the minor to the major axis of the ellipse; i.e., $\tan\theta = (E_R - E_L)/(E_R + E_L)$. The two mentioned parameters $[\theta]$ (molar ellipticity measured in $\deg \times cm^2/dmol$) and ΔE are related via the following equation: $[\theta] = 3298.2\Delta\varepsilon = 100 \times \theta/lC$, where l is the cell pathlength, and C is the protein concentration.

Since typical CD spectra report on the $\pi \rightarrow \pi^*$ and $n \rightarrow \pi^*$ transitions of the same structural motifs (Woody 1968), CD spectroscopy is sensitive to global secondary structure and has long been used to monitor the amount of α-helix, β-sheet and random coil in proteins. In fact, because of different chromophore arrays interact with light differently, different elements of protein secondary structure have characteristic CD spectra. For example, stable α-helices have negative bands at 222 nm and 208 nm and a positive band at 193 nm, β-structural proteins have a negative band at 218 nm and a positive band at 195 nm, whereas disordered proteins have very low ellipticity above 210 nm and are characterized by a negative band near 195 nm (Greenfield 2006). Furthermore, a CD spectrum of any protein can be considered as a simple weighted sum of these reference spectra (Greenfield and Fasman 1969). A more accurate approach is based on assumption that a CD spectrum can be analyzed as a linear combination of the CD spectra of proteins whose secondary structure is known from X-ray crystallography (Provencher and Glockner 1981). This analysis is particularly sensitive for the evaluation of α-helical and β-sheet structures, whereas the β-turn and remainder structures are determined with essentially lower accuracies (Provencher and Glockner 1981). Although classical Provencher-Glockner type of analysis is heavily dependent on accurate evaluation of protein concentration, a more recent study showed that protein secondary structure can be estimated rather well independently of protein concentration (Raussens et al. 2003). In this approach, a single-wavelength normalization procedure is made first, and then a quadratic model equation including one or two wavelength intensities is applied. As a result, protein secondary structure is estimated based on the information on the values of CD signal at 193.0 nm, 196.0 nm, 207.0 nm (the point of normalization), 211.0 nm, and 234.0 nm (Raussens et al. 2003).

4.4.1.2 Circular Dichroism of Extended IDPs

Despite the fact that CD spectroscopy cannot provide the secondary structure of specific residues and therefore belongs to the category of the low-resolution structural techniques, this approach is definitely a tool of choice for the reliable and relatively fast evaluation of the overall secondary structure content in a given protein or peptide. Among obvious advantages of CD is the ability of this method to be used for the analysis of diluted samples (e.g., containing 20 μg or less of protein) in aqueous solutions under physiological conditions.

One should keep in mind that even extended IDPs (which are typically considered as the most disordered species of natural proteins) are not random coil polypeptides since their different regions possess specific conformational preferences, exhibiting dynamic secondary structure elements or residual secondary structure (Uversky 2002b). These dynamic secondary structure elements can be stabilized or perturbed by different chemical (solvent, ionic strength, pH) or physical (temperature) agents, by post-translational modifications, and by ligands (Chemes et al. 2012).

Figure 7.1a represents the far-UV CD spectra of several extended IDPs (α-synuclein, prothymosin α, phosphodiesterase γ-subunit, and caldesmon 636–771 fragment) in comparison with the far-UV CD spectra of typical ordered proteins. One can see that extended IDPs possess distinctive far-UV CD spectra with characteristic deep minima in the vicinity of 200 nm, and relatively low ellipticity at 220 nm. The analysis of these spectra yields a low content of ordered secondary structure (α-helices and β-sheets). It has been emphasized that distinctive features of far-UV CD spectra can be used to differentiate native coils and native pre-molten globules (Uversky 2002a). In fact, Fig. 7.1b represents a "double wavelength" plot, $[\theta]_{222}$ vs. $[\theta]_{200}$, that was proposed as a tool to assort extended IDPs into two non-overlapping groups characterized by different levels of residual secondary structure (Uversky 2002a). Figure 7.1b shows that approximately half of the ~100 proteins analysed by far-UV CD spectroscopy possessed the far-UV CD spectra characteristic of almost completely unfolded polypeptide chains: with $[\theta]_{200} = -(18{,}900 \pm 2{,}800)$ deg·cm²·dmol⁻¹ and $[\theta]_{222} = -(1700 \pm 700)$ deg·cm²·dmol⁻¹. On the other hand, the other half of these proteins possessed CD spectra consistent with the existence of some residual secondary structure, thereby resembling the pre-molten globule states of globular proteins (with $[\theta]_{200} = -(10700 \pm 1{,}300)$ $deg \cdot cm^2 \cdot dmol^{-1}$ and $[\theta]_{222} = -(3{,}900 \pm 1{,}100)$ $deg \cdot cm^2 \cdot dmol^{-1}$) (Uversky 2002a).

It was also emphasized that the difference in the shape of far-UV CD spectra alone does not provide exclusive grounds for an unambiguous discrimination between these two classes of extended IDPs. However, among more than 100 proteins with a coil-like shape of far-UV CD spectrum (Uversky 2002a), 23 proteins

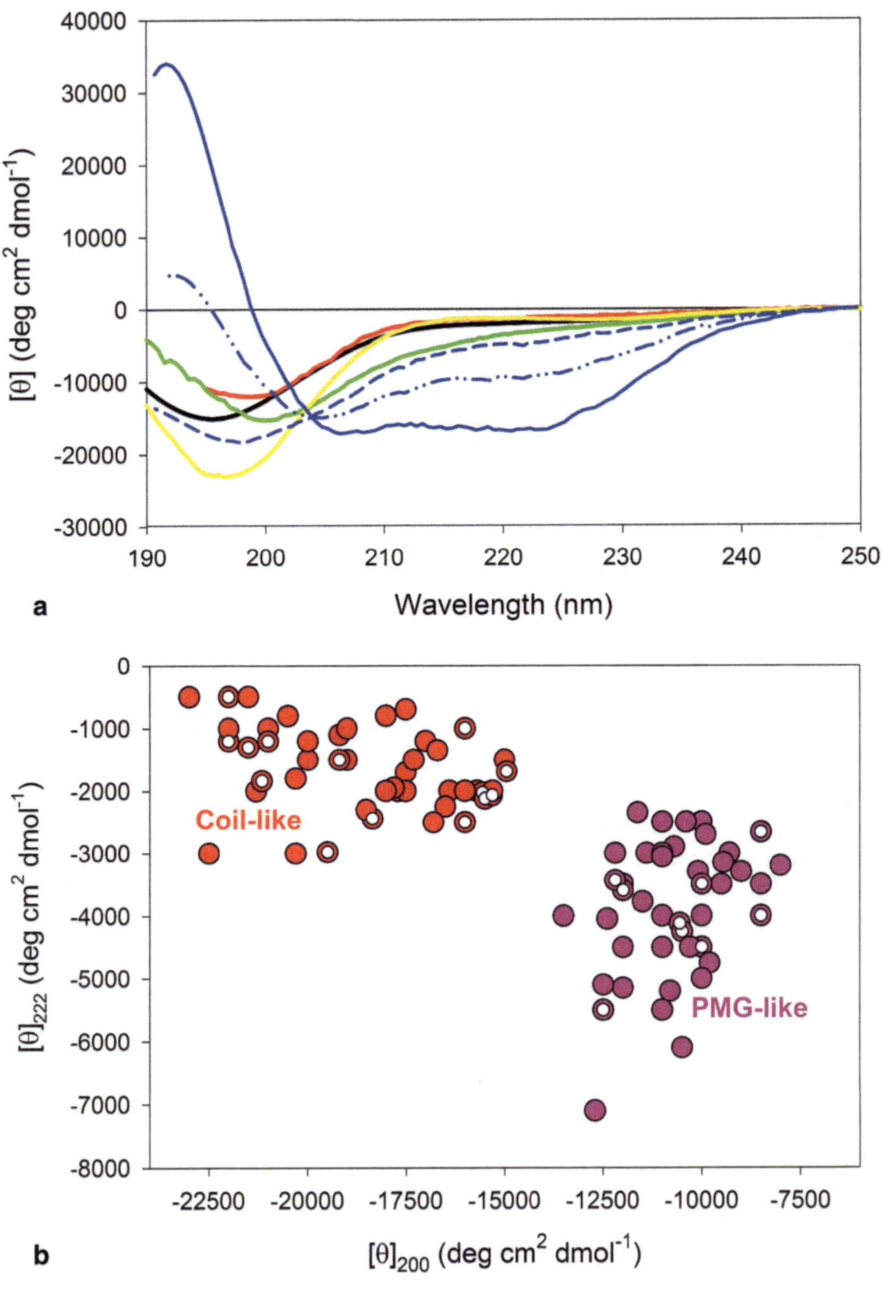

Fig. 7.1 Evaluating the secondary structure of extended IDPs by CD spectroscopy. **a** Far-UV CD spectra of extended IDPs, α-synuclein (*black line*), prothymosin α(*yellow line*), caldesmon 636–771 fragment (*green line*), and phosphodiesterase γ-subunit (*red line*). Far-UV CD spectra of a typical ordered protein (apomyoglobin) in different conformational states are shown for comparison as a set of *blue lines*. Spectra for folded, acid unfolded (pH 2.0) and a partially folded form stabilized at pH 2 by salt (50 mM Na_2SO_4) are shown as solid, short dash, and dash-dot-

were simultaneously characterized by CD and hydrodynamic methods (see below), making classification more certain. Intrinsic pre-molten globules and intrinsic coils studied by both techniques are indicated in Fig. 7.1b as white-dotted and black-dotted symbols, respectively. These data clearly showed that the more compact polypeptides (with the pre-molten globule-like hydrodynamic characteristics) possessed larger amounts of residual secondary structure than the less compact, coil-like IDPs (Uversky 2002a). This situation represents a nice illustration of the importance of the multiparametric approach, since it shows that the simultaneous application of CD spectroscopy and hydrodynamic analysis provides very strong evidence for the notion that extended IDPs should be subdivided into two structurally distinct groups: intrinsic coils and intrinsic pre-molten globules (Uversky 2002a).

4.4.1.3 Optical Rotatory Dispersion (ORD) in Analysis of Protein Secondary Structure

In addition to CD, a useful polarization spectroscopy technique for the analysis of protein secondary structure is optical rotation (OR) (Chen and Kliger 2012) in the form of optical rotatory dispersion (ORD) (Keston and Lospalluto 1953) or time-resolved optical rotatory dispersion (TRORD) (Lewis et al. 1985; Milder et al. 1990; Chen et al. 2010). ORD and CD measurements are analytically interconvertible by use of the Kramers-Kronig transform (Moscowitz 1962), and therefore they report the same secondary structure information. There are important practical differences between the two methods. In comparison to CD, ORD measurements have the advantage that the signals can be monitored away from the absorption bands (Gratzer and Cowburn 1969). This enhances the amount of light transmitted by the solution and renders it a sensitive probe for changes in protein structure. However, for this very reason, which offers a signal-to-noise advantage to ORD measurements, there are also more difficulties in the interpretation of ORD due to multiple Cotton contributions to the tail of the signal.

Already early studies revealed the usefulness of ORD for structural and conformational analysis of proteins and polypeptides (Jirgensons and Hnilica 1965; Jirgensons 1965). In fact, it has been shown that the ORD spectrum of polyglutamic acid in a disordered form is very different from the α-helical form of this polypeptide, that 20 of 28 globular proteins have ORD spectra closely resembling those of α-helical polyglutamic acid, and that denaturation of enzymes by acid, alkali, or the detergent sodium dodecyl sulfate was not accompanied by complete disorganization of their structure (Jirgensons 1965). These and other studies revealed that protein secondary structure can be efficiently evaluated the analysis of their ORD spectra (Chen et al. 1972).

---

dot lines, respectively. **b** Analysis of the far-UV CD spectra in terms of double wavelength plot, $[\theta]_{222}$ vs. $[\theta]_{200}$, provides a means for division of the extended IDPs into coil-like (*red circles*) and pre-molten globule-like (*PMG-like*) subclasses (*pink circles*). Intrinsic pre-molten globules and intrinsic coils for which the hydrodynamic parameters were measured are marked by *white-dotted* symbols. Data from (Uversky 2002a) were used to make this plot

Far-UV time-resolved ORD (TRORD) was shown to give significant secondary structure insights into the kinetics of many protein function-related reactions (e.g. phytochrome, myoglobin, photoactive yellow protein, and light-, oxygen-, or voltage-regulated domains (Chen et al. 1993; Chen et al. 1997; Chen et al. 2007b; Chen et al. 2003b; Chen and Kliger 1996)) and folding reactions (e.g. cytochrome *c* and azobenzene cross-linked peptides (Chen et al. 1998; Chen et al. 1999; Chen et al. 2003c, 2004; Chen et al. 2003d; Chen et al. 2007a; Chen et al. 2008)).

4.4.1.4 Fourier-Transform Infrared Spectroscopy (FTIR)

FTIR spectroscopy is one of the most powerful tools for the study of protein conformation, dynamics, and aggregation (Arrondo et al. 1993; Arrondo and Goni 1999; Seshadri et al. 1999; Barth and Zscherp 2002; Barth 2007; Goormaghtigh et al. 1994c, b, a; Natalello and Doglia 2010). Infrared light can be absorbed by a molecular vibration when the frequencies of light and vibration coincide. The frequency of the vibration and the probability of absorption depend on the strength and polarity of the vibrating bonds and are therefore influenced by intra- and intermolecular effects. In first approximation, the relationships between the vibrational spectrum of a molecule and its structure and environment can be illustrated reasonably well by the simple two-atomic oscillator. According to this approximation, the frequency of such a two-atomic oscillator increases when the force constant increases, that is when the bond strength increases. As a result, double and triple bonds absorb at higher wavenumber than single bonds since for double and triple bonds the force constant is approximately twice or three times that of a single bond. The second factor significantly influencing the frequency of oscillations is related to the masses of the vibrating atoms, with smaller atoms being characterized by the faster vibration. Also, different types of vibrations are typically distinguished in a molecule with several atoms: stretching vibrations (where the bond length elongates and contracts periodically), bending vibrations (where the bond angle of a 3-atomic fragment of the molecule changes in plane or out of plane), and torsional vibrations (where the torsional angle of a 4-atomic fragment of the molecule oscillates). Bond lengths and bond angles are known as internal coordinates of a molecule. In a typical situation, several of such internal coordinates are coupled and oscillate together with the same frequency and pass through their equilibrium position at the same time. Such type of motion, termed a normal mode of vibration, corresponds to a vibrational degree of freedom. Each normal mode is independent from the others.

There are several normal modes describing the behaviour of a polypeptide chain, whose structural repeat unit, the peptide group, generates up to nine characteristic bands named amide A, B, I, II, III, IV, V, VI, and VII, with amide I and amide II being the two major bands of protein infrared spectrum. Some of these modes are briefly described below.

N-H stretching vibrations—amides A and B (~3300 and~3170 cm^{-1}*).* The N-H stretching vibration gives rise to the amide A band which is located between 3310 and 3270 cm^{-1}. This type of vibration exists exclusively within the NH group and is therefore insensitive to the conformation of the polypeptide backbone, being, however, strongly dependent on the strength of the hydrogen bond. Both the amide A and amide B bands (which corresponds to the weak absorption between 3100 and 3030 cm^{-1}) are the parts of the Fermi resonance between the first overtone of amide II and the N-H stretching vibration. In α-helical polypeptides, the N-H stretching vibration is resonant with an overtone of the amide II vibration, in β-sheets with an amide II combination mode (Barth 2007).

C=O stretching—amide I (~1650 cm^{-1}). The amide I vibration corresponds to absorption near 1650 cm^{-1}. This band arises mainly from the C=O stretching vibration with minor contributions from the out-of-phase C-N stretching vibration, the C-C-N deformation, and the N-H in-plane bend (Barth 2007). The latter is responsible for the sensitivity of the amide I band to N-deuteration of the backbone. The extent to which the several internal coordinates contribute to the amide I normal mode depends on the backbone structure (Krimm and Bandekar 1986).

Although the amide I vibration is known to be affected by the nature of the side chain (Measey et al. 2005), the strong dependence of this vibration mode on secondary structure of the backbone determines the wide usage of the amide vibration for secondary structure analysis in protein structural studies. In fact, major types of protein secondary structure have specific contributions to the overall infrared spectrum of a protein. Although these secondary structure-related bands are broadened, essentially overlapped, and typically cannot be directly distinguished in the amide envelope, they can be efficiently resolved using a Fourier self-deconvolution protocol (Susi and Byler 1986; Byler and Susi 1986; Susi and Byler 1987). For example, β-turns absorb in the 1682–1662 cm^{-1} region, whereas β-sheet structure is characterized by the peaks at the 1689–1682 and 1637–1613 cm^{-1}, with intramolecular antiparallel β-sheets being ascribed to the 1630–1640 cm^{-1} region, α-helices and 3_{10}-helices have peaks at 1648–1658 cm^{-1} and 1660–1666 cm^{-1} respectively, whereas unordered structure has a broad band at within the 1645–1637 cm^{-1} region (Barth 2007; Barth and Zscherp 2002).

Amide II (~1550 cm^{-1}). The amide II mode is the out-of-phase combination of the N-H in plane bend and the C-N stretching vibration with smaller contributions from the C=O in plane bend and the C-C and N-C stretching vibrations (Barth 2007). Similar to the amide I vibration, the amide II vibration is hardly affected by side chain vibrations. Although the correlation between secondary structure types and frequency of the amide II vibration is less straightforward than that for the amide I vibration (Barth 2007; Jackson and Mantsch 1991), this band can be used for obtaining valuable structural information and secondary structure evaluation (Oberg et al. 2004). In fact, the very useful application of the amide II band in protein structure analysis is based on the observation that N-deuteration converts the amide II mode to largely a C-N stretching vibration at 1490–1460 cm^{-1} (named amide II′ mode) (Barth 2007). Since the N-^{2}H bending vibration has a considerably lower frequency than the N-^{1}H bending vibration, it does not couple with the C-N stretching vibration anymore, but is mixed with some other modes in the 1070−900 cm^{-1} region (Barth 2007). Because N-H bending contributes to the amide II mode but not to the amide II′ mode, both modes are differently affected by backbone conformation and the environment of the amide group. For example, hydrogen bonding will be sensed predominantly by the NH bending vibration, which contributes to the amide

II but not to the amide II′ vibration. As a result, the effect of a hydrogen bond will be larger on the amide II vibration than on the amide II′ vibration (Barth 2007).

Amide III (1400 – 1200 cm^{-1}). The amide III mode is the in-phase combination of the N-H bending and the C-N stretching vibration with small contributions from the C = O in plane bending and the C-C stretching vibration. In polypeptides and proteins, the composition of this mode is more complex, since it depends on side chain structure and since N-H bending contributes to several modes in the 1400 to 1200 cm^{-1} region (Barth 2007). Despite the noticeable side chain contributions to the amide III, this mode can be used for secondary structure prediction (Cai and Singh 1999, 2004). Upon N-deuteration, the N-^{2}H bending vibration separates out and the other coordinates become redistributed into other modes (Barth 2007).

The mentioned correlation between the vibrational spectra and helix, coil, sheet and turn structural motifs allows conformations that affect the C = O and N-H stretches of the amide bands to be monitored (Callender and Dyer 2002; Callender et al. 1998). IR spectra are sensitive to backbone secondary structures and to the formation of intermolecular β-sheets in protein aggregates. Furthermore, FTIR spectroscopy allows the examination of not only proteins in solution but also highly scattering protein aggregates (Natalello et al. 2012). This technique has been successfully applied for the characterization of IDP native secondary structure, induced folding, and aggregation (Natalello and Doglia 2010).

4.4.1.5 Raman Spectroscopy

Raman spectroscopy has emerged as a valuable tool for exploring the structure and dynamics of biomolecules over the last 40 years (Small et al. 1970). This technique relies on inelastic scattering, also known as Raman scattering, of the monochromatic light by matter, where the laser light interacts with molecular vibrations, phonons or other excitations in the system, resulting in the energy of the laser photons being shifted up or down. This shift in energy gives information about the vibrational modes in the system. Therefore, similar to infrared (IR) absorption, Raman scattering spectroscopy provides information on the vibrational, rotational, and other low-frequency modes in a system. Therefore these two techniques provide complementary information. Importantly, although both techniques depend on the same types of transitions, the technique-specific selection rules are somewhat different. As a result, weak bands in the IR may be strong in the Raman and vice versa (Pelton and McLean 2000). Typical Raman spectra reflect the difference between the incident and scattered radiation frequencies which depend on the types of bonds and their modes of vibration (Pelton and McLean 2000).

While resonance Raman, which involves the excitation of molecules within the wavenumber range of optical absorption of chromophores, is very frequently utilized to probe site-specific structural changes, non-resonance Raman spectroscopy can still be used to explore the secondary structure changes of proteins and peptides in a way similar to IR spectroscopy, i.e. by measuring and analysing the band profiles of amide backbone modes (Bandekar 1992).

In terms of application to IDPs, Raman spectroscopy can be used for the structural analysis and description of conformational changes of peptide backbones (Schweitzer-Stenner et al. 2012a). It has been emphasized that Raman spectroscopy has the ability to provide a variety of probes to look into the structure of disordered polypeptides (Maiti et al. 2004). Raman studies of protein secondary structure have followed the approach taken by CD studies and focused on correlating the positions of the amide I and amide III vibrations with the crystallographically determined fraction of each secondary structural element present in globular proteins (Williams 1986; Berjot et al. 1987; Thomas 2002; Sane et al. 1999). In relation to structural analysis of IDPs, Raman spectroscopy has several important advantages for characterization of their vibrational spectra and secondary structural tendencies (Maiti et al. 2004). An illustrative example of the power of Raman spectroscopy in conformational analysis of IDPs is given by the work of Maiti *et al.*, who showed that this technique provides a reliable structural description of α-synuclein in the presence and absence of methanol, sodium dodecyl sulfate (SDS), and hexafluoro-2-propanol (HFIP) (Maiti et al. 2004).

Another important Raman scattering-based technique is Raman optical activity (ROA), which is able to determine the vibrational optical activity by measuring a small difference in the intensity of Raman scattering from chiral molecules in right- and left-circularly polarized incident laser light, or, equivalently, as the intensity of a small circularly polarized component in the scattered light using incident light of fixed polarization (Barron et al. 2000). Contrarily to the conventional Raman spectrum of a protein, which is dominated by bands arising from the amino acid side chains that often obscure the peptide backbone bands, the largest signals in a ROA spectrum are often associated with vibrational coordinates that sample the most rigid and chiral parts of the structure; i.e., the protein backbone (Barron et al. 2000). As a result, protein ROA spectra contain information on the secondary and tertiary structure of the polypeptide backbone, backbone hydration and side chain conformation. In addition to structural description of ordered proteins, ROA can also be used to analyze the structural elements present in unfolded and partially unfolded states of globular proteins, as well as for the structural analysis of IDPs (Smyth et al. 2001; Syme et al. 2002; Zhu et al. 2005). Surprisingly, the Raman optical activity spectra of several IDPs (such bovine β- and κ-caseins, recombinant human α-, β-, and γ-synuclein, together with the A30P and A53T mutants of α-synuclein associated with familial cases of Parkinson's disease, and recombinant human tau46 protein together with the tau46 P301 L mutant associated with inherited frontotemporal dementia) were shown to be very similar, being dominated by a strong positive band centred at approximately 1318 cm^{-1} that may be due to the polyproline II (PPII) helical conformation (Syme et al. 2002).

4.4.2 Vibrational and Electron Paramagnetic Resonance Spectroscopy

A very useful development of IR spectroscopy is isotope-edited IR spectroscopy, which is a powerful tool for studying the structural and dynamic properties of peptides and proteins with site-specific resolution (Buchner and Kubelka 2012). Here, isotopically labelled amino acids are introduced at specific locations within the protein of interest (Tadesse et al. 1991; Decatur 2006). Since amide I vibrations of ^{13}C labelled residues are decoupled from ^{12}C vibrations and result in a separate signal shifted to lower wavenumbers, this technique provides information on the local structure within the labelled protein segment (Buchner and Kubelka 2012) and can be used for the analysis of the backbone conformation (secondary structure) of the labelled region (Huang et al. 2004), as well as to follow changes in tertiary structure that can be probed through the solvent exposure of labelled amides (Manas et al. 2000; Walsh et al. 2003; Brewer et al. 2007; Fesinmeyer et al. 2005).

Besides FTIR spectroscopy, vibrational circular dichroism (VCD) spectroscopy has emerged as a very powerful tool for the structural analysis of peptides and proteins (Keiderling 1996; Keiderling and Xu 2002). In the case of VCD, infrared rather than UV or visible light is used to probe the dichroisms of bands associated with transitions between the different vibrational energy levels of the electronic ground state of a molecule (Schweitzer-Stenner et al. 2012b). This serves as a measure of chirality in the vicinity of the respective chromophore as defined by the normal mode composition of the mode probed by the IR-absorption (Schweitzer-Stenner et al. 2012b).

The fast intrinsic time scale of infrared absorption and the sensitivity of molecular vibrational frequencies to their environments can be applied with site-specificity by introducing the artificial amino acid β-thiocyanatoalanine, or cyanylated cysteine, into chosen sites within IDPs (Yang et al. 2012). This cyanylation of cysteine, which converts the native cysteine thiol group to a covalently attached thiocyanate, represents a simple and useful approach for the structural analysis of IDPs based on the detection of a vibrationally active artificial amino acid. As described in (Yang et al. 2012): "The CN stretching band of aliphatic thiocyanate appears in an otherwise clear spectral window near 2160 cm^{-1} and is a reasonably strong infrared absorber (Choi et al. 2008; Maienschein-Cline and Londergan 2007). The CN stretching absorption frequency is sensitive to the local presence or absence of water (McMahon et al. 2010), as well as to the local electric field presented by the nearby structure (Fafarman et al. 2006). Its line-width is dynamically sensitive (Edelstein et al. 2010) to fluctuations of the local solvent and structure on the fs-ps time scale (which is the time scale of H-bond formation and breaking and dipolar reorientation)." It has been shown that this technique can be used in the analysis of partner-induced structural transitions in an IDP (Bischak et al. 2010).

Electron paramagnetic resonance (EPR) spectroscopy studies chemical species that have unpaired electrons, such as paramagnetic metalloproteins. EPR-sensitive reporter groups (spin labels or spin probes) can also be introduced into biological systems via site-directed spin-labelling (SDSL). SDSL is usually accomplished by cysteine-substitution mutagenesis followed by covalent modification of the unique sulfhydryl group with a selective nitroxide reagent (Feix and Klug 1998; Hubbell et al. 1996; Hubbell et al. 1998; Hubbell et al. 2000; Biswas et al. 2001; Columbus and Hubbell 2002; Hubbell et al. 2003; Fanucci and Cafiso 2006; Klare and Steinhoff 2009; Altenbach et al. 1989; Altenbach et al. 1990). SDSL EPR spectroscopy has been shown to be a sensitive and powerful method for studying structural transitions within IDPs (Morin et al. 2006; Belle et al. 2008; Pirman et al. 2011; Habchi et al. 2012).

4.4.3 Fluorescence-Based Techniques

Various fluorescence characteristics such as the shape and position of the intrinsic fluorescence spectrum, fluorescence anisotropy and lifetime, accessibility of the chromophore groups to external quenchers, steady-state and time-resolved parameters of the fluorescent dyes, and fluorescence resonance energy transfer can be used to describe the intramolecular mobility and compactness of an IDP (Neyroz and Ciurli 2012). In addition to intrinsic fluorescence (e.g. tryptophan fluorescence), useful information on the conformational behaviour of IDPs can be obtained by using the extrinsic fluorescence of labels covalently bound to IDPs, or probes that bind to proteins by weak non-covalent coupling. For example, binding of 8-anilino-1-naphthalenesulfonate (ANS) fluorescent probe to proteins is accompanied by a dramatic increase in the quantum yield of ANS fluorescence coupled with significant blue shift (Sulatskaya et al. 2012). As a result, this fluorescent probe is widely used for testing the presence of hydrophobic clusters and hydrophobic "pockets" in the structure of target objects (Stryer 1965; Peters Jr 1996), as well as for the analysis of partially folded intermediates (Semisotnov et al. 1991; Semisotnov et al. 1987; Goto and Fink 1989; Rodionova et al. 1989; Ptitsyn et al. 1990) and IDPs (Neyroz et al. 2006; Uversky et al. 2001; Bailey et al. 2001; Lavery and McEwan 2008).

The lifetimes of fluorescent states are very sensitive to the environment of the fluorophores. The advantage of measuring lifetimes comes from the fact that this can reveal the heterogeneity and dynamic properties of this environment (Schreurs et al. 2012). In fact, lifetime analysis can be used to characterize static and dynamic conformational properties and the heterogeneity of fluorescent groups in different areas of a protein and as a function of time for an evolving protein. The fluorescent groups involved are either natural amino acid side chains, such as tryptophan (Trp) or tyrosine (Tyr), or fluorescent labels covalently engineered into the protein. The phenomena that determine the lifetime of a fluorophore are its intrinsic properties, dynamic quenching by neighbouring groups, and exposure to the solvent, as well as FRET between different groups (Schreurs et al. 2012).

The distance dependence of FRET is a very useful tool for the characterization of polypeptide dynamics (Flory 1969; Stryer and Haugland 1967). The fluorescent donor and acceptor are attached to the polypeptide through synthesis or mutagenesis, with a defined separation in the primary sequence. In a dynamic polypeptide chain, FRET efficiency can be related to the root mean squared (rms) displacement in any direction, $R_{rms} = \sqrt{\langle R \rangle^2}$, or absolute distance in stable structures (Choi et al. 2012). Such methods are used to examine polymer models of the denatured state of a protein (Chen and Rhoades 2008) and IDPs under native conditions (Ferreon et al. 2009; Weninger et al. 2008). Although ensemble FRET can provide valuable insights into IDPs, single molecule FRET (smFRET) can resolve sample heterogeneity and to some extent probe dynamics (Choi et al. 2012). FRET can measure distances and conformational fluctuations on the scale of 2–8 nm (Stryer and Haugland 1967). For a highly disordered polypeptide chain, this corresponds to fluorophore separations of 50 to 175 residues in the primary sequence (Choi et al. 2012).

Measurements of dynamic FRET (Forster 1948, 1959, 1965; Steinberg 1971; Stryer et al. 1982; Van Der Meer 1994; Haas 2004), which is based on the distance dependent interactions between the excited state dipoles, can be applied for the determination of long-range distances between the labelled sites in disordered or partially folded proteins (Haas 2012). Ensemble time resolved FRET (trFRET) and smFRET measurements of double labelled protein samples can be used for the analysis of distributions of intramolecular segmental end-to-end distances and for the monitoring and analysis of fast fluctuations and fast and slow conformational transitions within selected sections of the molecule (Haas 2012). FRET measurements can also be used for the detection and analysis of intermolecular interactions (Haas 2012).

Fluorescence correlation spectroscopy (FCS) can be used for an accurate determination of the diffusion coefficient of fluorescently labelled subjects (Petrasek and Schwille 2008) including IDPs. This technique uses a confocal microscope to analyse the intensity fluctuations that appear over time, when fluorescent molecules are diffusing in and out of the confocal volume (Nath et al. 2012). This method can also uncover heterogeneity of diffusion coefficients and the formation of spikes when bright objects are formed. Such an analysis of heterogeneity over time can be quantified to study protein-DNA interactions (Vercammen et al. 2002) or complex formation (Buyens et al. 2008). Importantly, this technique can be used for various analyses of proteins *in vitro*, in cellular extracts and inside cells. For example, using a combination of FCS and FRET it was shown that early aggregation steps of a classical IDP, α-synuclein, are characterized by a contagious conformational change (Nath et al. 2010).

4.4.4 Single Molecule Techniques

The basic idea of a single molecule experiment using FRET is very simple: a donor dye and an acceptor dye are attached to specific residues of a protein. If a protein molecule resides in the volume illuminated by the focused laser beam, excitation of the donor dye can result in energy transfer to the acceptor. The efficiency of energy transfer, which can be determined from the rates of detected donor and acceptor photons, depends on the distance between the fluorophores and can thus be related to the separation of the dyes (Schuler et al. 2012).

smFRET has been used to address a wide range of questions in protein folding and dynamics (Schuler and Eaton 2008; Schuler and Haran 2008). Due to the possibility for resolving conformational heterogeneity, the method is uniquely suited to probing the properties of proteins in their denatured and other non-native states (Schuler and Eaton 2008). The application of smFRET to IDPs has increased notably in the last few years (Ferreon et al. 2010). In addition to the analysis of immobilized proteins, smFRET can be used within confocal microscopy settings (Schuler et al. 2012).

Immobilization of single fluorescently labelled molecules under conditions that retain their biological activity allows for extended smFRET-based observation of the same molecule for tens of seconds. This approach can capture slow conformational transitions or protein binding and unbinding cycles. Using an open geometry for immobilisation allows for direct observation of the response to changing solution conditions or ligand binding (Choi et al. 2012).

Atomic force microscopy (AFM)-based single molecule force spectroscopy (SMFS) is one very useful technique for gaining information on the biophysical properties of IDPs (Oroz et al. 2012). In SMFS the protein of interest is stretched, typically in order to measure its mechanical resistance. This resistance is usually unique and characteristic in folded proteins. However, the different conformations of IDPs may also exhibit diverse mechanical properties. Therefore, SMFS provides a way to analyse the conformational plasticity of these proteins (Oroz et al. 2012). For example, SMFS measurements were used for the unique conformational analysis of an IDP (such as α-synuclein) inserted as a domain in a chimeric multimodular polyprotein (Sandal et al. 2008). The insertion of an IDP into a chimeric polyprotein in which it is flanked by several globular domains, which act as both "molecular handles" and as internal mechanical gauges, is absolutely necessary to perform sufficiently clean SMFS experiments (Sandal et al. 2008). This requirement limits the general usefulness of the technique, since the needed flanking modules unavoidably influence the energy landscape of the IDP domain via steric and electrostatic effects, and its resulting conformational equilibria are thus different from those of the free IDP in the same conditions (Brucale et al. 2009). However, this technique can be used to assess the impact of a single factor of choice on the (perturbed) conformational distribution of a disordered domain (Brucale et al. 2012). For example, this approach was recently used to show that α-synuclein point mutations linked to familial Parkinson's disease increase the propensity of the α-synuclein domain to acquire compact structures under the tested conditions (Brucale et al. 2009).

High-speed atomic force microscopy (HS-AFM) can not only visualize IDPs and IDPRs, but also generate molecular movies that can be used to gain important information on the mechanical properties of IDPRs (Ando and Kodera 2012). HS-AFM can visualize IDPs that are weakly attached to a highly flat surface in solution, without chemical immobilization (Ando et al. 2003; Ando et al. 2001; Yamamoto et al. 2010; Ando et al. 2007). This technique was successfully applied for the observation of the dynamic behaviour of several proteins in action (Kodera et al. 2010; Shibata et al. 2010; Yamamoto et al. 2008; Milhiet et al. 2010). In application to IDPs, HS-AFM was shown to report the location of ordered and disordered regions in the molecules, mechanical properties of IDPRs, and the dynamic of order-disorder transitions of IDPRs (Ando and Kodera 2012).

4.5 Hydrodynamic Techniques and other Tools for the Evaluation of the Hydrodynamic Dimensions and Shapes of IDPs

Detailed information on the hydrodynamic dimensions of a polypeptide provide very important constraints for IDP structural characterization. IDPs have increased hydrodynamic volumes relative to ordered proteins of similar molecular mass. The degree of protein compaction can be evaluated by various techniques such as gel filtration, viscometry, small angle X-ray scattering (SAXS), small angle neutron scattering (SANS), sedimentation, and dynamic and static light scattering.

One of the most unambiguous characteristics of the conformational state of a globular protein is its hydrodynamic dimensions. It was noted long ago that hydrodynamic techniques may help to recognize when a protein has lost all of its non-covalent structure, i.e. when it becomes unfolded (Tanford 1968). This is because an essential increase in the hydrodynamic volume is associated with the unfolding of a protein molecule. It is known that globular proteins may exist in at least four different conformations—folded, molten globule, pre-molten globule and unfolded (Ptitsyn 1995; Uversky and Ptitsyn 1994, 1996b; Uversky 2002a, b, 2003)—which may easily be discriminated by the degree of compactness of the polypeptide chain. In fact, Fig. 7.2a shows that the hydrodynamic volume of a polypeptide chain in the molten globule, the pre-molten globule and the unfolded states, in comparison with that of the completely folded state, increases 1.5, ~3 and ~12 times, respectively. Finally, it has been established that folded and unfolded conformations of globular proteins possess very different molecular mass dependencies of their hydrodynamic radii, R_S (Tanford 1968; Tcherkasskaya et al. 2003; Tcherkasskaya and Uversky 2001, 2003).

Based on these considerations, hydrodynamic techniques were used to clarify the physical nature of extended IDPs. To illustrate a point, Fig. 7.2b compares log(R_S) vs. log(M) curves for extended IDPs with the same dependencies for the folded, molten globule, pre-molten globule, and urea- or guanidinium chloride (GdmCl) unfolded globular proteins (Uversky 2002a). The log(R_S) *vs.* log(M) dependencies for different conformations of globular proteins can be described by straight lines:

$$\log(R_S{}^F) = -(0.204 \times 0.024) + (0.357 \times 0.005) \times \log(M) \quad (7.1)$$

$$\log(R_S{}^{MG}) = -(0.053 \times 0.094) + (0.334 \times 0.021) \times \log(M) \quad (7.2)$$

$$\log(R_S{}^{PMG}) = -(0.21 \times 0.18) + (0.392 \times 0.041) \times \log(M) \quad (7.3)$$

$$\log(R_S{}^{U(urea)}) = -(0.649 \times 0.024) + (0.521 \times 0.004) \times \log(M) \quad (7.4)$$

$$\log(R_S{}^{U(GdmCl)}) = -(0.723 \times 0.024) + (0.543 \times 0.007) \times \log(M) \quad (7.5)$$

Here F, MG, PMG, U(urea) and U(GdmCl) correspond to the folded, molten globule, pre-molten globule, urea-, and GdmCl-unfolded globular proteins, respectively.

As for the extended IDPs, Fig. 7.2b clearly shows that they can be divided into two groups based on their log(R_S) vs. log(M) dependence. One group of the extended IDPs behaved as polymeric coils in poor solvent (denoted as IDP(coil)), whereas proteins from the other group were noticeably more compact, being close with respect to their hydrodynamic characteristics to pre-molten globules (denoted as IDP(PMG)) (Uversky 2002a):

$$\log(R_S{}^{IDP(PMG)}) = -(0.239 \times 0.055) + (0.403 \times 0.012) \times \log(M) \quad (7.6)$$

$$\log(R_S{}^{IDP(coil)}) = -(0.551 \times 0.032) + (0.493 \times 0.008) \times \log(M) \quad (7.7)$$

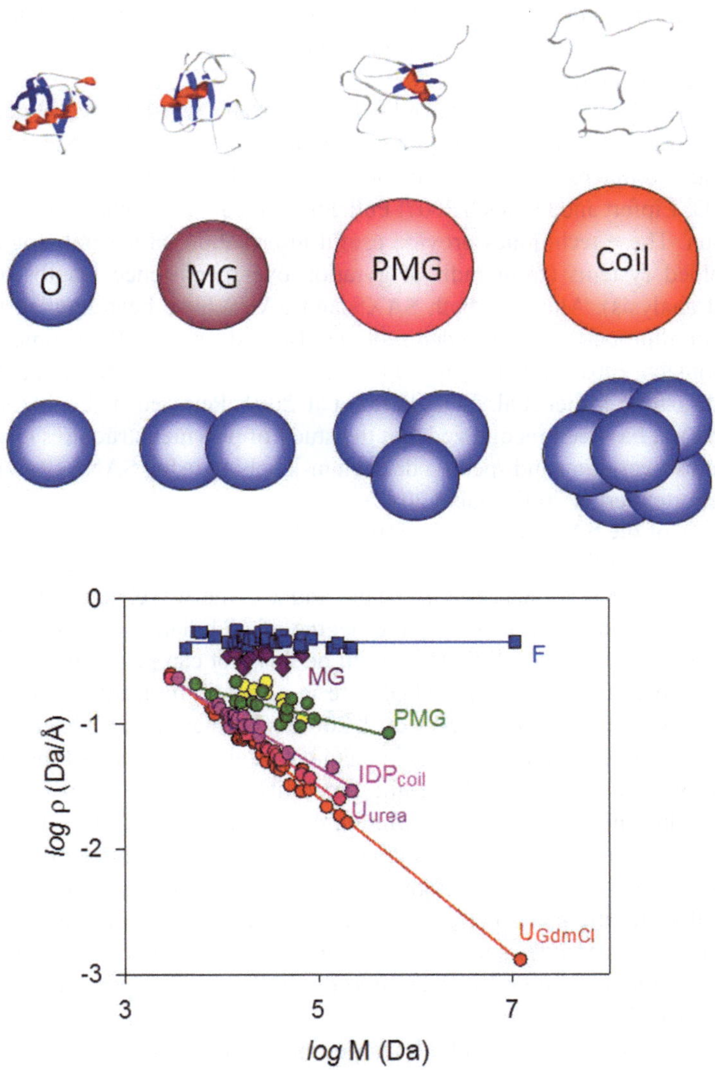

Fig. 7.2 Hydrodynamic dimensions of variously disordered IDPs. **a** *Top line*: Collapsed (molten globule-like, *MG*) disorder; extended (pre-molten globule-like, *PMG*) disorder; (coil-like, *coil*) disorder. An ordered globular protein of the same length is also shown for comparison. The figure represents model structures of a 100 residue-long polypeptide chain. *Middle line*: Relative hydrodynamic volumes occupied by a 100 residue-long polypeptide chain in these four conformations. Spheres in the *middle* lines show an increase in the hydrodynamic volume relative to the volume of the corresponding ordered protein. Modified from (Uversky and Dunker 2010). *Bottom line*: Apparent oligomerization states evaluated assuming that volumes occupied by differently disordered forms of a 100 residue-long polypeptide chain correspond to the oligomers of a 100 residue-long folded/ordered protein. **b** Variation of the density of protein molecules ρ with protein molecular weight *M* for folded/ordered (*blue squares*), molten globular (*dark pink diamonds*), pre-molten globular (*yellow circles*), 8 M urea-unfolded (*blue-pink circles*), and 6 M GdmCl-unfolded (*red circles*) conformational states of globular proteins and extended IDPs with coil-like (*pink circles*) or pre-molten globule-like properties (*green circles*). The data used to plot dependencies for native, molten globule, pre-molten globule and GdmCl-unfolded states of globular proteins are taken from (Tcherkasskaya and Uversky 2001)

This dramatic dependence of the hydrodynamic dimensions of an IDP on the degree of disorderedness should definitely be taken into account to avoid misinterpretation of the experimental data.

SANS and SAXS (Chap. 8) have been used extensively for several decades to study the structural properties of polymers (Chu and Hsiao 2001; Schurtenberger 2002) and bio-macromolecules (Doniach 2001; Heller 2010; Jacrot 1976; Koch et al. 2003; Lipfert and Doniach 2007; Putnam et al. 2007; Svergun and Koch 2002) in solution. These techniques provide useful information on several length-scales for unfolded systems (from radii of gyration over persistence lengths to cross-sectional analysis). Although both SAXS and SANS have been applied to study the structural properties of unfolded proteins (Bernadó et al. 2005; Calmettes et al. 1994; Doniach 2001; Gabel et al. 2009; Kataoka et al. 1995; Kirste et al. 1969; Kohn et al. 2004; Millet et al. 2002; Perez et al. 2001; Petrescu et al. 1997; Petrescu et al. 1998), it has been recognized that the study of polymer structures can benefit from contrast variation and specific deuterium-labelling using SANS (Rawiso et al. 1987; Schurtenberger 2002; Gabel 2012).

Analysis of the SAXS data in the form of a Kratky plot (a plot of $I(S)*S^2$ versus S, with $I(S)$ being the scattering intensity, and S being the scattering vector given by $2sin\theta/\lambda$, where θ is the scattering angle and λ is the wavelength of the X-ray) can provide crucial information on the degree of globularity of a given protein. Here, the lack of a characteristic maximum on this plot can be taken as the indication of the absence of a tightly packed core in a protein molecule (Uversky et al. 2000b; Uversky et al. 1999; Uversky et al. 2001b). Furthermore, SAXS can provide quantitative characterization of IDPs in solution via the utilization of the ensemble optimization method (EOM). Here, the flexibility of IDPs is taken into account by considering the coexistence of different conformations that are selected using a genetic algorithm from a pool containing a large number of randomly generated models covering the protein configurational space and that contribute to the experimental scattering pattern (Bernadó et al. 2007). A combination of SAXS with high-resolution techniques such as NMR can be used to generate reliable models and to gain unique structural insights into the IDP over multiple structural scales (Receveur-Brechot and Durand 2012).

The hydrodynamic size of an IDP is large compared to a folded protein of similar molecular mass and the resulting mass-to-size ratio is unusual (Receveur-Brechot et al. 2006; Uversky 2002b). The sedimentation coefficient, which can be obtained from sedimentation velocity (SV) analytical ultracentrifugation (AUC), is directly related to this ratio and can be easily interpreted in terms of frictional ratio (Salvay et al. 2012). In fact, AUC-based SV experiments provide information on the molar mass (M) and hydrodynamic (Stokes) radius (R_H) of macromolecules in solution (Ebel 2007; Lebowitz et al. 2002; Ebel 2004; Howlett et al. 2006). AUC is complementary to the methods based on size determination (e.g. size exclusion chromatography, dynamic light scattering) that probe R_H. As an illustration, it is difficult to distinguish an IDP (monomer) that appears large in size-exclusion chromatography (SEC) from a dimer. The same IDP will sediment more slowly than a folded monomer while a dimer would sediment faster, leading to an easy experimental diagnostic of the extended shape and association state of the protein (Manon and Ebel 2010).

It has been indicated that the combination of dynamic light scattering (DLS) and static light scattering (SLS) is a unique technique for the analysis of the molecular dimensions of IDPs coupled with thorough control of their monomeric state (Gast and Fiedler 2012).

As was already pointed out, gel filtration chromatography (or SEC) is a useful tool for structural and conformational analyses of IDPs (Uversky 2012). SEC can be used for the estimation of the hydrodynamic dimensions of a given IDP, evaluation of its association state, and analysis of interactions with binding partners, and for induced folding studies. It also can be used to physically separate IDP conformers based on their hydrodynamic dimensions, thus providing a unique possibility for the independent analysis of their physicochemical properties (Uversky 2012). SEC is very useful in ascertaining the degree of compactness of a protein, and can distinguish between partially and fully unfolded states, since an increase in hydrodynamic volume is associated with unfolding. The transformation of a typical globular protein into a molten globule state results in a ~15–20% increase in its hydrodynamic radius (Ptitsyn 1995; Ptitsyn et al. 1995; Uversky 1994, 1993, 2003; Tcherkasskaya and Uversky 2003). The relative increase in the hydrodynamic volume of less folded intermediates is even larger (Uversky 2003; Tcherkasskaya et al. 2003; Tcherkasskaya and Uversky 2003; Uversky and Ptitsyn 1996b; Ptitsyn et al. 1995; Uversky and Ptitsyn 1994). Furthermore, equilibrium conformations of any given IDP (coil-like, PMG-like, or molten globule-like) can easily be discriminated by the degree of compactness of the polypeptide chain (Uversky 2012) (see Fig. 7.2).

In the protein field, the most frequent uses of SEC are separation of proteins based on their size and estimation of their molecular masses. Formally, SEC is a separation technique based on hydrodynamic radius and not molecular mass. However, as follows from Eqs. 6.1–6.7, for similarly shaped/disordered molecules hydrodynamic radius is proportional to molecular mass. We can therefore think of SEC as a mass-based separation even though this is not strictly true. A few words must be added here to illustrate how ignoring the physical principles of SEC can lead to the misinterpretation of SEC data and reaching incorrect conclusions if SEC is used to estimate protein molecular mass and not its hydrodynamic radius. A polypeptide of 300 residues with a molecular mass of 35 kDa in its folded, molten globular, pre-molten globular IDP, and coil-like IDP forms has an R_S of 26.2, 29.2, 39.1, and 48.9 Å, respectively. However, if SEC column is calibrated in molecular masses of globular proteins and is used to evaluate the molecular mass of an unknown protein, the same 300 residue-long polypeptide chain in the aforementioned structural states would be found to have molecular masses of 35.0, 47.4, 107.5, and 201.1 kDa, respectively. These data would then suggest that the query protein exists as a monomer, dynamic dimer, very stable trimer or very stable hexamer, instead of a differently disordered polypeptide of 300 residues.

Obviously, a combination of multiple techniques provides a better characterization of a subject under study. Size exclusion chromatography coupled, AUC and quasi-elastic light scattering (QELS) coupled online to a Tetra Detector Array (which combines right and low angle light scattering (RALS and LALS) detectors, a spectrophotometer (UV), a refractometer (refractive index (RI)), and pressure transducers) represents a useful platform for characterizing the hydrodynamic properties of macromolecules, including IDPs (Karst et al. 2012). The combined application of these techniques evaluates the molecular mass, intrinsic viscosity, and hydrodynamic radius of a protein and provides information on its time-averaged apparent hydration and shape factor (Karst et al. 2012). This unique technology has been successfully applied for studies of IDPs, protein/ligand interactions (Chenal et al. 2009; Sotomayor Perez et al. 2010), protein/protein interactions, and proteins exhibiting anomalous behaviour (Bourdeau et al. 2009).

4.6 Evaluating the Conformational Stability of IDPs

IDPs, being highly dynamic, are characterized by low conformational stability, which is reflected in the low steepness of the transition curves describing their unfolding induced by strong denaturants or even with the complete lack of the sigmoidal shape of the unfolding curves. This is in strict contrast to the solvent-induced unfolding of ordered globular proteins, which is known to be a highly cooperative process (Uversky 2009). Based on the analysis of the shapes of unfolding transitions in ordered globular proteins it has been concluded that the steepness of urea- or GdmCl-induced unfolding curves depends strongly on whether a given protein has a rigid tertiary structure (i.e. is ordered) or is already denatured and exists as a molten globule (Ptitsyn and Uversky 1994; Uversky and Ptitsyn 1996a). In application to IDPs, it has been proposed that this type of analysis can be used to differentiate whether a given protein has ordered (rigid) structure or exists as a native molten globule (Uversky 2002a). Although the denaturant-induced unfolding of a native molten globule can be described by a shallow sigmoidal curve (e.g. see (Neyroz et al. 2006)), urea- or GdmCl-induced structural changes in native pre-molten globules or native coils are non-cooperative and typically seen as monotonous featureless changes in the studied parameters, which is due to the low content of the residual structure in these species.

Using the high thermal resistance of extended IDPs and their insensitivity to urea-induced unfolding, a useful two-dimensional electrophoresis technique was elaborated for the *de novo* recognition and characterization of IDPs (Szollosi et al. 2008; Tantos and Tompa 2012). This approach represents a combination of a native gel electrophoresis of heat-treated proteins followed by a second, denaturing gel containing 8 M urea and was shown to be a straightforward technique to separate IDPs from globular proteins in a cellular extract. The rationale for the first dimension is that IDPs are very often heat-stable, and thus heat treatment results in a good initial separation from globular proteins, most of which aggregate and precipitate. In the native gel, IDPs and rare heat-stable globular proteins will then be separated according to their charge/mass ratios. As urea is uncharged and IDPs are just as "denatured" in 8 M urea as under native conditions, they are expected to run the same distance in the second dimension and end up along the diagonal. Heat-stable globular proteins, on the other hand, will unfold in urea, slow down in the second gel, and arrive above the diagonal (Szollosi et al. 2008; Tantos and Tompa 2012).

Since typical extended IDPs contain relatively few hydrophobic residues and are enriched in amino acids carrying a net charge at physiological pH, these proteins are characterized by a "turned out" response to changes in pH (Uversky et al. 1999; Uversky et al. 2001; Konno et al. 1997; Lynn et al. 1999; Johansson et al. 1998), where a decrease (or increase) in pH induces partial folding of extended IDPs due to the minimization of their large net charge present at neutral pH, thereby decreasing charge/charge intramolecular repulsion and permitting hydrophobic-driven collapse to the partially-folded conformation (Uversky et al. 2001; Smith and Jelokhani-Niaraki 2012).

The analysis of temperature effects on the structural properties of several extended IDPs revealed that native coils and native pre-molten globules possess a so-called "turned out" response to heat (Uversky et al. 2001; Permyakov et al. 2003; Uversky et al. 2002; Timm et al. 1992; Kim et al. 2000), where an increase in temperature induces partial folding of IDPs, rather than the unfolding typical of ordered globular proteins. The effects of elevated temperatures were attributed to the increased strength of the hydrophobic interaction at higher temperatures, leading to a stronger hydrophobic attraction, which is the major driving force for folding (Uversky et al. 2001).

4.7 *Mass Spectrometry-Based Approaches*

4.7.1 Analysing IDPs with Electro-Spray Ionization Mass Spectrometry (ESI-MS)

Since the charge acquired by biomolecules during ESI is strongly influenced by their three-dimensional structure in the sprayed solution, ion charge-state distribution (CSD) analysis in ESI-MS represents a robust and fast technique for the direct detection and characterization of co-existing protein conformations in solution (Abzalimov et al. 2012). For example, tightly-folded protein conformations, with minimal solvent exposure, give rise to ESI generated ions carrying a small number of charges, whereas less compact conformers will accommodate larger number of charges upon the electro-spray ionization process depending on the extent of their unfolding. Therefore, an ESI-MS spectrum of multiple protein conformers represents a linear combination of ionic contributions from individual conformers (Gumerov et al. 2002). This allows the direct detection and characterization of distinct conformers, including transiently populated ones, and transitions between them. This feature of ESI-MS is widely employed in studies of protein dynamics (Konermann and Douglas 1998; Frimpong et al. 2007; Invernizzi and Grandori 2007) and protein-ligand (Zhang et al. 2004) and protein-protein interactions (Griffith and Kaltashov 2003; Simmons et al. 2004).

4.7.2 Finding IDPRs by Hydrogen/Deuterium Exchange Mass Spectrometry

Another useful application of mass spectrometry (MS) is the monitoring of the incorporation of deuterium into a protein's backbone amide. Analysis of this hydrogen/deuterium exchange (HDX) in proteins by MS (HDX-MS) coupled to high-performance liquid chromatography (HPLC) can provide invaluable structural information about the protein of interest (Smith et al. 1997; Zhang and Smith 1993). The rate of exchange of the backbone amide hydrogen is determined by its chemical environment and solvent accessibility. In fact, hydrogen deeply buried within the protein core or involved in hydrogen bonding would require a substantial local or global conformational change before exchange is possible and would therefore be considered to have a high level of protection from HDX. Conversely, hydrogen located on the surface of the protein or not constrained by hydrogen bonding would be unprotected and readily undergo exchange. This ability to readily distinguish protein regions by their level of protection makes HDX-MS an ideal technique for detecting intrinsic disorder (Bobst and Kaltashov 2012).

4.7.3 Finding Binding Domains Involved in Protein-Protein Interactions

MS is indispensable for proteomic studies as it allows the identification of thousands of proteins in a single experiment (Yates et al. 2009). The combination of chemical cross-linking, proteolysis, and MS analysis represents a powerful tool for the identification of the binding domains found in protein-protein interactions (Back et al. 2003; Eyles and Kaltashov 2004; Farmer and Caprioli 1998; Kalkhof et al. 2005; Schulz et al. 2004; Sinz 2003; Trester-Zedlitz et al. 2003). This approach has multiple advantages (Sinz 2006; Auclair et al. 2012). In fact, cross-linkers with different space arms and chemistries have inherently different masses and can therefore give insight into distance constraints and highlight networks of nearby residues. The mass of the protein complex is not limiting because proteolysis precedes mass measurement, resulting in peptides that are tractable for liquid chromatography-mass spectrometry (LC-MS)/MS analysis. The quantities of protein required for MS are very small, since only nanomoles, and in some cases even femtomoles, of protein are needed. Cross-linking is conducted in solution, and therefore flexible regions of proteins can undergo any necessary disorder-to-order transitions in order to form intermolecular interactions (Sinz 2006; Auclair et al. 2007; Auclair et al. 2012).

4.8 Other Useful Techniques for the Structural Analysis of IDPs

4.8.1 Immunochemical Analyses of IDPs

Immunochemical methods may also be applied toward the elucidation of protein disorder. The immunoglobulins obtained against a given protein may be specific for different macromolecular levels: the primary structure (Amit et al. 1985; Wilson et al. 1985), secondary structure (Fujio et al. 1985), or tertiary structure (Amit et al. 1985; Wilson et al. 1985). In the latter case, the antigenic determinants may reside on either the neighbouring residues in the chain (loops) (Amit et al. 1985; Wilson et al. 1985) or on spatially distant residues (Fujio et al. 1985). Furthermore, it has been shown that antibodies in the immune serum may possess a high affinity for the internal elements of an antigen (Fujio et al. 1985). Thus, antibodies may be successfully used to study the structural changes that a protein-immunogen undergoes upon changes to experimental conditions. For example, antibodies obtained against the Ca^{2+}-saturated F_1-fragment of prothrombin did not interact with the calcium-free apo-form of this protein (Furie and Furie 1979). An analogous effect was also observed in the case of osteocalcin (Delmas et al. 1984).

4.8.2 Abnormal Electrophoretic Mobility

Due to their distinctive, significantly polar amino acid compositions, IDPs bind less sodium dodecyl sulphate (SDS) than ordered proteins. Because of this, IDPs and hybrid proteins with long IDPRs possess abnormal mobility in SDS polyacrylamide gel electrophoresis experiments, giving rise to the apparent molecular masses that are noticeably higher than molecular masses calculated from sequence data or measured by mass spectrometry (Tompa 2002; Receveur-Brechot et al. 2006).

4.8.3 Limited Proteolysis of IDPs and Hybrid Proteins with IDPRs

The extent of proteolytic digestion by specific proteases, such as trypsin, correlates with flexibility in the region of the cut site and not just to surface exposure (Hubbard 1998). In fact, the structure and dynamics of a substrate protein play a crucial role in determining the efficiency of proteolysis (Hubbard et al. 1998; Hubbard et al. 1994). It has been established that sites of limited proteolysis are structurally different from the inhibitor loops, suggesting that they must undergo a conformational change in order to enter the proteinase active site (Hubbard et al. 1991). Furthermore, it has been shown for several proteins that local unfolding of at least 13 residues is needed for a set of observed cut-sites to properly fit into trypsin's active site (Hubbard et al. 1998; Hubbard et al. 1994). It has also been established that limited proteolytic sites are typically found within flexible solvent-exposed loop regions (as indicated by crystallographic temperature factors or B-values) (Hubbard et al. 1991; Fontana et al. 1986; Novotny and Bruccoleri 1987), and are notably absent in regions of regular secondary structure, especially within β-sheets (Hubbard et al. 1994; Fontana et al. 1997a, b). These proteolytic sites protrude from the protein surface and are expected to be found at regions where local packing does not inhibit the local unfolding (Hubbard et al. 1991). It has also been established that computational indications of order and disorder were perfectly correlated with protease digestion profiles (Iakoucheva et al. 2001a, b); regions predicted to be ordered were generally not cut at all, while regions predicted to be disordered were rapidly cut.

Therefore, proteolytic digestion happens much faster in IDPRs and can be used to map ordered and disordered regions in a protein (Fontana et al. 2012). Depending on the amount of intrinsic disorder in a given protein, three major scenarios for proteolytic digestion are expected (Johnson et al. 2012). Highly disordered proteins are expected to be digested quickly, typically without accumulation of stable fragments. However, some of these proteins might produce semi-stable fragments, the number and protease resistance of which are expected to be dependent on the amount and stability of partially ordered structure. Highly ordered proteins are expected be digested slowly, typically with accumulation of one or several stable fragments. Partially disordered proteins (proteins with long IDRs) are expected to show intermediate proteolytic behaviour. As accessible cut sites in disordered regions are cleaved, stable fragments may emerge if enough structural stability is maintained. Therefore, by comparing the rates of digestion as determined by the disappearance of the initial protein band on SDS-PAGE and by the analysis of the digestion patterns, one can loosely characterize the degree of disorder in protein samples (Johnson et al. 2012).

4.8.4 Measuring the Mean Net Charge of IDPs

The electrophoretic mobility of a protein sample is measured by applying an electric field between two electrodes. Although charged macromolecules dispersed in the solvent will migrate toward the electrode of opposite sign with a velocity that is proportional to the applied voltage, electrophoretic mobility is also dependent on the charge, mass, hydration and shape of the macromolecule (Sotomayor-Perez et al. 2012). It is believed that the electrophoretic mobility is proportional to the number of charges and inversely correlated to the protein molecular mass and frictional ratio $\left(f/f_0\right)$ (Basak and Ladisch 1995). Therefore, the mean net charge of an IDP is evaluated from the experimentally determined parameters, hydrodynamic radius and electrophoretic mobility (Sotomayor-Perez et al. 2011). Furthermore, an accurate determination of the mean net charge of an IDP in both the ligand-free and ligand-bound states allows one to estimate the number of ligands bound to the protein in the holo-state (Sotomayor-Perez et al. 2012).

4.8.5 Evaluating Tertiary Structure of IDPs

There are two types of optically active chromophores in proteins—side groups of aromatic amino acid residues and peptide bonds (Adler et al. 1973; Fasman 1996). Far-UV CD is used for the evaluation of protein secondary structure, whereas CD spectra in the near-UV region (250–350 nm), also known as the aromatic region, reflect the symmetry of the environment of aromatic amino acid residues and, consequently, are characteristic of protein tertiary structure. The absorption and CD bands of proteins in the near-UV region are weaker by an order of magnitude or more relative to those in the far-UV region. Therefore, near-UV CD measurements require higher concentrations and/or longer path lengths than those used in the far-UV CD analyses. However, this wavelength region has several advantages, such as high signal-to-noise ratio and lack of noticeable contribution from the absorption of various solvent components, such as most buffer components, and chemical denaturants such as urea and guanidinium chloride. Due to the relatively small number of aromatic residues in most proteins combined with vibronic fine structure characteristic of each type of aromatic amino acid, it is possible to assign specific features in the near-UV CD spectrum to particular residue types. On the other hand, the lack of rigid tertiary structure in a protein containing aromatic residues may be easily detected by the simplified near-UV CD spectrum with low intensity.

5 Concluding Remarks: Looking at an Elephant with Multifaceted Eyes

The accurate characterization of the multifaceted phenomenon of protein intrinsic disorder clearly requires a multiparametric approach. The use of this approach for the structural and dynamic characterization of IDPs provides a number of important advantages. In essence, multiparametric analysis resembles the compound or multifaceted eyes of insects, which, compared to simple eyes, possess a very large view angle, and can detect fast movement (Völkel et al. 2003; Uversky and Dunker 2012c). Detailed biophysical studies utilizing a wide spectrum of techniques sensitive to the different levels of protein structure and to dynamic behaviour at different time scales are crucial for the structural characterization of IDPs and for the clarification of the relationship between their highly dynamic structure and biological functions.

References

Abzalimov RR, Frimpong AK, Kaltashov IA (2012) Detection and characterization of large-scale protein conformational transitions in solution using charge-state distribution analysis in ESI-MS. Methods Mol Biol 896:365–373

Adler AJ, Greenfield NJ, Fasman GD (1973) Circular dichroism and optical rotatory dispersion of proteins and polypeptides. Methods Enzymol 27:675–735

Altenbach C, Flitsch SL, Khorana HG et al (1989) Structural studies on transmembrane proteins. 2. Spin labeling of bacteriorhodopsin mutants at unique cysteines. Biochemistry 28(19):7806–7812

Altenbach C, Marti T, Khorana HG et al (1990) Transmembrane protein-structure: spin labeling of bacteriorhodopsin mutants. Science 248(4959):1088–1092

Ami D, Natalello A, Zullini A et al (2004) Fourier transform infrared microspectroscopy as a new tool for nematode studies. FEBS lett 576(3):297–300

Ami D, Neri T, Natalello A et al (2008) Embryonic stem cell differentiation studied by FT-IR spectroscopy. Biochim Biophys Acta 1783(1):98–106
Ami D, Natalello A, Doglia SM (2012) Fourier transform infrared microspectroscopy of complex biological systems: from intact cells to whole organisms. Methods Mol Biol 895:85–100
Amit AG, Mariuzza RA, Phillips SE et al (1985) Three-dimensional structure of an antigen-antibody complex at 6 A resolution. Nature 313(5998):156–158
Ando T, Kodera N (2012) Visualization of mobility by atomic force microscopy. Methods Mol Biol 896:57–69
Ando T, Kodera N, Takai E et al (2001) A high-speed atomic force microscope for studying biological macromolecules. Proc Natl Acad Sci U S A 98(22):12468–12472
Ando T, Kodera N, Naito Y et al (2003) A high-speed atomic force microscope for studying biological macromolecules in action. ChemPhysChem 4(11):1196–1202
Ando T, Uchihashi T, Kodera N et al (2007) High-speed atomic force microscopy for observing dynamic biomolecular processes. J Mol Recognit 20(6):448–458
Arnone A, Bier CJ, Cotton FA et al (1971) A high resolution structure of an inhibitor complex of the extracellular nuclease of staphylococcus aureus. I. Experimental procedures and chain tracing. J Biol Chem 246(7):2302–2316
Arrondo JL, Goni FM (1999) Structure and dynamics of membrane proteins as studied by infrared spectroscopy. Prog Biophys Mol Biol 72(4):367–405
Arrondo JL, Muga A, Castresana J et al (1993) Quantitative studies of the structure of proteins in solution by fourier-transform infrared spectroscopy. Prog Biophys Mol Biol 59(1):23–56
Auclair JR, Green KM, Shandilya S et al (2007) Mass spectrometry analysis of HIV-1 Vif reveals an increase in ordered structure upon oligomerization in regions necessary for viral infectivity. Proteins 69(2):270–284
Auclair JR, Somasundaran M, Green KM et al (2012) Mass spectrometry tools for analysis of intermolecular interactions. Methods Mol Biol 896:387–398
Back JW, de Jong L, Muijsers AO et al (2003) Chemical cross-linking and mass spectrometry for protein structural modeling. J Mol Biol 331(2):303–313
Bailey RW, Dunker AK, Brown CJ et al (2001) Clusterin, a binding protein with a molten globule-like region. BioChemistry 40(39):11828–11840
Bandekar J (1992) Amide modes and protein conformation. Biochim Biophys Acta 1120(2):123–143
Barron LD, Hecht L, Blanch EW et al (2000) Solution structure and dynamics of biomolecules from Rraman optical activity. Prog Biophys Mol Biol 73(1):1–49
Barth A (2007) Infrared spectroscopy of proteins. Biochim Biophys Acta 1767(9):1073–1101
Barth A, Zscherp C (2002) What vibrations tell us about proteins. Q Rev Biophys 35(4):369–430
Basak SK, Ladisch MR (1995) Correlation of electrophoretic mobilities of proteins and peptides with their physicochemical properties. Anal Biochem 226(1):51–58
Belle V, Rouger S, Costanzo S et al (2008) Mapping α-helical induced folding within the intrinsically disordered C-terminal domain of the measles virus nucleoprotein by site-directed spin-labeling EPR spectroscopy. Proteins 73(4):973–988
Berjot M, Marx J, Alix AJP (1987) Determination of the secondary structure of proteins from the Raman amide-I band—the reference intensity profiles method. J Raman Spectrosc 18(4):289–300
Bernadó P, Blanchard L, Timmins P et al (2005) A structural model for unfolded proteins from residual dipolar couplings and small-angle x-ray scattering. Proc Natl Acad Sci U S A 102 (47):17002–17007
Bernadó P, Mylonas E, Petoukhov MV et al (2007) Structural characterization of flexible proteins using small-angle X-ray scattering. J Am Chem Soc 129(17):5656–5664
Binolfi A, Theillet FX, Selenko P (2012) Bacterial in-cell NMR of human alpha-synuclein: a disordered monomer by nature? Biochem Soc Trans 40(5):950–954
Bischak CG, Longhi S, Snead DM et al (2010) Probing structural transitions in the intrinsically disordered C-terminal domain of the measles virus nucleoprotein by vibrational spectroscopy of cyanylated cysteines. Biophys J 99:1676–1683
Biswas R, Kuhne H, Brudvig GW et al (2001) Use of EPR spectroscopy to study macromolecular structure and function. Sci Prog 84(Pt 1):45–67

Bloomer AC, Champness JN, Bricogne G et al (1978) Protein disk of tobacco mosaic virus at 2.8 A resolution showing the interactions within and between subunits. Nature 276(5686):362–368
Bobst CE, Kaltashov IA (2012) Localizing flexible regions in proteins using hydrogen-deuterium exchange mass spectrometry. Methods Mol Biol 896:375–385
Bodart JF, Wieruszeski JM, Amniai L et al (2008) NMR observation of Tau in Xenopus oocytes. J Magn Reson 192(2):252–257
Bode W, Schwager P, Huber R (1978) The transition of bovine trypsinogen to a trypsin-like state upon strong ligand binding. The refined crystal structures of the bovine trypsinogen-pancreatic trypsin inhibitor complex and of its ternary complex with Ile-Val at 1.9 A resolution. J Mol Biol 118(1):99–112
Bourdeau RW, Malito E, Chenal A et al (2009) Cellular functions and X-ray structure of anthrolysin O, a cholesterol-dependent cytolysin secreted by Bacillus anthracis. J Biol Chem 284(21):14645–14656
Brewer SH, Song BB, Raleigh DP et al (2007) Residue specific resolution of protein folding dynamics using isotope-edited infrared temperature jump spectroscopy. Biochemistry 46(11):3279–3285
Brucale M, Sandal M, Maio SD et al (2009) Pathogenic mutations shift the equilibria of alpha-synuclein single molecules towards structured conformers. Chembiochem 10(1):176–183
Brucale M, Tessari I, Bubacco L et al (2012) Single-molecule force spectroscopy of chimeric polyprotein constructs containing intrinsically disordered domains. Methods Mol Biol 896:47–56
Buchner GS, Kubelka J (2012) Isotope-edited infrared spectroscopy. Methods Mol Biol 895:347–358
Buyens K, Lucas B, Raemdonck K et al (2008) A fast and sensitive method for measuring the integrity of siRNA-carrier complexes in full human serum. J Control Release 126(1):67–76
Byler DM, Susi H (1986) Examination of the secondary structure of proteins by deconvolved FTIR spectra. Biopolymers 25(3):469–487
Cai S, Singh BR (1999) Identification of beta-turn and random coil amide III infrared bands for secondary structure estimation of proteins. Biophys Chem 80(1):7–20
Cai S, Singh BR (2004) A distinct utility of the amide III infrared band for secondary structure estimation of aqueous protein solutions using partial least squares methods. Biochemistry 43(9):2541–2549
Callender R, Dyer RB (2002) Probing protein dynamics using temperature jump relaxation spectroscopy. Curr Opin Struct Biol 12(5):628–633
Callender RH, Dyer RB, Gilmanshin R et al (1998) Fast events in protein folding: the time evolution of primary processes. Annu Rev Phys Chem 49:173–202
Calmettes P, Durand D, Desmadril M et al (1994) How random is a highly denatured protein? Biophys Chem 53(1–2):105–113
Chemes LB, Alonso LG, Noval MG et al (2012) Circular dichroism techniques for the analysis of intrinsically disordered proteins and domains. Methods Mol Biol 895:387–404
Chen EF, Kliger DS (1996) Time-resolved near UV circular dichroism and absorption studies of carbonmonoxymyoglobin photolysis intermediates. Inorg Chim Acta 242(1–2):149–158
Chen E, Kliger DS (2012) Deconstructing time-resolved optical rotatory dispersion kinetic measurements of cytochrome c folding: from molten globule to the native state. Methods Mol Biol 895:405–419
Chen H, Rhoades E (2008) Fluorescence characterization of denatured proteins. Curr Opin Struct Biol 18(4):516–524
Chen YH, Yang JT, Martinez HM (1972) Determination of the secondary structures of proteins by circular dichroism and optical rotatory dispersion. BioChemistry 11(22):4120–4131
Chen EF, Parker W, Lewis JW et al (1993) Time-resolved UV circular dichroism of phytochrome a: folding of the N-terminal region. J Am Chem Soc 115(21):9854–9855
Chen EF, Lapko VN, Song PS et al (1997) Dynamics of the N-terminal alpha-helix unfolding in the photoreversion reaction of phytochrome A. BioChemistry 36(16):4903–4908
Chen EF, Wood MJ, Fink AL et al (1998) Time-resolved circular dichroism studies of protein folding intermediates of cytochrome c. BioChemistry 37(16):5589–5598

Chen EF, Wittung-Stafshede P, Kliger DS (1999) Far-UV time-resolved circular dichroism detection of electron-transfer-triggered cytochrome c folding. J Am Chem Soc 121(16):3811–3817
Chen E, Kumita JR, Woolley GA et al (2003a) The kinetics of helix unfolding of an azobenzene cross-linked peptide probed by nanosecond time-resolved optical rotatory dispersion. J Am Chem Soc 125(41):12443–12449
Chen EF, Gensch T, Gross AB et al (2003b) Dynamics of protein and chromophore structural changes in the photocycle of photoactive yellow protein monitored by time-resolved optical rotatory dispersion. BioChemistry 42(7):2062–2071
Chen EF, Goldbeck RA, Kliger DS (2003c) Earliest events in protein folding: submicrosecond secondary structure formation in reduced cytochrome *c*. J Phys Chem A 107(40):8149–8155
Chen EF, Kumita JR, Woolley GA et al (2003d) The kinetics of helix unfolding of an azobenzene cross-linked peptide probed by nanosecond time-resolved optical rotatory dispersion. J Am Chem Soc 125(41):12443–12449
Chen EF, Goldbeck RA, Kliger DS (2004) The earliest events in protein folding: a structural requirement for ultrafast folding in cytochrome *c*. J Am Chem Soc 126(36):11175–11181
Chen E, Abel CJ, Goldbeck RA et al (2007a) Non-native heme-histidine ligation promotes microsecond time scale secondary structure formation in reduced horse heart cytochrome c. BioChemistry 46(43):12463–12472
Chen EF, Swartz TE, Bogomolni RA et al (2007b) A LOV story: the signaling state of the Phot1 LOV2 photocycle involves chromophore-triggered protein structure relaxation, as probed by far-UV time-resolved optical rotatory dispersion spectroscopy. Biochemistry 46(15):4619–4624
Chen E, Van Vranken V, Kliger DS (2008) The folding kinetics of the SDS-induced molten globule form of reduced cytochrome c. BioChemistry 47(19):5450–5459
Chen E, Goldbeck RA, Kliger DS (2010) Nanosecond time-resolved polarization spectroscopies: tools for probing protein reaction mechanisms. Methods 52(1):3–11
Chenal A, Guijarro JI, Raynal B et al (2009) RTX calcium binding motifs are intrinsically disordered in the absence of calcium: implication for protein secretion. J Biol Chem 284(3):1781–1789
Choi JH, Oh KI, Cho MH (2008) Azido-derivatized compounds as IR probes of local electrostatic environment: theoretical studies. J Chem Phys 129(17):11
Choi UB, Weninger KR, Bowen ME (2012) Immobilization of proteins for single-molecule fluorescence resonance energy transfer measurements of conformation and dynamics. Methods Mol Biol 896:3–20
Choo LP, Wetzel DL, Halliday WC et al (1996) In situ characterization of beta-amyloid in Alzheimer's diseased tissue by synchrotron Fourier transform infrared microspectroscopy. Biophys J 71(4):1672–1679
Choy WY, Forman-Kay JD (2001) Calculation of ensembles of structures representing the unfolded state of an SH3 domain. J Mol Biol 308(5):1011–1032
Chu B, Hsiao BS (2001) Small-angle X-ray scattering of polymers. Chem Rev 101(6):1727–1761
Columbus L, Hubbell WL (2002) A new spin on protein dynamics. Trends Biochem Sci 27(6): 288–295
Daughdrill GW, Pielak GJ, Uversky VN et al (2005) Natively disordered proteins. In: Buchner J, Kiefhaber T (eds) Handbook of protein folding. Wiley-VCH Verlag GmbH & Co KGaA, Weinheim, pp 271–353
Decatur SM (2006) Elucidation of residue-level structure and dynamics of polypeptides via isotope-edited infrared spectroscopy. Acc Chem Res 39(3):169–175
Dedmon MM, Patel CN, Young GB et al (2002) FlgM gains structure in living cells. Proc Natl Acad Sci U S A 99(20):12681–12684
Delmas PD, Stenner DD, Romberg RW et al (1984) Immunochemical studies of conformational alterations in bone γ-carboxyglutamic acid containing protein. BioChem 23(20):4720–4725
Dhar A, Gruebele M (2011) Fast relaxation Imaging in living cells. Curr Protoc Protein Sci (editorial board, john E Coligan et al.) Chap. 28:Unit 28 21
Dhar A, Prigozhin M, Gelman H et al (2012) Studying IDP stability and dynamics by fast relaxation imaging in living cells. Methods Mol Biol 895:101–111

Di Domenico T, Walsh I, Martin AJ et al (2012) MobiDB: a comprehensive database of intrinsic protein disorder annotations. Bioinformatics 28(15):2080–2081

Diomede L, Cassata G, Fiordaliso F et al (2010) Tetracycline and its analogues protect Caenorhabditis elegans from β amyloid-induced toxicity by targeting oligomers. Neurobiol Dis 40(2):424–431

Djerassi C (1960) Optical rotatory dispersion: applications to organic chemistry. McGraw-Hill, New York

Doglia SM, Ami D, Natalello A et al (2008) Fourier transform infrared spectroscopy analysis of the conformational quality of recombinant proteins within inclusion bodies. Biotechnol J 3(2):193–201

Doniach S (2001) Changes in biomolecular conformation seen by small angle X-ray scattering. Chem Rev 101:1763–1778

Dunker AK, Obradovic Z (2001) The protein trinity–linking function and disorder. Nat Biotechnol 19(9):805–806

Dunker AK, Uversky VN (2010) Drugs for 'protein clouds': targeting intrinsically disordered transcription factors. Curr Opin Pharmacol 10(6):782–788

Dunker AK, Garner E, Guilliot S et al (1998) Protein disorder and the evolution of molecular recognition: theory, predictions and observations. Pac Symp Biocomput:473–484

Dunker AK, Obradovic Z, Romero P et al (2000) Intrinsic protein disorder in complete genomes. Genome Inform Ser Workshop Genome Inform 11:161–171

Dunker AK, Lawson JD, Brown CJ et al (2001) Intrinsically disordered protein. J Mol Graph Model 19(1):26–59

Dunker AK, Brown CJ, Lawson JD et al (2002) Intrinsic disorder and protein function. Bio Chem 41(21):6573–6582

Dunker AK, Cortese MS, Romero P et al (2005) Flexible nets: the roles of intrinsic disorder in protein interaction networks. FEBS J 272(20):5129–5148

Dyson HJ, Wright PE (2002) Coupling of folding and binding for unstructured proteins. Curr Opin Struct Biol 12(1):54–60

Ebbinghaus S, Dhar A, McDonald D et al (2010) Protein folding stability and dynamics imaged in a living cell. Nat Methods 7(4):319–323

Ebel C (2004) Analytical ultracentrifugation for the study of biological macromolecules. Progr Colloid Polym Sci 127:73–82

Ebel C (2007) Analytical ultracentrifugation. State of the art and perspectives. In: Uversky VN, Permyakov EA (eds) Protein structures: methods in protein structure and stability analysis, vol Chap. 2.2. vol methods in protein structure and stability analysis. Part C. Conformational stability, size, shape and surface of protein molecules. Nova Science Publishers, New York, pp 229–260

Edelstein L, Stetz MA, McMahon HA et al (2010) The effects of Alpha-Helical structure and cyanylated cysteine on each other. J Phys Chem B 114(14):4931–4936

Eliezer D (2007) Characterizing residual structure in disordered protein states using nuclear magnetic resonance. Methods Mol Biol 350:49–67

Eliezer D (2009) Biophysical characterization of intrinsically disordered proteins. Curr Opin Struct Biol 19(1):23–30

Eyles SJ, Kaltashov IA (2004) Methods to study protein dynamics and folding by mass spectrometry. Methods 34(1):88–99

Fafarman AT, Webb LJ, Chuang JI et al (2006) Site-specific conversion of cysteine thiols into thiocyanate creates an IR probe for electric fields in proteins. J Am Chem Soc 128(41):13356–13357

Fanucci GE, Cafiso DS (2006) Recent advances and applications of site-directed spin labeling. Curr Opin Struct Biol 16(5):644–653

Farmer TB, Caprioli RM (1998) Determination of protein-protein interactions by matrix-assisted laser desorption/ionization mass spectrometry. J Mass Spectrom 33(8):697–704

Fasman GD (1996) Circular dichroism and the conformational analysis of biomolecules. Plenem Press, New York

Feix JB, Klug CS (1998) Site-directed spin-labeling of membrane proteins and peptide-membrane interactions. In: Berliner L (ed) Biological magnetic resonance, vol spin labeling: the next millenium. Plenum Press, New York, pp 251–281
Ferreon AC, Gambin Y, Lemke EA et al (2009) Interplay of alpha-synuclein binding and conformational switching probed by single-molecule fluorescence. Proc Natl Acad Sci U S A 106(14):5645–5650
Ferreon AC, Moran CR, Gambin Y et al (2010) Single-molecule fluorescence studies of intrinsically disordered proteins. Methods Enzymol 472:179–204
Fesinmeyer RM, Peterson ES, Dyer RB et al (2005) Studies of helix fraying and solvation using C-13' isotopomers. Protein Sci 14(9):2324–2332
Flory JP (1969) Statistical mechanics of chain molecules. Interscience, New York
Fontana A, Fassina G, Vita C et al. (1986) Correlation between sites of limited proteolysis and segmental mobility in thermolysin. Biochem 25(8):1847–1851
Fontana A, Polverino de Laureto P et al (1993) Molecular aspects of proteolysis of globular proteins. In: van den Tweel W, Harder A, Buitelear M (eds) Protein stability and stabilization. Elsevier Science, Amsterdam, pp 101–110
Fontana A, Polverino de Laureto P, De Filippis V et al (1997a) Probing the partly folded states of proteins by limited proteolysis. Fold Des 2(2):R17–R26
Fontana A, Zambonin M, Polverino de Laureto P et al (1997b) Probing the conformational state of apomyoglobin by limited proteolysis. J Mol Biol 266(2):223–230
Fontana A, de Laureto PP, Spolaore B et al (2012) Identifying disordered regions in proteins by limited proteolysis. Methods Mol Biol 896:297–318
Forster TH (1948) Zwischen Molekulare Energie Wanderung und Fluoreszenz. Ann Phys (Leipzig) 2:55–75
Forster TH (1959) Transfer mechanisms of electroinc excitation. Discuss Fraday Soc 27:7–17
Forster TH (1965) Delocalized excitation and excitation transfer. In: Sinaonglu O (ed) Modern quantum Chemistry, Istnabul lectures part III: action of light and organic crystals. Academic Press, New York, pp 93–137
Freedberg DI, Selenko P (2014) Live cell NMR. Annu Rev Biophysics 43:171–192
Frimpong AK, Abzalimov RR, Eyles SJ et al (2007) Gas-phase interference-free analysis of protein ion charge-state distributions: detection of small-scale conformational transitions accompanying pepsin inactivation. Anal Chem 79(11):4154–4161
Fujio H, Takagaki Y, Ha YM et al (1985) Native and non-native conformation-specific antibodies directed to the loop region of hen egg-white lysozyme. J Biochem 98(4):949–962
Furie B, Furie BC (1979) Conformation-specific antibodies as probes of the gamma-carboxyglutamic acid-rich region of bovine prothrombin. Studies of metal-induced structural changes. J Biol Chem 254(19):9766–9771
Gabel F (2012) Small angle neutron scattering for the structural study of intrinsically disordered proteins in solution: a practical guide. Methods Mol Biol 896:123–135
Gabel F, Jensen MR, Zaccai G et al (2009) Quantitative modelfree analysis of urea binding to unfolded ubiquitin using a combination of small angle X-ray and neutron scattering. J Am Chem Soc 131(25):8769–8771
Garner E, Cannon P, Romero P et al (1998) Predicting disordered regions from amino acid sequence: common themes despite differing structural characterization. Genome Inform Ser Workshop Genome Inform 9:201–213
Gast K, Fiedler C (2012) Dynamic and static light scattering of intrinsically disordered proteins. Methods Mol Biol 896:137–161
Gerstein M (1998) How representative are the known structures of the proteins in a complete genome? A comprehensive structural census. Fold Des 3(6):497–512
Gonzalez-Montalban N, Natalello A, Garcia-Fruitos E et al (2008) In situ protein folding and activation in bacterial inclusion bodies. Biotechnol Bioeng 100(4):797–802
Goormaghtigh E, Cabiaux V, Ruysschaert JM (1994a) Determination of soluble and membrane protein structure by fourier transform infrared spectroscopy. I. Assignments and model compounds. Subcell biochem 23:329–362

Goormaghtigh E, Cabiaux V, Ruysschaert JM (1994b) Determination of soluble and membrane protein structure by fourier transform infrared spectroscopy. II. Experimental aspects, side chain structure, and H/D exchange. Subcell biochem 23:363–403

Goormaghtigh E, Cabiaux V, Ruysschaert JM (1994c) Determination of soluble and membrane protein structure by fourier transform infrared spectroscopy. III. Secondary structures. Subcell biochem 23:405–450

Goto Y, Fink AL (1989) Conformational states of beta-lactamase: molten-globule states at acidic and alkaline pH with high salt. BioChem 28(3):945–952

Gratzer WB, Cowburn DA (1969) Optical activity of biopolymers. Nature 222(5192):426–431

Greenfield NJ (2006) Using circular dichroism spectra to estimate protein secondary structure. Nat Protoc 1(6):2876–2890

Greenfield N, Fasman GD (1969) Computed circular dichroism spectra for the evaluation of protein conformation. BioChem 8(10):4108–4116

Griffith WP, Kaltashov IA (2003) Highly asymmetric interactions between globin chains during hemoglobin assembly revealed by electrospray ionization mass spectrometry. BioChem 42(33):10024–10033

Gumerov DR, Dobo A, Kaltashov IA (2002) Protein-ion charge-state distributions in electrospray ionization mass spectrometry: distinguishing conformational contributions from masking effects. Eur J Mass Spectrom 8(2):123–129

Haas E (2004) Fluorescence resonance energy transfer (FRET) and single molecule fluorescence detection studies of the mechanism of protein folding and unfolding. In: Kiefhaber JBa T (ed) Protein Folding Handbook. Part I, vol I. WILEY-VCH Verlag GmbH & Co.KGaA, Weinheim, pp 573–633

Haas E (2012) Ensemble FRET methods in studies of intrinsically disordered proteins. Methods Mol Biol 895:467–498

Habchi J, Martinho M, Gruet A et al (2012) Monitoring structural transitions in IDPs by site-directed spin labeling EPR spectroscopy. Methods Mol Biol 895:361–386

He B, Wang K, Liu Y et al (2009) Predicting intrinsic disorder in proteins: an overview. Cell Res 19(8):929–949

Heller WT (2010) Small-angle neutron scattering and contrast variation: a powerful combination for studying biological structures. Acta Crystallogr D Biol Crystallogr 66(Pt 11):1213–1217

Heraud P, Tobin MJ (2009) The emergence of biospectroscopy in stem cell research. Stem cell research 3(1):12–14

Howlett GJ, Minton AP, Rivas G (2006) Analytical ultracentrifugation for the study of protein association and assembly. Curr Opin Chem Biol 10(5):430–436

Huang A, Stultz CM (2008) The effect of a DeltaK280 mutation on the unfolded state of a microtubule-binding repeat in Tau. PLoS Comput Biol 4(8):e1000155

Huang R, Kubelka J, Barber-Armstrong W et al (2004) Nature of vibrational coupling in helical peptides: an isotopic labeling study. J Am Chem Soc 126(8):2346–2354

Hubbard SJ (1998) The structural aspects of limited proteolysis of native proteins. Biochim Biophys Acta 1382(2):191–206

Hubbard SJ, Campbell SF, Thornton JM (1991) Molecular recognition. Conformational analysis of limited proteolytic sites and serine proteinase protein inhibitors. J Mol Biol 220(2):507–530

Hubbard SJ, Eisenmenger F, Thornton JM (1994) Modeling studies of the change in conformation required for cleavage of limited proteolytic sites. Protein Sci 3:757–768

Hubbard SJ, Beynon RJ, Thornton JM (1998) Assessment of conformational parameters as predictors of limited proteolytic sites in native protein structures. Protein Eng 11:349–359

Hubbell WL, McHaourab HS, Altenbach C et al (1996) Watching proteins move using site-directed spin labeling. Structure 4(7):779–783

Hubbell WL, Gross A, Langen R et al (1998) Recent advances in site-directed spin labeling of proteins. Curr Opin Struct Biol 8(5):649–656

Hubbell WL, Cafiso DS, Altenbach C (2000) Identifying conformational changes with site-directed spin labeling. Nat Struct Biol 7(9):735–739

Hubbell WL, Altenbach C, Hubbell CM et al (2003) Rhodopsin structure, dynamics, and activation: a perspective from crystallography, site-directed spin labeling, sulfhydryl reactivity, and disulfide cross-linking. Adv Protein Chem 63:243–290
Iakoucheva LM, Kimzey AL, Masselon CD et al (2001a) Identification of intrinsic order and disorder in the DNA repair protein XPA. Protein Sci 10(3):560–571
Iakoucheva LM, Kimzey AL, Masselon CD et al (2001b) Aberrant mobility phenomena of the DNA repair protein XPA. Protein Sci 10(7):1353–1362
Iakoucheva LM, Brown CJ, Lawson JD et al (2002) Intrinsic disorder in cell-signaling and cancer-associated proteins. J Mol Biol 323(3):573–584
Iakoucheva LM, Radivojac P, Brown CJ et al (2004) The importance of intrinsic disorder for protein phosphorylation. Nucleic Acids Res 32(3):1037–1049
Invernizzi G, Grandori R (2007) Detection of the equilibrium folding intermediate of beta-lactoglobulin in the presence of trifluoroethanol by mass spectrometry. Rapid Commun Mass Spectrom 21(6):1049–1052
Jackson M, Mantsch HH (1991) Protein secondary structure from FT-IR spectroscopy: correlation with dihedral angles from three-dimensional Ramachandran plots. Can J Chem 69:1639–1642
Jacrot B (1976) The study of biological structures by neurtron scattering from solution. Rep Prog Phys 39(10):911–953
Jensen MR, Markwick PR, Meier S et al (2009) Quantitative determination of the conformational properties of partially folded and intrinsically disordered proteins using NMR dipolar couplings. Structure 17(9):1169–1185
Jensen MR, Salmon L, Nodet G et al (2010) Defining conformational ensembles of intrinsically disordered and partially folded proteins directly from chemical shifts. J Am Chem Soc 132(4):1270-1272
Jirgensons B (1965) The cotton effects in the optical rotatory dispersion of proteins as new criteria of conformation. J Biol Chem 240:1064–1071
Jirgensons B, Hnilica LS (1965) The conformational changes of calf-thymus histone fractions as determined by the optical rotary dispersion. Biochim Biophys Acta 109(1):241–249
Johansson J, Gudmundsson GH, Rottenberg ME et al (1998) Conformation-dependent antibacterial activity of the naturally occurring human peptide LL-37. J Biol Chem 273(6):3718–3724
Johnson WC Jr (1988) Secondary structure of proteins through circular dichroism spectroscopy. Annu Rev Biophys Biophys Chem 17:145–166
Johnson DE, Xue B, Sickmeier MD et al (2012) High-throughput characterization of intrinsic disorder in proteins from the Protein Structure Initiative. J Struct Biol 180(1):201–215
Kalkhof S, Ihling C, Mechtler K et al (2005) Chemical cross-linking and high-performance fourier transform ion cyclotron resonance mass spectrometry for protein interaction analysis: application to a calmodulin/target peptide complex. Anal Chem 77(2):495–503
Karplus PA, Schulz GE (1985) Prediction of chain flexibility in proteins. Naturwissenschaften 72:212–213
Karst JC, Sotomayor-Perez AC, Ladant D et al (2012) Estimation of intrinsically disordered protein shape and time-averaged apparent hydration in native conditions by a combination of hydrodynamic methods. Methods Mol Biol 896:163–177
Kataoka M, Nishii I, Fujisawa T et al (1995) Structural characterization of the molten globule and native states of apomyoglobin by solution X-ray scattering. J Mol Biol 249(1):215–228
Keiderling TA (1996) Vibrational circular dichroism: application to conformational analysis of biomolecules. In: Fasman GD (ed) Circular dichroism and the conformational analysis of biomolecules. Plenumn Press, New York, pp 555
Keiderling TA, Xu Q (2002) Unfolded proteins studied with IR and VCD spectra. Adv Protein Chem 62:111–161
Kelly SM, Price NC (1997) The application of circular dichroism to studies of protein folding and unfolding. Biochim Biophys Acta 1338(2):161–185
Kelly JG, Singh MN, Stringfellow HF et al (2009) Derivation of a subtype-specific biochemical signature of endometrial carcinoma using synchrotron-based fourier-transform infrared microspectroscopy. Cancer Lett 274(2):208–217

Keston A, Lospalluto J (1953) Simple ultrasensitive spectropolarimeters. Fed Proc 12:229
Kim TD, Ryu HJ, Cho HI et al (2000) Thermal behavior of proteins: heat-resistant proteins and their heat-induced secondary structural changes. BioChemistry 39(48):14839–14846
Kirste RG, Schulz GV, Stuhrmann HB (1969) Die Konformationsaenderung des Pottwal-Mesmyoglobins bei der reversiblen Denaturierung im pH-Bereich 7 bis 1. Z Naturforsch B 24:1385–1392
Klare JP, Steinhoff HJ (2009) Spin labeling EPR. Photosynth Res 102(2–3):377–390
Kneipp J, Miller LM, Joncic M et al (2003) In situ identification of protein structural changes in prion-infected tissue. Biochim Biophys Acta 1639(3):152–158
Koch MH, Vachette P, Svergun DI (2003) Small-angle scattering: a view on the properties, structures and structural changes of biological macromolecules in solution. Q Rev Biophys 36(2):147–227
Kodera N, Yamamoto D, Ishikawa R et al (2010) Video imaging of walking myosin V by high-speed atomic force microscopy. Nature 468(7320):72–76
Kohn JE, Millet IS, Jacob J et al (2004) Random-coil behavior and the dimensins of chemically unfolded proteins. Proc Natl Acad Sci U S A 101(34):12491–12496
Konermann L, Douglas DJ (1998) Equilibrium unfolding of proteins monitored by electrospray ionization mass spectrometry: distinguishing two-state from multi-state transitions. Rapid Commun Mass Spectrom 12(8):435–442
Konno T, Tanaka N, Kataoka M et al (1997) A circular dichroism study of preferential hydration and alcohol effects on a denatured protein, pig calpastatin domain I. Biochim Biophys Acta 1342(1):73–82
Kretlow A, Wang Q, Kneipp J et al (2006) FTIR-microspectroscopy of prion-infected nervous tissue. Biochim Biophys Acta 1758(7):948–959
Krimm S, Bandekar J (1986) Vibrational spectroscopy and conformation of peptides, polypeptides, and proteins. Adv Protein Chem 38:181–364
Kundu S, Melton JS, Sorensen DC et al (2002) Dynamics of proteins in crystals: comparison of experiment with simple models. Biophys J 83(2):723–732
Lavery DN, McEwan IJ (2008) Structural characterization of the native NH2-terminal transactivation domain of the human androgen receptor: a collapsed disordered conformation underlies structural plasticity and protein-induced folding. BioChemistry 47(11):3360–3369
Le Gall L, Romero PR, Cortese MS et al (2007) Intrinsic disorder in the protein data bank. J Biomol Struct Dyn 24(4):325–342
Lebowitz J, Lewis MS, Schuck P (2002) Modern analytical ultracentrifugation in protein science: a tutorial review. Protein Sci 11(9):2067–2079
Lewis JW, Tilton RF, Einterz CM et al (1985) New technique for measuring circular dichroism changes on a nanosecond time scale—application to (carbonmonoxy)myoglobin and (carbonmonoxy)hemoglobin. J Phys Chem 89(2):289–294
Li C, Charlton LM, Lakkavaram A et al (2008) Differential dynamical effects of macromolecular crowding on an intrinsically disordered protein and a globular protein: implications for in-cell NMR spectroscopy. J Am Chem Soc 130(20):6310–6311
Lipfert J, Doniach S (2007) Small-angle X-ray scattering from RNA, proteins, and protein complexes. Annu Rev Biophys Biomol Struct 36:307–327
Longhi S, Uversky VN (eds) (2010) Instrumental analysis of intrinsically disordered proteins: assessing structure and conformation. The Wiley series in protein and peptide science. Wiley, Hoboken
Lynn A, Chandra S, Malhotra P et al (1999) Heme binding and polymerization by plasmodium falciparum histidine rich protein II: influence of pH on activity and conformation. FEBS Lett 459(2):267–271
Maienschein-Cline MG, Londergan CH (2007) The CN stretching band of aliphatic thiocyanate is sensitive to solvent dynamics and specific solvation. J Phys Chem A 111(40):10020–10025
Maiti NC, Apetri MM, Zagorski MG et al (2004) Raman spectroscopic characterization of secondary structure in natively unfolded proteins: alpha-synuclein. J Am Chem Soc 126(8):2399–2408

Manas ES, Getahun Z, Wright WW et al (2000) Infrared spectra of amide groups in alpha-helical proteins: evidence for hydrogen bonding between helices and water. J Am Chem Soc 122:9883–9890
Manon F, Ebel C (2010) Analytical ultracentrifugation, a useful tool to probe intrinsically disordered proteins. In: Uversky VN, Longhi S (eds) Instrumental analysis of intrinsically disordered proteins: assessing structure and conformation. Wiley series in protein and peptide science. Wiley, Hoboken, pp 433–449
McMahon HA, Alfieri KN, Clark KAA et al (2010) Cyanylated cysteine: a covalently attached vibrational probe of protein-lipid contacts. J Phys Chem Lett 1(5):850–855
McNulty BC, Young GB, Pielak GJ (2006) Macromolecular crowding in the escherichia coli periplasm maintains alpha-synuclein disorder. J Mol Biol 355(5):893–897
Measey T, Hagarman A, Eker F et al (2005) Side chain dependence of intensity and wavenumber position of amide I' in IR and visible Raman spectra of XA and AX dipeptides. J Phys Chem B 109(16):8195–8205
Milder SJ, Gold JS, Kliger DS (1990) Assignments of ground-state and excited-state spectra from time-resolved absorption and circular dichroism measurements of the ^2E state of (D)-$Cr(Bpy)_3^{3+}$. Inorg Chem 29(13):2506–2511
Milhiet PE, Yamamoto D, Berthoumieu O et al (2010) Deciphering the structure, growth and assembly of amyloid-like fibrils using high-speed atomic force microscopy. PLoS ONE 5(10):e13240
Millet IS, Doniach S, Plaxco KW (2002) Toward a taxonomy of the denatured state: small angle studies of unfolded proteins. Adv Protein Chem 62:241–262
Mittag T, Forman-Kay JD (2007) Atomic-level characterization of disordered protein ensembles. Curr Opin Struct Biol 17(1):3–14
Mittag T, Orlicky S, Choy WY et al (2008) Dynamic equilibrium engagement of a polyvalent ligand with a single-site receptor. Proc Natl Acad Sci U S A 105(46):17772–17777
Mittag T, Marsh J, Grishaev A et al (2010) Structure/function implications in a dynamic complex of the intrinsically disordered Sic1 with the Cdc4 subunit of an SCF ubiquitin ligase. Structure 18(4):494–506
Morin B, Bourhis JM, Belle V et al (2006) Assessing induced folding of an intrinsically disordered protein by site-directed spin-labeling EPR spectroscopy. J Phys Chem B 110(41):20596–20608
Moscowitz A (1962) Theoretical aspects of optical activity.1. Small molecules. Adv Chem Phys 4:67–112
Nash P, Tang X, Orlicky S et al (2001) Multisite phosphorylation of a CDK inhibitor sets a threshold for the onset of DNA replication. Nature 414(6863):514–521
Natalello A, Doglia SM (2010) Intrinsically disordered proteins and induced folding studied by fourier transform infrared spectroscopy. In: Uversky VN, Longhi S (eds) Instrumental analysis of intrinsically disordered proteins: assessing structure and conformation. (Wiley Series on Protein and Peptide Science). Wiley, Hoboken
Natalello A, Ami D, Doglia SM (2012) Fourier transform infrared spectroscopy of intrinsically disordered proteins: measurement procedures and data analyses. Methods Mol Biol 895:229–244
Nath S, Meuvis J, Hendrix J et al (2010) Early aggregation steps in alpha-synuclein as measured by FCS and FRET: evidence for a contagious conformational change. Biophys J 98(7):1302–1311
Nath S, Deng M, Engelborghs Y (2012) Fluorescence correlation spectroscopy to determine the diffusion coefficient of α-synuclein and follow early oligomer formation. Methods Mol Biol 895:499–506
Neyroz P, Ciurli S (2012) Intrinsic fluorescence of intrinsically disordered proteins. Methods Mol Biol 895:435–440
Neyroz P, Zambelli B, Ciurli S (2006) Intrinsically disordered structure of Bacillus pasteurii UreG as revealed by steady-state and time-resolved fluorescence spectroscopy. BioChemistry 45(29):8918–8930
Novotny J, Bruccoleri RE (1987) Correlation among sites of limited proteolysis, enzyme accessibility and segmental mobility. FEBS Lett 211(2):185–189
Oberg KA, Ruysschaert JM, Goormaghtigh E (2004) The optimization of protein secondary structure determination with infrared and circular dichroism spectra. Eur J Biochem/FEBS 271(14):2937–2948

Oldfield CJ, Cheng Y, Cortese MS et al (2005) Coupled folding and binding with α-helix-forming molecular recognition elements. Biochemistry 44(37):12454–12470
Oldfield CJ, Meng J, Yang JY et al (2008) Flexible nets: disorder and induced fit in the associations of p53 and 14-3-3 with their partners. BMC Genomics 9(Suppl 1):S1
Oroz J, Hervas R, Valbuena A et al (2012) Unequivocal single-molecule force spectroscopy of intrinsically disordered proteins. Methods Mol Biol 896:71–87
Orsini F, Ami D, Villa AM et al (2000) FT-IR microspectroscopy for microbiological studies. J Microbiol Methods 42(1):17–27
Pelton JT, McLean LR (2000) Spectroscopic methods for analysis of protein secondary structure. Anal Biochem 277(2):167–176
Peng K, Obradovic Z, Vucetic S (2004) Exploring bias in the protein data bank using contrast classifiers. Pac Symp Biocomput:435–446
Perez J, Vachette P, Russo D et al (2001) Heat-induced unfolding of neocarzinostatin, a small all-beta protein investigated by small-angle X-ray scattering. J Mol Biol 308(4):721–743
Permyakov SE, Millett IS, Doniach S et al (2003) Natively unfolded C-terminal domain of caldesmon remains substantially unstructured after the effective binding to calmodulin. Proteins 53(4):855–862
Pervushin K, Vamvaca K, Vogeli B et al (2007) Structure and dynamics of a molten globular enzyme. Nat Struct Mol Biol 14(12):1202–1206
Peters T Jr (1996) All about albumin: biochemistry, genetics, and medical application. Academic Press, New York
Petrasek Z, Schwille P (2008) Precise measurement of diffusion coefficients using scanning fluorescence correlation spectroscopy. Biophys J 94(4):1437–1448
Petrescu AJ, Receveur V, Calmettes P et al (1997) Small-angle neutron scattering by a strongly denatured protein: analysis using random polymer theory. Biophys J 72(1):335–342
Petrescu AJ, Receveur V, Calmettes P et al (1998) Excluded volume in the configurational distribution of a strongly-dentaured protein. Protein Sci 7(6):1396–1403
Pirman NL, Milshteyn E, Galiano L et al (2011) Characterization of the disordered-to-alpha-helical transition of IA by SDSL-EPR spectroscopy. Protein Sci 20(1):150–159
Provencher SW, Glockner J (1981) Estimation of globular protein secondary structure from circular dichroism. BioChem 20(1):33–37
Ptitsyn OB (1995) Molten globule and protein folding. Adv Protein Chem 47:83–229
Ptitsyn OB, Uversky VN (1994) The molten globule is a third thermodynamical state of protein molecules. FEBS Lett 341(1):15–18
Ptitsyn OB, Pain RH, Semisotnov GV et al (1990) Evidence for a molten globule state as a general intermediate in protein folding. FEBS lett 262(1):20–24
Ptitsyn OB, Bychkova VE, Uversky VN (1995) Kinetic and equilibrium folding intermediates. Philos Trans R Soc Lond B Biol Sci 348(1323):35–41
Putnam CD, Hammel M, Hura GL et al (2007) X-ray solution scattering (SAXS) combined with crystallography and computation: defining accurate macromolecular structures, conformations and assemblies in solution. Q Rev Biophys 40(3):191–285
Radivojac P, Obradovic Z, Smith DK et al (2004) Protein flexibility and intrinsic disorder. Protein Sci 13(1):71–80
Radivojac P, Iakoucheva LM, Oldfield CJ et al (2007) Intrinsic disorder and functional proteomics. Biophys J 92(5):1439–1456
Raussens V, Ruysschaert JM, Goormaghtigh E (2003) Protein concentration is not an absolute prerequisite for the determination of secondary structure from circular dichroism spectra: a new scaling method. Anal Biochem 319(1):114–121
Rawiso M, Duplessix R, Picot C (1987) Scattering function of polysterene. Macromolecules 20:630–648
Receveur-Brechot V, Durand D (2012) How random are intrinsically disordered proteins? A small angle scattering perspective. Curr Protein Pept Sci 13(1):55–75
Receveur-Brechot V, Bourhis JM, Uversky VN et al (2006) Assessing protein disorder and induced folding. Proteins 62(1):24–45
Rhodes G (1993) Crystallography made crystal clear: a guide for users of macromolecular models. Academic Press, San Diego

Rodionova NA, Semisotnov GV, Kutyshenko VP et al (1989) Staged equilibrium of carbonic anhydrase unfolding in strong denaturants. Mol Biol (Mosk) 23(3):683–692

Romero P, Obradovic Z, Li X et al (2001) Sequence complexity of disordered protein. Proteins 42(1):38–48

Salvay AG, Communie G, Ebel C (2012) Sedimentation velocity analytical ultracentrifugation for intrinsically disordered proteins. Methods Mol Biol 896:91–105

Sandal M, Valle F, Tessari I et al (2008) Conformational equilibria in monomeric α-synuclein at the single-molecule level. Plos Biology 6(1):99–108

Sane SU, Cramer SM, Przybycien TM (1999) A holistic approach to protein secondary structure characterization using amide I band Raman spectroscopy. Anal Biochem 269(2):255–272

Schreurs S, Kluba M, Meuvis J et al (2012) Fluorescence lifetime measurements of intrinsically unstructured proteins: application to alpha-synuclein. Methods Mol Biol 895:461–466

Schuler B, Eaton WA (2008) Protein folding studied by single-molecule FRET. Curr Opin Struct Biol 18(1):16–26

Schuler B, Haran G (2008) Protein folding and dynamics from optical single molecule spectroscopy. In: Rigler R, Vogel H (eds) Single molecules and nanotechnology, vol 12. Springer, Berlin, pp 181–216 (Springer Series in Biophysics)

Schuler B, Muller-Spath S, Soranno A et al (2012) Application of confocal single-molecule FRET to intrinsically disordered proteins. Methods Mol Biol 896:21–45

Schultz CP, Liu KZ, Johnston JB et al (1997) Prognosis of chronic lymphocytic leukemia from infrared spectra of lymphocytes. J Mol Struct 408:253–256

Schulz DM, Ihling C, Clore GM et al (2004) Mapping the topology and determination of a low-resolution three-dimensional structure of the calmodulin-melittin complex by chemical cross-linking and high-resolution FTICRMS: direct demonstration of multiple binding modes. BioChem 43(16):4703–4715

Schurtenberger P (2002) Static properties of polymers. In: Lindner P, Zemb T (eds) Neutrons, X-rays and light. Delta series, North Holland

Schweitzer-Stenner R, Soffer JB, Toal S et al (2012a) Structural analysis of unfolded peptides by Raman spectroscopy. Methods Mol Biol 895:315–346

Schweitzer-Stenner R, Soffer JB, Verbaro D (2012b) Structure analysis of unfolded peptides I: vibrational circular dichroism spectroscopy. Methods Mol Biol 895:271–313

Selenko P, Wagner G (2007) Looking into live cells with in-cell NMR spectroscopy. J Struct Biol 158(2):244–253

Semisotnov GV, Rodionova NA, Kutyshenko VP et al (1987) Sequential mechanism of refolding of carbonic anhydrase B. FEBS lett 224 (1):9–13

Semisotnov GV, Rodionova NA, Razgulyaev OI et al (1991) Study of the "molten globule" intermediate state in protein folding by a hydrophobic fluorescent probe. Biopolymers 31(1):119–128

Seshadri S, Khurana R, Fink AL (1999) Fourier transform infrared spectroscopy in analysis of protein deposits. Methods Enzymol 309:559–576

Shaw RA, Mantsch HH (1999) Vibrational biospectroscopy: from plants to animals to humans. A historical perspective. J Mol Struct 480–481:1–13

Shaw RA, Guijon FB, Paraskevas M et al (1999) Infrared spectroscopy of exfoliated cervical cell specimens. Proceed with caution. Anal Quant Cytol Histol 21(4):292–302

Shibata M, Yamashita H, Uchihashi T et al (2010) High-speed atomic force microscopy shows dynamic molecular processes in photoactivated bacteriorhodopsin. Nat Nanotechnol 5(3):208–212

Simmons DA, Wilson DJ, Lajoie GA et al (2004) Subunit disassembly and unfolding kinetics of hemoglobin studied by time-resolved electrospray mass spectrometry. Biochem 43(46):14792–14801

Sinz A (2003) Chemical cross-linking and mass spectrometry for mapping three-dimensional structures of proteins and protein complexes. J Mass Spectrom 38(12):1225–1237

Sinz A (2006) Chemical cross-linking and mass spectrometry to map three-dimensional protein structures and protein-protein interactions. Mass Spectrom Rev 25(4):663–682

Small EW, Fanconi B, Peticolas WL (1970) Raman spectra and the phonon dispersion of polyglycine. J Chem Phys 52(9):4369–4379

Smith MD, Jelokhani-Niaraki M (2012) pH-induced changes in intrinsically disordered proteins. Methods Mol Biol 896:223–231

Smith DL, Deng Y, Zhang Z (1997) Probing the non-covalent structure of proteins by amide hydrogen exchange and mass spectrometry. J Mass Spectrom 32(2):135–146

Smyth E, Syme CD, Blanch EW et al (2001) Solution structure of native proteins with irregular folds from Raman optical activity. Biopolymers 58(2):138–151

Sotomayor Perez AC, Karst JC, Davi M et al (2010) Characterization of the regions involved in the calcium-induced folding of the intrinsically disordered RTX motifs from the bordetella pertussis adenylate cyclase toxin. J Mol Biol 397(2):534–549

Sotomayor-Perez AC, Ladant D, Chenal A (2011) Calcium-induced folding of intrinsically disordered repeat-in-toxin (RTX) motifs via changes of protein charges and oligomerization states. J Biol Chem 286(19):16997–17004

Sotomayor-Perez AC, Karst JC, Ladant D et al (2012) Mean net charge of intrinsically disordered proteins: experimental determination of protein valence by electrophoretic mobility measurements. Methods Mol Biol 896:331–349

Steinberg IZ (1971) Long-range nonradiative transfer of electronic excitation energy in proteins and polypeptides. Annu Rev Biochem 40:83–114

Stryer L (1965) The interaction of a naphthalene dye with apomyoglobin and apohemoglobin. A fluorescent probe of non-polar binding sites. J Mol Biol 13(2):482–495

Stryer L, Haugland RP (1967) Energy transfer: a spectroscopic ruler. Proc Natl Acad Sci U S A 58(2):719–726

Stryer L, Thomas DD, Meares CF (1982) Diffusion-enhanced fluorescence energy transfer. Annu Rev Biophys Bioeng 11:203–222

Sulatskaya AI, Povarova OI, Kuznetsova IM et al (2012) Binding stoichiometry and affinity of fluorescent dyes to proteins in different structural states. Methods Mol Biol 895:441–460

Susi H, Byler DM (1986) Resolution-enhanced fourier transform infrared spectroscopy of enzymes. Methods Enzymol 130:290–311

Susi H, Byler DM (1987) Fourier transform infrared study of proteins with parallel beta-chains. Arch Biochem Biophys 258(2):465–469

Svergun DI, Koch MHJ (2002) Small-angle scattering studies of biological macromolecules in solution. Rep Prog Phys 66(10):1735–1782

Syme CD, Blanch EW, Holt C et al (2002) A Raman optical activity study of rheomorphism in caseins, synucleins and tau. New insight into the structure and behaviour of natively unfolded proteins. Eur J Biochem/FEBS 269(1):148–156

Szollosi E, Bokor M, Bodor A et al (2008) Intrinsic structural disorder of DF31, a *Drosophila* protein of chromatin decondensation and remodeling activities. J Proteome Res 7(6):2291–2299

Tadesse L, Nazarbaghi R, Walters L (1991) Isotopically enhanced infrared spectroscopy: a novel method for examining secondary structure at specific sites in conformationally heterogeneous peptides. J Am Chem Soc 113:7036–7037

Takaoka Y, Kioi Y, Morito A et al (2013) Quantitative comparison of protein dynamics in live cells and in vitro by in-cell (19)F-NMR. Chem Commun 49(27):2801–2803

Tanford C (1968) Protein denaturation. Adv Protein Chem 23:121–282

Tanthanuch W, Thumanu K, Lorthongpanich C et al (2010) Neural differentiation of mouse embryonic stem cells studied by FTIR spectroscopy. J Mol Struct 967:189–195

Tantos A, Tompa P (2012) Identification of intrinsically disordered proteins by a special 2D electrophoresis. Methods Mol Biol 896:215–222

Tcherkasskaya O, Uversky VN (2001) Denatured collapsed states in protein folding: example of apomyoglobin. Proteins 44(3):244–254

Tcherkasskaya O, Uversky VN (2003) Polymeric aspects of protein folding: a brief overview. Protein Pept Lett 10(3):239–245

Tcherkasskaya O, Davidson EA, Uversky VN (2003) Biophysical constraints for protein structure prediction. J Proteome Res 2(1):37–42

Theillet FX, Binolfi A, Frembgen-Kesner T et al (2014) Physicochemical properties of cells and their effects on intrinsically disordered proteins (IDPs). Chem Rev 114(13):6661–6714
Thomas GJ (2002) New structural insights from Raman spectroscopy of proteins and their assemblies. Biopolymers 67(4–5):214–225
Timm DE, Vissavajjhala P, Ross AH et al (1992) Spectroscopic and chemical studies of the interaction between nerve growth factor (NGF) and the extracellular domain of the low affinity NGF receptor. Protein Sci 1(8):1023–1031
Tompa P (2002) Intrinsically unstructured proteins. Trends Biochem Sci 27(10):527–533
Trester-Zedlitz M, Kamada K, Burley SK et al (2003) A modular cross-linking approach for exploring protein interactions. J Am Chem Soc 125(9):2416–2425
Tsvetkov P, Shaul Y (2012) Determination of IUP based on susceptibility for degradation by default. Methods Mol Biol 895:3–18
Tsvetkov P, Asher G, Paz A et al (2008) Operational definition of intrinsically unstructured protein sequences based on susceptibility to the 20 S proteasome. Proteins 70(4):1357–1366
Uversky VN (1993) Use of fast protein size-exclusion liquid chromatography to study the unfolding of proteins which denature through the molten globule. Biochem 32(48):13288–13298
Uversky VN (1994) Gel-permeation chromatography as a unique instrument for quantitative and qualitative analysis of protein denaturation and unfolding. Int J Bio-Chromatography 1: 103–114
Uversky VN (2002a) Natively unfolded proteins: a point where biology waits for physics. Protein Sci 11(4):739–756
Uversky VN (2002b) What does it mean to be natively unfolded? Eur J Biochem/FEBS 269(1): 2–12
Uversky VN (2003) Protein folding revisited. A polypeptide chain at the folding-misfolding-nonfolding cross-roads: which way to go? Cell Mol Life Sci 60(9):1852–1871
Uversky VN (2009) Intrinsically disordered proteins and their environment: effects of strong denaturants, temperature, pH, counter Ions, membranes, binding partners, osmolytes, and macromolecular crowding. Protein J 28(7–8):305–325
Uversky VN (2010) The mysterious unfoldome: structureless, underappreciated, yet vital part of any given proteome. J Biomed Biotechnol 2010:568068
Uversky VN (2011) Multitude of binding modes attainable by intrinsically disordered proteins: a portrait gallery of disorder-based complexes. Chem Soc Rev 40(3):1623–1634
Uversky VN (2012) Size-exclusion chromatography in structural analysis of intrinsically disordered proteins. Methods Mol Biol 896:179–194
Uversky VN, Dunker AK (2010) Understanding protein non-folding. Biochim Biophys Acta 1804(6):1231–1264
Uversky VN, Dunker AK (eds) (2012a) Experimental tools for the analysis of intrinsically disordered protein: volume I methods in molecular biology. Humana Press, Totowa
Uversky VN, Dunker AK (eds) (2012b) Experimental tools for the analysis of intrinsically disordered protein: volume II methods in molecular biology. Humana Press, Totowa
Uversky VN, Dunker AK (2012c) Multiparametric analysis of intrinsically disordered proteins: looking at intrinsic disorder through compound eyes. Anal Chem 84(5):2096–2104
Uversky VN, Ptitsyn OB (1994) "Partly folded" state, a new equilibrium state of protein molecules: four-state guanidinium chloride-induced unfolding of β-lactamase at low temperature. BioChem 33(10):2782–2791
Uversky VN, Ptitsyn OB (1996a) All-or-none solvent-induced transitions between native, molten globule and unfolded states in globular proteins. Fold Des 1(2):117–122
Uversky VN, Ptitsyn OB (1996b) Further evidence on the equilibrium "pre-molten globule state": four-state guanidinium chloride-induced unfolding of carbonic anhydrase B at low temperature. J Mol Biol 255(1):215–228
Uversky VN, Kutyshenko VP, Protasova N et al (1996) Circularly permuted dihydrofolate reductase possesses all the properties of the molten globule state, but can resume functional tertiary structure by interaction with its ligands. Protein Sci 5(9):1844–1851

Uversky VN, Gillespie JR, Millett IS et al (1999) Natively unfolded human prothymosin α adopts partially folded collapsed conformation at acidic pH. BioChemistry 38(45):15009–15016

Uversky VN, Gillespie JR, Fink AL (2000a) Why are "natively unfolded" proteins unstructured under physiologic conditions? Proteins 41(3):415–427

Uversky VN, Gillespie JR, Millett IS et al (2000b) Zn(2+)-mediated structure formation and compaction of the "natively unfolded" human prothymosin alpha. Biochem Biophys Res Commun 267(2):663–668

Uversky VN, Li J, Fink AL (2001) Evidence for a partially folded intermediate in alpha-synuclein fibril formation. J Biol Chem 276(14):10737–10744

Uversky VN, Permyakov SE, Zagranichny VE et al (2002) Effect of zinc and temperature on the conformation of the gamma subunit of retinal phosphodiesterase: a natively unfolded protein. J Proteome Res 1(2):149–159

Uversky VN, Oldfield CJ, Dunker AK (2005) Showing your ID: intrinsic disorder as an ID for recognition, regulation and cell signaling. J Mol Recognit 18(5):343–384

Uversky VN, Oldfield CJ, Dunker AK (2008) Intrinsically disordered proteins in human diseases: introducing the D2 concept. Annu Rev Biophys 37:215–246

Vacic V, Uversky VN, Dunker AK et al (2007) Composition profiler: a tool for discovery and visualization of amino acid composition differences. BMC Bioinformatics 8:211

Vamvaca K, Jelesarov I, Hilvert D (2008) Kinetics and thermodynamics of ligand binding to a molten globular enzyme and its native counterpart. J Mol Biol 382(4):971–977

Van Der Meer WB, Coker G III et al (1994) Resonance energy transfer theory and sata. VCH Publishers, Inc., New York

Vassilenko KS, Uversky VN (2002) Native-like secondary structure of molten globules. Biochim Biophys Acta 1594(1):168–177

Vercammen J, Maertens G, Gerard M et al (2002) DNA-induced polymerization of HIV-1 integrase analyzed with fluorescence fluctuation spectroscopy. J Biol Chem 277(41):38045–38052

Vihinen M, Torkkila E, Riikonen P (1994) Accuracy of protein flexibility predictions. Proteins 19(2):141–149

Völkel R, Eisner M, Weible KJ (2003) Miniaturized imaging systems. Microelectron Eng 67–68 (1):461–472

Walsh STR, Cheng RP, Wright WW et al (2003) The hydration of amides in helices; a comprehensive picture from molecular dynamics, IR, and NMR. Protein Sci 12(3):520–531

Walsh MJ, Hammiche A, Fellous TG et al (2009) Tracking the cell hierarchy in the human intestine using biochemical signatures derived by mid-infrared microspectroscopy. Stem Cell Res 3(1):15–27

Ward JJ, Sodhi JS, McGuffin LJ et al (2004) Prediction and functional analysis of native disorder in proteins from the three kingdoms of life. J Mol Biol 337(3):635–645

Weninger K, Bowen ME, Choi UB et al (2008) Accessory proteins stabilize the acceptor complex for synaptobrevin, the 1:1 syntaxin/SNAP-25 complex. Structure 16(2):308–320

Williams RW (1986) Protein secondary structure-analysis using Raman amide-I and Amide-Iii spectra. Methods Enzymol 130:311–331

Williams RM, Obradovi Z, Mathura V et al (2001) The protein non-folding problem: amino acid determinants of intrinsic order and disorder. Pac Symp Biocomput:89–100

Wilson IA, Haft DH, Getzoff ED et al (1985) Identical short peptide sequences in unrelated proteins can have different conformations: a testing ground for theories of immune recognition. Proc Natl Acad Sci U S A 82(16):5255–5259

Wood BR, Chernenko T, Matthaus C et al (2008) Shedding new light on the molecular architecture of oocytes using a combination of synchrotron fourier transform-infrared and Raman spectroscopic mapping. Anal Chem 80(23):9065–9072

Woody RW (1968) Improved calculation of the n-pi rotational strength in polypeptides. J Chem Phys 49(11):4797–4806

Woody RW (1995) Circular dichroism. Methods Enzymol 246:34–71

Woycechowsky KJ, Choutko A, Vamvaca K et al (2008) Relative tolerance of an enzymatic molten globule and its thermostable counterpart to point mutation. Biochemistry 47(51):13489–13496

Wright PE, Dyson HJ (1999) Intrinsically unstructured proteins: re-assessing the protein structure-function paradigm. J Mol Biol 293(2):321–331
Xu M, Ermolenkov VV, He W et al (2005) Lysozyme fibrillation: deep UV Raman spectroscopic characterization of protein structural transformation. Biopolymers 79(1):58–61
Xu M, Ermolenkov VV, Uversky VN et al (2008) Hen egg white lysozyme fibrillation: a deep-UV resonance Raman spectroscopic study. J Biophotonics 1(3):215–229
Yamamoto D, Uchihashi T, Kodera N et al (2008) Anisotropic diffusion of point defects in a two-dimensional crystal of streptavidin observed by high-speed atomic force microscopy. Nanotechnology 19(38):384009
Yamamoto D, Uchihashi T, Kodera N et al (2010) High-speed atomic force microscopy techniques for observing dynamic biomolecular processes. Methods Enzymol 475:541–564
Yang H, Habchi J, Longhi S et al (2012) Monitoring structural transitions in IDPs by vibrational spectroscopy of cyanylated cysteine. Methods Mol Biol 895:245–270
Yates JR, Ruse CI, Nakorchevsky A (2009) Proteomics by mass spectrometry: approaches, advances, and applications. Annu Rev Biomed Eng 11:49–79
Zhang Z, Smith DL (1993) Determination of amide hydrogen exchange by mass spectrometry: a new tool for protein structure elucidation. Protein Sci 2 (4):522–531
Zhang M, Gumerov DR, Kaltashov IA et al (2004) Indirect detection of protein-metal binding: interaction of serum transferrin with In^{3+} and Bi^{3+}. J Am Soc Mass Spectrom 15(11):1658–1664
Zhu F, Isaacs NW, Hecht L et al (2005) Raman optical activity: a tool for protein structure analysis. Structure 13(10):1409–1419

Chapter 8
Application of SAXS for the Structural Characterization of IDPs

Michael Kachala, Erica Valentini and Dmitri I. Svergun

Abstract Small-angle X-ray scattering (SAXS) is a powerful structural method allowing one to study the structure, folding state and flexibility of native particles and complexes in solution and to rapidly analyze structural changes in response to variations in external conditions. New high brilliance sources and novel data analysis methods significantly enhanced resolution and reliability of structural models provided by the technique. Automation of the SAXS experiment, data processing and interpretation make solution SAXS a streamline tool for large scale structural studies in molecular biology. The method provides low resolution macromolecular shapes ab initio and is readily combined with other structural and biochemical techniques in integrative studies. Very importantly, SAXS is sensitive to macromolecular flexibility being one of the few structural techniques applicable to flexible systems and intrinsically disordered proteins (IDPs). A major recent development is the use of SAXS to study particle dynamics in solution by ensemble approaches, which allow one to quantitatively characterize flexible systems. Of special interest is the joint use of SAXS with solution NMR, given that both methods yield highly complementary structural information, in particular, for IDPs. In this chapter, we present the basics of SAXS and also consider protocols of the experiment and data analysis for different scenarios depending on the type of the studied object. These include *ab initio* shape reconstruction, validation of available high resolution structures and rigid body modelling for folded macromolecules and also characterisation of flexible proteins with the ensemble methods. The methods are illustrated by examples of recent applications and further perspectives of the integrative use of SAXS with NMR in the studies of IDPs are discussed.

M. Kachala (✉) · E. Valentini · D. I. Svergun
Hamburg Outstation, European Molecular Biology Laboratory,
c/o DESY, Notkestrasse 85, 22603 Hamburg, Germany
e-mail: mkachala@embl-hamburg.de

D. I. Svergun
e-mail: svergun@embl-hamburg.de

M. Kachala · E. Valentini
Department of Chemistry, Hamburg University,
Martin-Luther-King Platz 6, 20146, Hamburg, Germany

I. C. Felli, R. Pierattelli (eds.), *Intrinsically Disordered Proteins Studied by NMR Spectroscopy,* Advances in Experimental Medicine and Biology,
DOI 10.1007/978-3-319-20164-1_8

Keywords Small-angle X-ray scattering · Solution scattering · Hybrid methods in structural biology · Intrinsically disordered proteins · Ensemble description of flexible proteins · Ab initio shape reconstruction · Rigid body modelling

1 Introduction

Small angle X-ray and neutron scattering (SAXS and SANS) are powerful and rapid techniques for characterising biological macromolecules in solution at a broad range of sizes ranging from a few kDa to GDa. In small angle scattering (SAS), a solution of macromolecules is irradiated by a monochromatic X-ray or neutron beam and the scattered intensity I is recorded by a two-dimensional detector (see Fig. 8.1). The two techniques, SAXS and SANS, differ in the type of radiation utilized—X-rays in the former and neutrons in the latter. The X-rays are scattered by electrons and neutrons are scattered by nuclei, and in both cases only elastic scattering is considered where the energy (and thus the wavelength) of the incoming beam are the same as in the scattered beam.

The potential of SAXS for studies of disperse systems was discovered by the French physicist Andre Guinier in the 1930's when he observed that the intensity of the scattered beam depends on the grain size in the material (Guinier 1939). This technique was later also found to be useful for the analysis of the size and shapes of biological macromolecules in solutions (Kratky 1963).

The radiation source for a SAXS experiment can be a synchrotron storage ring (e.g. PETRA III (Hamburg) or ESRF (Grenoble); all major synchrotrons in the world currently offer SAXS beamlines) or an in-house camera (laboratory X-ray instruments are produced by several companies). Because of the high brilliance at modern (so-called third generation) synchrotrons, very little sample is needed (10–30 μl) and the exposure time can be down to the sub-second range. SAXS experiments using in-house machines (and also SANS measurements, which can only be performed at large scale facilities, fission reactors or spallation sources) usually take much longer times, i.e. minutes to several hours. More about SAXS/SANS instrumentation can be read in (Svergun et al. 2013).

Short measurement times and recent advances in automation allow the performance of high-throughput SAXS experiments in which changes in the macromolecule conformations are analysed under different environmental conditions such as pH or temperature. The latest developments in SAXS also allow for measurement times in the microsecond range for the time-resolved analysis of processes such as protein folding kinetics. Radiation damage caused by X-rays is a serious problem with high-brilliance X-ray beamlines, but can be combatted by flowing the sample while measuring or adding compounds such as reducing agents or glycerol in low quantities.

During a SAXS or SANS experiment each sample measurement must be accompanied by a measurement of the buffer (pure solvent scattering). The buffer scattering will then be subtracted from the sample scattering in order to obtain the net scattering from the dissolved particles. The main sample requirements for SAS experiments are (a) the sample should be pure (ideally, 95 % monodisperse

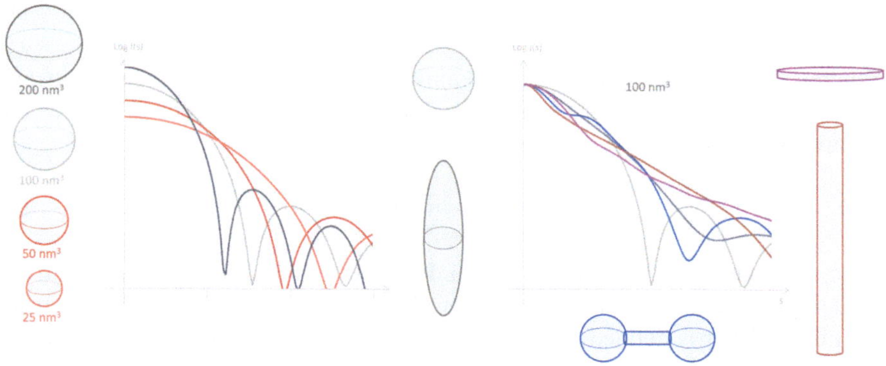

Fig. 8.1 Scattering curve size and shape (courtesy of Al Kikhney)

or better); (b) unspecific aggregates must be absent; and (c) the sample should be measured at different solute concentrations. Relatively dilute solutions are studied such that a typical concentration series looks like 0.5, 1, and 2 mg/ml, and, if available, higher concentrations (5 mg/ml and higher). The concentration series is used to extrapolate the sample signal to infinite dilution, i.e. to simulate a concentration of 0 mg/ml. The useful signal coming from the solute depends on the number of macromolecules in the illuminated volume, i.e. is proportional to the concentration. However, higher solute concentrations can lead to interparticle interactions that alter the signal at lower angles and make the data more difficult to analyse. At lower concentrations the interparticle interaction effects are usually negligible but the data is much noisier. Extrapolation to infinite dilution is therefore an important step in the scattering experiments and subsequent data analysis.

The formation of unspecific aggregates due to strong attractive interparticle interactions must be avoided because these aggregates significantly alter the scattering patterns and the data from such samples can usually not be meaningfully analysed. Therefore, prior to the SAXS/SANS experiments it is necessary to check the sample purity using other techniques such as gel filtration chromatography, dynamic light scattering (DLS) or analytical ultracentrifugation (AUC).

SANS experiments require larger sample quantities (about 300 μl) and there are fewer neutron facilities available then SAXS ones. Also, the major advantage of SANS—the use of deuteration of the sample or solvent—is a very powerful approach for characterising multicomponent macromolecular complexes (Svergun 2010), but this is of less importance for studies of intrinsically disordered proteins (IDPs). For these reasons the rest of the chapter will focus mostly on SAXS.

2 Overview of the SAXS Theoretical Background

Scattering by the ensemble of macromolecules randomly oriented in the solvent is isotropic because of the average over their orientations. The intensity $I(s)$ can be defined as a function of the modulus of the scattering vector s (also called q in some publications), which is defined as:

$$s = 4\pi \sin(\theta) / \lambda \quad (1)$$

Where λ is the wavelength of the incoming radiation and 2θ is the angle between the incident and the scattered beam. The scattering intensity can be defined as the squared amplitude,

$$I(s) = < I(s) > = < A(s) A^*(s) > \quad (2)$$

Here, < > stands for the average over all directions of the scattering vector and the scattering amplitude $A(\boldsymbol{s})$ is a Fourier transformation of the excess electron density $\Delta\rho(\mathbf{r})$. The latter is defined as the difference between the scattering density of the solute $\rho(\mathbf{r})$ and that of the solvent $\rho_s : \Delta\rho(\boldsymbol{r}) = \rho(\boldsymbol{r}) - \rho_s$

$$A(\boldsymbol{s}) = \int_V \Delta\rho(\boldsymbol{r}) \exp(i\boldsymbol{sr})\, d\boldsymbol{r} \quad (3)$$

Two types of samples can be analysed with SAS:

1. monodisperse, where all the particles are identical,
2. polydisperse, where the particles are different from each other.

In the case of monodisperse samples the intensity is proportional to that of a single particle averaged over all orientations; several parameters defining the size and shape of the particle can therefore be extracted from the distribution $p(r)$ of the distances r within the particle.

$$p(r) = \frac{r^2}{2\pi^2} \int_0^\infty s^2 I(s) \frac{\sin(sr)}{sr} ds \quad (4)$$

The maximum distance inside a particle (D_{max}) corresponds to the distance r beyond which the $p(r)$ distribution is equal to zero. The $p(r)$ distribution is not only used to calculate $D_{\max}$ but also gives information about the overall shape of the particle (see Fig. 8.2).

The inverse Fourier transform of the $p(r)$ distribution function on the interval [0, D_{max}] gives the scattering intensity $I(s)$ (see Fig. 8.3).

$$I(s) = 4\pi \int_0^{D_{\max}} p(r) \frac{\sin(sr)}{sr} dr \quad (5)$$

At very small angles the SAXS data can be represented by two parameters, the forward scattering intensity $I(0)$ and the radius of gyration R_g using the classical Guinier approximation $I(s) \approx I(0)\exp(-(sR_g)^2)$, which is valid in the range $s < 1.3/R_g$ (Guinier 1939). These parameters are calculated from the Guinier plot, which should be a linear function in the coordinates $\ln(I(s))$ *vs.* s^2. From the forward scattering to $s = 0$ it is possible to calculate the molecular weight (MW) and therefore assess the oligomerization state of the particle. In SAXS one of the possible ways to calculate

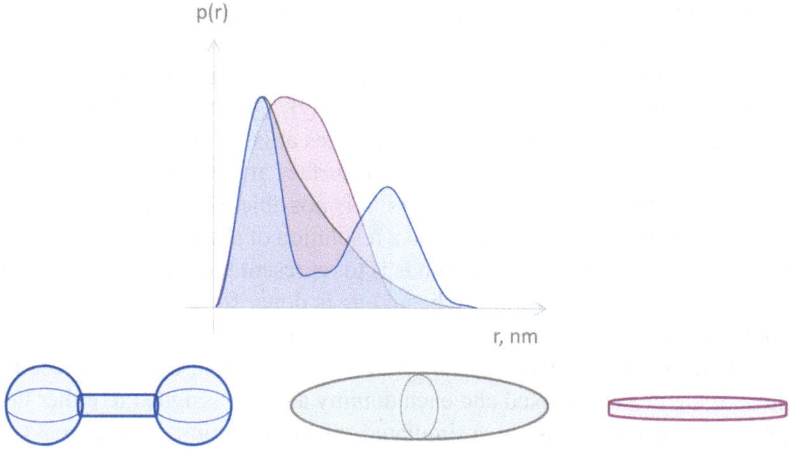

Fig. 8.2 The $p(r)$ distribution depends on the dimension and shape of the macromolecule (courtesy of Al Kikhney, European Molecular Biology Laboratory-Hamburg Outstation, unpublished data)

the MW is based on a separate measurement of a known standard protein (for example bovine serum albumin), for which the MW is known, through the proportion:

$$\frac{I(0)_{\text{standard}}}{MW_{\text{standard}}} = \frac{I(0)_{\text{molecule}}}{MW_{\text{molecule}}} \tag{6}$$

where both *I(0)* values are normalized for the solute concentration.

The slope of the Guinier plot gives R_g, which is defined as the average of squared distances to the centre-of-mass of the molecule weighted by the contrast. R_g provides indications about the overall shape of the particle; for the given volume, lower R_g values correspond to more compact particles, with the lowest R_g being that of a spherical shape.

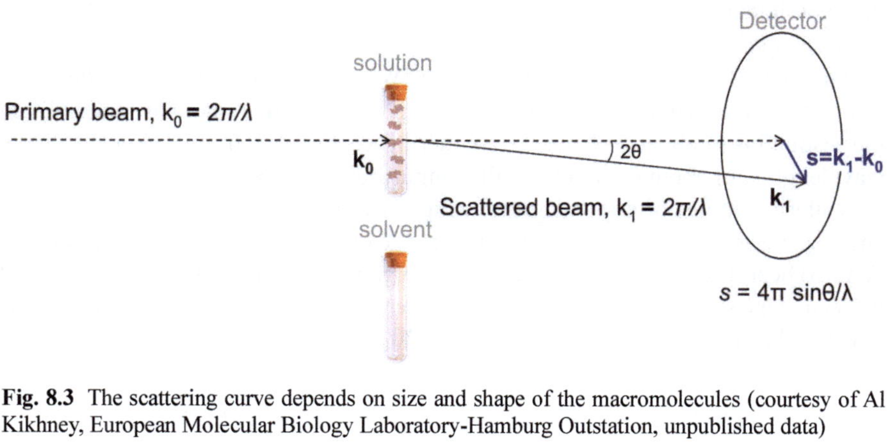

Fig. 8.3 The scattering curve depends on size and shape of the macromolecules (courtesy of Al Kikhney, European Molecular Biology Laboratory-Hamburg Outstation, unpublished data)

The hydrated particle volume and its specific surface can be derived from the scattering data using Porod's invariant and Porod asymptotics (Porod 1982). The Porod invariant Q is proportional to the integrated intensity weighted with s^2 and the particle volume is expressed as $V_p = 2\pi^2 I(Q)/Q$. The asymptotic behaviour of the scattering curve for homogeneous particles at large angles indicates that the intensity is proportional to the product of the surface area of the particles and s^{-4}.

In addition to the structural parameters, it is possible to obtain an *ab initio* low-resolution shape of the macromolecule at a resolution of about 1–2 nm from a SAXS experiment. The idea of *ab initio* methods is to represent the particle as an assembly of densely-packed spheres (dummy beads) as is done, for example, in the DAMMIN program (Svergun 1999). The radius of the dummy beads depends on the size of the particle and, in particular, on D_{max} (a few 1000 beads are typically used). The positions of the beads are fixed and each dummy atom is assigned to either the solute (protein) or solvent phase by a simulated annealing optimization process aimed at finding the shape that gives the curve I_{calc} (s) fitting the experimental data I_{exp} (s) with the minimum discrepancy.

$$\chi^2 = \frac{1}{N-1}\sum_{j=1}^{N}\left[\frac{I_{\exp}(s_j) - cI_{calc}(s_j)}{\sigma(s_j)}\right] \qquad (7)$$

where c is a scaling factor, N is the number of points and σ is the experimental error (Blanchet and Svergun 2013). Penalties such as looseness and disconnectivity are also taken into account in order to build a physically sensible model.

Furthermore, with SAXS it is also possible to reconstruct the quaternary structure of complexes, when the high-resolution structures of the single subunits are available, using a rigid body modelling approach (Petoukhov and Svergun 2005). One can compute a theoretical scattering from a given high-resolution structure (e.g. with CRYSOL (Svergun et al. 1995)); the computed curve can then be compared to the experimental one to minimize the discrepancy value χ^2 (Eq. 7.7).

Most of the modelling software applications in SAXS use algorithms aimed at the minimization of the discrepancy between the theoretical scattering curve and the experimental one. Some of these procedures will be explained in detail later on in this chapter.

The overall parameters (R_g, D_{max}, $I(0)$, MW, and Porod volume) and the shape information are extracted from the scattering intensity in an angular range within the resolution to about 1–2 nm, i.e. scattering vector $s \approx 3$–6 nm^{-1}. For the X-ray wavelength λ of about 0.1 nm typically employed in SAXS, this range corresponds to scattering angles of a few degrees, which is why the technique is called "small angle scattering". Even higher angles (the range of "wide angle X-ray scattering" or WAXS) bear information about the internal organization of the macromolecule and its secondary structure (see Fig. 8.4); however, interpretation of the WAXS data is not straightforward (Graewert and Svergun 2013).

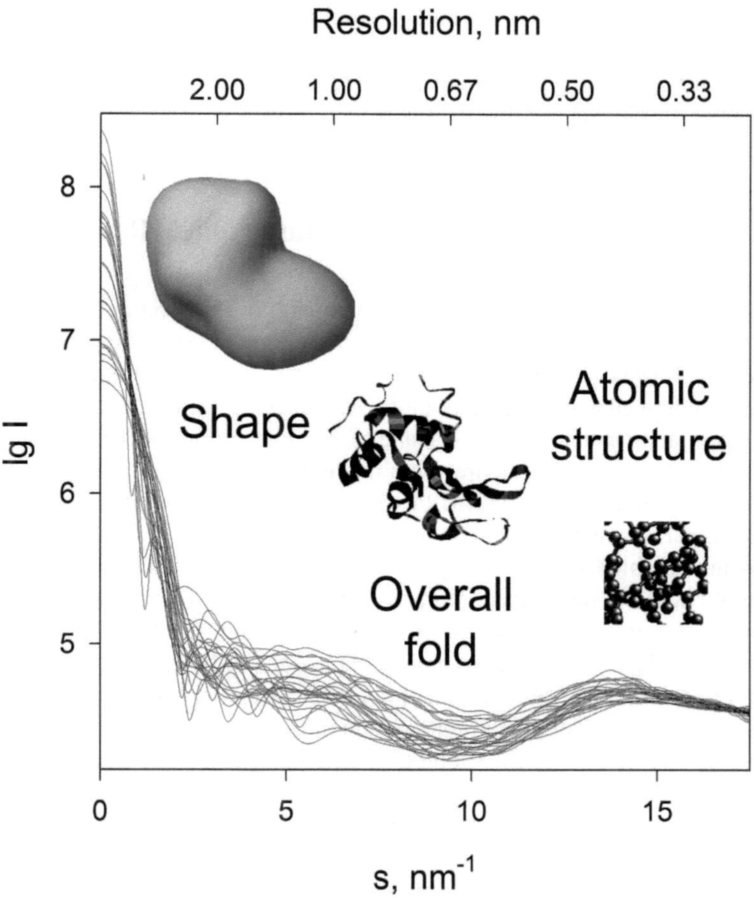

Fig. 8.4 SAXS-WAXS resolution (Svergun and Koch 2002)

For monodisperse systems, the overall parameters such as R_g, D_{max} and MW directly describe those of a single particle. In the case of polydisperse systems, however, one deals with different species in solution and these parameters become averages over the distribution of these species. The equation describing the scattering intensity of a polydisperse system can be written as:

$$I(s) = \sum_k v_k I_k(s) \tag{8}$$

Where k is the number of species in solution, v_k is the volume fraction of a single species and $I_k(s)$ is its normalized intensity. In most cases, the task of the analysis lies in reconstructing the volume fractions given (or assuming) the intensities of the single components.

Solutions of IDPs and flexible modular multi-domain proteins belong to polydisperse samples containing astronomic numbers of components, and are therefore not easy to structurally analyse. Several developments that will be addressed further in this chapter have advanced this field in recent years.

3 SAXS as a Complementary Technique to NMR

SAXS can be extremely useful when coupled with other structural techniques, especially with high-resolution techniques such as X-ray crystallography and nuclear magnetic resonance (NMR). SAXS and NMR are rather complementary techniques for the structural characterisation of biological macromolecules. Indeed, SAXS provides information about the overall shape of the molecule and NMR yields high-resolution data on the local structure. Both methods use solutions in the mM range as samples, but in many cases the NMR experiment requires isotope labelling, making sample preparation expensive and time-consuming. The range of molecular masses of proteins suitable for a SAXS experiment goes from a few kDa to GDa, while for structural NMR the range spans from 10 Da to 100 kDa. This means that for large proteins or biomolecular complexes (> 100 kDa), a high-resolution structure of individual subunits or components can be obtained by NMR and the overall shape of the entire construct as well as of the individual components may be assessed with SAXS. Further, when studying multidomain proteins or complexes, SAXS is more sensitive to movements of the subunits or domains (which alter the overall shape) while NMR is more sensitive to their rotations through residual dipolar coupling (RDC) and interfaces (chemical shifts and spin labels). A combination of the two methods is therefore a powerful approach to characterize such complexes (Mattinen et al. 2002). Another difference between the techniques is the acquisition time: a single NMR experiment can take days or even weeks, while SAXS measurements of one sample are done in seconds on modern synchrotrons and in hours using in-house sources. Furthermore, the interpretation of an NMR spectrum usually takes days or weeks while the SAXS data analysis may be completed in minutes (e.g. validation of a given high-resolution structure or *ab initio* shape analysis) or hours (e.g. rigid body modelling or multi-phase *ab initio* modelling). The dynamic capabilities of these techniques also differ significantly, with SAXS data averaged over the exposure time while NMR yields data on the time scale of the spin relaxation. All the mentioned similarities and differences show the complementarity of SAXS and NMR and a joint use of the techniques is one of the most widely applied strategies for the characterization of biological macromolecules.

Several protocols for SAXS data collection and analysis for different scenarios of its combined use with NMR are outlined later in this section. Only globular proteins are considered here, while approaches to the characterization of flexible biomolecular systems are presented in the next sections.

3.1 *Validation of High-Resolution Models*

The amount of available high-resolution structures is rapidly increasing and the validation of these structures in solution, i.e. in conditions close to native, is becoming more important. SAXS is among the main techniques to rapidly perform this validation, and the procedure typically consists of the following steps:

1. The SAXS data is collected on dilute solutions using several concentrations of purified sample to avoid aggregation and radiation damage effects.
2. Possible effects caused by intermolecular interaction are eliminated by extrapolation of experimental data from various concentrations to infinite dilution. This can be done, for example, by either using the program PRIMUS (Konarev et al. 2003) with a graphical interface or with the command line program ALMERGE (Franke et al. 2012).
3. Evaluation of R_g and the forward scattering $I(0)$ is performed using the Guinier approximation, for example interactively with PRIMUS (Konarev et al. 2003) or by the automated program AutoRG (Petoukhov et al. 2007). The forward scattering is needed for the further estimation of the MW and consequently the oligomeric state.
4. The maximal dimension of the molecule D_{max} is defined from the distance distribution function $p(r)$, which can be obtained either interactively (e.g. using the program GNOM (Svergun 1992)) or automatically (with DATGNOM (Petoukhov et al. 2007)).
5. An *ab initio* reconstruction of the overall shape is performed using one of the bead modelling programs (e.g. DAMMIN (Svergun 1999), DAMMIF (Franke and Svergun 2009), GASBOR (Svergun et al. 2001)). The modelling is usually performed several times and the obtained models are clustered and averaged using DAMAVER (Volkov and Svergun 2003). The final model is then overlaid with the available high-resolution structure to see if they match each other. This can be done automatically with the program SUPCOMB, which also provides a measurement of the agreement, a so-called normalized spatial discrepancy (Kozin and Svergun 2001).
6. As a final or additional step the direct calculation of the scattering pattern of the given high-resolution structure and its comparison with the experimental data can be performed, for example using CRYSOL (Svergun et al. 1995). If several alternative models are considered, the one with the smallest χ^2 is the one most likely to be present in solution.

3.1.1 Rigid Body Modelling

The determination of the high-resolution structure of large biomolecules and complexes using NMR or X-ray crystallography can be difficult because of problems with assignment of the resonances with the first technique and potential difficulties in growing crystals with the second one. In such cases SAXS provides an alternative by building hybrid models. As explained in the introduction, if the atomic structures are available they can be used as building blocks to reconstruct the quaternary structure of the protein or complex. In order to obtain more accurate models, the scattering data of individual subunits and, if possible, of partial constructs should be collected in addition to the scattering pattern for the entire construct. Assuming that the arrangement of subunits is the same in the entire complex and in partial constructs, the simultaneous fit of the corresponding datasets adds extra information for a more reliable determination of the overall structure. The initial analysis of all scattering curves used for additional modelling is described in steps 1–5 of the previous section. The next stages include the following steps:

6. Calculate the scattering amplitudes for each subunit in a reference orientation with CRYSOL (Svergun et al. 1995). If scattering data for individual components is available, the fit obtained by CRYSOL can be used for the validation of their given high-resolution structures.
7. Employ SASREF (Petoukhov and Svergun 2005) to build an interconnected complex from rigid subunits without steric clashes that fits the experimental scattering profile. If the tentative model of the complex is known *a priori* it could be refined by this procedure. As in the case of *ab initio* modelling, rigid body reconstruction should be run multiple times to discover the variability of the possible quaternary structures corresponding to the experimental data.
   In addition to high-resolution structures, other types of NMR data can be used in rigid body modelling:

8a. The information about distances and interfaces obtained via chemical shifts, nuclear Overhauser effect (NOE) and paramagnetic relaxation enhancement (PRE) can be used in SASREF to set restraints on relative positions of the subunits as distances between certain residues or loops.

8b. Data about the mutual orientation of the subunits can be derived from RDCs and pseudocontact chemical shifts(PCSs) and also used as SASREF restrictions. If the studied system consists of two components, their relative orientations are defined with a four-fold degeneracy (the rotations of one of the subunits by 180 degrees around the orthogonal axes of the reference frame). Taking this into account, rigid body modelling can be done only with translations probing the four possible relative orientations determined based on the RDC data.

The models obtained by the modelling with either distance or orientation constrictions should be compared with the unrestrained models to check if the imposed restraints introduced any bias.

3.2 SAXS and IDPs

SAXS is traditionally used to analyse the shape of globular proteins or complexes in purified monodisperse solutions, when all particles can be considered identical and the scattering pattern reflects the properties of a single particle. In this case a low-resolution reconstruction of the overall shape of particles in solution using either the *ab initio* approach (when no *a priori* structural information is available) or hybrid modelling (when high-resolution structures of domains or subunits is given) is possible and successfully applied even in the most complicated situations.

These approaches are not applicable to polydisperse solutions because the scattering pattern is averaged over different types of particles. In this case, the structure of the individual components cannot be determined solely from the scattering profile. However, if the scattering curves of individual components are given it is possible to obtain their volume fractions through Eq. 7.8.This approach is typically used to estimate the volume fractions of components in oligomeric mixtures, for instance monomer-dimer equilibrium (Konarev et al. 2003), but it can also be suitable for the characterisation of more complicated systems, and even for flexible systems such as IDPs or multidomain proteins with flexible linkers.

For flexible systems, each component is a single conformation of the protein and the number of conformations can be astronomical, and therefore plain decomposition into the volume fractions is not feasible. Still, the scattering pattern of a flexible system is an average of patterns of individual conformation existing in solution. This average causes the distinct behaviour of the SAXS data of IDPs shown in Fig. 8.5 (Bernado and Svergun 2012), where scattering profiles of 10 random conformations of a synthetic 100 amino acid-long polypeptide extracted from a pool of 10,000

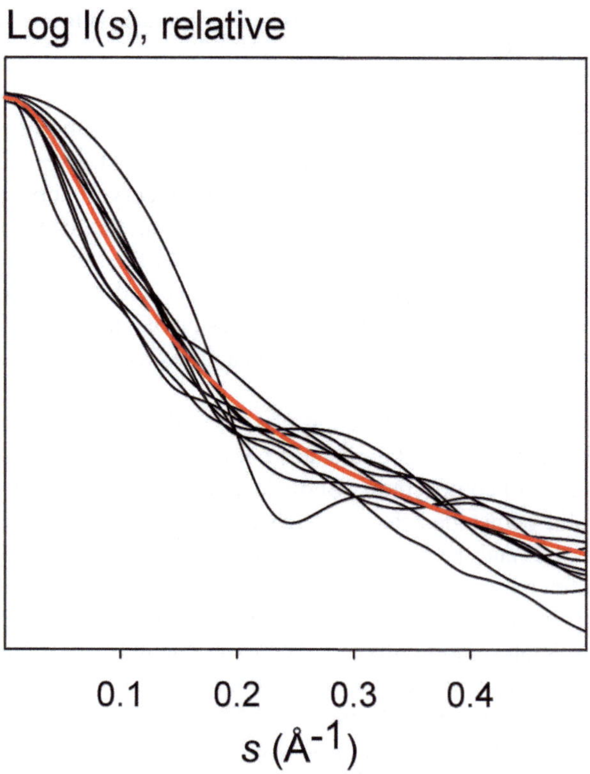

Fig. 8.5 Individual SAXS profiles (*black*) of 10 randomly selected chains and averaged curves of 10,000 conformations (*red*) (Bernado and Svergun 2012)

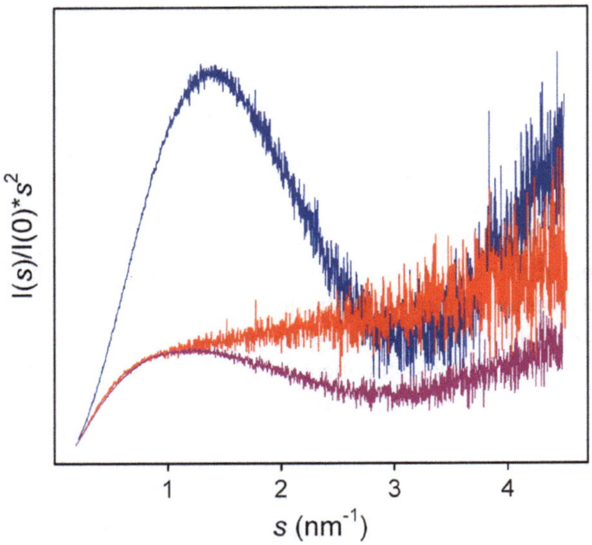

Fig. 8.6 Kratky plot for three constructs of Src kinase. The globular SH3 domain (*blue*), the fully disordered unique domain (*red*), and a construct joining both domains (*purple*). The prototypical features of globular and disordered domains are combined in the partially folded construct. Courtesy of Yolanda Pérez and Miquel Pons (Institute for Research in Biomedicine, Barcelona). (Bernado and Svergun 2012)

conformations are exhibited together with the averaged scattering profile of the entire pool. Each individual conformation displays distinct features in the simulated momentum transfer range and the initial parts of the curves, corresponding to the largest intramolecular distances, vary greatly, indicating that conformations of unfolded proteins are very diverse in size and shape. At the same time the averaged SAXS profile of 10,000 conformations is smoother and has almost no features. Such behaviour of the scattering curve is often a sign of a disordered state, although a simple visual analysis is not sufficient to unambiguously determine if the protein is unfolded.

A Kratky plot ($I(s) \cdot s^2$ as a function of s) is traditionally used to qualitatively distinguish unfolded and structured states because the appearance of this plot visualizes the degree of compactness (Bernadó and Blackledge 2009). The scattering intensity of solid bodies decreases at higher angles with momentum transfer approximately as $1/s^4$ and the Kratky plot for globular proteins has a bell-like shape with a well-defined maximum. As another limiting case, the scattering curve of an ideal Gaussian chain has an asymptotic behaviour as $1/s^2$ and has a plateau at large s values in the Kratky representation. The experimental scattering curves of unfolded proteins display a plateau over a specific range of s followed by a monotonic increase. Typical experimental Kratky plots for globular, partially unfolded, and completely disordered proteins are presented in Fig. 8.6.

Another way to detect unfolded proteins is based on the comparison of the experimentally obtained R_g of the sample with theoretical estimates for globular and disordered proteins. IDPs usually demonstrate larger overall parameters due to the presence of extended conformations. The most used quantitative description for R_g is based on Flory's equation (Flory 1953), which postulates the dependence between R_g and the MW of the protein as a power law:

$$R_g = R_0 N^{\nu} \tag{9}$$

Where N is the number of residues in the chain, R_0 is a constant, and ν is an exponential scaling factor. Flory's equation estimated ν to be approximately 0.6 for the Gaussian chain and further calculations determined a refined value of 0.588 (Le Guillou and Zinn-Justin 1977). The measurements of R_g for 26 chemically denatured proteins containing from 8 to 549 amino acids established the following values:$\nu = 0.598 \pm 0.028$ and $R_0 = 1.927 \pm 0.27$ (Kohn et al. 2004). The obtained values indicate that denatured proteins behave similarly to random coils in terms of R_g.

The comparison of measured R_g with the corresponding theoretical estimate for unfolded proteins obtained using Flory's equation is widely applied for the detection of disordered conformations, and yet it is not clear whether chemically denatured and intrinsically disordered proteins have equivalent conformational properties. It has been reported that chemical denaturation agents such as urea or guanidinium chloride interact with backbone and/or side chain atoms, possibly causing an alteration of Ramachandran populations (Stumpe and Grubmüller 2007). The observed perturbations at the residue level could result in changes in the overall properties. An NMR study based on several RDC measurements along the backbone of ubiquitin showed that the chemically denatured samples have larger populations of extended conformations as compared to IDPs (Meier et al. 2007). The ensemble approach (see next section) has also been applied to compare R_g values for chemically denatured and natively unfolded proteins, proving a 15% increase in extended conformations in ensembles describing denatured proteins as compared to IDPs (Bernadó and Blackledge 2009). The same study derived new estimates for the Flory's equation parameters for IDPs:$\nu = 0.522 \pm 0.010$ and $R_0 = 2.54 \pm 0.01$.

Although the traditional methods of modelling used in the SAXS analysis are not applicable for flexible systems, they can be used to identify disorder. For example, an attempt at *ab initio* modelling performed using the scattering curve of an unfolded protein will result in highly elongated shapes because of the extended conformations. The most effective method for the quantitative characterization of flexible structures with SAXS is currently an ensemble approach, which is reviewed in the next section.

3.3 Ensemble Representation of IDPs in SAXS Data Analysis and its Application

The main problem for a quantitative SAXS analysis in flexible systems lies in the tremendous number of coexisting conformers in solution. A short time ago the only approach to obtain a qualitative description of such proteins was the use of Kratky plots to distinguish between globular and unfolded structures. Today, several methods based on an ensemble representation of flexible structures in solution are making quantitative analysis possible. In this section the theoretical basis of the ensemble approach is briefly presented, complemented with an overview of existing software implementations and examples of their practical application (also in combination with NMR).

There are different types of flexibility present in proteins: fast fluctuations in globular proteins, slow molecular reconfigurations on a large scale, and constitutive disorder of IDPs and intrinsically disordered regions (IDRs). In all of these cases the dynamics of the protein is essential for the biological functionality of the molecules, e.g. recognition, regulation or catalysis. The widely accepted concept used for the quantitative description of dynamic systems is an *ensemble of conformations* (Bernadó et al. 2005; Bernadó et al. 2007; Bernadó and Blackledge 2009; von Ossowski et al. 2005) also called the *supertertiary structure* (Tompa 2012). In order to be reliable the properties of the ensembles must correspond to available experimental data obtained by experimental methods (NMR, SAXS, circular dichroism (CD), bioinformatics methods such as structure prediction, etc.). In SAXS data analysis an ensemble represents a polydisperse mixture of various conformations, each with its own scattering pattern. Based on these ideas several methods for the SAXS data analysis of flexible molecules have been developed, following a strategy consisting of three major steps:

1. Generation of a large pool of various structures covering the conformational space of the studied protein;
2. Calculation of the scattering properties of each individual conformation;
3. Selection of an ensemble of structures that fits the experimental data using one or another optimization method.

In the following paragraphs we describe different approaches to implement this strategy and their distinct features.

3.3.1 Ensemble Optimization Method

The first method developed to analyse SAXS data of flexible structures using the ensemble approach is the ensemble optimization method (EOM) (Bernadó et al. 2007). The underlying idea of this method is the reconstruction of the experimental SAXS curve as the average scattering profile of coexisting conformations in solution. The number of conformations is unknown beforehand and can be determined during the optimization procedure. A sub-ensemble is selected from pre-computed scattering patterns of a large pool, which represents maximum possible flexibility according to the protein topology. EOM is not a completely model-independent approach, but instead relies on the structural limitations coming from the known information about the system and being imposed during the pool generation.

The potential solution considered by EOM at each step of optimization is an ensemble of N different conformers of the same molecule or, in more complicated cases, of a mixture of conformers with different oligomeric states. The scattering profile of the ensemble is calculated using the individual scattering patterns and assuming equal populations of each conformer:

$$I^{EOM}(s) = \frac{1}{N}\sum_{n=1}^{N} I_n(s) \qquad (10)$$

Where $I_n(s)$ is the scattering from the n-th conformer. This principle requires that the pool adequately covers the conformational space of the molecule and contains $M >> N$ structures. The optimization is performed through the genetic algorithm (GA) to select an ensemble that fits the experimental data assuming that all structures are equally represented in the ensemble.

EOM is mostly applied to estimate parameters of highly flexible structures such as IDPs, which have tremendous number of conformations in solution. The generation of complete pools containing such an amount of structures is practically impossible, which is why EOM results are reported not as distinct structures but as distributions of low resolution parameters including radius of gyration (R_g), maximum dimension (D_{max}), interdomain distances and anisotropy. The distributions of the selected ensemble and of the pool are compared with each other to outline the overall properties of studied biomolecules. Although EOM results are of low resolution they have a significant advantage over traditional methods for characterizing flexible proteins because they provide distributions of structural parameters instead of just the averaged values. Additionally, the resolution of EOM can be improved by using scattering data from deletion mutants and analysing them together with the full length constructs (see the example on tau protein below).

A notable example of the application of EOM together with NMR data is the study of the two-domain ribosomal L12 protein. L12 forms dimers in solution via its N-terminal while the C-terminal is connected through a highly flexible 20-amino acid long linker expected to move freely in solution (Bernado et al. 2010). The scenario for investigation of such systems is similar to an IDP case, because even with the short flexible linker the number of possible conformations in solution is astronomical. The application of EOM for multidomain proteins provides, in addition to the distribution of the standard parameters R_g and D_{max}, the distribution of the interdomain distances, which are determined by the structure of the linker between domains. In the case of L12 the optimized ensemble was obtained by fitting scattering data in correspondence with the correlation times of orientation eigenmodes compatible with ^{15}N relaxation. The use of EOM demonstrated that the overall shape of L12 is longer and anisotropy is more pronounced than it would be with a random linker; consequently, the flexible linker is in an extended conformation (Fig. 8.7). Moreover, the analysis also showed an asymmetry of both linkers, which may be relevant for the biological function of L12, providing efficient translation.

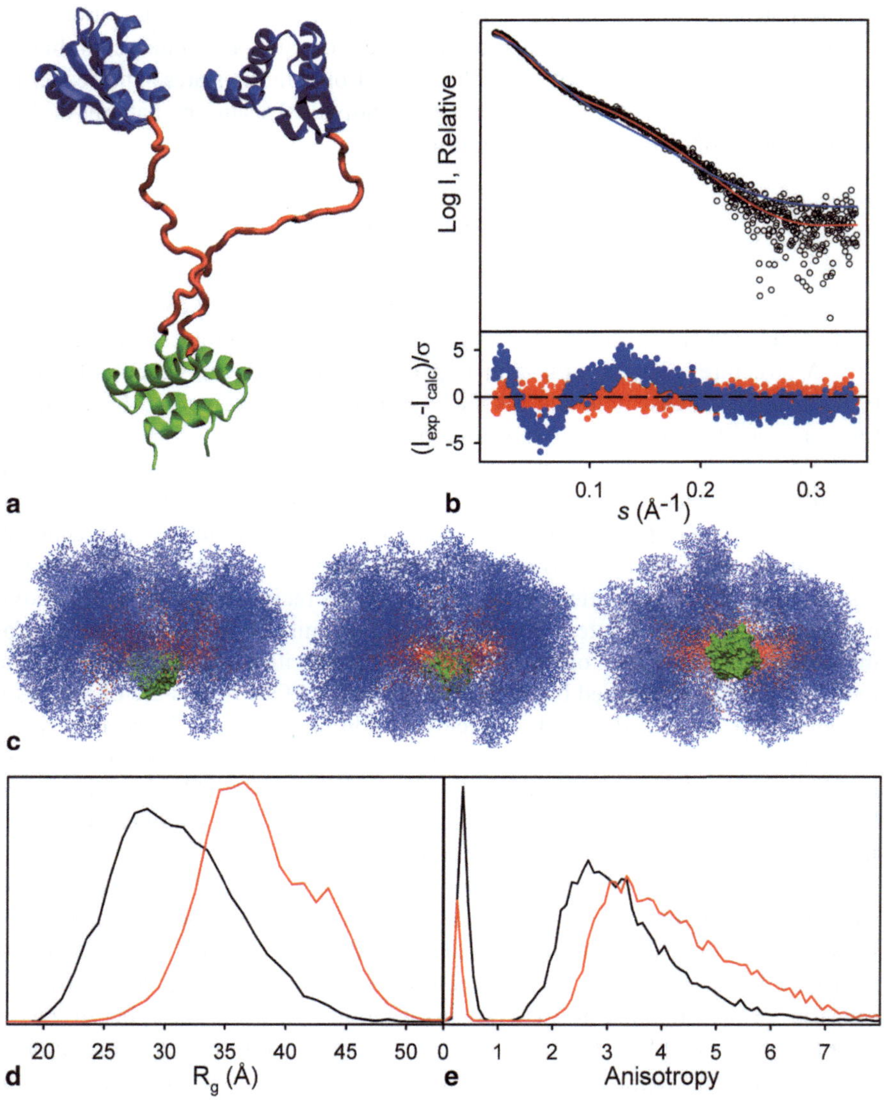

Fig. 8.7 Ensemble optimization analysis of the SAXS profile measured for L12. **a** Cartoon of a single L12 conformation, 1rqu (16), showing the NTD dimer (*green*), the CTD (*blue*), and the linker (*red*). **b** Logarithm of the scattering intensity (black dots) as a function of the momentum transfer, $s = 4\pi \sin(\theta) / \lambda$. The fitted scattering profile of the optimized ensemble (OE) obtained by the EOM approach is shown in *red*. The theoretical scattering curve of the random ensemble (RE, *green line*) is shown for comparison. The *bottom* panel displays the point-by-point error function for the two ensembles using the same colour code. Both ensembles contain 10,000 independent conformers. **c** Three orthogonal views of a random subset (N=50) of the OE; colour code as in panel A. The orientation in the side view (*left*) is the same as in panel **a**. **d** Radius of gyration (Rg) and (**e**) anisotropy (**a**) distributions for the RE (*black lines*) and the OE (*red lines*). The sharp peaks at A<1 correspond to oblate conformers with populations of 4.8 % and 14.2 % for the OE and RE, respectively. (Bernado et al. 2010)

3.3.2 Minimal Ensemble Search

The approach employed in the minimal ensemble search (MES) (Pelikan et al. 2009) method is similar to that of EOM, with a focus on avoiding overfitting the experimental data. The chosen strategy to achieve this goal is to determine the minimal number of conformations and the corresponding curves which, being treated as an ensemble, describe the given data. During the process of determination the number of used conformations is increased from two to five. The relative populations of each selected structure may also change in the course of optimization. The change of the discrepancy χ^2 with the increase in the number of conformations is used to determine a minimal ensemble that fits the experimental data. A comparative analysis of the final ensemble and the initially generated pool is used to draw conclusions about the flexibility of the studied macromolecule in a similar way as in EOM. Due to its nature MES is more suitable in cases of biomolecules with limited flexibility, where the conformational space can be covered with a few structures.

3.3.3 Basic-Set Supported SAXS

Basic-set supported SAXS (BSS-SAXS) was developed for the analysis of conformational transitions of Hck tyrosine kinase (Yang et al. 2010). In the same way as the aforementioned approaches BSS-SAXS starts with covering the conformational space and pre-computing the scattering profiles of each structure, and then selecting conformations/curves that collectively describe the input data. However, unlike other methods, the investigators pre-select the conformations and associated scattering curves before the optimization process; for example, in the original study of Hck tyrosine kinase, this was done in two steps. First, a large number of residue-level coarse-grinded structures were clustered into 25 groups considered to be distinct conformational states of the kinase, and a scattering profile was calculated for each group. Second, the number of clusters was reduced to nine by clustering groups with similar theoretical SAXS properties, a step that can result in conformationally different structures being grouped in the same cluster. BSS-SAXS employs a Bayesian-based Monte Carlo algorithm yielding not only the fraction of each conformation but also their uncertainties. BSS-SAXS, similarly to MES, is especially effective for macromolecular systems with a number of discrete conformations. The technique employs an accurate statistical method to analyse the data, but it can only be used with a realistic representation of the conformational sampling of the biomolecule, which is not always available.

3.3.4 Ensemble Refinement of SAXS

As in all the other mentioned methods, the first step of the ensemble refinement of SAXS (EROS) approach (Rozycki et al. 2011) is an extensive sampling of the conformational space. Similarly to BSS-SAXS, a coarse-grained approach is applied to generate the structures, which are later clustered to determine independent states. The relative population of the clusters is an optimization parameter that is varied to fit the experimental data. The initial values are defined by the number of conformations in each cluster and the refinement is based on minimisation of a pseudo-free energy function consisting of two terms. The first term is the discrepancy χ^2 that compares the experimental and computed scattering profiles. The second term proposed to avoid overfitting is effective entropy, which assigns a penalty to variations in the relative cluster populations as compared to the initial values. The inventors of this method therefore presume that the initial cluster populations are close to reality and the algorithm just refines the values, treating large changes as overfitting. In the subsequent publication, a simple term increasing with the number of conformation is used in the pseudo-free energy function instead of maximum entropy (Francis et al. 2011). EROS was applied to investigate properties of the ESCRT-III CHMP3 protein, which has a long terminal tail with two small helical regions (Rozycki et al. 2011). Six conformations selected by EROS using a SAXS curve corresponding to low salt concentration conditions are very similar to the crystallographic structure with one of the small helices bound to the globular part of the protein. At high salt concentration the small helices are much more flexible and the ensemble is represented by 60 conformations.

3.3.5 ENSEMBLE

One example of the ensemble approach that uses many types of experimental data including SAXS profiles and several NMR parameters (RDCs, J-couplings, chemical shifts, PREs, NOEs, etc.) is ENSEMBLE (Krzeminski et al. 2013). The pool generation in this approach proceeds in a different way than the previously described methods. First, a large pool of 100,000 conformations called the "initial soup" is populated by structures provided by the user and/or conformers generated by the TraDES program (Feldman and Hogue 2000). For each structure the different parameters are back-calculated either by ENSEMBLE or external software, for example CRYSOL (Svergun et al. 1995) in the case of SAXS data. During the next step 5000 conformations from the initial soup are randomly selected to create the "initial pool". Here, each conformer can be selected several times, but the algorithm prefers structures that have been chosen fewer times. The pool content is periodically updated by removing structures that are not involved in the fitting of experimental data and by adding slightly modified conformers from the selected ensemble. In the selection phase ENSEMBLE selects structures from the initial pool. Each type of experimental data is assigned a weight defining its importance in the scoring function. The ensemble is selected by the"switching Monte-Carlo algorithm" and is considered to fit the experimental data when the discrepancy is lower than a threshold automatically determined by ENSEMBLE for each type of input data. During the optimization process these thresholds as well as the weights of the scoring function may change in order to fit the experimental data. The ENSEMBLE approach provides an opportunity to incorporate data of various natures, for example NMR and SAXS data, and fit them simultaneously. The underrestraining and overfitting problem is approached in ENSEMBLE by finding the smallest ensemble that explains all of the experimental observations. The method was applied in the structural characterisation of the Sic1 protein and its hexaphosphorylated variant pSic1 (Paoletti et al. 2009) by following the NMR parameters chemical shifts, PREs, RDCs, and ^{15}N R_2 with SAXS data.

3.3.6 Flexible-Meccano

The program Flexible-Meccano (FM) is another example of software that generates ensembles to simultaneously describe SAXS and NMR data and has thus far been the most tested structural model for IDPs. FM samples conformational space and consecutively assembles rigid peptidic units according to the residue-specific Ramachandran space and a coarse-grained description of the side-chains to avoid clashes within the chain. FM has been tested for a large number of IDPs and has successfully described several NMR observables and SAXS data measured for these proteins (Jensen et al. 2009).

3.3.7 Maximum Occurrence

The maximum occurrence (MO) method employs a different approach to integrate NMR and SAXS data. MO is defined as the maximum fraction of time a system can spend in a given conformation and still be compatible with the experimental observations. To determine this fraction, the weight of a conformation in the ensemble is increased until the experimental data cannot be fitted by the best possible ensemble of other conformations. The procedure is performed for each conformer and the ones with the largest occurrence are selected. One successful example of its application is the aforementioned combined use of SAXS data with RDCs and PCSs obtained by NMR. This was done to investigate the structural disorder of calmodulin, a two-domain flexible protein (Bertini et al. 2010). RDC and PCS data were obtained in three measurements with Tb^{3+}, $Tm^{3+}$and Dy^{3+} loaded in the N-terminal domain of calmodulin and demonstrated the presence of extensive interdomain mobility. The computationally expensive procedure of MO calculation was applied to 400 calmodulin conformations of a large pool of 56,000 structures that exhaustively samples the translational and orientational interdomain space. Comparison of the MO for each conformation clearly shows that moderately extended conformations are the most populated (35 %) whereas closed and fully extended structures only reach 5 and 15 % MO, respectively.

3.4 *Protocols for the Characterization of Flexible Systems Using SAXS (EOM)*

A typical analysis protocol of SAXS data for flexible bimolecular systems using EOM includes several steps: the first three are common to most data analysis approaches for any type of molecule and are presented in the previous section. The other steps are specific to the ensemble-based analysis of unfolded or flexible proteins and are listed here for the case of EOM.

4. The initial step of the EOM procedure is the generation of the random pool,where the protein sequence is used as input. Here the user may select the following parameters:

   a. *Degree of flexibility*, which can be either random or native. The former corresponds to the case of IDPs or flexible linkers of multidomain proteins, while the latter is more appropriate for regions that are expected to be more structured, but whose exact structure is unknown.
   b. The generated structures may incorporate *high-resolution elements* if they are available. The EOM aligns the sequences of the whole construct and the provided fragment and places the fragment at the first suitable position.
   c. Pools consisting of oligomeric assemblies can be generated using the *symmetry* parameter of EOM. The user can choose a symmetry of the assembly core from P1 to P19 or P*n*2 (where *n* can be from 1 to 12). The flexible part of the structure may be generated using the same symmetry as the high-resolution core or in an asymmetric manner.
   d. The optimal *number of structures* that can cover the conformational space depends on the size of the unfolded protein or disordered region. To represent short disordered regions (e.g. flexible linkers),5000 conformations may suffice; for large, completely unfolded proteins the pool size should not be less than 20,000. An important feature of EOM is the ability to incorporate high-resolution models into the generated structures; in this way, proteins with flexible loops or interdomain linkers can be modelled.

5. In the second step, EOM selects the ensemble that fits the experimental data using GA. The selection procedure is repeated several times by using different random sequences and averaging the results. The users can choose the number of runs of the genetic algorithm(more runs will increase accuracy, but take longer time) and the number of input curves to fit (e.g. several curves with different concentrations).
6. The last step is the analysis of the results of the genetic algorithm selection. Distributions of the properties of the chosen ensemble such as R_g and D_{max} are compared to the corresponding distributions of the initial pool. For example, if the mean R_g of the selected ensemble is larger than that of the pool, the solute particles are more extended then randomly generated structures. The correspondence between the mean R_g of the ensemble and the one obtained using the Guinier approximation (step 3) is a good way to control the consistency of the genetic algorithm selection. The width (standard deviation) of the distribu-

tions shows the variability of the selected conformers and therefore provides a measure of flexibility;the ratio of the standard deviations of the ensemble and the pool is denoted R_{ratio}. Another metric used for the quantitative characterization of the ensemble properties is R_{flex}, which is based on the information entropy of the obtained distribution and varies from 0 % (no flexibility) to 100 % (maximum flexibility). In some cases EOM can also be used to discriminate between two subpopulations of the same protein in solution, for example open and closed states.

One of the examples of EOM application is the structural characterization of the N-terminal region of the phosphoprotein of vesicular stomatitis virus (VSV-P_{60}), whose interaction with the nucleoprotein is crucial for the encapsidation of the viral RNA genome (Leyrat et al. 2011).The protein is 68 amino acids long and contains the recognition element of the nucleoprotein. Large values of R_g and D_{max}, the poor dispersion of ^{1}H NMR signals, and CD spectroscopy and size-exclusion chromatography (SEC) data indicate that the protein is intrinsically disordered. Other NMR methods such as chemical shifts and relaxation rates showed the presence of two regions in the protein with transient helical conformations. The analysis of SAXS data performed with EOM displayed a bimodal distribution of R_g values with two subpopulations corresponding to compact and extended conformers (Fig. 8.8a, b). This distribution is considered to be robust because it was observed regardless of the parameters of the genetic algorithm. A series of experiments with different additives affecting the structure of the protein were completed to discover the structural nature of the two subpopulations. The first measurements were performed in a 50 mM Arg/Glu buffer that induced compaction of the protein thus increasing its solubility and stability (Blobel et al. 2011). When the concentration of Arg/Glu was decreased to 10 mM the amount of compact conformations decreased (Fig. 8.8c, d). In the case of the addition of 6 M of destabilizing consolute GdmCl, the unimodal distribution of extended conformers was observed (Fig. 8.8d, e). Conversely the addition of the stabilizing agent trimethylamine N-oxide (TMAO) increased the subpopulation consisting of compact structures (Fig. 8.8f, g). The authors assumed that the existence of transient helical elements and the distribution of charged residues plays a role in the preconfiguration of certain VSV-P_{60} conformations to enable the recognition of the nucleoprotein. The bimodal distribution of overall parameters is not unique and was observed in other IDPs (Paz et al. 2008; Boze et al. 2010), and EOM's ability to distinguish between two subpopulations makes it a helpful tool to explore such properties of unfolded proteins.

3.4.1 Multiple Constructs (Deletion Mutants)

Due to the low-resolution nature of SAXS data, EOM provides overall information but not the exact conformation of a certain region. More structural details, even with a low-resolution method, can be obtained using deletion mutants. The protocol of experiments in this case includes measurements of multiple constructs according to steps 1–5 of the single construct scenario. For each mutant, a pool

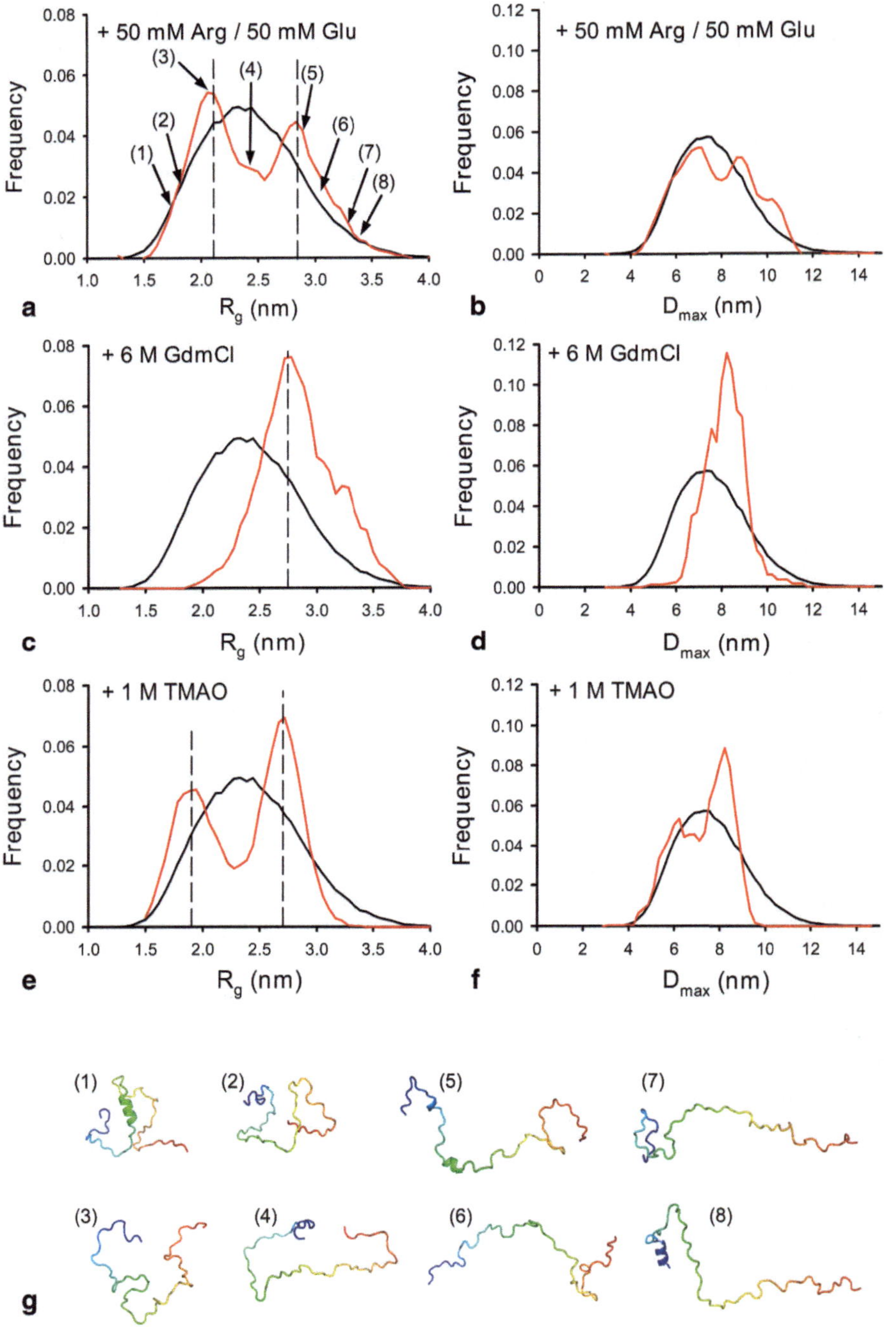

Fig. 8.8 Conformational ensemble selection and effects of stabilizing and destabilizing cosolutes. Ensembles of 20 conformers that collectively reproduce the experimental curves were selected from the initial ensemble. In each figure, the *black* curve shows the R_g and D_{max} distributions calculated for the initial ensemble of conformers, while the red curve shows the R_g and D_{max} distributions of the selected ensemble that fit the experimental SAXS data. **a** R_g and **b** D_{max} distribution

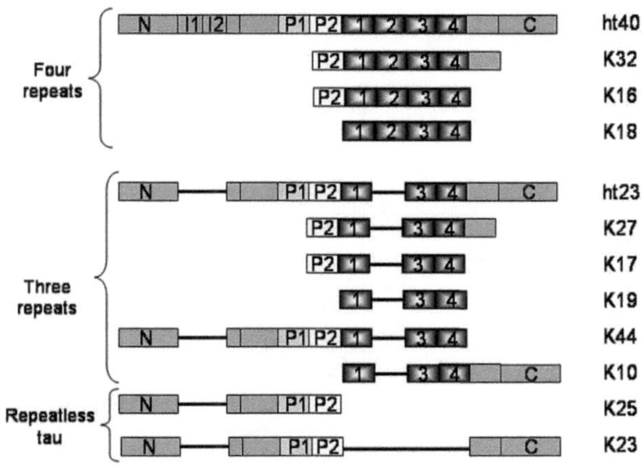

Fig. 8.9 Bar diagrams of isoforms and mutants of tau protein. All residue numbers refer to the sequence of ht40, the longest of the human tau isoforms in the central nervous system [441 residues]. Constructs with four repeats: ht40 (441 amino acids); K32, residues (M)S198-Y394; K16, residues (M)S198-E372; and K18, residues (M)Q244-E372. Constructs with three repeats: ht23 (352 amino acids); K27, residues (M)S198-Y394 without the second repeat; K17, residues (M) S198-E372 without the second repeat; K19, three repeats where the second repeat is missing (M) Q244-E372; K44, residues M1-E372 without the second repeat and where two N-terminal inserts (E45-T102) are missing; K10, residues (M)Q244-L441 without the second repeat. For repeatless tau, K25 contains the amino-terminal domain of ht23 and consists of residues M1-L243 (residues E45-T102, representing the amino-terminal inserts in ht40, are missing) and K23 represents the ht23 molecule without repeats (residues Q244-N368 are missing). (Mylonas et al. 2008)

of conformations is taken from the appropriate portions of the full length proteins generated in the main pool using the corresponding amino acid sequence. The GA is then applied to select ensembles that simultaneously fit the available experimental data to distributions of R_g and D_{max} for all the constructs. A comparison of these distributions is used to determine the contribution of different protein regions to overall flexibility and compactness and therefore to obtain clues of the local structural properties of the regions. An example of the implementation of this idea is the study of tau protein (Mylonas et al. 2008), which plays a role in stabilizing the neuronal microtubule and is found in abnormal deposits in the brains of Alzheimer's disease patients (Mandelkow and Mandelkow 1998). Three isoforms of this protein with four, three and zero repeats (Fig. 8.9) were investigated and for each isoform, SAXS data for the full-length constructs and several different deletion mutants were collected (Fig. 8.10). The EOM results (Fig. 8.11)

from the EOM analysis in 50 mM Glu and 50 mM Arg. **c** R_g and **d** D_{max} distribution from the EOM analysis in 10 mM Glu and 10 mM Arg. **e** R_g and **f** D_{max} distributions from the EOM analysis in 6 M GdmCl. **g** R_g and **h** D_{max} distributions from the EOM analysis in 1 M TMAO. (**i**) Members of the conformational ensemble. The cartoon models show some of the selected models at varying Rg values, highlighting the presence of residual helical structures. The chains are coloured from the N-terminal (*blue*) to the C-terminal (*red*). The numbers refer to their position in the distribution shown in (**a**). (Leyrat et al. 2011)

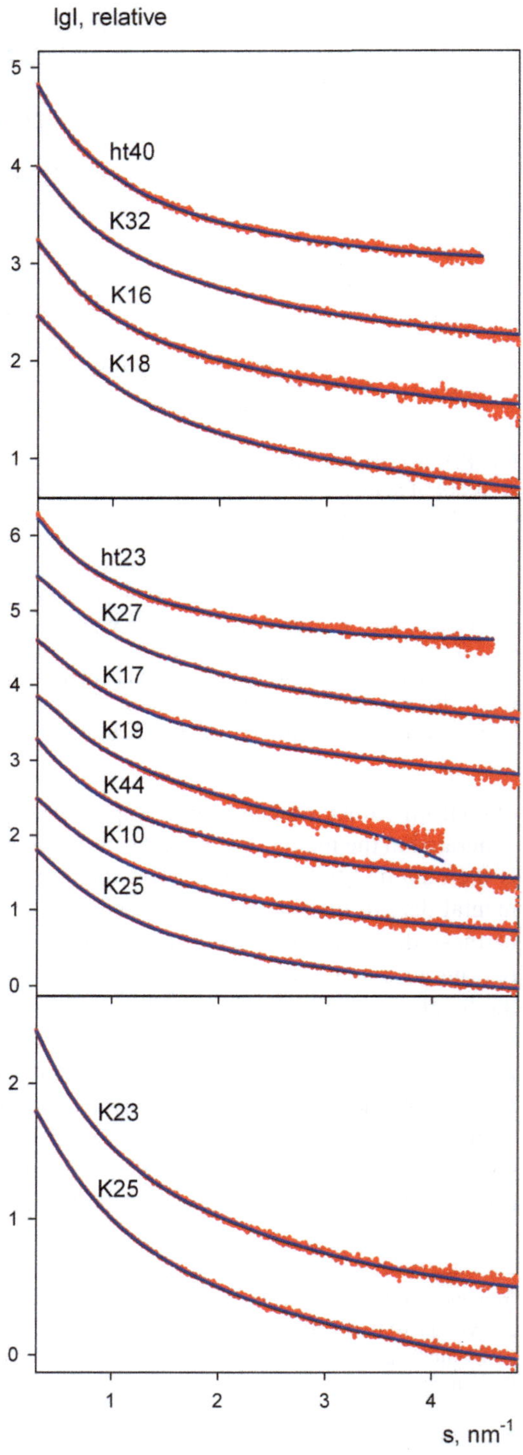

Fig. 8.10 Experimental SAXS data (O) with the corresponding ensemble fit (s). Constructs were grouped into three categories: **a** ht40 (containing four repeats), **b** ht23 (containing three repeats with K25, which is the N-terminal part of ht23), and **c** K23 (repeatless tau equivalent to ht23 and the N-terminus of ht23). (Mylonas et al. 2008)

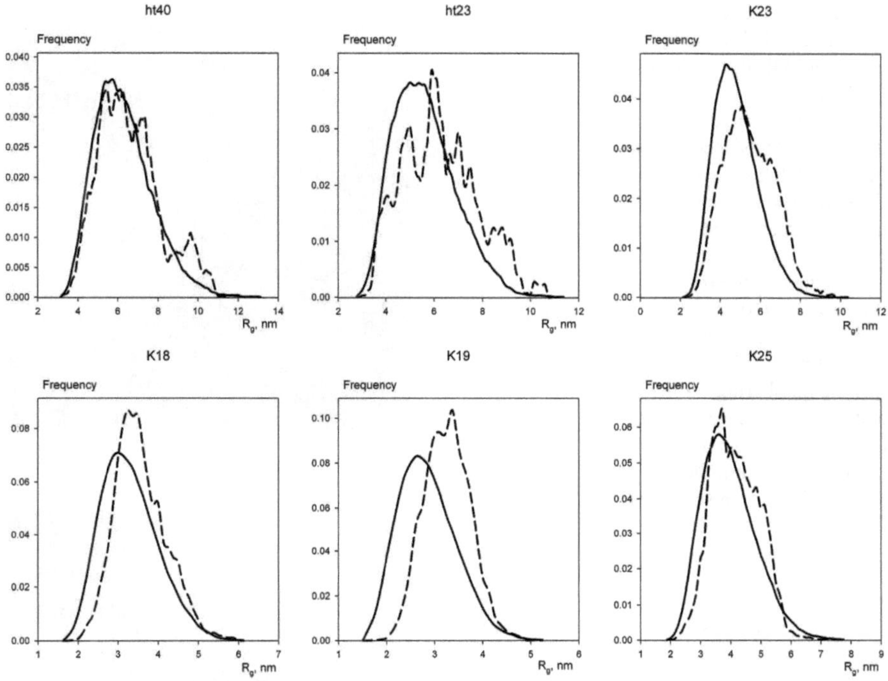

Fig. 8.11 Radius of gyration distributions of the pools (s) vs. the selected structures (—) using EOM (the histograms were smoothed using a sliding average). The integral of the area defined by the histograms equals 1. (Mylonas et al. 2008)

determined that the so-called repeat region is the source of residual secondary structure in tau, which agrees with previously obtained NMR data indicating the presence of turns and extended fragments in this region. The averaged Cα-Cα interresidue distance matrix (Fig. 8.12) obtained as the result of multiple curve fitting identifies a distinct conformational behaviour depending on the number of repeats in the isoforms. In the three-repeat isoform (ht23) the maximum separation is located within the repeat domain itself, while for the full-length isoform (ht40) with four domains an enhanced separation is found between the repeat domain and the preceding region. These results imply that the global arrangement of the chains may depend on the number of turns (one per repeat), resulting in the extension or shortening of the average interdomain distances as compared to a random coil.

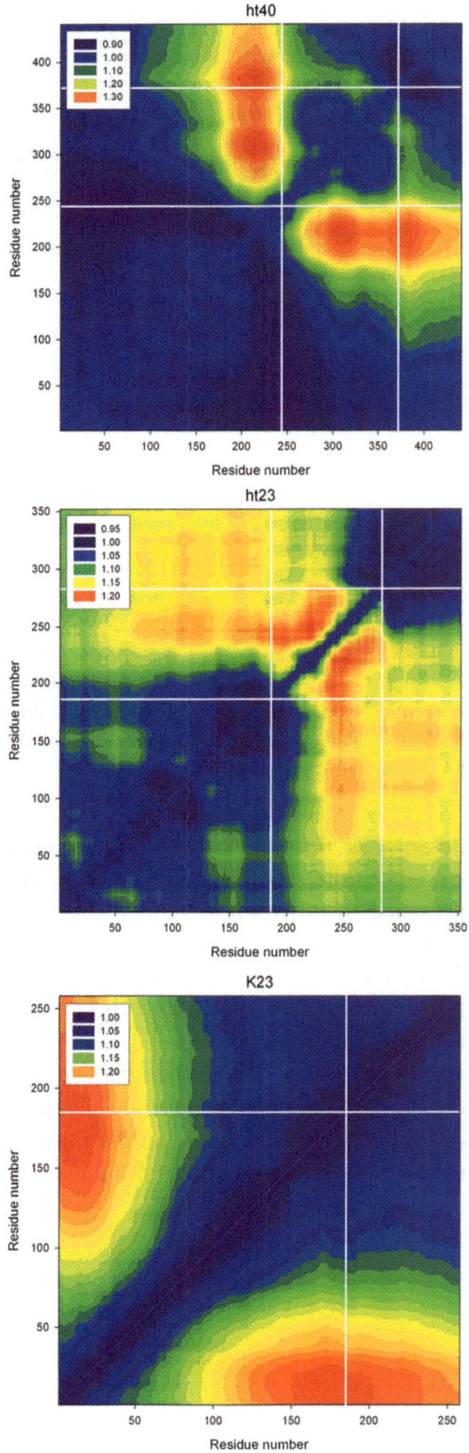

Fig. 8.12 CR-CR distance plot of ht40, ht23, and K23 using multiple curve fitting. Each plot point shows the ratio of the average CR distance of the selected structures to all the structures from the pool. The legend shows the ratio represented by each colour. *White lines* indicate the residues of the repeat domain for ht40 and ht23 and the N- and C-terminal connections for K23. Note that the numbering for ht23 and K23 is not based on the ht40 construct, but gives the ordinal residue numbers of the two former constructs. (Mylonas et al. 2008)

4 Conclusions

This chapter focused on the applications of SAXS to IDPs and there was only a brief description of the basics of the technique and its numerous other applications in structural biology. A more comprehensive description of SAXS/SANS and its practical applications can be found in recent reviews (e.g. Graewert and Svergun 2013) and monographs (Svergun et al. 2013). SAXS experienced a recent renaissance in structural biology thanks to recent advances in instrumentation and methods development, from a new generation of synchrotrons to the automation of data analysis software and novel methods of structure analysis. With the advent of the ensemble approach, the last decade also brought long-awaited progress in applications to IDPs, allowing for the quantitative description of these rather complicated systems.

We have briefly described experimental data collection, computational methods and analysis protocols, and presented practical examples to demonstrate that SAXS is a powerful technique for the structural characterization of unfolded and disordered proteins. The ensemble approach has proven to be a powerful tool for the quantitative analysis of IDP data as evidenced by a number of various implementations of this idea and an increasing number of studies employing this approach.

SAXS is inherently a low-resolution method and therefore profits from joint use with high-resolution techniques. The combination of SAXS and NMR provides an opportunity to analyse protein properties on multiple levels, ensuring a comprehensive and accurate description especially with flexible systems such as IDPs. The study of IDPs is a rapidly developing field, and SAXS, which is able to characterize complex systems in solution, is expected to further contribute to a better understanding of the problems of "unstructural biology".

References

Bernadó P, Blackledge M (2009) A self-consistent description of the conformational behavior of chemically denatured proteins from NMR and small angle scattering. Biophys J 97(10):2839–2845. doi:10.1016/j.bpj.2009.08.044

Bernadó P, Svergun DI (2012) Structural analysis of intrinsically disordered proteins by small-angle X-ray scattering. Mol Biosyst 8(1):151–167. doi:10.1039/c1mb05275f

Bernadó P, Blanchard L, Timmins P et al (2005) A structural model for unfolded proteins from residual dipolar couplings and small-angle x-ray scattering. Proc Natl Acad Sci U S A 102(47):17002–17007

Bernadó P, Mylonas E, Petoukhov MV et al (2007) Structural characterization of flexible proteins using small-angle X-ray scattering. J Am Chem Soc 129(17):5656–5664. doi:10.1021/ja069124n

Bernadó P, Modig K, Grela P et al (2010) Structure and dynamics of ribosomal protein L12: An ensemble model based on SAXS and NMR relaxation. Biophys J 98(10):2374–2382. doi:S0006-3495(10)00263-8 (10.1016/j.bpj.2010.02.012)

Bertini I, Giachetti A, Luchinat C et al (2010) Conformational space of flexible biological macromolecules from average data. J Am Chem Soc 132(38):13553–13558. doi:10.1021/ja1063923

Blanchet CE, Svergun DI (2013) Small-angle X-ray scattering on biological macromolecules and nanocomposites in solution. Annu Rev Phys Chem 64:37–54. doi:10.1146/annurev-physchem-040412-110132

Blobel J, Brath U, Bernadó P et al (2011) Protein loop compaction and the origin of the effect of arginine and glutamic acid mixtures on solubility, stability and transient oligomerization of proteins. Eur Biophys J 40(12):1327–1338. doi:10.1007/s00249-011-0686-3

Boze H, Marlin T, Durand D et al (2010) Proline-rich salivary proteins have extended conformations. Biophys J 99(2):656–665. doi:10.1016/j.bpj.2010.04.050

Feldman HJ, Hogue CW (2000) A fast method to sample real protein conformational space. Proteins 39(2):112–131

Flory PJ (1953) Principles of polymer chemistry. Cornell University Press, Ithaca, United States

Francis DM, Rozycki B, Koveal D et al (2011) Structural basis of p38alpha regulation by hematopoietic tyrosine phosphatase. Nat Chem Biol 7(12):916–924. doi:10.1038/nchembio.707

Franke D, Svergun DI (2009) DAMMIF, a program for rapid ab-initio shape determination in small-angle scattering. J Appl Cryst 42:342–346. doi:10.1107/S0021889809000338

Franke D, Kikhney AG, Svergun DI (2012) Automated acquisition and analysis of small angle X-ray scattering data. Nucl Instrum Methods Phys Res 689:52–59

Graewert MA, Svergun DI (2013) Impact and progress in small and wide angle X-ray scattering (SAXS and WAXS). Curr Opin Struct Biol 23(5):748–754. doi:10.1016/j.sbi.2013.06.007

Guinier A (1939) La diffraction des rayons X aux tres petits angles; application a l'etude de phenomenes ultramicroscopiques. Ann Phys 12:161–237

Jensen MR, Markwick PRL, Meier S et al (2009) Quantitative determination of the conformational properties of partially folded and intrinsically disordered proteins using NMR dipolar couplings. Structure 17(9):1169–1185. doi:10.1016/j.str.2009.08.001

Kohn JE, Millett IS, Jacob J et al (2004) Random-coil behavior and the dimensions of chemically unfolded proteins. Proc Natl Acad Sci U S A 101(34):12491–12496

Konarev PV, Volkov VV, Sokolova AV et al (2003) PRIMUS–a Windows-PC based system for small-angle scattering data analysis. J Appl Crystallogr 36:1277–1282

Kozin MB, Svergun DI (2001) Automated matching of high- and low-resolution structural models. J Appl Crystallogr 34:33–41

Kratky O (1963) X-ray small angle scattering with substances of biological interest in diluted solutions. Progress in biophysics and molecular biology 13:105–173

Krzeminski M, Marsh JA, Neale C et al (2013) Characterization of disordered proteins with ENSEMBLE. Bioinformatics 29(3):398–399. doi:10.1093/bioinformatics/bts701

Le Guillou JC, Zinn-Justin J (1977) Critical exponents for the n-vector model in three dimensions from field theory. Phys Rev Lett 39(2):95–98

Leyrat C, Jensen MR, Ribeiro EA Jr et al (2011) The N(0)-binding region of the vesicular stomatitis virus phosphoprotein is globally disordered but contains transient alpha-helices. Protein Sci 20(3):542–556. doi:10.1002/pro.587

Mandelkow EM, Mandelkow E (1998) Tau in Alzheimer's disease. Trends Cell Biol 8(11):425–427

Mattinen ML, Paakkonen K, Ikonen T et al (2002) Quaternary structure built from subunits combining NMR and small-angle x-ray scattering data. Biophys J 83(2):1177–1183

Meier S, Grzesiek S, Blackledge M (2007) Mapping the conformational landscape of urea-denatured ubiquitin using residual dipolar couplings. J Am Chem Soc 129(31):9799–9807. doi:10.1021/ja0724339

Mylonas E, Hascher A, Bernadó P et al (2008) Domain conformation of tau protein studied by solution small-angle X-ray scattering. Biochemistry 47(39):10345–10353

Paoletti F, Covaceuszach S, Konarev PV et al (2009) Intrinsic structural disorder of mouse proNGF. Proteins 75(4):990–1009. doi:10.1002/prot.22311

Paz A, Zeev-Ben-Mordehai T, Lundqvist M et al (2008) Biophysical characterization of the unstructured cytoplasmic domain of the human neuronal adhesion protein neuroligin 3. Biophys J 95(4):1928–1944. doi:S0006-3495(08)70151-6 (10.1529/biophysj.107.126995)

Pelikan M, Hura GL, Hammel M (2009) Structure and flexibility within proteins as identified through small angle X-ray scattering. Gen Physiol Biophys 28(2):174–189

Petoukhov MV, Svergun DI (2005) Global rigid body modeling of macromolecular complexes against small-angle scattering data. Biophys J 89(2):1237–1250. doi:10.1529/biophysj.105.064154

Petoukhov MV, Konarev PV, Kikhney AG et al (2007) ATSAS 2.1–towards automated and web-supported small-angle scattering data analysis. J Appl Cryst 40(s1):s223–s228

Porod G (1982) General theory. In: Glatter O, Kratky O (eds) Small-angle X-ray scattering. Academic, London, pp 17–51

Rozycki B, Kim YC, Hummer G (2011) SAXS ensemble refinement of ESCRT-III CHMP3 conformational transitions. Structure 19(1):109–116. doi:S0969-2126(10)00395-3 (10.1016/j.str.2010.10.006)

Stumpe MC, Grubmüller H (2007) Interaction of urea with amino acids: implications for urea-induced protein denaturation. J Am Chem Soc 129(51):16126–16131

Svergun DI (1992) Determination of the regularization parameter in indirect-transform methods using perceptual criteria. J Appl Crystallogr 25:495–503

Svergun DI (1999) Restoring low resolution structure of biological macromolecules from solution scattering using simulated annealing. Biophys J 76(6):2879–2886

Svergun DI (2010) Small-angle X-ray and neutron scattering as a tool for structural systems biology. Biol Chem 391(7):737–743. doi:10.1515/BC.2010.093

Svergun DI, Koch MHJ (2002) Advances in structure analysis using small-angle scattering in solution. Curr Opin Struct Biol 12(5):654–660

Svergun DI, Barberato C, Koch MHJ (1995) CRYSOL–a program to evaluate X-ray solution scattering of biological macromolecules from atomic coordinates. J Appl Crystallogr 28:768–773

Svergun DI, Petoukhov MV, Koch MHJ (2001) Determination of domain structure of proteins from X-ray solution scattering. Biophys J 80(6):2946–2953. doi:10.1016/S0006-3495(01)76260-1

Svergun DI, Koch MHJ, Timmins PA et al (2013) Small angle X-ray and neutron scattering from solutions of biological macromolecules. OUP, Oxford

Tompa P (2012) On the supertertiary structure of proteins. Nat Chem Biol 8(7):597–600. doi:10.1038/nchembio.1009

Volkov VV, Svergun DI (2003) Uniqueness of ab initio shape determination in small angle scattering. J Appl Crystallogr 36:860–864

von Ossowski I, Eaton JT, Czjzek M et al (2005) Protein disorder: conformational distribution of the flexible linker in a chimeric double cellulase. Biophys J 88(4):2823–2832

Yang SC, Blachowicz L, Makowski L et al (2010) Multidomain assembled states of Hck tyrosine kinase in solution. Proc Natl Acad Sci U S A 107(36):15757–15762. doi:10.1073/pnas.1004569107

Chapter 9
Bioinformatics Approaches for Predicting Disordered Protein Motifs

Pallab Bhowmick, Mainak Guharoy and Peter Tompa

Abstract Short, linear motifs (SLiMs) in proteins are functional microdomains consisting of contiguous residue segments along the protein sequence, typically not more than 10 consecutive amino acids in length with less than 5 defined positions. Many positions are 'degenerate' thus offering flexibility in terms of the amino acid types allowed at those positions. Their short length and degenerate nature confers evolutionary plasticity meaning that SLiMs often evolve convergently. Further, SLiMs have a propensity to occur within intrinsically unstructured protein segments and this confers versatile functionality to unstructured regions of the proteome. SLiMs mediate multiple types of protein interactions based on domain-peptide recognition and guide functions including posttranslational modifications, subcellular localization of proteins, and ligand binding. SLiMs thus behave as modular interaction units that confer versatility to protein function and SLiM-mediated interactions are increasingly being recognized as therapeutic targets. In this chapter we start with a brief description about the properties of SLiMs and their interactions and then move on to discuss algorithms and tools including several web-based methods that enable the discovery of novel SLiMs (*de novo* motif discovery) as well as the prediction of novel occurrences of known SLiMs. Both individual amino acid sequences as well as sets of protein sequences can be scanned using these methods to obtain statistically overrepresented sequence patterns. Lists of putatively functional SLiMs are then assembled based on parameters such as evolutionary sequence conservation, disorder scores, structural data, gene ontology terms and other contextual information that helps to assess the functional credibility or significance of these motifs. These bioinformatics methods should certainly guide experiments aimed at motif discovery.

P. Tompa (✉) · M. Guharoy · P. Bhowmick
VIB Department of Structural Biology, Vrije Universiteit Brussel (VUB), Building E, Pleinlaan 2, 1050 Brussels, Belgium
e-mail: mainak.guharoy@vib-vub.be

P. Tompa
Institute of Enzymology, Research Center of Natural Sciences, Hungarian Academy of Sciences, Budapest, Hungary
e-mail: ptompa@vub.ac.be

I. C. Felli, R. Pierattelli (eds.), *Intrinsically Disordered Proteins Studied by NMR Spectroscopy*, Advances in Experimental Medicine and Biology,
DOI 10.1007/978-3-319-20164-1_9

Keywords Protein sequence · Short linear motifs · Motif prediction · Intrinsic disorder · Post-translational modification · PTM · Multiple sequence alignments · Position-specific weight matrix · PWM · Protein interaction · Evolutionary conservation

1 Introduction

Protein modularity is a central and recurrent theme in our understanding of protein function. The basic functioning of almost all proteins occurs by the interaction of its modules with various other partners (proteins, nucleic acids, small molecules, etc.). Each module has a defined set of function(s) (eg, interactions with specific partners) that is linked to its surface characteristics, shape and structural dynamics and the variety of functions that a protein can carry out is closely linked to the number and types of modules it contains (Bhattacharyya et al. 2006). These modules include globular domains, *S*hort *Li*near *M*otifs (SLiMs) or other *Mo*lecular *R*ecognition *F*eatures (MoRFs). The presence of these elements in a given protein will determine its function by specifying its set of interaction partners.

Protein domains possess well-defined three dimensional structures with the members of any given domain family sharing strong and clearly visible evolutionary relationships; domain signatures are therefore comparatively easy to detect from protein primary sequence using information contained in databases such as Pfam (Finn et al. 2014) and Prosite (Sigrist et al. 2013). Domain structures can also be predicted reliably using *in silico* methods such as homology modelling based on sequence-structure alignments and this is now done routinely in protein structure prediction competitions like CASP (*C*ritical *A*ssessment of protein *S*tructure *P*rediction) (Moult et al. 2014). The Protein Data Bank (PDB) currently has more than 100,000 deposited structures that have accumulated rapidly over the past few decades (Berman et al. 2013), and most of the domain types are now thought to have been discovered.

At present, scientists are focusing not only on structured regions of the proteome but also on the disordered regions in search of functional modules (Tompa 2012; Habchi et al. 2014). In eukaryotes, up to 33% of the proteome may have putative long disordered segments (defined as > 30 consecutive disordered residues) (Ward et al. 2004). Contained within these disordered regions, there may be a million or more estimated peptide motifs (SLiMs) existing in the proteome (Tompa et al. 2014) although relatively few of them have been discovered and experimentally validated so far. Work over the past decade has brought to the forefront the importance of sequence (peptide) motifs in protein function. These motifs are typically found at functional sites of proteins like cleavage sites, binding sites, sites for post-translational modifications and sub-cellular targeting sites. Some of the functions mediated by peptide motifs include specific protein-protein interactions, regulatory functions and signal transduction (Van Roey et al. 2014). The large number of annotated motifs in the *E*ukaryotic *L*inear *M*otif (ELM) database (Dinkel et al. 2014) provide overwhelming evidence of the fact that linear motifs are a ubiquitous and essential part of cellular biology.

Although clearly very abundant, true positive (ie, functional) linear motif instances are difficult to predict *de novo* from protein sequences due to the difficulty associated with obtaining robust statistical assessments (Gould et al. 2010). It is therefore of great interest to discover (using both computational and experimental techniques) new functional motifs that may form the basis of future drug discovery, by disrupting or regulating important interactions.

2 Short Linear Motifs (SLiMs) and Molecular Recognition Features (MoRFs)

In this chapter we focus on the characteristic features of SLiMs and on the various algorithms that have been developed to aid in their identification. Protein sequence motifs (SLiMs) have been described as functional microdomains that are short and flexible in length (between 2 to 11 consecutive residues). These are thought to arise by convergent evolution (Davey et al. 2009; Dinkel et al. 2014), thus the same SLiM may be found within otherwise unrelated proteins. They form compact functional modules and mainly occur within intrinsically disordered regions and surface accessible regions of proteins (Fuxreiter et al. 2007). Of the residues that constitute a SLiM, only a certain fraction are invariant (ie, fully conserved) across multiple instances of the motif. Usually these residues confer functional specificity, for binding interactions and/or undergo posttranslational modifications (PTMs). Other positions may tolerate conservative substitutions (eg, residues with similar size and/or physicochemical characteristics may be used interchangeably). Finally, some positions are not under selective constraints (wildcard positions). Thus, SLiMs have well-defined sequence patterns that are usually represented graphically using sequence logos (Schneider and Stephens 1990) or by machine-readable regular expressions (REs), that constitute position-specific definitions of allowed residue types and/or certain wildcard or ambiguous positions. Regular Expressions (REs) will be explained and elaborated upon later in the chapter.

*Mo*lecular *R*ecognition *F*eatures (MoRFs) are so-called because these protein segments form a specific class of intrinsically disordered regions (IDRs) that exhibit specific molecular recognition and binding functions. MoRFs are short (usually 20 residues or fewer) segments that are located within longer IDRs and are very interaction-prone (Vacic et al. 2007). MoRFs undergo characteristic disorder-to-order transitions upon binding to their partners (Mohan et al. 2006); based upon their bound state structures, they have been classified into α-MoRFs, β-MoRFs and ι-MoRFs (the latter class forms non-regular structures without regular backbone hydrogen bonding patterns). Unlike SLiMs, MoRFs are not defined on the basis of a sequence pattern (RE), but as interaction-prone disordered segments that (are predicted to) form ordered secondary structures upon binding to a protein partner. However, MoRF segments may themselves contain SLiMs, such as demonstrated in Fig. 9.1a (see next section).

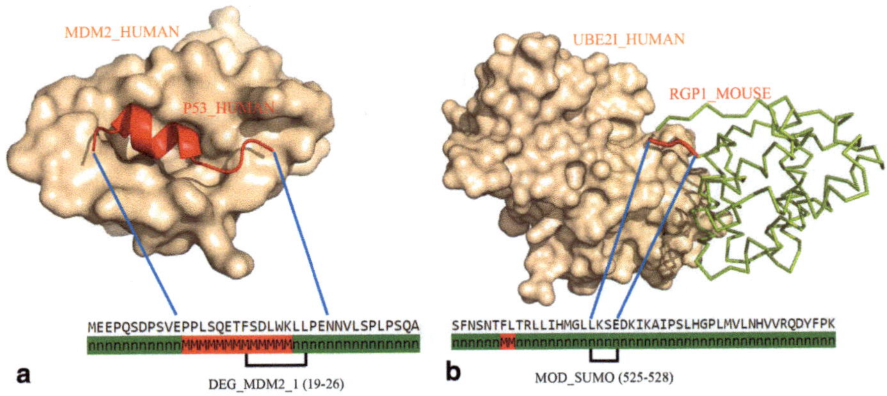

Fig. 9.1 Examples of SLiM-mediated interactions. **a** The p53 peptide (*red* cartoon) that is recognized by the folded SWIB domain (surface representation) of MDM2 (PDB code: 1YCR) is a MoRF that attains a helical bound state conformation. This MoRF region also contains a SLiM (degron) as indicated on the figure. **b** Interaction between the mammalian SUMO E2 enzyme (UBE2I, in surface representation) and its SUMOylation substrate RanGAP1 (*green* ribbon) mediated by a modification motif (shown in *red*) (PDB code: 1KPS). In both the figures, the amino acid sequence of the peptide motif segments and their sequence neighborhood are shown below their respective molecular diagrams along with the MoRFPred predictions (the letter 'M' on a *red* background indicates the segments that are predicted to be a MoRF, whereas 'n' against a *green* background indicates non-MoRF residues). The SLiM segments and their corresponding ELM identifiers are also indicated

3 Motif (SLiM)-Mediated Interactions and Their Biological Importance

Our current understanding of protein-protein interactions has changed significantly with the knowledge of how IDRs play crucial roles in enabling protein interactions ('domain-peptide' interactions) (Dinkel et al. 2014; Petsalaki and Russell 2008; Edwards et al. 2012). Interactions mediated by SLiMs have been shown to function in diverse processes, such as in the control of cell cycle progression, substrate selection for proteasomal degradation, targeting proteins to specific subcellular locations and for stabilizing scaffolding complexes. Figure 9.1a shows an example of a motif-mediated interaction (a p53 peptide bound to the folded SWIB domain of MDM2) (Schon et al. 2002). The region of p53 present in the crystal structure contains an 8-residue SLiM (the ELM degradation motif 'DEG_MDM2_1'). The motif is disordered in the unbound state, but forms an α-helical secondary structure in the complex with MDM2, thus conforming to the classical definition of a MoRF. In this example, the SLiM overlaps with a larger MoRF segment that can be detected by the MoRFPred predictor (Disfani et al. 2012). Figure 9.1b illustrates recognition of the ELM SUMOylation motif 'MOD_SUMO' present on the C-terminal domain of RanGAP1 by the mammalian SUMO E2 enzyme UBE2I (Bernier-Villamor et al. 2002). Note that in this case the peptide motif is not classified as a MoRF by the predictor.

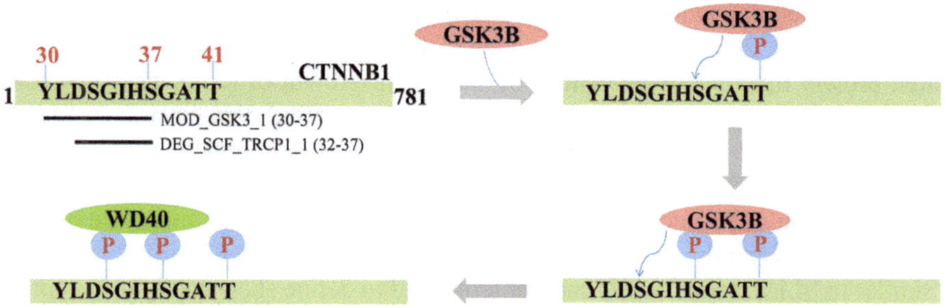

Fig. 9.2 Schematic illustration of the use of multiple overlapping SLiMs (ELM identifiers MOD_GSK3_1 and DEG_SCF_TRCP1) in beta-catenin (CTNNB1) that allows the recognition and relay (sequential) phosphorylation of beta-catenin by glycogen synthase kinase-3 beta (GSK3B) resulting in the activation of a degradation motif (degron) that is recognized by the WD40 repeat domain of the substrate adaptor subunit of a multi-subunit E3 ubiquitin ligase, resulting in the ubiquitination of beta-catenin and its 26 S proteasome-mediated degradation. Phospho groups are shown in *blue circles* and 'P' written in *red*

Interface areas in peptide-protein complexes observed in the PDB average about 500 Å^2 (London et al. 2012), significantly smaller than the size of an average protein-protein hetero-interface (1900 Å^2) or homodimer interface (3900 Å^2) (Janin et al. 2008). The limited size of SLiM-mediated interfaces often results in micromolar binding affinity for these interactions, whereas globular protein-protein complexes formed via domain-domain interactions can be much stronger (nano-molar or lower Kd). This permits transient and reversible interactions that are necessary for many dynamic cellular binding events, such as those required for the rapid transmission of intracellular signals (Neduva and Russell 2005; Gibson 2009).

A further advantage is the 'switching' behaviour that can be achieved by the use of PTMs within SLiMs to regulate interactions. Phosphorylation/dephosphorylation is widely used to enhance (or disrupt) interactions for example, and this enables direct cross-talk between multiple signaling pathways (Akiva et al. 2012). Multiple SLiMs can also form more complex switches by co-operating with each other and acting in synergy with post-translational modifications to assist switching between different functional states of proteins (Dinkel et al. 2014). In the example illustrated in Fig. 9.2, the phosphorylation of beta-catenin (CTNNB1) at Thr41 generates a docking site for Glycogen synthase kinase-3 beta (GSK3B) which phosphorylates Ser37 and generates a new docking site for GSK3B. Subsequent phosphorylation of Ser33 by GSK3B switches CTNNB1 binding specificity to the F-box/WD40 repeat containing protein BTRC which functions as a substrate recognition component of a SCF (SKP1-CUL1-F-box protein) multi-subunit E3 ubiquitin-protein ligase. This results in the recruitment of β-catenin to the SCF E3 ligase complex followed by ubiquitination and proteasome-dependent degradation of β-catenin (Wu et al. 2003; Hagen and Vidal-Puig 2002; Van Roey et al. 2013).

SLiMs represent an important target for diseases, both in terms of causal mutations and potential therapeutics (Uyar et al. 2014). Further, many pathogens have taken advantage of the plasticity of SLiMs by mimicking host motifs to dysregulate and rewire cellular pathways of the host to their own advantage (Davey et al. 2011b; Kadaveru et al. 2008). Our growing appreciation of the importance of motif-mediated protein functions is evidenced by the recent growth of motif databases. The eukaryotic linear motif (ELM) resource maintains curated data on protein SLiMs whose functional validity has been demonstrated experimentally (Dinkel et al. 2014). MiniMotifMiner (MnM) (Mi et al. 2012) is another resource dedicated to the annotation and detection of a broad spectrum of motifs from a large number of species and currently contains 880 consensus minimotifs and 294,053 instances. Similar to SLiM, minimotif is another term used to define short contiguous peptide sequences that possess a demonstrated function (including post translation modifications, binding to a target protein or molecule and protein trafficking) in at least one protein. Another database ScanSite (Obenauer et al. 2003) stores data for 65 motifs in 12 different groups (functionally similar motifs have been grouped together). Similarly, Prosite (Sigrist et al. 2013) contains data for 1308 patterns or regular expressions although it contains domain signatures in addition to SLiMs. However, in spite of their immense functional importance in eukaryotic cell regulation, detailed information regarding the majority of SLiMs are still limited, and at present only a small proportion of human motifs have been discovered (Tompa et al. 2014). This highlights the pressing need to develop and further enhance computational methods that can efficiently predict novel SLiMs in protein sequences and thereby serve as a useful guide for experimental motif discovery efforts.

4 Representing Motifs: *R*egular *E*xpressions (REs), *P*osition *W*eighted *M*atrices (PWMs) and *P*osition-*S*pecific *S*coring *M*atrices (PSSMs)

SLiMs are commonly represented by RE-patterns and PWMs. SLiMs are comprised of both defined amino acid positions as well as wildcard positions which may be occupied by any amino acid type. Defined positions may be (i) *fixed* or *invariant*, in which only a single amino acid type is permitted at that position, or (ii) *ambiguous*, in which case multiple amino acids (often of similar size and/or physicochemical properties) may occupy that site and still result in a functional SLiM. Thus, a RE describes a sequence of letters that may match at each position in a given motif. The simplest RE is just a string of letters, such as the "RGD" motif present in extracellular matrix proteins that is recognised by different members of the integrin family (Corti and Curnis 2011). This regular expression matches only one defined amino acid sequence: Arg-Gly-Asp (RGD). To allow variable positions in a RE, additional symbols are used. For example, [KR] specifies that either K or R may be present; {min, max} specifies a range of minimum and maximum numbers of residues allowed (eg. M{0,1}) indicates that Met can either be absent (0) or can be present but only once (1); the '.' (dot symbol) at a given position indicates that any amino acid is allowed at that position. One disadvantage of REs is that residue-specific frequency information is lost: [KR] does not indicate the relative occurrence frequency

Table 9.1 Description of the different types of symbols used to construct Regular Expressions (REs) for peptide motif representation

Character	Name	Description
.	Dot	Any amino acid allowed
[…]	Allowed character class	Amino acids listed are allowed
[^..]	Disallowed character class	Amino acids listed are not allowed
X{min, max}	Allowed range (number) of consecutive specified character 'X'	Min required, max allowed
^	Caret	Matches the amino terminal
$	Dollar	Matches the carboxy terminal
?	Question	One amino acid is allowed but is optional
*	Star	Any number of amino acids are allowed but are optional
+	Plus	One amino acid is allowed, additional are optional
|	Alternation	Matches either expression it separates

of K vs R. Table 9.1 provides an overview of how regular expressions are used to represent sequence motifs.

Unlike REs, PWMs indicate the probability of each residue type occurring at each position in a motif. PWMs are widely used for characterizing and predicting sequence motifs (Bailey 2008). A PWM is an 'n' by 'w' matrix where 'n' is the number of letters in the sequence alphabet (20 amino acids for proteins) and 'w' is the number of motif positions. $P_{a,i}$ represents the probability of letter 'a' at the ith position in the motif. A PWM can be used to define an occurrence probability for any possible sequence containing 'w' characters (calculated as the product of the corresponding entries in the PWM), based on the assumption that each motif position is statistically independent. The relationship between a RE and the corresponding PWM is shown in Fig. 9.3 for the KEN-box motif. The 16 validated occurrences (sites) from which this motif was constructed (data from ELM entry DEG_APCC_KENBOX_2) are shown aligned with each other on the left-hand panel. The corresponding RE is shown in the middle panel along with the observed counts of each letter in the corresponding alignment columns (frequency table). The PWM is shown on the right-hand panel. Finally, the figure represents the KEN-box sequence logo (Schneider and Stephens 1990).

Motif discovery algorithms also output a position-specific scoring matrix (PSSM) which takes the background probabilities of different letters into account (Bailey 2008). The PSSM entries are calculated as a log likelihood: $S_{a,j} = \log_2 (P_{a,j}/f_a)$, where f_a is the overall (background) probability of letter 'a' in the set of input sequences that will be scanned for motif occurrences, and $P_{a,j}$ represents the frequency of letter 'a' at the jth position as explained earlier. Sequences are assigned scores by summing up (rather than multiplying position specific probabilities as with a PWM) the appropriate numbers from the PSSM table. PSSM scores are more useful for scanning sequences as compared to PWM probabilities because they allow scaling by background probability: this reduces false positive rates caused by non-uniform distribution of letters in sequences (Xia 2012; Bailey 2008).

Aligned sites					
Site1	N	K	E	N	E
Site2	D	K	E	N	G
Site3	D	K	E	N	E
Site4	D	K	E	N	Q
Site5	E	K	E	N	E
Site6	K	K	E	N	G
Site7	Q	K	E	N	S
Site8	S	K	E	N	V
Site9	D	K	E	N	M
Site10	N	K	E	N	Y
Site11	V	K	E	N	L
Site12	N	K	E	N	D
Site13	S	K	E	N	Q
Site14	N	K	E	N	A
Site15	N	K	E	N	Y
Site16	A	K	E	N	E

Regular Expression (RE)					
XKENX					
G	0	0	0	0	2
A	1	0	0	0	1
V	1	0	0	0	1
S	2	0	0	0	1
T	0	0	0	0	0
C	0	0	0	0	0
M	0	0	0	0	1
L	0	0	0	0	1
I	0	0	0	0	0
K	1	16	0	0	0
R	0	0	0	0	0
E	1	0	16	0	4
D	4	0	0	0	1
Q	1	0	0	0	2
N	5	0	0	16	0
F	0	0	0	0	0
Y	0	0	0	0	2
W	0	0	0	0	0
P	0	0	0	0	0
H	0	0	0	0	0

Position-specific Weight Matrix(PWM)					
G	0	0	0	0	0.125
A	0.062	0	0	0	0.062
V	0.062	0	0	0	0.062
S	0.125	0	0	0	0.062
T	0	0	0	0	0
C	0	0	0	0	0
M	0	0	0	0	0.062
L	0	0	0	0	0.062
I	0	0	0	0	0
K	0.062	1	0	0	0
R	0	0	0	0	0
E	0.062	0	1	0	0.25
D	0.25	0	0	0	0.062
Q	0.062	0	0	0	0.125
N	0.312	0	0	1	0
F	0	0	0	0	0
Y	0	0	0	0	0.125
W	0	0	0	0	0
P	0	0	0	0	0
H	0	0	0	0	0

Fig. 9.3 Converting a multiple sequence alignment of known motif instances into a RE and PWM. The alignment of motif sites (validated instances of the KEN-box (Dinkel et al. 2014)) is shown on the *left*. The RE is shown at the *top* of the *middle* panel. The counts of each amino acid type in each alignment column (the position specific count matrix, PSCM) are shown beneath the RE. The PWM is shown on the *right* hand side. The last figure shows the information content sequence logo for the motif (generated by http://weblogo.berkeley.edu/logo.cgi)

5 Overview of Functionally Specialized SLiM Categories in ELM

The latest published ELM release contained 197 classes and 2404 instances (Dinkel et al. 2014). SLiMs in ELM have been classified into six categories based on their function: proteolytic cleavage sites ('CLV'), sub-cellular targeting sites ('TRG'), ligand binding sites ('LIG'), post-translational modification sites ('MOD'), destruction motifs or degrons ('DEG') and finally, docking sites ('DOC') (Table 9.2). Figure 9.4 shows representative examples of SLiM-mediated interactions from each ELM class (except 'CLV' sites for which none of the entries had a corresponding PDB entry).

Table 9.2 Summary of data stored in the ELM database (as of September 2013) (reprinted with permission from Dinkel et al. 2014). Breakup of ELM data according to (1) the six ELM class types (LIG, MOD, TRG, DEG, DOC and CLV motifs) and the number of ELM classes corresponding to each class, (2) ELM instances by organism type, (3) the number of ELMs that are represented in the PDB, and finally, (4) the number of GO terms associated with the data in ELM

Functional sites	ELM classes		ELM instances		PDB structures	GO terms	
Total	197			2404	290	419	
By category	LIG	103	Human	1391		Biological process	217
	MOD	30	Mouse	211			
	TRG	23	Rat	115		Cell compartment	95
	DEG	15	Yeast	86			
	DOC	15	Fly	77		Molecular function	107
	CLV	11	Other	524			

Cleavage ‘CLV’ sites are recognised by proteases for the processing of predecessor proteins into their active biological products (eg, N-arginine dibasic convertase is an endopeptidase that recognizes (.RK)|(RR[^KR]) dibasic cleavage sites for processing secreted proteins (Hospital et al. 2000)). ‘TRG’ motifs are used for protein recognition and targeting to diverse sub-cellular compartments: for example, the ‘tyrosine-based sorting signal’ (Y..[LMVIF] motif) is found in the cytosolic tails of some membrane proteins and is responsible for deciding the traffic flow in endosomal and secretory pathways (Fig. 9.4a). Motifs that mediate binding to globular protein domains form the ‘LIG’ class: for example, the AP2 (Adaptor Protein) α subunit recognizes and binds to accessory endocytic proteins such as amphiphysin, AP180 and synaptojanin170 via their F.D.F motifs resulting in their recruitment to the site of clathrin coated vesicle formation and thereby assists and regulates vesicle assembly (Brett et al. 2002) (Fig. 9.4b). SLiMs located at post-translational modification sites constitute the ‘MOD’ class (eg, the Protein kinase B substrate phosphorylation site has residue preferences as shown in Fig. 9.4c).

Earlier ELM versions contained only these four motif categories (‘CLV’, ‘TRG’, ‘LIG’, and ’MOD’) (Gould et al. 2010). Recently however with the increase in the number of ELM classes, two additional but functionally specialized ‘LIG’ (ligand-binding) categories were introduced—‘DEG’ (degron) motifs and ‘DOC’ (docking) motifs. Degrons are motif sequences embedded within proteins that enable their specific recognition by E3 ubiquitin ligases, normally resulting in the channeling of these substrates into the ubiquitin-proteasomal degradation pathway (Glickman and Ciechanover 2002). For example, the [IL]A(P).{6,8}[FLIVM].[FLIVM] motif present in the α subunit of the heterodimeric transcription factor Hif-1 (hypoxia-inducible factor 1) is an oxygen-dependent degron that is hydroxylated by prolyl hydroxylases under conditions of normal oxygen availability (Masson and Ratcliffe 2003). Prolyl hydroxylation confers degron recognition and binding by the von Hippel-Lindau tumor suppressor protein (pVHL) (Fig. 9.4d) which forms a multi-subunit E3 ubiquitin ligase complex with elongin C, elongin B, Cul-2, and Rbx1 leading to the ubiquitination and proteasomal degradation of Hif-1α (Min et al. 2002).

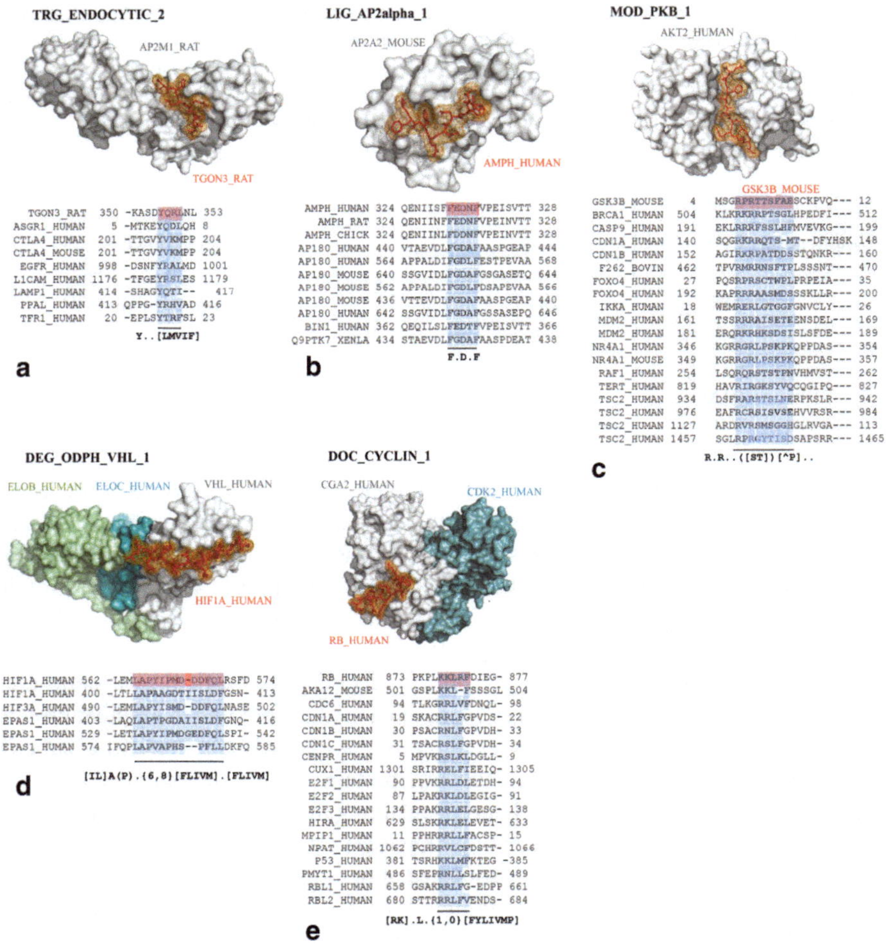

Fig. 9.4 PDB structures corresponding to representative examples from each ELM class showing the SLiM peptide (drawn using stick representation, *colored red* and surrounded by a surface *mesh*) in complex with their globular protein partners (displayed using *light grey* surface representation). SLiM-containing sequence segments of all the experimentally validated vertebrate instances (data from ELM) are shown in the multiple sequence alignments. The first sequence in each alignment corresponds to the SLiM-containing protein shown in the PDB structure (SLiM residues are shown in *red*). SLiM residues for the other instances are highlighted using *light blue color*. Consensus motif patterns are shown in *bold* under each alignment. **a** Targeting motif derived from the trans-Golgi network integral membrane protein (TGN38) interacting with the mu subunit of the adaptor protein complex 2, Ap2m1 (PDB code: 1BXX). **b** Ligand binding motif from human Amphiphysin interacting with the alpha-2 subunit of the adaptor protein complex 2, Ap2a2 (PDB code: 1KY7). **c** Modification motif from Glycogen synthase kinase-3 beta (Gsk3b) in complex with the kinase domain from RAC-beta serine/threonine-protein kinase (AKT2) (PDB code: 1O6K). **d** Degradation (*degron*) motif of human hypoxia-inducible factor 1-α protein (HIF1A) interacting with the Von Hippel-Lindau (VHL) component of the multi-subunit VHL ubiquitination complex (PDB code: 1LM8). **e** Docking motif derived from human Retinoblastoma-associated protein, RB1 interacting with the cyclin A2/CDK2 complex (PDB code: 1H25). Figures were drawn using PyMol

Finally, docking('DOC')motifs are used to recruit modifying enzymes onto their target substrates. However, 'DOC' sites are distinct from 'MOD' sites that are targeted for the actual enzymatic modification; initial binding to docking motifs on the substrate helps to direct and enhance enzyme specificity for the modification site (the two motifs together can be considered to possess a bi-partite architecture). For example, the docking motif DOC_CYCLIN_1 ([RK].L.{0,1}[FYLIVMP]) initiates substrate interactions with cyclin (Fig. 9.4e) resulting in increased specificity of phosphorylation (at the associated MOD_CDK_1 phosphorylation sites) by cyclin/Cdk complexes (Takeda et al. 2001).

6 Motif Discovery Algorithms and Tools

Given the diverse gamut of functions that are mediated by SLiMs, the development of methods and algorithms that will aid in (1) the discovery of new motifs (*de novo* motif prediction), and (2) filtering functional motif instances from the background of stochastic occurrences, is expected to be useful for identifying functional sites in proteins, especially within the unstructured segments. Usually motif discovery algorithms fall into three categories: enumeration, deterministic optimization and probabilistic optimization (D'Haeseleer 2006).

Enumeration is an exhaustive search based word counting method. The target sequences are broken up into shorter fragments (words of length 'n') and by counting the occurrence frequencies of all 'n-mers', the method attempts to identify statistically overrepresented short motifs. The highest occurrence frequency within the target sequences does not necessarily indicate a specific motif; statistical overrepresentation can be more reliably estimated by searching for motif patterns that appear more frequently than the random expectation (this random expectation is based on a background model that takes into account compositional biases). These steps need to be repeated several times until it finds statistically significant motifs. Further, by allowing mismatches and degeneracy in certain positions, consensus motifs can be defined in a more flexible and realistic manner. Alternatively, multiple overrepresented motifs that exhibit similarity may be combined into a single, more flexible motif. However, this method is computationally expensive because it requires the generation and storage of large numbers of short segments in memory.

Deterministic optimization is based on Expectation Maximization (EM). In the first step of EM, a PWM is initialized with a single n-mer segment of user-defined length ('n') along with some amount of background frequencies (nucleotides or amino acids). Next the input sequences are split into substrings (n-mers) and each substring then matched against the PWM. A probability value is calculated that indicates whether the substring was generated by the motif (PWM) model or by the background sequence distribution. Taking a weighted average of the current probabilities for each substring, the PWM is refined and the probabilities for the substrings then recalculated based on the updated PWM. The steps are repeated iteratively until a maximum likelihood motif model (PWM) is obtained. A well-known implementation of EM is the Multiple EM for Motif Elicitation (MEME) software (Bailey et al. 2006).

Finally, probabilistic optimization is based on Gibbs sampling. Briefly one motif from each input sequence is randomly selected to determine an initial model and a PSSM is built from those sub-strings. Then the PSSM is used to scan each input sequence to find a motif that better contributes to improve the PSSM quality; this new motif with higher PSSM score is then added to the model and the old motif is removed. This process is repeated until the PSSM reaches convergence. The algorithm assumes that most of the target sequences will contain the motif. *Align*s Nucleic *A*cid *C*onserved *E*lements (AlignACE) (Chen et al. 2008) is a program based on the Gibbs sampling approach and is used to discover motifs from sets of DNA sequences.

Many *de novo* motif discovery tools are currently available that are dedicated to discover motifs present in disordered protein regions. De novo discovery methods take as input the protein primary sequence and utilize features such as disordered structural environment and evolutionary context as pointers to reduce false positive matches (Davey et al. 2012b). Functional SLiMs have been characterized to be enriched within disordered regions of the proteome, motif residues can be distinguished from their sequence neighborhood on the basis of higher evolutionary conservation, and furthermore, SLiMs often exhibit a propensity to form ordered secondary structures upon partner binding (Davey et al. 2012b). These additional layers of information are therefore used to enhance the filtering and removal of false positive hits.

Additional strategies to improve true positive motif detection include: removal prior to input of sequence segments that are spurious for motif discovery (eg, masking repeat sequences and low complexity regions), and sequence regions that are poorly represented in SLiMs (such as well structured domains, transmembrane segments and poorly conserved segments). Furthermore, the use of multiple motif predictors that cover a range of motif descriptions and search algorithms, followed by a comparison of results is always recommended. Optimizing the runtime details such as motif width, expected number of motif occurrences, deciding cutoffs for various parameters also require careful consideration. Sometimes it may be useful to combine similar motifs into a smaller set of (more) flexible motif descriptions. Users should also consider multiple high scoring motifs as the top hit may not necessarily be the most biologically relevant. Finally, the chances of detecting a true functional motif are also maximized if one can reduce (based on available evidence) the number of sequences that are not likely to possess that functionality ("noise").

The Discovery@Bioware portal (http://bioware.ucd.ie/~compass/biowareweb/) and MEME Suite (http://meme.nbcr.net) contain a host of useful resources pertaining to the discovery, characterization and analysis of SLiMs (Table 9.3). The Eukaryotic Linear Motif (ELM) resource (http://elm.eu.org) has an extensive collection of curated SLiM instances, and is a useful tool for sequence annotation to identify protein segments that match known functional SLiMs. Regular expressions

Table 9.3 A list of commonly used motif discovery resources that enable motif prediction, discovery and analysis

Name	Description
SLiMProb	Searches for occurrences of pre-defined motifs (REs) in protein sequences (http://bioware.ucd.ie/~compass/biowareweb/)
SLiMSearch 3	Searches for occurrences of pre-defined motifs proteome wide (http://bioware.ucd.ie/~compass/biowareweb/)
SLiMPred	Predicts potential SLiMs in a protein sequence (http://bioware.ucd.ie/~compass/biowareweb/)
SLiMPrints	Predicts potential motifs by searching for clusters of locally conserved residues present in intrinsically disordered regions (http://bioware.ucd.ie/~compass/biowareweb/)
SLiMFinder	Identify SLiMs in a group of proteins (http://bioware.ucd.ie/~compass/biowareweb/)
GLAM2	Identify DNA or protein motifs using gapped local alignment (http://meme.nbcr.net)
MEME	Identify DNA or protein motifs using EM (http://meme.nbcr.net)
ELM	Database of experimentally validated SLiMs in eukaryotic proteins and a resource for investigating candidate functional SLiMs (http://elm.eu.org/)
MnM	Examines query protein for presence of short contiguous peptide sequences that have a known function in at least one protein (http://mnm.engr.uconn.edu/MNM/SMSSearchServlet)

representing the ELM classes are used by ELM's motif detection pipeline to scan proteins for putative SLiM instances (Davey et al. 2012a; Dinkel et al. 2012). Minimotif Miner (MnM, http://mnm.engr.uconn.edu/MNM/SMSSearchServlet) is also widely used for motif searches and analysis.

7 Details of Usage and Functionality of Some Selected Motif Discovery Tools

SLiMPrints (short linear motif fingerprints, currently at version 3.0) attempts to identify putative functional motifs from the input amino acid sequence on the basis of evolutionary conservation as a discriminatory feature for SLiM discovery (Davey et al. 2012a). Residue conservation statistics are analyzed and their significance estimated by comparison against the background conservation of neighboring residues. The method identifies relatively conserved (overconstrained) proximal residue clusters present within disordered regions; such "islands of conservation" located inside structurally unconstrained and mutation-prone disordered regions have been shown to be indicative of putatively functional SLiMs. The reader is referred to the original publication for a detailed description of the methodology (Davey et al. 2012a).

We demonstrate here how the user can provide input to the SLiMPrints web application (http://bioware.ucd.ie/~compass/biowareweb/Server_pages/slimprints.php), provide a brief overview of the methodology involved and finally, describe how the output is displayed and its contents. The user can analyse a protein of interest by providing the UniProt Accession of the protein into the search box (Fig. 9.5, "Query protein"). SLiMPrints contains pre-computed multiple sequence alignments of least divergent orthologs selected using the GOPHER algorithm, following a BLAST search for homologs against a database of EnsEMBL metazoan (plus *Saccharomyces cerevisiae*) genomes (Flicek et al. 2011). The alignments have been processed to increase their quality by the removal of potential biases (for example, low complexity regions in highly divergent proteins were removed from the alignments and alignments with identified orthologs in <10 metazoan species were not considered further) (Davey et al. 2012a). Further, regions shown to be deficient in motifs (annotated domains, transmembrane segments, extracellular regions and highly structured residues) are masked before the motif discovery step. Because the algorithm aims to identify regions of functional constraint (proximal clusters of strongly conserved residues) against a backdrop of evolutionary drift especially within disordered segments, relative local conservation (RLC) statistics (that measures residue conservation against the background conservation of a neighboring sequence window) are employed to obtain better information about the putative functionality of a motif region. SLiMPrints combines RLC and disorder predictions to identify putative SLiMs in the input sequence. Figure 9.5 illustrates an example SLiMPrints output using human p53 as the input sequence. The output contains the identified motifs ranked by their significance score (Sig_{motif} is a metric that represents the likelihood/significance of the observed grouping of highly conserved residues that form a putative sequence motif (Davey et al. 2012a)). The underlying alignment(s) corresponding to the respective motif regions can be visualized by clicking on the "view" links. The RE of the obtained motifs and their sequence context (with the motif start and end residue positions in the input sequence) are also printed. The average IUPred (Dosztanyi et al. 2005) disorder score of the motif is also output. Finally, if the obtained motif matches an annotated ELM identifier, the ELM entry is also shown.

SLiMFinder (Short, Linear Motif Finder) software/web server (http://bioware.ucd.ie/~compass/biowareweb/Server_pages/slimfinder.php) is intended to allow researchers to *de novo* discover novel SLiMs from a set of input sequences (Davey et al. 2010). The purpose is to identify shared motifs among a set of unrelated proteins that possess a common function suspected to be SLiM-mediated (eg, binding to a common protein partner). SLiMFinder accounts for evolutionary relationships amongst the input sequences by clustering them into unrelated protein clusters (UPCs), such that proteins separated into different clusters do not share any BLAST-detectable similarity (Altschul et al. 1990). An explicit model of convergent evolution is used whereby the method searches for SLiMs that are statistically overrepresented in a maximum number of proteins from the different UPCs. SLiMFinder combines two algorithms: (i) SLiMBuild, which performs the actual task of identifying recurring motifs, and, (ii) SLiMChance estimates the statistical significance of returned motifs. We refer the reader to the original publication for full details of the methodology involved (Edwards et al. 2007).

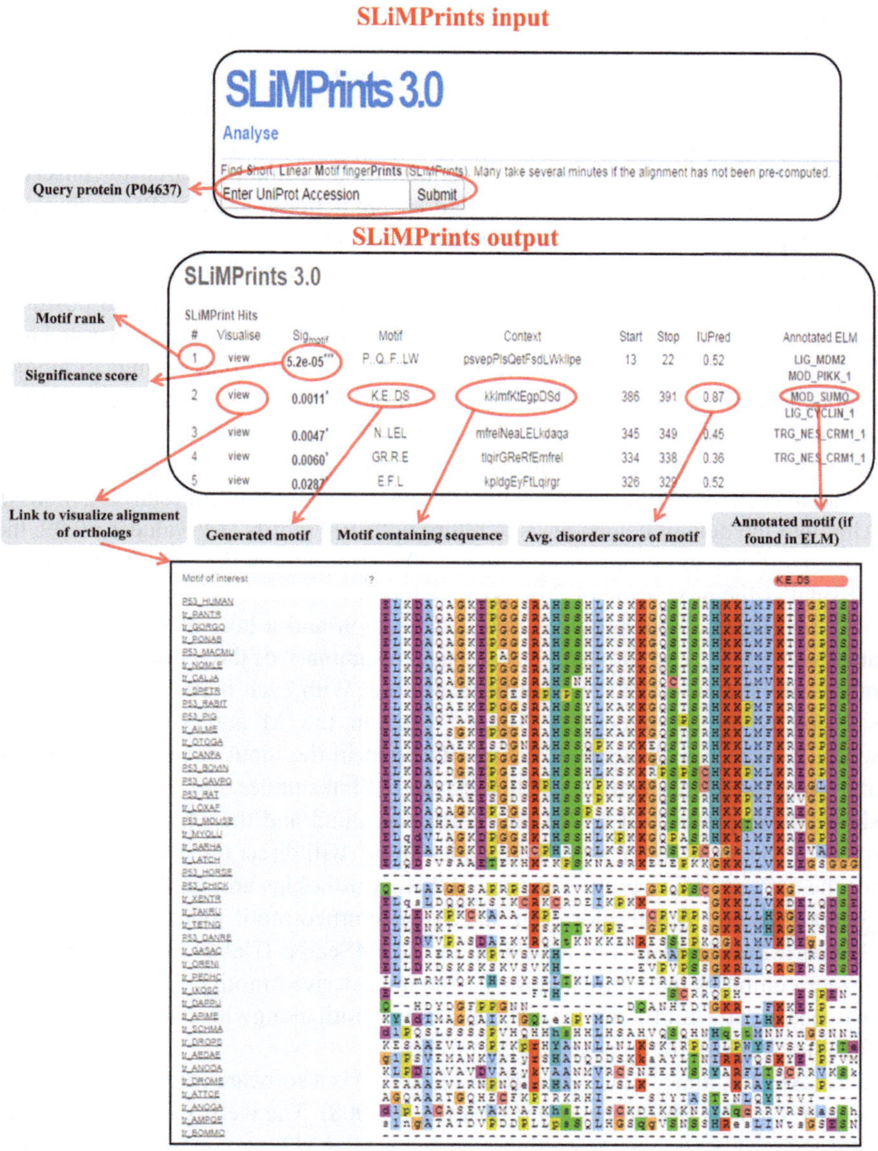

Fig. 9.5 SLiMPrints input and output. Input options: SLiMPrints takes as input a UniProt accession number (shown on the *top* panel). Output options: summarized results of SLiMPrint hits are initially displayed as shown below the input options panel. This section provides a summary of the identified motifs along with their main features (highlighted using the *red* ovals and the *red arrows*). The results specify the motif rank, a "Visualize" option (link to visualize alignment of orthologs, an example of which is shown in the *bottom* panel), "Sig_{motif}" (Significance score of the identified motif), "Motif" (Regular Expression of the observed motif), "Context" (motif containing sub-sequence), "IUPred" (average disorder score of the motif) and "Annotated ELM" (if the motif is found in ELM)

Figure 9.6 shows the SLiMFinder web server input page. The input may be a list of UniProt IDs or user-built sequence files in UniProt or FASTA format. Next to the input box are the lists of options (separately for 'Masking', 'SLiMBuild', 'SLiMChance' and 'Output') that the user can employ to fine tune searches. First, there are multiple options to mask out regions (from the input sequences) known to be depleted in SLiMs: users can exclude from the motif search unconserved residues, ordered regions (based on IUPred predictions) such as Pfam domains, low complexity regions as well as certain amino acid types. Next, SLiMBuild has options that specify the minimum and maximum number of consecutive wildcard positions that are to be permitted, the total number of allowed wildcard positions and the minimum number of input sequences that must contain each generated motif for it to be returned as a putative SLiM. Users will also find settings to modify residue groupings based on physicochemical or other parameters: these groupings are used to define ambiguous SLiM positions. Once a set of motifs is generated by SLiMBuild, the SLiMChance algorithm assigns a statistical significance score (P-value) to each motif (the user can select the significance cutoff for returning motifs). Although the default behaviour is to return upto 100 motifs at P-value $<=0.99$, the most significant motifs are those with $P<=0.05$ (the stricter the significance cutoff, the smaller the proportion of false positive hits).

SLiMFinder output provides rich visualization and a host of options for data analysis (Fig. 9.6). In the main output page, a summary of the returned (predicted) motifs are shown ranked by significance score. With each motif hit there are associated hyperlinks: under the "Aligned" column, the 'M' and 'A' alignment links will allow the visualization of the motif region in the input sequences ('masked' and 'unmasked', respectively). Clicking the red links under the "Proteins" column shows those proteins in which the motif was found and their position in the sequence. The small thumbnail figure under "Plot" will direct the user to alignments for the corresponding protein and its GOPHER orthologs around the region of the generated motif. Finally, for each putatively returned motif there are links to run CompariMotif (Edwards et al. 2008) and SLiMSearch (Davey et al. 2011a): the former compares the motif to known, literature-derived motifs, whereas the latter searches for all UniProt entries that contain this motif alongwith statistical estimates about the validity of the observed occurrence.

GLAM2 (*G*apped *L*ocal *A*lignment of *M*otifs) is a software for finding motifs in input (protein or DNA) sequences (Frith et al. 2008). The web version is located at http://meme.nbcr.net/meme/cgi-bin/glam2.cgi. GLAM2 examines the set of input sequences for common motifs and finds a motif alignment with maximum score. GLAM2 enables the detection of gapped (*ie*, with indels) motifs. The algorithm starts from an initial random alignment constructed from the input sequences and uses simulated annealing to make repetitive changes to it. These changes are random and they affect the motif score (which can either increase or decrease), the idea being to prevent the system from being trapped in local optima. The changes are applied iteratively until the score fails to improve further even after 'n' successive changes ($n=10,000$ by default). The types of changes that are possible and their details are beyond the scope of this chapter and the reader is referred to the original

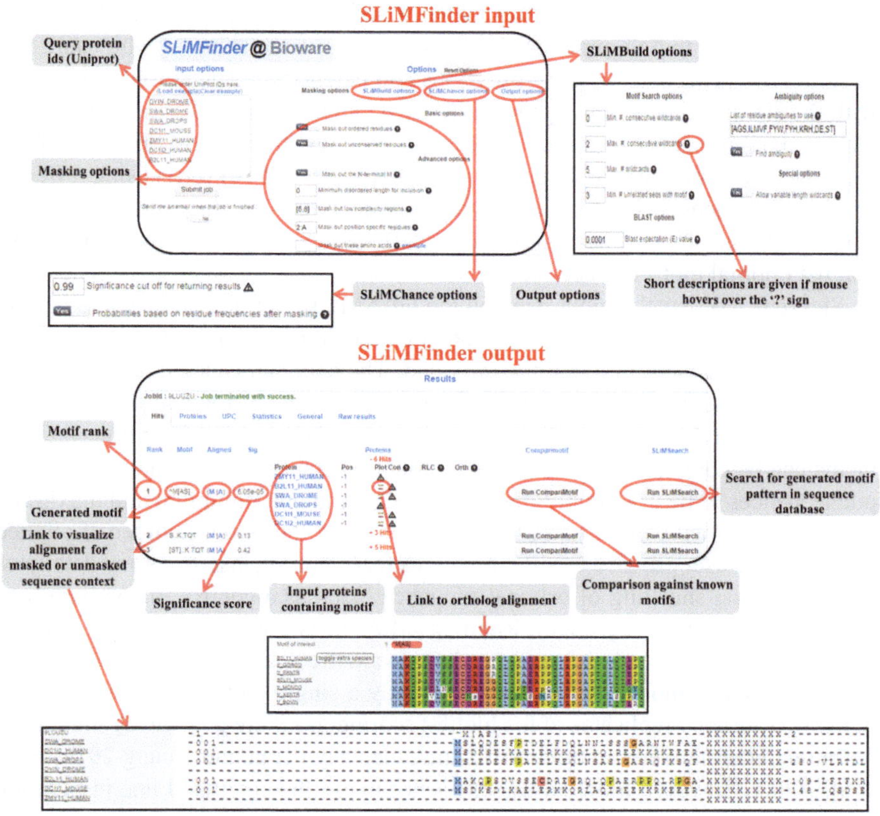

Fig. 9.6 SLiMFinder input and output. Input options are shown on the *top*. Input is a list of UniProt identifiers corresponding to the set of proteins in which we want to discover common (shared) motifs. Options are categorized into the following sub-sections: "Masking", "SLiMBuild", "SLiMChance" and "Output" options (shown using the red ovals and *arrows*). The web server provides short descriptions for each option if the user hovers the mouse over the '?' sign next to each option. Output: summarized results are initially displayed (shown in the panel below the input options). This section outputs the "Rank" (motif rank), "Motif" (RE of the generated motif), "Aligned (M|A)" (links to visualize motif alignments for masked or unmasked sequence context, an example is shown on the bottom most panel), "Sig" (motif significance score), "Proteins" (list of input proteins that contain the motif). Under the "Proteins" header, the user will see in red the number of proteins containing each predicted motif. By clicking on the number of hits, the output will expand to show the names of those proteins from the input list that contain the motif in question. Each protein can then be further analyzed for that motif based on the conservation statistics (for example by clicking on "Link to ortholog alignment"). Finally, "Run CompariMotif" (comparison against known motifs) and "Run SLiMSearch" (search for generated motif pattern in sequence databases) functions are also available for each predicted motif

publication (Frith et al. 2008). Essentially, GLAM2 builds on the idea that motifs contain a certain number of "key positions" defined by strict residue preferences at highly conserved and therefore presumed to be functional sites. The algorithm optimizes the number of key positions and then searches for an alignment of substrings (one from each input sequence) to match a series of key positions. Thus in the scoring scheme, the alignments of identical or similar residues in the same key positions are rewarded, whereas insertions and deletions are penalised. Ultimately with the simulated annealing approach GLAM2 attempts to find a motif alignment with maximum score. To cross-check that a reproducible, high-scoring motif has been identified, the steps are repeated multiple (by default 10) times using different starting alignments selected randomly by the program. The algorithm then checks whether similar (but not necessarily identical) alignments recur. This is suggestive that the optimal motif has been found.

Figure 9.7 shows the input page on the GLAM2 server and an example output. Input can be either in the form of a text file containing the input sequences or by pasting the sequences into the box provided. The user can check details about the input formatting by clicking on the links (colored cyan) just above the input box. There are several parameters that can be customized (Fig. 9.7). The allowed alignments can be constrained by specifying variables such as: minimum number of input sequences to be used in building the motif alignment, minimum and maximum number of aligned columns (*ie*, key positions), and the initial number of aligned columns. The user can also modify the scores for tolerating insertions and deletions, and turn off/on shuffling of original sequence (used as a control to compare with the score of original sequence). Running GLAM2 is computationally heavy and the analysis time depends on sequence length and the size of the input dataset. One feature of this method is that it can detect only a single motif at a given time (by default 10 variants/replicates of the highest scoring motif are generated) and it does not model alternative binding motifs simultaneously (Tran and Huang 2014; Frith et al. 2008). However, more advanced users can use the command line installation to detect alternate (weaker) motifs, by first masking the strongest identified motif region (using the program 'glam2mask') and then re-running GLAM2.

The output is provided in three different formats: html, text and MEME text format. Figure 9.7 (*bottom*) shows a screenshot from the html output page. Because GLAM2 attempts to find the strongest motif in the set of input sequences using a 'replication strategy', if the top ranking motifs are very similar to each other, it is an indication that a successful replication has been achieved. Thus, by default GLAM2 outputs 10 variations of the strongest motif shared by the input sequences (this value "number of alignment replicates" can be changed by the user). Thus the topmost/first alignment is the interesting one: the purpose of the others is to indicate the reproducibility of the first motif. The output contains the list of motifs with maximum score and corresponding alignments of the motif containing segments (only the first one is shown in Fig. 9.7), their start and end positions, marginal score for each motif segment (this reflects the amount by which the total alignment score would decrease if that segment were to be removed from the alignment; thus, higher scores reflect better matches to the motif), and finally, the motif sequence logo. For each candidate motif, GLAM2 has additional options including, for example, scanning the motif against sequence databases (using GLAM2SCAN). The HTML output page also provides a link to view the *P*osition *S*pecific *P*robability *M*atrix (PSPM).

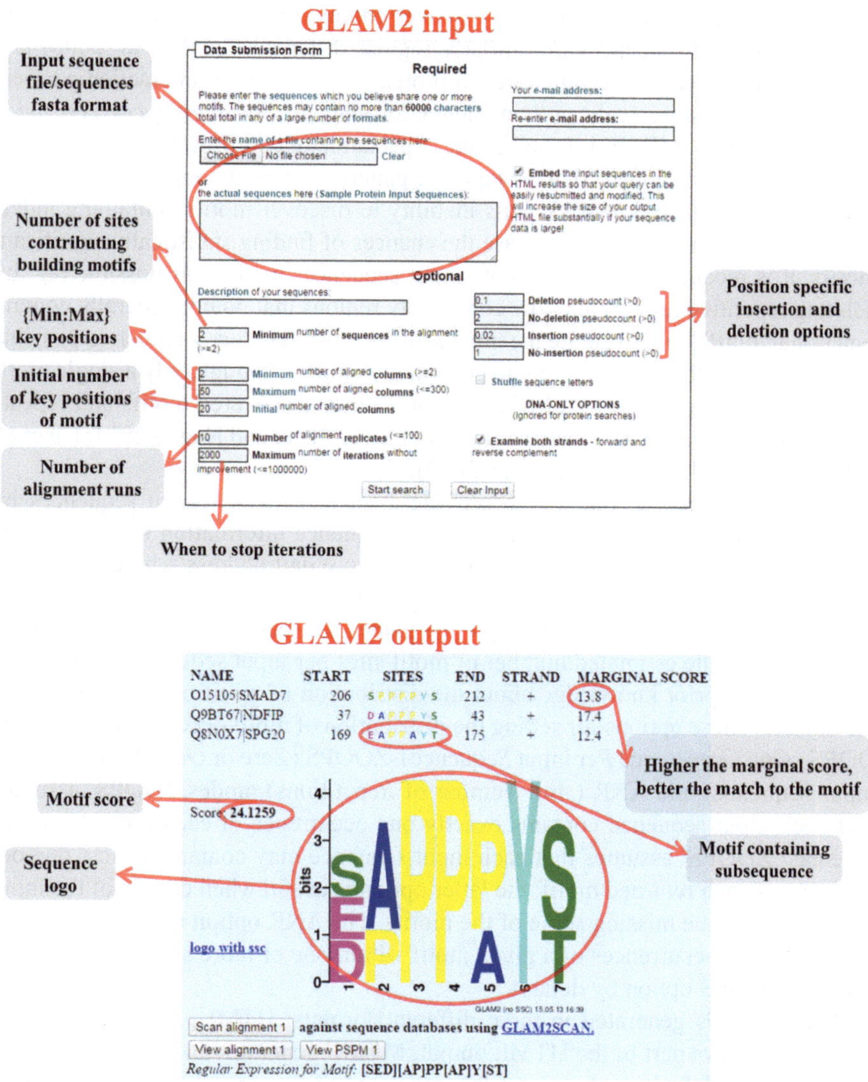

Fig. 9.7 GLAM2 input and output. Input (*top* panel) is accepted in fasta format. The available input options are shown using *red arrows*. These include options to specify the number of sites contributing to the motif (if known), number of key positions (maximum, minimum and initial number), maximum number of iterations and position specific insertion and deletion penalty scores. Output (*bottom* panel) showing the best statistically significant motif and a list of motif occurrences in the input dataset, their start and end positions, and marginal score followed by the motif logo. Hyperlinked buttons ("Scan alignment", "View alignment" and "View PSPM") that allow the motif to be analysed are shown at the bottom

MEME *M*ultiple *E*M for *M*otif *E*licitation (MEME) is a widely used tool for searching novel 'signals' in sets of biological sequences (Bailey et al. 2006); the webserver version is available on MEME suite (http://meme.nbcr.net/meme/cgi-bin/meme.cgi).

MEME has been used previously to discover common transcription factor binding sites in promoter sequences of similarly regulated genes (Lyons et al. 2000) and to identify novel sequence signatures in proteins with common interaction partners identified from large scale protein interaction data in *Saccharomyces cerevisiae* (Fang et al. 2005). MEME is based on the expectation maximization (EM) algorithm and it looks for ungapped, shared sequence patterns within the input (DNA or protein) sequences. One drawback is its inability to discover motifs containing indels as it does not allow gaps. To increase the chances of finding statistically significant motifs, it is recommended to keep the input sequences as short as possible (eg, by deleting repetitive regions and low complexity regions that do not generally contain functional motifs) and to curate the input sequence list to reduce as much as possible those sequences that are not likely to contain the motif. Although only a single motif can be modeled at a time, MEME erases previously discovered motifs and repeats the search, this enables new patterns to be extracted (Tran and Huang 2014; Hu et al. 2005; Bailey et al. 2006; Bailey et al. 2009).

For web server use, one has to provide a set of FASTA format sequences by either uploading a text file or by pasting the sequence information into the box as shown in Fig. 9.8. The other required input is an email address where the results will be sent. MEME searches for motifs ranging from 6 and 50 residues in length by default, although the user can specify other values between $\{2,300\}$. There is an option to specify the estimated number of motif sites per input sequence, particularly if there is any prior knowledge about the distribution of motif occurrences within the dataset. These options for setting the distribution of motif occurrences are called OOPS (*O*ne *O*ccurrence *P*er input *S*equence), ZOOPS (*Z*ero or *O*ne *O*ccurrence *P*er input *S*equence) and ANR (*A*ny *N*umber of *R*epetitions) modes. 'OOPS' assumes that each input sequence contains exactly one occurrence of each returned motif, whereas 'ZOOPS' assumes that each input sequence may contain at most one occurrence of each returned motif; the latter option is useful when certain of the input sequences may be missing some of the motifs. The ANR option can be used to explore multiple occurrences of a given motif within one or more sequences. MEME uses the ZOOPS option by default.

The output is generated in three different formats: HTML, TEXT and XML. Figure 9.8 shows part of the HTML output. MEME generates up to three top-ranking motifs by default, and each of the generated motifs may be present in either a subset of sequences or in all the input sequences (this refers to the number of occurrences). Every output motif is assigned an 'E-value'. The E-value refers to the probability of finding an equally well-conserved sequence pattern in random sequences; thus, the lower the E-value, the greater the statistical significance of the observed motif. The output overview shows the rank of the motif, its E-value and number of occurrences (sites) and the sequence logo for the motif. Below the "Motif Overview" section, further details about each of the identified motifs are available. This includes the multiple alignments showing the identified motif region in the

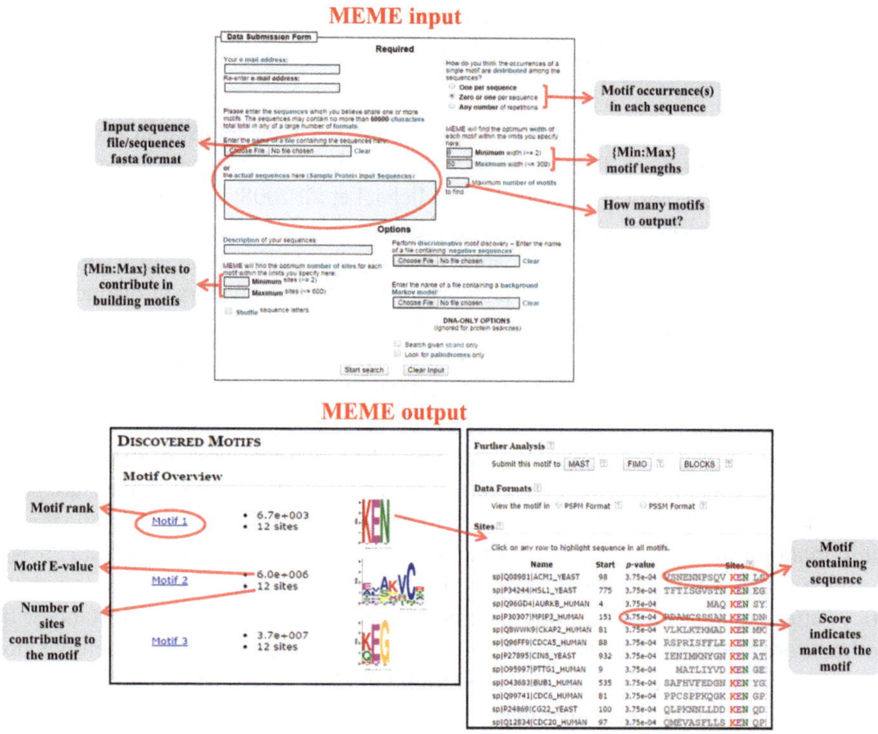

Fig. 9.8 MEME input and output. Input options are shown in the *top* panel. There are options to include the number of sites for each motif (if there is prior knowledge about the number of occurrences), and options to specify motif length. Output (*bottom left*) showing a list of protein motifs (by default 3 motifs) that MEME has discovered in the input sequences. Some of the hyperlinked buttons that allow the motif to be analysed further are shown at the *bottom right*

input sequences (Fig. 9.8, bottom right panel). Below the alignments are so-called "Block diagrams" showing the relative positions of the motifs within the input sequences (not displayed in the figure). Clickable buttons allow each motif to be analysed by other programs. Clicking on the 'MAST' (*Motif Alignment* and *Search Tool*) button will send the motif to the MAST web server where various sequence databases (or sets of user-uploaded sequences) can be searched for sequences that contain matches to that motif. Similarly, the button 'FIMO' (*Find Individual Motif Occurrences*) (Grant et al. 2011) will also trigger searches of sequence databases for hits to the motif patterns. Finally, these motifs may be compared against entries in the BLOCKS database of protein motifs (Henikoff et al. 1999) by clicking on the 'BLOCKS' button.

8 Prediction Performance on Disordered Motifs: Case Study on the KEN-box Motif

KEN-box mediated target selection is one of the mechanisms used in proteasomal destruction of mitotic cell cycle regulatory proteins via the Anaphase-promoting complex (APC/C complex) (Peters 2006; Michael et al. 2008; Pfleger and Kirschner 2000). 'KEN' motifs are significantly enriched in proteins with cell cycle keywords and further the KEN-box is significantly conserved throughout the eukaryotic taxon (Michael et al. 2008). Cdh1 and Cdc20 act as APC/C co-activators at distinct stages of the cell cycle. Cdc20 interacts with the APC complex during the M phase and is later replaced by Cdh1 (late M/G1 transition). Whereas both Cdh1 and Cdc20 can recognise target proteins via the Destruction Box (D-box) motif, the KEN-box is only recognised by Cdh1. Interestingly Cdc20 itself contains a KEN-box that is identified by Cdh1 and undergoes temporal degradation; Cdh1 then replaces Cdc20 as the adaptor of the APC complex. However Cdh1 contains two D-box motifs that ensure self-degradation of Cdh1 via APC/C in an auto-regulatory feedback mechanism; this is important for tuning the levels of active Cdh1 throughout G1 (Listovsky et al. 2004).

Motif discovery algorithms have to deal with the problem of spurious (stochastic) pattern matches that turn out to be non-functional (false positive) instances. In other words, merely observing a KEN pattern within a protein sequence does not necessarily indicate a functional degradation targeting motif. Many factors including protein cellular compartmentalization, tertiary structure and motif accessibility, etc regulate interaction of the KEN-containing protein with APC/C. All the functional KEN-box motifs discovered so far have been found within natively unfolded (disordered) regions of proteins; however, certain proteins (eg, HIPK4) carry a KEN-motif within a globular domain although their role in proteasomal degradation is unknown (Michael et al. 2008).

KEN-box instances were collected from the ELM database: 16 instances from 14 proteins were found classified as true positives (Dinkel et al. 2014). Table 9.4 shows their prediction performance using the 4 motif discovery algorithms discussed in the previous section. Whereas SLiMPrints analyzes every protein individually, the other methods (SLiMFinder, GLAM2 and MEME) take a set of sequences as input. Thus the complete set of 14 sequences carrying validated KEN motifs were supplied as input. With each method, we always tried the default settings first to evaluate how well these parameters performed. Any modifications that were necessary are mentioned at the appropriate places in the following description.

Of the 16 known instances, *SLiMPrints* returned 9 instances as significant hits ($P<0.05$) that either completely or partially overlapped with the known KEN box and were recognized as being similar to the ELM entry LIG_APCC_KENbox_2. For two proteins ('CIN8_YEAST' and 'VE1_BPV1') it completely failed to predict the KEN-boxes. In case of the viral protein 'VE1_BPV1', this failure may have been due to the fact that SLiMPrints has been trained on the EnsEMBL (Flicek et al. 2014) metazoan and *Saccharomyces cerevisiae* genomes, and therefore it is unable to predict for viral proteins. For 'CIN8_YEAST' the program resulted in an error message.

Table 9.4 Prediction accuracy on the KEN-box (.KEN.) motif using four motif discovery algorithms ('Yes' indicates that the motif was successfully identified, 'No' that the method failed to identify the motif; '*' indicates that the KEN motif was returned by the algorithm as a significant hit; (Number) indicates the rank obtained for the predicted motif)

KEN-box containing proteins			Motif discovery methods used			
Protein name	Gene name	Start,End	SLiMPrints[a]	SLiM-Finder[b]	GLAM2[c]	MEME[c]
ACM1_YEAST	ACM1	97,101	Yes*(2)	Yes*(10)	Yes(1)	Yes(1)
AURKB_HUMAN	AURKB	3,7	Yes(5)	Yes*(10)	Yes(1)	Yes(1)
BUB1_HUMAN	BUB1	534,538	Yes*(3)	Yes*(10)	No	Yes(1)
BUB1_HUMAN	BUB1	624,628	Yes(16)	Yes*(10)	Yes(1)	No
BUB1B_HUMAN	BUB1B	25,29	Yes(25)	Yes*(10)	Yes(1)	Yes(1)
BUB1B_HUMAN	BUB1B	303,307	Yes*(9)	Yes*(10)	No	No
CDC20_HUMAN	CDC20	96,100	Yes(20)	Yes*(10)	Yes(1)	Yes(1)
CDC6_HUMAN	CDC6	80,84	Yes*(1)	Yes*(10)	Yes(1)	Yes(1)
CDCA5_HUMAN	CDCA5	87,91	Yes*(2)	Yes*(10)	Yes(1)	Yes(1)
CG22_YEAST	CLB2	99,103	Yes*(1)	Yes*(10)	Yes(1)	Yes(1)
CIN8_YEAST	CIN8	931,935	No	Yes*(10)	Yes(1)	Yes(1)
CKAP2_HUMAN	CKAP2	80,84	Yes*(1)	Yes*(10)	Yes(1)	Yes(1)
HSL1_YEAST	HSL1	774,778	Yes*(8)	Yes*(10)	Yes(1)	Yes(1)
MPIP3_HUMAN	CDC25C	150,154	Yes(19)	Yes*(10)	Yes(1)	Yes(1)
PTTG1_HUMAN	PTTG1	8,12	Yes*(3)	Yes*(10)	Yes(1)	Yes(1)
VE1_BPV1	E1	27,31	No	Yes*(10)	Yes(1)	Yes(1)

[a] SLiMPrints accepts a single protein sequence at a time and provides the score for the identified motif

[b] SLiMFinder can take multiple sequences simultaneously as input. SLiMFinder can either use the complete set of input sequences or automatically selects a subset thereof such that a high confidence motif can be generated. For this example SLiMFinder returned a list of 11 significant motifs, KEN motif was found in 10th position. Two similar motifs ('KEN..D' and 'KEN.{1,2}P') were ranked at 4th and 6th positions respectively

[c] Although GLAM2 and MEME can optimize how many sequences to use in order to obtain significant candidate motifs, in this case study both methods were controlled to use all 14 input sequences simultaneously. This was meaningful because in this particular example we knew beforehand that all the input sequences contained a true positive KEN motif

SLiMFinder performed significantly well on the dataset using default parameters. SLiMFinder outputs a list of candidate motifs identified from the set of input sequences ranked by their significance score. We found a KEN motif (with a significance score of 0.002) at rank 10 that contained all 16 KEN instances. Interestingly, two higher ranking motifs that closely resembled the KEN were also found: KEN.P ranked #4 (Sigscore=6.96E-5) and KEN.{1,2}P ranked #6 (Sigscore=9.53E-5). These two motifs contained 9 and 10 respectively of the total KEN instances present in the input dataset.

GLAM2 initially failed to detect the KEN-motif in the input set. The following parameters were used (all default settings, except for the number of motif containing sequences, which we knew beforehand to be 14): –z 14 (number of sequences), –a 2 (minimum width of motif), –b 50 (maximum width of motif), –w 20 (initial number of 'key positions'), and –n 2000 (number of iterations). On reflection, we felt that there was a mismatch between the length of the KEN motif and the value used for the "initial number of key positions" parameter; accordingly, we modified this to a low value consistent with the length of the motif being searched (*ie*, w=2). This enabled GLAM2 to successfully identify 14 out of the 16 motif instances (Table 9.4). BUB1_HUMAN and BUB1B_HUMAN each contain 2 validated KEN-boxes, however only one from each protein was identified (since GLAM2 assumes that every input sequence may contain at most one occurrence of each motif). Further, we tested different values of 'w', and all values in the range [2, 15] were successfully able to recover 14 instances (one from each input sequence).

MEME also did an excellent job of discovering KEN-box motifs in the ELM benchmark dataset. It successfully identified 14 of the 16 instances using the following parameters: –minw 6 (minimum width of motif), –maxw 50 (maximum width of motif), –minsites 14 (minimum number of motifs), –maxsites 14 (maximum number of motifs), and –mod zoops (zero or one occurrences). The 'minsites' and 'maxsites' values were set to 14 since the number of motif occurrences in the dataset were already known (default values were used for all the other parameters). However, MEME failed to identify the second motifs of 'BUB1_HUMAN' (624, 628) and 'BUB1B_HUMAN' (303, 307) because the 'zoops' mode assumes that each input sequence may contain at most one occurrence of each motif. Although we knew that these two sequences contained 2 KEN-boxes each, there is no parameter setting on the input page where we could set the number of motif occurrences exactly to 2. We did however use the ANR (*A*ny *N*umber of *R*epetitions) option to try and detect the multiple motifs. However, this option resulted in a large number of false positive hits and even so the multiple KEN's in both BUB1 and BUB1B remained unidentified.

9 Limitations of Motif Discovery Algorithms

Although motif discovery algorithms have improved considerably over the past years, considerable challenges remain. For example, since a large majority of motif types have been characterized to be preferentially located in disordered protein segments, one main challenge will be to design effective multiple sequence alignment tools that can efficiently align intrinsically disordered regions. However, it can also be argued that by focusing mostly on IDRs and by routinely masking out structured domains we might miss finding (some) novel SLiMs. On the other hand, another level of complexity is introduced if we include domain sequences in the alignments used for motif discovery. The strong similarities between domain sequences would hide the weak SLiM signals. Although it is difficult to estimate how frequently functional SLiMs may occur within domains (eg, on their surface regions), this might be an avenue to explore in the future. Another limitation of motif discovery algorithms is their unsuitability to take entire genomes as input to discover motifs. Especially with short length motifs, their statistical significance in the context of the entire proteome is difficult to establish. Therefore, motif discovery tools need to be improved further to be able to discover the full complement of short linear motifs in the proteome.

References

Akiva E, Friedlander G, Itzhaki Z et al (2012) A dynamic view of domain-motif interactions. PLoS Comput Biol 8(1):e1002341. doi:10.1371/journal.pcbi.1002341

Altschul SF, Gish W, Miller W et al (1990) Basic local alignment search tool. J Mol Biol 215(3):403–410. doi:10.1016/S0022-2836(05)80360-2

Bailey TL (2008) Discovering sequence motifs. Methods in Mol Biol 452:231–251. doi:10.1007/978-1-60327-159-212

Bailey TL, Williams N, Misleh C et al (2006) MEME: discovering and analyzing DNA and protein sequence motifs. Nucleic Acids Res 34(Web Server issue):W369–373. doi:10.1093/nar/gkl198

Bailey TL, Boden M, Buske FA et al (2009) MEME SUITE: tools for motif discovery and searching. Nucleic Acids Res 37(Web Server issue):W202–208. doi:10.1093/nar/gkp335

Berman HM, Kleywegt GJ, Nakamura H et al (2013) The future of the protein data bank. Biopolymers 99(3):218–222. doi:10.1002/bip.22132

Bernier-Villamor V, Sampson DA, Matunis MJ et al (2002) Structural basis for E2-mediated SUMO conjugation revealed by a complex between ubiquitin-conjugating enzyme Ubc9 and RanGAP1. Cell 108(3):345–356

Bhattacharyya RP, Remenyi A, Yeh BJ et al (2006) Domains, motifs, and scaffolds: the role of modular interactions in the evolution and wiring of cell signaling circuits. Annu Rev Biochem 75:655–680. doi:10.1146/annurev.biochem.75.103004.142710

Brett TJ, Traub LM, Fremont DH (2002) Accessory protein recruitment motifs in clathrin-mediated endocytosis. Structure 10(6):797–809

Chen X, Guo L, Fan Z et al (2008) W-AlignACE: an improved Gibbs sampling algorithm based on more accurate position weight matrices learned from sequence and gene expression/ChIP-chip data. Bioinformatics 24(9):1121–1128. doi:10.1093/bioinformatics/btn088

Corti A, Curnis F (2011) Isoaspartate-dependent molecular switches for integrin-ligand recognition. J Cell Sci 124(Pt 4):515–522. doi:10.1242/jcs.077172

Davey NE, Shields DC, Edwards RJ (2009) Masking residues using context-specific evolutionary conservation significantly improves short linear motif discovery. Bioinformatics 25(4):443–450. doi:10.1093/bioinformatics/btn664

Davey NE, Haslam NJ, Shields DC et al (2010) SLiMFinder: a web server to find novel, significantly over-represented, short protein motifs. Nucleic Acids Res 38(Web Server issue):W534–539. doi:10.1093/nar/gkq440

Davey NE, Haslam NJ, Shields DC et al (2011a) SLiMSearch 2.0: biological context for short linear motifs in proteins. Nucleic Acids Res 39(Web Server issue):W56–60. doi:10.1093/nar/gkr402

Davey NE, Trave G, Gibson TJ (2011b) How viruses hijack cell regulation. Trends Biochem Sci 36(3):159–169. doi:10.1016/j.tibs.2010.10.002

Davey NE, Cowan JL, Shields DC et al (2012a) SLiMPrints: conservation-based discovery of functional motif fingerprints in intrinsically disordered protein regions. Nucleic Acids Res 40(21):10628–10641. doi:10.1093/nar/gks854

Davey NE, Van Roey K, Weatheritt RJ et al (2012b) Attributes of short linear motifs. Mol Biosyst 8(1):268–281. doi:10.1039/c1mb05231d

D'Haeseleer P (2006) How does DNA sequence motif discovery work? Nat Biotechnol 24(8):959–961. doi:10.1038/nbt0806-959

Dinkel H, Michael S, Weatheritt RJ et al (2012) ELM—the database of eukaryotic linear motifs. Nucleic Acids Res 40(Database issue):D242–D251. doi:10.1093/nar/gkr1064

Dinkel H, Van Roey K, Michael S et al (2014) The eukaryotic linear motif resource ELM: 10 years and counting. Nucleic Acids Res 42(Database issue):D259–D266. doi:10.1093/nar/gkt1047

Disfani FM, Hsu WL, Mizianty MJ et al (2012) MoRFpred, a computational tool for sequence-based prediction and characterization of short disorder-to-order transitioning binding regions in proteins. Bioinformatics 28(12):i75–i83. doi:10.1093/bioinformatics/bts209

Dosztanyi Z, Csizmok V, Tompa P et al (2005) IUPred: web server for the prediction of intrinsically unstructured regions of proteins based on estimated energy content. Bioinformatics 21(16):3433–3434. doi:10.1093/bioinformatics/bti541

Edwards RJ, Davey NE, Shields DC (2007) SLiMFinder: a probabilistic method for identifying over-represented, convergently evolved, short linear motifs in proteins. PloS ONE 2(10):e967. doi:10.1371/journal.pone.0000967

Edwards RJ, Davey NE, Shields DC (2008) CompariMotif: quick and easy comparisons of sequence motifs. Bioinformatics 24(10):1307–1309. doi:10.1093/bioinformatics/btn105

Edwards RJ, Davey NE, O'Brien K et al (2012) Interactome-wide prediction of short, disordered protein interaction motifs in humans. Mol Biosyst 8(1):282–295. doi:10.1039/c1mb05212h

Fang J, Haasl RJ, Dong Y et al (2005) Discover protein sequence signatures from protein-protein interaction data. BMC Bioinformatics 6:277. doi:10.1186/1471-2105-6-277

Finn RD, Bateman A, Clements J et al (2014) Pfam: the protein families database. Nucleic Acids Res 42(Database issue):D222–D230. doi:10.1093/nar/gkt1223

Flicek P, Amode MR, Barrell D et al (2011) Ensembl 2011. Nucleic Acids Res 39(Database issue):D800–D806. doi:10.1093/nar/gkq1064

Flicek P, Amode MR, Barrell D et al (2014) Ensembl 2014. Nucleic Acids Res 42(Database issue):D749–D755. doi:10.1093/nar/gkt1196

Frith MC, Saunders NF, Kobe B et al (2008) Discovering sequence motifs with arbitrary insertions and deletions. PLoS Comput Biol 4(4):e1000071. doi:10.1371/journal.pcbi.1000071

Fuxreiter M, Tompa P, Simon I (2007) Local structural disorder imparts plasticity on linear motifs. Bioinformatics 23(8):950–956. doi:10.1093/bioinformatics/btm035

Gibson TJ (2009) Cell regulation: determined to signal discrete cooperation. Trends Biochem Sci 34(10):471–482. doi:10.1016/j.tibs.2009.06.007

Glickman MH, Ciechanover A (2002) The ubiquitin-proteasome proteolytic pathway: destruction for the sake of construction. Physiol Rev 82(2):373–428. doi:10.1152/physrev.00027.2001

Gould CM, Diella F, Via A et al (2010) ELM: the status of the 2010 eukaryotic linear motif resource. Nucleic Acids Res 38(Database issue):D167–D180. doi:10.1093/nar/gkp1016

Grant CE, Bailey TL, Noble WS (2011) FIMO: scanning for occurrences of a given motif. Bioinformatics 27(7):1017–1018. doi:10.1093/bioinformatics/btr064

Habchi J, Tompa P, Longhi S et al (2014) Introducing protein intrinsic disorder. Chem Rev 114(13):6561–6588. doi:10.1021/cr400514h

Hagen T, Vidal-Puig A (2002) Characterisation of the phosphorylation of beta-catenin at the GSK-3 priming site Ser45. Biochem Biophys Res Commun 294(2):324–328. doi:10.1016/S0006-291×(02)00485-0

Henikoff JG, Henikoff S, Pietrokovski S (1999) New features of the Blocks Database servers. Nucleic Acids Res 27(1):226–228
Hospital V, Chesneau V, Balogh A et al (2000) N-arginine dibasic convertase (nardilysin) isoforms are soluble dibasic-specific metalloendopeptidases that localize in the cytoplasm and at the cell surface. Biochem J 349(Pt 2):587–597
Hu J, Li B, Kihara D (2005) Limitations and potentials of current motif discovery algorithms. Nucleic Acids Res 33(15):4899–4913. doi:10.1093/nar/gki791
Janin J, Bahadur RP, Chakrabarti P (2008) Protein-protein interaction and quaternary structure. Q Rev Biophys 41(2):133–180. doi:10.1017/S0033583508004708
Kadaveru K, Vyas J, Schiller MR (2008) Viral infection and human disease–insights from minimotifs. Front Biosci: A J Virt Lib 13:6455–6471
Listovsky T, Oren YS, Yudkovsky Y et al (2004) Mammalian Cdh1/Fzr mediates its own degradation. EMBO J 23(7):1619–1626. doi:10.1038/sj.emboj.7600149
London N, Raveh B, Schueler-Furman O (2012) Modeling peptide-protein interactions. Methods Mol Biol 857:375–398. doi:10.1007/978-1-61779-588-617
Lyons TJ, Gasch AP, Gaither LA et al (2000) Genome-wide characterization of the Zap1p zinc-responsive regulon in yeast. Proc Natl Acad Sci U S A 97(14):7957–7962
Masson N, Ratcliffe PJ (2003) HIF prolyl and asparaginyl hydroxylases in the biological response to intracellular O(2) levels. J Cell Sci 116(Pt 15):3041–3049. doi:10.1242/jcs.00655
Mi T, Merlin JC, Deverasetty S et al (2012) Minimotif Miner 3.0: database expansion and significantly improved reduction of false-positive predictions from consensus sequences. Nucleic Acids Res 40(Database issue):D252–D260. doi:10.1093/nar/gkr1189
Michael S, Trave G, Ramu C et al (2008) Discovery of candidate KEN-box motifs using cell cycle keyword enrichment combined with native disorder prediction and motif conservation. Bioinformatics 24(4):453–457. doi:10.1093/bioinformatics/btm624
Min JH, Yang H, Ivan M et al (2002) Structure of an HIF-1alpha -pVHL complex: hydroxyproline recognition in signaling. Science 296(5574):1886–1889. doi:10.1126/science.1073440
Mohan A, Oldfield CJ, Radivojac P et al (2006) Analysis of molecular recognition features (MoRFs). J Mol Biol 362(5):1043–1059. doi:10.1016/j.jmb.2006.07.087
Moult J, Fidelis K, Kryshtafovych A et al (2014) Critical assessment of methods of protein structure prediction (CASP)–round x. Proteins 82(Suppl 2):1–6. doi:10.1002/prot.24452
Neduva V, Russell RB (2005) Linear motifs: evolutionary interaction switches. FEBS lett 579 (15):3342–3345. doi:10.1016/j.febslet.2005.04.005
Obenauer JC, Cantley LC, Yaffe MB (2003) Scansite 2.0: Proteome-wide prediction of cell signaling interactions using short sequence motifs. Nucleic Acids Res 31(13):3635–3641
Peters JM (2006) The anaphase promoting complex/cyclosome: a machine designed to destroy. Nat Rev Mol Cell Biol 7 (9):644–656. doi:10.1038/nrm1988
Petsalaki E, Russell RB (2008) Peptide-mediated interactions in biological systems: new discoveries and applications. Curr Opin Biotechnol 19(4):344–350. doi:10.1016/j.copbio.2008.06.004
Pfleger CM, Kirschner MW (2000) The KEN box: an APC recognition signal distinct from the D box targeted by Cdh1. Genes Dev 14 (6):655–665
Van Roey K, Dinkel H, Weatheritt RJ et al (2013) The switches. ELM resource: a compendium of conditional regulatory interaction interfaces. Sci Signal 6(269):rs7. doi:10.1126/scisignal.2003345
Van Roey K, Uyar B, Weatheritt RJ et al (2014) Short linear motifs: ubiquitous and functionally diverse protein interaction modules directing cell regulation. Chem Rev 114(13):6733–6778. doi:10.1021/cr400585q
Schneider TD, Stephens RM (1990) Sequence logos: a new way to display consensus sequences. Nucleic Acids Res 18(20):6097–6100
Schon O, Friedler A, Bycroft M et al (2002) Molecular mechanism of the interaction between MDM2 and p53. J Mol Biol 323(3):491–501
Sigrist CJ, de Castro E, Cerutti L et al (2013) New and continuing developments at PROSITE. Nucleic Acids Res 41(Database issue):D344–D347. doi:10.1093/nar/gks1067

Takeda DY, Wohlschlegel JA, Dutta A (2001) A bipartite substrate recognition motif for cyclin-dependent kinases. J Biol Chem 276(3):1993–1997. doi:10.1074/jbc.M005719200
Tompa P (2012) Intrinsically disordered proteins: a 10-year recap. Trends Biochem Sci 37(12):509–516. doi:10.1016/j.tibs.2012.08.004
Tompa P, Davey NE, Gibson TJ et al (2014) A million peptide motifs for the molecular biologist. Mol Cell 55(2):161–169. doi:10.1016/j.molcel.2014.05.032
Tran NT, Huang CH (2014) A survey of motif finding Web tools for detecting binding site motifs in ChIP-Seq data. Biology Direct 9:4. doi:10.1186/1745-6150-9-4
Uyar B, Weatheritt RJ, Dinkel H et al (2014) Proteome-wide analysis of human disease mutations in short linear motifs: neglected players in cancer? Mol Biosyst 10(10):2626–2642. doi:10.1039/c4mb00290c
Vacic V, Oldfield CJ, Mohan A et al (2007) Characterization of molecular recognition features, MoRFs, and their binding partners. J Proteome Res 6(6):2351–2366. doi:10.1021/pr0701411
Ward JJ, Sodhi JS, McGuffin LJ et al (2004) Prediction and functional analysis of native disorder in proteins from the three kingdoms of life. J Mol Biol 337(3):635–645. doi:10.1016/j.jmb.2004.02.002
Wu G, Xu G, Schulman BA et al (2003) Structure of a beta-TrCP1-Skp1-beta-catenin complex: destruction motif binding and lysine specificity of the SCF(beta-TrCP1) ubiquitin ligase. Mol Cell 11(6):1445–1456
Xia X (2012) Position weight matrix, Gibbs sampler, and the associated significance tests in motif characterization and prediction. Scientifica 2012:917540. doi:10.6064/2012/917540

Chapter 10
Towards Understanding Protein Disorder *In-Cell*

Cesyen Cedeño, Hadas Raveh-Amit, András Dinnyés and Peter Tompa

Abstract Investigating the activity and structure of cellular biochemical machinery at atomic resolution has been a point of paramount significance for understanding health and disease over the decades. The underlying molecular mechanisms are primarily studied *in vitro*. Nuclear magnetic resonance (NMR) is a technique that allows to look into cells and study proteins and other constituents, thanks to careful experimental design and technological advances (spectrometer sensitivity and pulse sequence design). Here we outline current applications of the technique and propose a realistic future for the field.

Keywords In-cell NMR · Isotopic labeling · Cell types · Cell extracts

1 Introduction

The structural underpinning of enzymatic processes in solution has been extensively described. Despite technical sophistication, experimentalists have unveiled molecular mechanisms, and details of activities, and described structural features of isolated biological systems with relative accuracy. In these experiments, experimental conditions can usually be controlled due to dealing with an isolated, purified protein. However, living cells are extremely heterogeneous at the micro-environment level in which a protein/enzyme displays an optimum performance. That is, the exact conditions in which a protein exists cannot be fully reproduced by bench biologists, fueling discussions about the relevance of some physiological

A. Dinnyés (✉) · H. Raveh-Amit
BioTalentum Ltd, Aulich L. str. 26, 2100 Godollo, Hungary
e-mail: andras.dinnyes@biotalentum.hu

C. Cedeño · P. Tompa
VIB Department of Structural Biology, Vrije Universiteit Brussel,
Brussels 1050, Belgium

P. Tompa
Institute of Enzymology, Biological Research Center,
Hungarian Academy of Sciences, Budapest 1518, Hungary

I. C. Felli, R. Pierattelli (eds.), *Intrinsically Disordered Proteins Studied by NMR Spectroscopy,* Advances in Experimental Medicine and Biology,
DOI 10.1007/978-3-319-20164-1_10

parameters (Ellis 2001). A scientific debate regarding *in vitro* assays versus *in vivo* ones starts with a consideration of the protein environment, as well as the physiologically relevant concentration and the size of the protein under scrutiny. At the centre of these debates is nuclear magnetic resonance (NMR), which is the most powerful and versatile technique that allows the examination of proteins at atomic level. In this regard, there are basic, well-defined concepts that need to be addressed.

First of all, a disordered protein behaves very differently from a globular protein of similar size. Regardless of function and shape, a given protein freely diffusing in buffer (e.g. in a test tube) might not necessarily reflect its behaviour in a cell matrix, i.e., in the presence of other macromolecules, metabolites, membranes and local pH changes. Subtle shifts within a given population of interconverting structures can occur and binding events could also be affected from the kinetic and thermodynamic points of view.

The concept of protein crowding also complicates the cell's inner scenario. Crowding means that life inside a cell occurs in a non-continuous, dynamic and viscous environment, in which every molecule exerts an excluded-volume effect on other surrounding molecules, restricting both its rotational diffusion and internal motions.

Protein structural propensities and dynamics information can be deduced by the appropriate NMR experiments. However, the inner milieu of living cells poses a challenge for NMR measurements because of its inherent viscosity. As a general rule for spectroscopy purposes, macroscopic intracellular viscosity directly affects the rotational diffusion rate of a given protein inside the cell (D_r'). This magnitude differs from that of the same protein if observed in solution (D_r), and the cell components (primarily macromolecules) collectively contribute to these differences. The overall phenomena affecting rotational diffusion can be referred to as "molecular crowding" and the diffusional rotation of a molecule can be described as "tumbling".

It is not difficult to mimic molecular crowding by chemical agents, namely: polyethylene glycol (PEG), Ficoll, Dextrans or even protein matrices built up from bovine serum albumin (BSA) (Li et al. 2008). Under normal conditions, cells are estimated to have a protein concentration of ca. 400 mg/mL. Many other components, such as metabolites and membranes, are also present, and therefore mimicking agents do not reconstitute the entire system, although their use is widely accepted (Fig. 10.1). Due to the limitation of their use to gain physiological insight, they will never provide the wealth of information of an *in-cell* experiment due to the lack of biological relevance in terms of specific protein-protein interactions and signalling events. In the following section, we aim to address the most common questions to ask before embarking on an *in-cell* NMR experiment.

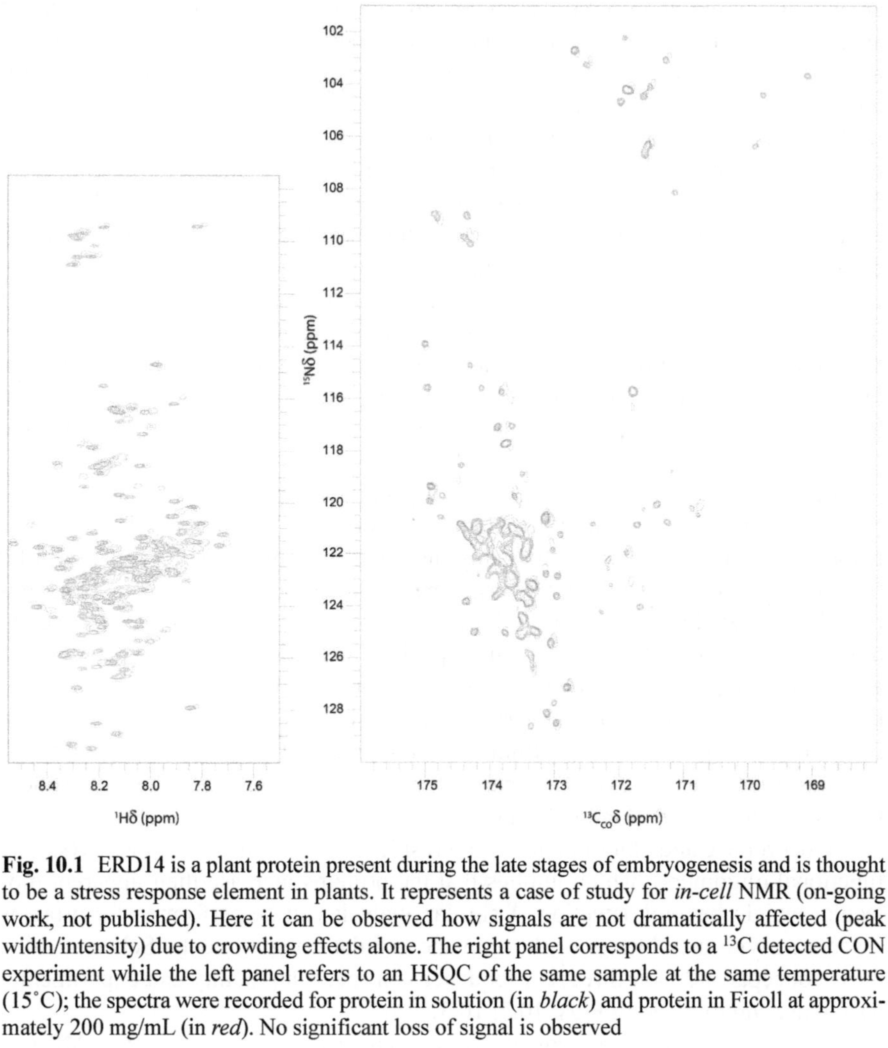

Fig. 10.1 ERD14 is a plant protein present during the late stages of embryogenesis and is thought to be a stress response element in plants. It represents a case of study for *in-cell* NMR (on-going work, not published). Here it can be observed how signals are not dramatically affected (peak width/intensity) due to crowding effects alone. The right panel corresponds to a ^{13}C detected CON experiment while the left panel refers to an HSQC of the same sample at the same temperature (15°C); the spectra were recorded for protein in solution (in *black*) and protein in Ficoll at approximately 200 mg/mL (in *red*). No significant loss of signal is observed

2 Is It Possible to Detect/Observe Proteins inside Cells using NMR?

The investigation of proteins by solution NMR depends on certain dynamics parameters (Chap. 3). The relaxation properties of the protein under investigation ($R_{1,2}$, also expressed as the reciprocal of $T_{1,2}$) are related to its global motions and affect the detection of signals by NMR. Intense cross peaks in correlation 2D spectra are usually recorded for either highly dynamic systems (intrinsically disordered proteins (IDP)) or for systems that are not so "dynamic" (globular proteins) but small in size; in these situations, the proteins can "tumble" freely and relaxation rates are favourable for NMR. This provides a rationale to why so many large IDP are actually observed by NMR, even though folded proteins of similar size are not amenable to solution state NMR. Among other factors, viscosity and temperature also play a role in the detectability of particular signals for certain regions of proteins or for the entire molecule.

The effects of peak broadening and loss of intensity are remarkable for large well-folded proteins, especially when embedded in viscous matrices, making them difficult to detect by NMR (Selenko and Wagner 2007). In contrast, IDPs experience little peak broadening and/or consequent loss of signal when embedded in a crowding environment, (Inomata et al. 2009). IDPs are therefore more suitable for conducting *in-cell* NMR experiments, whereas globularity and large size represent the most important limitations. On the one hand, globular proteins tend to "tumble" slowly when in solution, generating a particular set of intensities/peak-widths. On the other hand, increasing viscosity (crowding agents or cytosol) makes the protein tumble more slowly, decreasing intensity and therefore broadening peaks beyond detection.

Such limitations are found experimentally while aiming for *in-cell* NMR. Nonetheless, it is worth mentioning that signal intensity from a globular protein such as cytochrome c inside cells is remarkably diminished (Crowley et al. 2011). Similar observations, yet for different reasons, were obtained for ubiquitin, which engages in stable or transient interactions with other proteins inside human cells, affecting mobility and hence affecting tumbling. Specific interactions have also been reported to generate peak broadening in ubiquitin, although after mutating key residues that mediate the relevant protein-protein interactions, it is possible to recover the line width of signals and a spectrum comparable to that under *in vitro* conditions, meaning ubiquitin "recovers" its capacity to tumble almost freely even inside a viscous medium (Serber et al. 2001).

Cytochrome c and ubiquitin have long residence times within complexes (stable/functional oligomers), which implies that the proteins increase their apparent molecular weight beyond detection. A similar analysis can be performed for proteins associated with membranes or those inside cellular compartments. Schematically, these concepts and their direct consequences on spectrum quality and experimental design are summarized in Figs. 10.2, 10.3.

Regarding membrane proteins, the debate remains whether solid state NMR (ssNMR) could shed some light on their behaviour. However, the physiological relevance of such experiments is questionable due to the experimental design applied in ssNMR (spinning frequencies may not be compatible with normal living conditions for a cell).

3 Is it Possible to "Place" an Isotopically Labelled Sample inside a Living Cell?

The performance of *in-cell* NMR experiments also requires a critical assessment of the methods employed to deliver an isotopically labelled protein into the cell of interest. Depending on the type of cell and the availability of technological tools, it

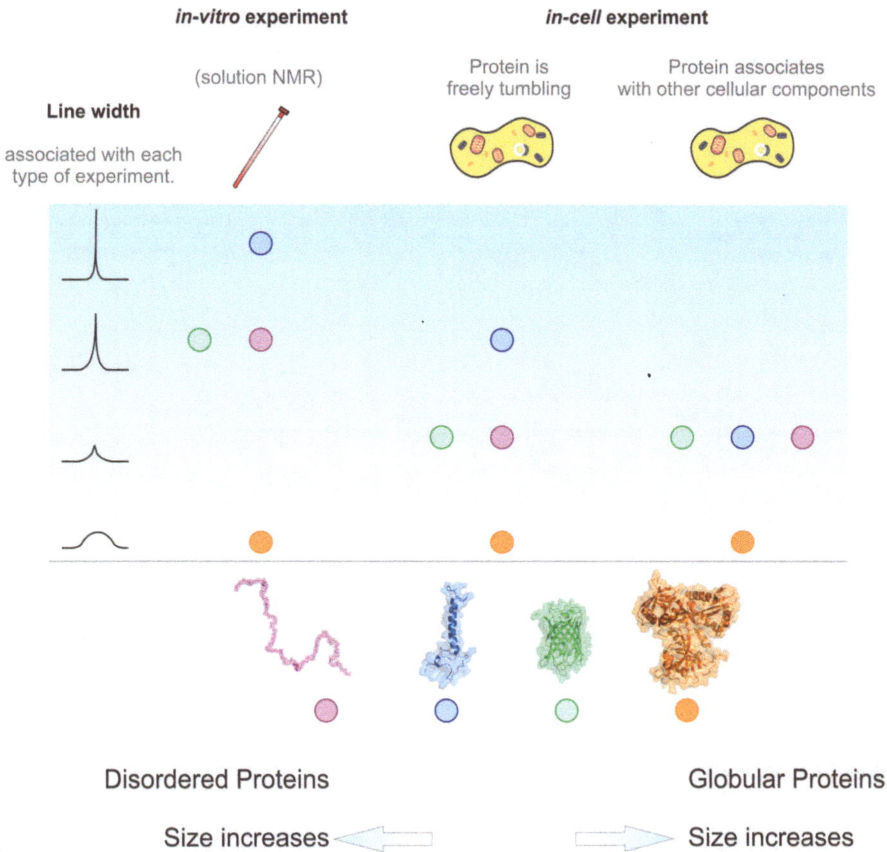

Fig. 10.2 Schematic representation of NMR experiments in terms of the intrinsic ability to detect signals. This highlights how feasible it is to carry out *in-cell* NMR studies for both globular and disordered proteins. Disordered proteins are good candidates in general (Theillet et al. 2014)

is possible to place proteins inside cells either by "production" of the sample using cellular endogenous machinery (e.g. aided by inducible systems) or by introducing them from outside (see Sect. 10.3).

Endogenous generation of labelled proteins is based, for example, induced via the T7 promoter of *E. coli*. After induction of the culture—and depending on the solubility of the translation product—it is possible to proceed to *in-cell* NMR studies. This has been demonstrated in *E. coli* under isotopic labelling conditions, upon IPTG addition, where it is possible to collect a HN-HSQC spectrum of overexpressed protein inside cells (Serber et al. 2001). The N-terminal domain (metal binding motif) of MerA was detected inside *E. coli* by this approach. Cells were grown in lysogeny broth (LB) and then harvested and re-suspended into labelling medium just before induction. This procedure ensures that the endogenous translation machinery will only produce the protein of interest upon induction, incorporating

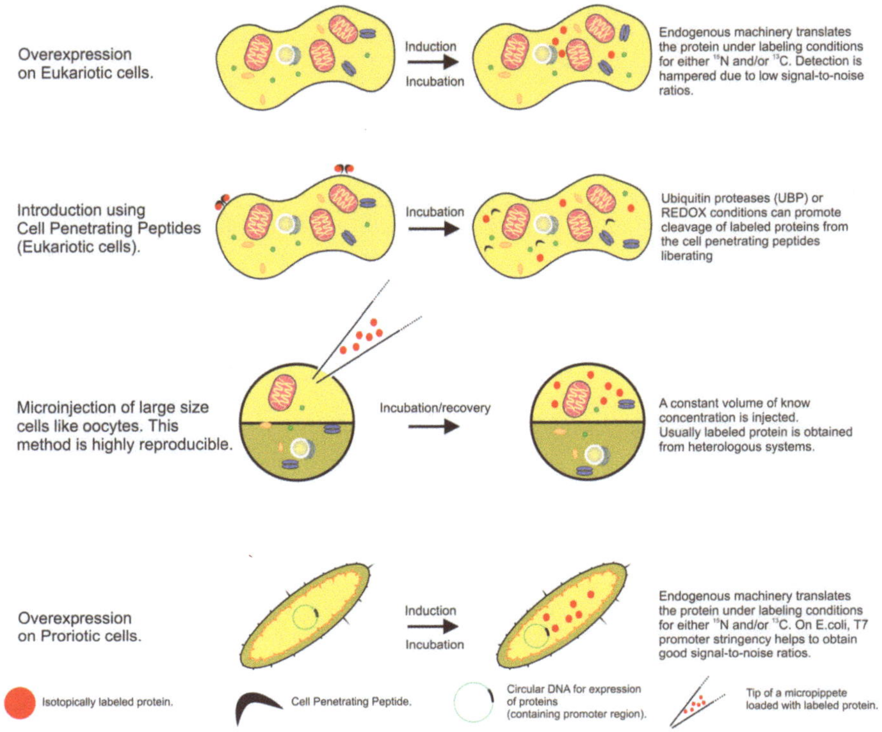

Fig. 10.3 Methods for protein delivery

isotopes. Due to the stringency of the T7 promoter, the translation of the desired product is tightly regulated, and only low concentrations of isotopically labelled by-products are formed; therefore, a working HSQC spectrum can be recorded after only 1 h of induction (Selenko et al. 2006).

The relevance of these studies might be strongly limited by cell viability. This is an important remark if one considers that the major advantage of *in-cell* NMR is not only the ability to study the protein under crowding conditions, but also that the protein is inside a living organism. Life in this sense refers to the full integrity and activity of the biochemical machinery inside a cell; therefore, cell survivability and the proper handling of cellular samples is critical when performing these experiments.

Early experiments demonstrated that it is possible to collect NMR data in *E. coli*, so the community interest rapidly moved towards more complicated systems, such as eukaryotic cells. Different technological tools have been adapted; microinjection and cell penetrating peptides were introduced at almost the same time in the *in-cell* NMR field.

Microinjection was already extensively used for DNA transfection using large cells such as oocytes and was soon customized to deliver isotopically labelled proteins. This technique was exploited for protein delivery allowing high-resolution correlation spectra not only for globular proteins (Selenko et al. 2006; Sakai et al. 2006), but also with IDPs (Bodart et al. 2008). Microinjection was introduced into this field as an attempt to produce the first eukaryotic *in-cell* NMR sample *Xenopus laevis* oocytes, selected for their size, cell cycle arrest capability and homogeneity (Selenko et al. 2006). Streptococcal GB1 domain (GB1, 56 residues, 7 kDa) was injected inside *Xenopus laevis* eggs (cell volume of ≈ 1 µL) using a commercially available micromanipulator and a microinjection system adjusted to deliver around 50 nL of purified labelled protein. Protein concentrations ranging between 0.5 and 10 mM yielded *in-cell* NMR samples (cells) with a protein concentration of about 25–500 µM. An exhaustive quality control was required to carry out a reproducible *in-cell* NMR experiment, primarily to prevent the protein from leaking out of the cells (Thongwichian and Selenko 2012). *In-cell* NMR experiments involving IDPs are more prone to yield reliable results even though this is not a rule of thumb (Barnes and Pielak 2011).

The introduction of proteins using cell-penetrating peptides (CPP) can also be achieved and many molecular tricks can be included to deliver untagged samples inside cells. For example, HIVtat peptide is a protein transduction domain derived from human immunodeficiency virus type 1 that is rapidly taken up by cells (Inomata et al. 2009). Constructs in which endogenous ubiquitin peptidases cleave the tat peptide from the protein of interest and generate the labelled free protein upon entry to cells, have been developed, as well as those which include an attachment of HIVtat via an S-S bond. In the latter, the CPP, which is not isotopically labelled, can be cleaved from the labelled protein so that the background from isotopically-labelled by-products is completely reduced inside cells. Bekei and collaborators demonstrated the amenability of this technique for IDPs to study the role of disorder under physiologically relevant conditions (Bekei et al. 2012a).

Alternatively, the formation of pores in cell membranes can also aid the delivery of labelled proteins inside living cells. Pore forming methods include the use of chemicals such as Ca^{2+}, toxins, or PEG, or physical forces such as electroporation and mechanical deformation (Sharei et al. 2013). Upon pore opening, diffusion drives the protein inside cells. Streptolysin, a known pore-forming toxin, has been used for IDP delivery into several cell lines, after which HSQC spectra with reasonable quality were obtained (Ogino et al. 2009; Bekei et al. 2012b).

Only a few studies using overexpression in eukaryotic cells and demonstrating its potential in the *in-cell* NMR field have been published (Banci et al. 2013; Hamatsu et al. 2013; Luchinat et al. 2014; Barbieri et al. 2014).

Even though it cannot be called in-cell in the strictest sense, there is also a modest study showing how peptides can bind a VEGF receptor on the surface of Porcine Aortic Endothelial Cells (Diana et al. 2015).

4 *In-Cell* NMR seems to be a very Useful Technique. What for?

Serber and collaborators showed how to manipulate bacteria (*E. coli*) in order to produce a good sample that can yield spectra of decent quality. Their work also demonstrated that specific methyl labelling is feasible for *in-cell* NMR samples (Serber et al. 2004). Even though no IDPs were used for these seminal experiments, they presented a strong proof-of-concept and physiologically relevant issues were soon addressed.

α-synuclein, a well-known IDP involved in neurodegeneration, has been studied by *in-cell* NMR almost from the very beginning. By measuring periplasmic α-synuclein in *E. coli* it was possible to demonstrate that there is no change in secondary structure propensity under cellular conditions. The disordered state of this protein prevails even when NMR experiments are carried out at a temperature slightly higher than in previous *in vitro* experiments. Equivalent *in vitro* experiments at high temperatures suggested a more helical structure (McNulty et al. 2006) however electrostatic contributions should be taken into account for each experiments either in solution or in-cell as was suggested by Croke (Croke et al. 2008). Pielak and collaborators concluded that α-synuclein and its disease-related variants remained disordered in both the periplasm of bacteria and under artificially induced crowding conditions using BSA (300 g/L). Dedmon, however, investigated FlgM inside *E. coli* and compared the cellular behaviour of the protein to when it is studied in solution. These experiments were validated by crowding agents, finally suggesting that crowding itself could trigger disorder-to-order transition in the C-terminal part of the protein while the N-terminal portion remains unfolded (Dedmon et al. 2002). The authors suggested that there could be two distinct types of IDPs, based on their ability to gain structure (or remain disordered) under physiologically relevant conditions. Taken together, these experiments also exemplify a persistent debate within the IDP field and they also demonstrate the unique potential that *in-cell* NMR has to address important questions without leaving biology aside.

Besides ^{13}C and ^{15}N, another "active" isotope for NMR is ^{19}F, also used for *in-cell* NMR purposes. ^{19}F 3-fluorotyrosine was initially observed inside *E. coli*, but its use was limited to small globular proteins (chymotrypsin inhibitor 2, calmodulin, ubiquitin) (Williams et al. 1997). Meanwhile, modifications with the synthetic amino acid trifluoromethyl-L-phenylalanine allowed the observation of signals from larger proteins (green fluorescent protein and histidinol dehydrogenase) (Li et al. 2010). In the latter, ^{19}F labelled α-synuclein was overexpressed in *E. coli* and the authors showed the suitability of the method for obtaining useful information about IDPs. In the same direction, ^{19}F NMR data helped to explore α-synuclein in the presence of mitochondria-like vesicles, evidencing regions that are more likely to interact with membranes of a given composition (Zigoneanu et al. 2012).

Many IDPs are predicted to be nuclear proteins, and it is therefore not surprising to find examples of nuclear events being analysed by *in-cell* NMR, including important post-translational modifications (PTM), such as phosphorylation (Selenko et al. 2008), acetylation (Liokatis et al. 2010) and methylation (Theillet et al. 2012) that regulate the cell cycle. Liokatis and collaborators reported both acetylations and phophorylations of the histone H3 N-terminal region (amino acids 1–33) using HeLa cell nuclear extracts (Liokatis et al. 2012). This disordered region of H3 histone was subject to detailed analyses *in vivo* as many relevant PTMs are tightly regulated and do not resemble previously reported patterns. Selenko later used the mentioned fragment of histone H3 and the N-terminal disordered regions of H4 and H2A to demonstrate the suitability of NMR for the observation of methylations. A tailored pulse sequence (^{1}H-^{13}C presat-SOFAST-HMQC) was required to detect $CH_2\varepsilon$ at the lysine residues and quantitatively assign the level of methylation in a given position by signal integration. An individual mapping of *de novo* methylation sites remained inaccessible by the latter approach, so a two-dimensional pulse sequence was developed, making it possible to differentiate the mono-, di- and trimethylated forms of lysine, particularly in the case of H2A. The approach was then tested using purified fragments of histone H3 reacting within cellular extracts of HeLa cells, demonstrating that the methodology has unique advantages and can be potentially useful during *in-cell* experiments, providing a wider range of functional assays using NMR (Theillet et al. 2012). In the context of developmental biology, these tools open the door towards a more realistic description of chromatin assembly in cell cycle progression in health and disease.

Using microinjected *X. laevis* eggs, it was possible to unveil a particular mechanism of adjacent phosphorylation in a disordered region of the SV40 regulatory domain (Ogino et al. 2009). The proposed mechanism highlights the advantage of continuously monitoring phosphorylation events within disordered regions of model substrates by *in-cell* NMR.

Tau, a protein related to Alzheimer's disease, represented a major challenge for both solution and *in-cell* NMR due to its size (45 kDa.). When tau was injected into oocytes, it was possible to confirm its association with microtubules and visualize some relevant phosphorylation events. These landmark experiments allowed the confirmation of two important facts: first, a largely disordered protein was visible inside cells despite the effect of crowding on its tumbling, and, second, that strong binding to microtubules inside cells did not hamper its analysis by *in-cell* techniques. Complementary studies were achieved using *in vitro* solution experiments in which tau was titrated with microtubules (Bodart et al. 2008). The contribution of these results is significant from a biological point of view and the authors have paved the way towards more comprehensive approaches regarding IDPs and related pathologies.

More detailed and accurate protein-protein interaction assays can be achieved by titrating interacting proteins inside *E. coli* (Burz and Shekhtman 2010). This method has been used to map interaction sites and recognition patterns in a residue-specific way inside the cell. It is relatively easy to manipulate labelling conditions and timing, which allows to visualize either only one protein at a time or multiple interactors. The IDP field awaits more in-depth studies on challenging cases in which a single protein can bind to several partners (e.g. moonlighting), as would be the case with hubs, adaptors and chaperones. Shekhtman and collaborators evaluated a library of peptides that can disrupt the complex of FKBP-FRB inside *E. coli* cells, showing that *in-cell* NMR can also aid in drug screening assays (Xie et al. 2009).

Bertini and collaborators demonstrated the feasibility of using ^{13}C direct detection *in-cell* to collect correlation spectra of folded proteins and of IDPs (Atx1 protein and α-synuclein) in *E. coli*. CON experiments showed a better signal dispersion for α-synuclein than the analogous HN correlation experiments, proving particularly useful for the study of IDPs. On the other hand, for folded proteins such as Atx1, the signal dispersion of 2D HN spectra was sufficient to acquire a snapshot of the protein in cell (Bertini et al. 2011; Felli et al. 2014; Binolfi et al. 2012).

5 Can We Describe the Ensemble Distribution of IDPs inside Cells?

There is a growing demand in the NMR community for describing reliable ensembles of proteins in cells, but such calculations require not only chemical shift data but also values of nuclear Overhauser effects (NOE). More recently, new protocols have been developed for IDPs based on *in vitro* data; these no longer require NOEs, which are not detectable in disordered systems. They are promising *in vitro* and are expected to be widely implemented for *in vivo* data as well in the near future (Ozenne et al. 2012; Granata et al. 2013).

Ensemble calculations of IDPs enable an accurate and meaningful approximation of naturally highly dynamic molecules. However, it is well accepted that the more experimental data points are used, the more accurate any given ensemble will be, regardless of the methodology employed. Residual dipolar coupling values (RDCs) are meaningful observables *in vitro* but they are not accessible for *in-cell* NMR due to difficulty of physical alignment required for the RDC. As paramagnetic relaxation enhancement (PRE) measurements *in vivo* would also provide important information, efforts are being made to make them accessible for *in vivo* structure calculation and other purposes.

Structure calculation requires complete sets of assigned chemical shifts, for which several multidimensional experiments are required. Such experiments involving long and time-consuming acquisitions are not amenable to *in-cell* NMR as cell viability can be dramatically compromised inside an NMR tube.

TTHA1718 is a globular protein from *Thermus thermophilus* that has been overexpressed inside *E. coli* for solving its "*in-cell*" structure (Sakakibara et al. 2009). This study compared structures of the protein *in vitro* and *in vivo*, and concluded that these structures do not differ, also implying that no protein-protein interactions occur. The challenge still remains for proteins that are exposed to their interacting partners and/or are being post-translationally modified in their native environments. Increasing the level of complexity of the system, Ito and co-workers prepared *in-cell* NMR samples using insect cells and a baculovirus protein expression system. NMR experiments involving 3D triple-resonance (HNCA, HN(CO)CA and HNCO) were collected using non-linear sampling (Hamatsu et al. 2013). Each experiment was collected for 3.5 h on average, which represents a significant reduction in time compared to normal sampling schemes, ensuring the viability of the cells. This study enabled chemical shift assignments directly for *in-cell* samples instead of transferring a previous *in vitro* assignment. The challenge stills remains regarding ensemble calculations; fast acquisition methods (Solyom et al. 2013) or protonless methods amenable to IDPs must still therefore be regularly employed for *in-cell* NMR (Gil et al. 2013).

6 Overall Dynamics of IDPs inside Cells: Pushing the Boundaries

In general, NMR can address the global motions and dynamics of macromolecules. Several *in-cell* NMR experiments addressed GB1. For example, Wang and collaborators (Wang et al. 2011) demonstrated that viscosity inside cells is a determinant of the dynamics and visibility of NMR signals, without limiting data acquisition. Furthermore, the authors showed the difference in line widths between TROSY and antiTROSY lines, namely ΔΔνTAT, which is a parameter that helps understand the rotational diffusion of molecules. ΔΔνTAT varies linearly for globular proteins in glycerol, and it indicates that GB1 rotates approximately 8–10 times slower inside *E. coli* cells. Nonetheless, a similar approach has not yet been taken for a disordered protein. Using ^{19}F NMR, Brindle and co-workers reached a similar conclusion regarding the reduction of rotation of proteins inside cells (Williams et al. 1997). It is difficult to separate rotational components that derive from high viscosity from the effect arising from protein-protein interactions; interaction-free methods should therefore be developed, requiring conceptual and technological advances (Li and Liu 2013).

Cino and collaborators have worked on IDPs under crowding conditions (Cino et al. 2012). By studying ProTα, TC-1 and α-synuclein, they concluded that globular parts of TC-1 experience dramatic to mild changes in R_1 (^{15}N) and R_2 (^{15}N), whereas disordered segments within the same protein do not exhibit major changes. ProTα exhibits increased R_2 (^{15}N) under crowding, while its R_1 (^{15}N) variation is small. In general, internal motion is restricted for IDPs under crowding conditions, suggesting motif stabilization towards binding of partners or substrates. The measured values of het-NOEs are unaffected, which indicates that backbone dynamics do not suffer variations even though structural diversity is still observed.

7 Perspective and Proposed Path

Biomolecular NMR provides adequate answers in many different fields, from a basic understanding of biochemistry at atomic detail all the way to complicated explanations of signalling in a cellular context (Amata et al. 2013). However, from the cell biology point of view, NMR techniques still often miss the link between detailed structural information and the processes of life such as protein regulation and metabolism.

There is limited information about structural changes and mechanisms of regulation mediated by PTMs other than those described above (namely small molecule modifications (SMM): phosphorylation, acetylation and methylation). Furthermore, there are only a handful of studies addressing a system in which several modifications are induced, even only considering *in vitro* studies. SUMOylation and ubiquitination, although essential modifications in every aspect of living processes, have not been thoroughly investigated (Cai et al. 2012).

Cell differentiation and regulation are processes in which several modifications work in a concerted way. In this context, a large body of information has been collected over the past years, translating into the so-called modern field of regenerative medicine. It is now well understood how differentiation is regulated, whereas the reverse process is still a matter of discussion. There are several alternative views in which protein disorder plays a key role (Xue et al. 2012). Novel means to revert differentiation and produce *induced* pluripotent stem cells (iPSC) currently represent the ultimate hope in terms of regenerative medicine.

In-cell NMR could provide the answers many scientists are awaiting to understand how certain transcription factors can revert the differentiation process when induced in somatic cells. The suggested proteins (transcription factors) listed in the table below share two defining features that make them highly interesting for *in-cell* NMR studies:

- they play crucial roles as transcription factors in regulating stem cell pluripotency and reprogramming somatic cells into iPSC, and
- they are predicted to contain intrinsically disordered domains (regions) as well as structured regions, whereas relevant experimentally determined three-dimensional structural data are sparse.

Not surprisingly—and as seen for many IDPs—the available structural data corresponds to either structured domains or a disordered region stabilized in a complex with binding proteins and DNA (see Table 10.1). Indeed, based primarily on bioinformatics analyses, a previous study suggested that most reprogramming factors are enriched with intrinsic disorder (Xue et al. 2012). It is of great interest to establish the following: a) the extensive 3D structural analysis of the reprogramming factors and b) the role that intrinsic disorder plays in regulation. Figure 10.4 shows the significant length of disordered regions missing within the structural data available for some selected proteins.

By the detailed analysis of specific cases (see Table 10.1 and Fig 10.4), it is possible to gather an idea about the role of disorder in the regulation of the activity/binding of these specific transcription factors. Such long disordered fragments are usually neglected/missing in previous *in vitro* studies, basically because of solubility, isolation/purification or crystallization issues. However, *in-cell* NMR may overcome these difficulties in an elegant way as the protein can be studied inside the most appropriate environment in terms of stability: the interior of the cell.

Expanding our knowledge in this sense is highly relevant as it helps bridge the gap between basic research and applied medicine.

Table 10.1 Suggested protein candidates for *in-cell* NMR studies: Sox2, Oct4, Klf, c-Myc and Nanog (Cai et al. 2012; Moretto-Zita et al. 2010)

Protein	Swissprot ID (Mouse/human)	Protein length (amino acids)	Known PTM	PDB structures available
Sox2	P48432/P48431	319	Sumoylation at Lys247 Phosphorylation of Ser246, Ser249, Ser250 and Ser251 Methylation at Arg113 Acetylation of Lys75	104X: Ternary complex DNA binding domains of, Oct1 Sox2 and DNA 1GT0: Crystal structure of a POU/HMG/DNA ternary complex. 2LE4: Solution structure of the HMG box DNA-binding domain of Sox2
Oct4	P20263/Q01860	352	Ubiquitination on Lys63 SUMOylation at Lys118 Phosphorylation at Ser229 (PKA) and Tyr327 (Abl)	3L1P: POU protein:DNA complex
Klf4	Q60793/Q43474	483	SUMOylation at Lys275 Phosphorylation at Ser123	2WBS: Crystal structure of the zinc finger domain of klf4 and its target DNA. 4M9E: Structure of Klf4 zinc finger in complex with methylated DNA
c-myc	P01108/P01106	439		1NKP: Crystal structure of Myc-Max recognising DNA
Nanog	Q80Z64/Q9H950	305	Phosphorylation at Ser52, Ser56 or 57, Ser65 and Ser77 or 78, Oct1	2VI6: Crystal structure of Nanog homodomain. 2KT0: Solution structure of human Nanog homodomain fragment

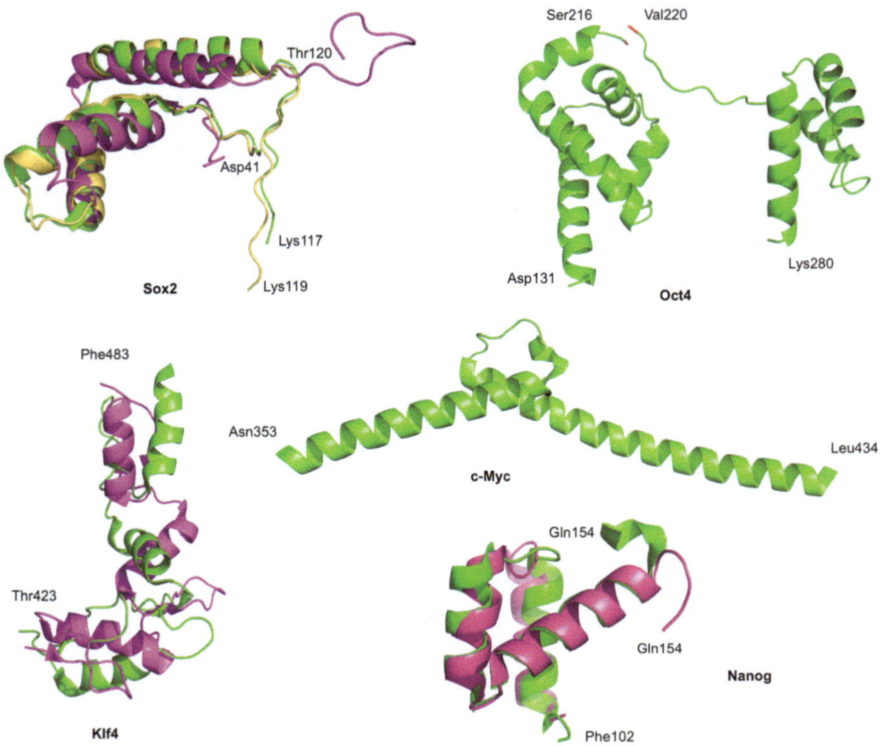

Fig. 10.4 Some structural features for proteins shown in Table 10.1. Although the structural data presented corresponds to folded regions, their missing N-terminal or C-terminal ends are predicted to be disordered. PDB codes for Sox2: 1O4X in *green*, 2LE4 in *purple*, and 1GT0 in *yellow*. In the case of Klf4: 2WBS in *green*, and 4M9E in *purple*. Finally for Nanog, 2VI6 is represented in *green* and 2KT0 is *purple*

Acknowledgements Cesyen Cedeño and Hadas Raveh-Amit were fellows in the IDPbyNMR Marie Curie project of the European Commission, 7th Framework Programme (contract no. 264257), and this work has been partially supported by this project. Peter Tompa acknowledges the Research Foundation Flanders (FWO) Odysseus grant G.0029.12.

References

Amata I, Maffei M, Igea A et al (2013) Multi-phosphorylation of the intrinsically disordered unique domain of c-Src studied by in-cell and real-time NMR spectroscopy. Chembiochem 14:1820–1827. doi:10.1002/cbic.201300139

Banci L, Barbieri L, Bertini I et al (2013) Atomic-resolution monitoring of protein maturation in live human cells by NMR. Nat Chem Biol 9:297–299. doi:10.1038/nchembio.1202

Barbieri L, Luchinat E, Banci L (2014) Structural insights of proteins in sub-cellular compartments: In-mitochondria NMR. Biochim Biophys Acta—Mol Cell Res 1843:2492–2496. doi:10.1016/j.bbamcr.2014.06.009

Barnes CO, Pielak GJ (2011) In-cell protein NMR and protein leakage. Proteins 79:347–351. doi:10.1002/prot.22906

Bekei B, Rose HM, Herzig M et al (2012a) In-cell NMR in mammalian cells: part 1. Methods Mol Biol 895:43–54. doi:10.1007/978-1-61779-927-3_4

Bekei B, Rose HM, Herzig M et al (2012b) In-cell NMR in mammalian cells: part 2. Methods Mol Biol 895:55–66. doi:10.1007/978-1-61779-927-3_5

Bertini I, Felli IC, Gonnelli L et al (2011) ^{13}C direct-detection biomolecular NMR spectroscopy in living cells. Angew Chem Int Ed Engl 50:2339–2341. doi:10.1002/anie.201006636

Binolfi A, Theillet F-X, Selenko P (2012) Bacterial in-cell NMR of human α-synuclein: a disordered monomer by nature? Biochem Soc Trans 40:950–954.

Bodart J-F, Wieruszeski J-M, Amniai L et al (2008) NMR observation of Tau in Xenopus oocytes. J Magn Reson 192:252–257. doi:10.1016/j.jmr.2008.03.006

Burz DS, Shekhtman A (2010) The STINT-NMR method for studying in-cell protein-protein interactions. Curr Protoc Protein Sci Chapter 17: Unit 17.11. doi:10.1002/0471140864.ps1711s61

Cai N, Li M, Qu J et al (2012) Post-translational modulation of pluripotency. J Mol Cell Biol 4:262–265. doi:10.1093/jmcb/mjs031

Cino EA, Karttunen M, Choy W-Y (2012) Effects of molecular crowding on the dynamics of intrinsically disordered proteins. PLoS One 7:e49876. doi:10.1371/journal.pone.0049876

Croke RL, Sallum CO, Watson E et al (2008) Hydrogen exchange of monomeric α-synuclein shows unfolded structure persists at physiological temperature and is independent of molecular crowding in *Escherichia coli*. Protein Sci 17:1434–1445. doi:10.1110/ps.033803.107

Crowley PB, Chow E, Papkovskaia T (2011) Protein interactions in the *Escherichia coli* cytosol: an impediment to in-cell NMR spectroscopy. ChemBioChem 12:1043–1048. doi:10.1002/cbic.201100063

Dedmon MM, Patel CN, Young GB et al (2002) FlgM gains structure in living cells. Proc Natl Acad Sci U S A 99:12681–12684. doi:10.1073/pnas.202331299

Diana D, Russomanno A, De Rosa L et al (2015) Functional binding surface of a β-Hairpin VEGF receptor targeting peptide determined by NMR spectroscopy in living cells. Chem—A Eur J 21:91–95. doi:10.1002/chem.201403335

Ellis RJ (2001) Macromolecular crowding: obvious but underappreciated. Trends Biochem Sci 26:597–604.

Felli IC, Gonnelli L, Pierattelli R (2014) In-cell ^{13}C NMR spectroscopy for the study of intrinsically disordered proteins. Nat Protoc 9:2005–2016.

Gil S, Hošek T, Solyom Z et al (2013) NMR spectroscopic studies of intrinsically disordered proteins at near-physiological conditions. Angew Chem Int Ed Engl 52:11808–11812. doi:10.1002/anie.201304272

Granata D, Camilloni C, Vendruscolo M et al (2013) Characterization of the free-energy landscapes of proteins by NMR-guided metadynamics. Proc Natl Acad Sci U S A 110:6817–6822. doi:10.1073/pnas.1218350110

Hamatsu J, O'Donovan D, Tanaka T et al (2013) High-resolution heteronuclear multidimensional NMR of proteins in living insect cells using a baculovirus protein expression system. J Am Chem Soc 135:1688–1691. doi:10.1021/ja310928u

Inomata K, Ohno A, Tochio H et al (2009) High-resolution multi-dimensional NMR spectroscopy of proteins in human cells. Nature 458:106–109. doi:10.1038/nature07839

Li C, Liu M (2013) Protein dynamics in living cells studied by in-cell NMR spectroscopy. FEBS Lett 587:1008–1011. doi:10.1016/j.febslet.2012.12.023

Li C, Charlton LM, Lakkavaram A et al (2008) Differential dynamical effects of macromolecular crowding on an intrinsically disordered protein and a globular protein: implications for in-cell NMR spectroscopy. J Am Chem Soc 130:6310–6311. doi:10.1021/ja801020z

Li C, Wang G-F, Wang Y et al (2010) Protein ^{19}F NMR in *Escherichia coli*. J Am Chem Soc 132:321–327. doi:10.1021/ja907966n

Liokatis S, Dose A, Schwarzer D et al (2010) Simultaneous detection of protein phosphorylation and acetylation by high-resolution NMR spectroscopy. J Am Chem Soc 132:14704–14705. doi:10.1021/ja106764y

Liokatis S, Stützer A, Elsässer SJ et al (2012) Phosphorylation of histone H3 Ser10 establishes a hierarchy for subsequent intramolecular modification events. Nat Struct Mol Biol 19:819–823. doi:10.1038/nsmb.2310

Luchinat E, Barbieri L, Rubino JT et al (2014) In-cell NMR reveals potential precursor of toxic species from SOD1 fALS mutants. Nat Commun 5:5502.

McNulty BC, Young GB, Pielak GJ (2006) Macromolecular crowding in the *Escherichia coli* periplasm maintains α-synuclein disorder. J Mol Biol 355:893–897. doi:10.1016/j.jmb.2005.11.033

Moretto-Zita M, Jin H, Shen Z et al (2010) Phosphorylation stabilizes Nanog by promoting its interaction with Pin1. Proc Natl Acad Sci U S A 107:13312–13317. doi:10.1073/pnas.1005847107

Ogino S, Kubo S, Umemoto R et al (2009) Observation of NMR signals from proteins introduced into living mammalian cells by reversible membrane permeabilization using a pore-forming toxin, streptolysin O. J Am Chem Soc 131:10834–10835. doi:10.1021/ja904407w

Ozenne V, Bauer F, Salmon L et al (2012) Flexible-meccano: a tool for the generation of explicit ensemble descriptions of intrinsically disordered proteins and their associated experimental observables. Bioinformatics 28:1463–1470. doi:10.1093/bioinformatics/bts172

Sakai T, Tochio H, Tenno T et al (2006) In-cell NMR spectroscopy of proteins inside Xenopus laevis oocytes. J Biomol NMR 36:179–188. doi:10.1007/s10858-006-9079-9

Sakakibara D, Sasaki A, Ikeya T et al (2009) Protein structure determination in living cells by in-cell NMR spectroscopy. Nature 458:102–105. doi:10.1038/nature07814

Selenko P, Wagner G (2007) Looking into live cells with in-cell NMR spectroscopy. J Struct Biol 158:244–253. doi:10.1016/j.jsb.2007.04.001

Selenko P, Serber Z, Gadea B et al (2006) Quantitative NMR analysis of the protein GB1 domain in Xenopus laevis egg extracts and intact oocytes. Proc Natl Acad Sci U S A 103:11904–11909. doi:10.1073/pnas.0604667103

Selenko P, Frueh DP, Elsaesser SJ et al (2008) In situ observation of protein phosphorylation by high-resolution NMR spectroscopy. Nat Struct Mol Biol 15:321–329. doi:10.1038/nsmb.1395

Serber Z, Ledwidge R, Miller SM et al (2001) Evaluation of parameters critical to observing proteins inside living *Escherichia coli* by in-cell NMR spectroscopy. J Am Chem Soc 123:8895–8901.

Serber Z, Straub W, Corsini L et al (2004) Methyl groups as probes for proteins and complexes in in-cell NMR experiments. J Am Chem Soc 126:7119–7125. doi:10.1021/ja049977k

Sharei A, Zoldan J, Adamo A et al (2013) A vector-free microfluidic platform for intracellular delivery. Proc Natl Acad Sci U S A 110:2082–2087. doi:10.1073/pnas.1218705110

Solyom Z, Schwarten M, Geist L et al (2013) BEST-TROSY experiments for time-efficient sequential resonance assignment of large disordered proteins. J Biomol NMR 55:311–321. doi:10.1007/s10858-013-9715-0

Theillet F-X, Liokatis S, Jost JO et al (2012) Site-specific mapping and time-resolved monitoring of lysine methylation by high-resolution NMR spectroscopy. J Am Chem Soc 134:7616–7619. doi:10.1021/ja301895f

Theillet F-X, Binolfi A, Frembgen-Kesner T et al (2014) Physicochemical properties of cells and their effects on intrinsically disordered proteins (IDPs). Chem Rev 114:6661–6714. doi:10.1021/cr400695p

Thongwichian R, Selenko P (2012) In-cell NMR in Xenopus laevis oocytes. Methods Mol Biol 895:33–41. doi:10.1007/978-1-61779-927-3_3

Wang Q, Zhuravleva A, Gierasch LM (2011) Exploring weak, transient protein-protein interactions in crowded in vivo environments by in-cell nuclear magnetic resonance spectroscopy. BioChemistry 50:9225–9236. doi:10.1021/bi201287e

Williams SP, Haggie PM, Brindle KM (1997) ^{19}F NMR measurements of the rotational mobility of proteins in vivo. Biophys J 72:490–498. doi:10.1016/S0006-3495(97)78690-9

Xie J, Thapa R, Reverdatto S et al (2009) Screening of small molecule interactor library by using in-cell NMR spectroscopy (SMILI-NMR). J Med Chem 52:3516–3522. doi:10.1021/jm9000743

Xue B, Oldfield CJ, Van Y et al (2012) Protein intrinsic disorder and induced pluripotent stem cells. Mol Biosyst 8:134–150. doi:10.1039/c1mb05163f

Zigoneanu IG, Yang YJ, Krois AS et al (2012) Interaction of α-synuclein with vesicles that mimic mitochondrial membranes. Biochim Biophys Acta 1818:512–519. doi:10.1016/j.bbamem.2011.11.024

Chapter 11
The Protein Ensemble Database

Mihaly Varadi and Peter Tompa

Abstract The scientific community's major conceptual notion of structural biology has recently shifted in emphasis from the classical structure-function paradigm due to the emergence of intrinsically disordered proteins (IDPs). As opposed to their folded cousins, these proteins are defined by the lack of a stable 3D fold and a high degree of inherent structural heterogeneity that is closely tied to their function. Due to their flexible nature, solution techniques such as small-angle X-ray scattering (SAXS), nuclear magnetic resonance (NMR) spectroscopy and fluorescence resonance energy transfer (FRET) are particularly well-suited for characterizing their biophysical properties. Computationally derived structural ensembles based on such experimental measurements provide models of the conformational sampling displayed by these proteins, and they may offer valuable insights into the functional consequences of inherent flexibility. The Protein Ensemble Database (http://pedb.vib.be) is the first openly accessible, manually curated online resource storing the ensemble models, protocols used during the calculation procedure, and underlying primary experimental data derived from SAXS and/or NMR measurements. By making this previously inaccessible data freely available to researchers, this novel resource is expected to promote the development of more advanced modelling methodologies, facilitate the design of standardized calculation protocols, and consequently lead to a better understanding of how function arises from the disordered state.

Keywords Database · Experimental validation · Calculation protocols · Ensembles of structures

P. Tompa (✉)
Institute of Enzymology, Research Centre for Natural Sciences, Hungarian Academy of Sciences, Budapest, Hungary
e-mail: ptompa@vub.ac.be

M. Varadi · P. Tompa
Department of Structural Biology, Vlaams Institute voor Biotechnologie (VIB), Brussels, Belgium

Vrije Universiteit Brussel (VUB), Structural Biology Brussel (SBB), Brussels, Belgium

I. C. Felli, R. Pierattelli (eds.), *Intrinsically Disordered Proteins Studied by NMR Spectroscopy,* Advances in Experimental Medicine and Biology,
DOI 10.1007/978-3-319-20164-1_11

1 Introduction to IDP Ensembles

The ultimate goal of structural biology is to understand the function of a protein in terms of its structural properties, as traditionally formulated in the structure-function paradigm. Connecting structure and function is never truly trivial, yet the phenomenon of the intrinsic structural disorder that can be observed in a large number of proteins has certainly added a layer of complexity that makes functional interpretation even more difficult. These proteins are termed intrinsically disordered (i.e. IDPs), and they are defined by the lack of a static tertiary structure under physiological conditions (Dyson and Wright 2005; Dunker et al. 2008; Tompa 2002). Instead of adopting a single well-defined fold, IDPs exist in a state of continuous fluctuation among conformations that are separated by low energy barriers. Disordered proteins and protein regions dominate biological processes such as cell-cycle regulation, transcription, translation and membrane fusion (Dunker et al. 2008; Dyson and Wright 2005; Tompa 2002; Gsponer and Babu 2009), and they are also often associated with pathological conditions such as Alzheimer's disease and Parkinson's disease (Chiti and Dobson 2006). Disordered regions may function either by serving as entropic chains, flexibly connecting folded protein domains, or by binding protein or nucleic acid partners. Binding of disordered regions might be transient, often modulated by post-translational modifications, or permanent, as in most cases where the protein acts as a scaffold. Disordered segments generally go through binding-induced folding or disorder-to-order transitions upon binding, but in a number of known cases the bound regions retain their flexibility, forming so-called 'fuzzy' complexes (Tompa and Fuxreiter 2008).

The presence of conformational flexibility may be inferred from either experimental observations or computational predictions. Missing residues in X-ray crystallographic structures, a distinct shape of a Kratky-plot representation of SAXS data (Chap. 8), or narrow dispersion of chemical shifts in NMR spectroscopy (Chap. 3) may all indicate structural disorder, as may a number of other techniques of lower resolution (Chap. 7) such as infra-red spectroscopy, electron microscopy (EM) or circular dichroism (CD) (Jensen et al. 2013; Salmon et al. 2010; Bernadó and Svergun 2012; Sethi et al. 2013). Residue-wise structural disorder propensities can also be predicted computationally from the primary amino acid sequence, since disordered regions are enriched in hydrophilic and charged residues (i.e. disorder-promoting amino acids) while being depleted in the hydrophobic residues that are otherwise responsible for organizing the hydrophobic core in most folded proteins (Uversky et al. 2000). Even though localizing conformationally flexible protein segments is relatively straightforward, gaining insights into their functions is a different matter altogether. In fact, the most exciting and biologically relevant current challenge in the field of 'unstructural biology' is elucidating the connection between structural disorder and the functions that may arise from the disordered states (Tompa 2011; Tompa and Varadi 2014).

The classical approach applied to well-folded proteins is to provide a high resolution, atomic-level structural description of the molecule, and then interpret its function by correlating this structural representation (and its many details) with the action and activity of the protein. Since disordered proteins are inherently flexible, they cannot be described by a single, static structure, but solution experiments yielding ensemble-averaged long- and short-range structural information can be used in synergy with computational methods to approximate the structural description of an IDP with an ensemble of conformations (Fisher and Stultz 2011; Bernadó et al. 2007; Marsh and Forman-Kay 2011). In recent years there has been an increase in the number of techniques that can derive ensemble models of disordered proteins from SAXS and NMR data. One of the major concerns regarding these models is that even though they are based on experimental data, and consequently do conform to experimental observations, they might still not be representative of the physical reality. Calculating the structural ensemble of a disordered protein is a typical ill-posed problem. The number of conceivable conformations of a flexible protein chain (and therefore the degree of freedom of the problem) is immense, while the number of currently available experimental observations (and therefore the amount of constraints that are used by recent modelling protocols for selecting relevant conformations) is rather limited. The consequence is that multiple different ensembles may fit the experimental observations equally well, and currently there is no way to differentiate between these models without cross-validation from an independent data source. Furthermore, the various computational procedures used in ensemble calculations are not yet standardized, and in fact have never been evaluated in a comparative manner. Of course, such comparisons have further been hindered by the inaccessibility of both the ensemble models and the underlying experimental data.

In order to offer remedies to some of these issues, we have developed the Protein Ensemble Database (Varadi et al. 2014), the first online repository of disordered/denatured protein ensembles and the experimental data that was used during the calculation procedures. Our aim with this contribution to the field of ensemble calculation is to facilitate the design of standardized protocols and a new generation of ensemble modelling techniques that can face the challenge posed by the conformational flexibility of IDPs and intrinsically disordered regions (IDRs). The field of ensemble calculation is a new field and one that is constantly evolving; next generation modelling tools and the incorporation of new data types will therefore lead to a significant increase in the predictive power of ensemble models (Table 11.1).

Table 11.1 The most commonly used ensemble calculation protocols

Protocol name	Constraints	Example protein	Reference
FM/ASTEROIDS	RDCs, PREs, CSs	Measles N-tail	(Jensen et al. 2011)
GAJOE/EOM	SAXS	CYNEX4 FRET probe	(Mertens et al. 2012)
ENSEMBLE	RDCs, PREs, NOEs, J-couplings, SAXS	Sic1	(Mittag et al. 2010)
REMD	SAXS, CSs	α-synuclein	(Allison et al. 2009)
BEGR	SAXS	p53 TAD	(Daughdrill et al. 2012)

2 The Techniques of Ensemble Modelling

Even though a high-resolution structural description of flexible proteins may not be achieved using X-ray crystallography or solid state NMR, other solution techniques are particularly well-suited for this task. The intricate interplay between biophysical information obtained from such experimental measurements and sophisticated computational modelling algorithms allows the structure of an IDP to be described as an ensemble of interchanging conformations. Experimental measurements on chemical shifts (CSs) (Jensen et al. 2011; Allison et al. 2009), residual dipolar couplings (RDCs) (Jensen et al. 2011; Mittag et al. 2010), paramagnetic relaxation enhancements (PREs) (Jensen et al. 2011; Mittag et al. 2010), and J-couplings (Mittag et al. 2010), as well as topological constraints derived from SAXS measurements (Mertens et al. 2012; Mittag et al. 2010; Allison et al. 2009) and distance restraints from FRET (Tompa and Varadi 2014) can all be combined into a set of constraints for the modelling procedure. These constraints are at the core of our attempt to discriminate between relevant and supposedly realistic conformations from randomly oriented chains by either driving the selection algorithms or by allowing the validation of the ensemble models.

Broadly speaking there are two families of approaches for calculating structural ensembles, both of which are aimed at providing ensembles that can explain the experimental data used during the calculation procedures (Fig. 11.1). Protocols in the first set all begin with the generation of a large pool of random or semi-random

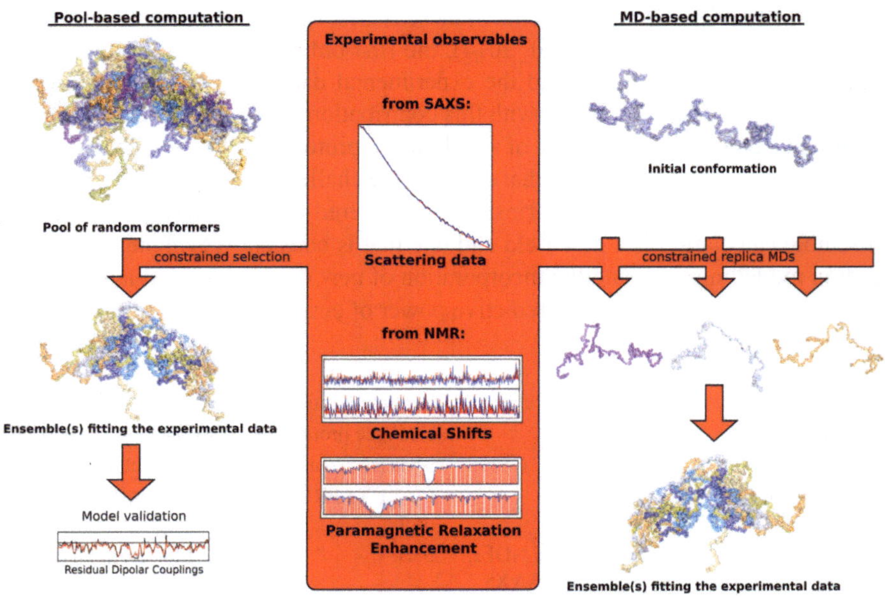

Fig. 11.1 Ensemble calculation procedures. There are two main families of ensemble modelling procedures. The pool-based computation (*left side*) is selecting from a large *pool of random or semi-random conformers*, while the MD-based computation (*right side*) is generating the ensembles by running short, parallel, constrained MD simulations. Both methods rely heavily on the input experimental data

conformations. Genetic algorithms are then deployed to subset the random pool into ensembles that fit to the experimental observations of SAXS and/or NMR. In contrast, the second approach starts from a single, randomly oriented protein chain. This random conformation is then driven through a special type of molecular dynamics (MD) simulation called replica-exchange meta-dynamics (REMD). A conformational sampling influenced by experimentally derived constraints is performed during the simulation.

2.1 Ensembles Selected from Random Pools

Calculation protocols falling into this family of ensemble modelling (Chap. 4) all start by generating a pool of a very large number of conformations. However, it is worth noting that even a pool of a million random conformers is merely a subset of all the conceivable conformations a protein chain might theoretically sample. The conformers in the pool are either randomly oriented or they might be forced to adhere to experimental (e.g. secondary chemical shift) data or theoretical constraints on Ψ/Φ angles, or secondary structure propensities. The most commonly used software for generating the pools are TraDES (Feldman and Hogue 2000, 2002), the Ensemble Optimization Method (EOM) (Bernadó et al. 2007; Bernadó and Svergun 2012) and Flexible-Meccano (FM) (Ozenne et al. 2012). In certain cases, for example when using FM, the conformers of the pool lack side-chains. Before proceeding further, the missing side-chains should be modelled using software such as SCCOMP (Eyal et al. 2004) or SCRWL (Canutescu et al. 2003). Once the pool is ready for the downstream analysis, theoretical biophysical parameters are calculated for each conformer. The calculation of theoretical values is crucial for comparing the random conformers to experimental data. If working with SAXS data, theoretical scattering curves need to be calculated, for example by the software CRYSOL (Bernadó et al. 2007). For NMR data, theoretical chemical shifts are calculated by ShiftX, SPARTA (Shen and Bax 2007) or other related chemical shift prediction approaches. Theoretical RDCs can be approximated by FM (Ozenne et al. 2012) or ENSEMBLE (Krzeminski et al. 2013), which use local alignment information, while also taking into account the long-range effects modulating the RDC baselines. Similarly, PREs, J-couplings, solvent accessibility, R_2 relaxation rates and nuclear Overhauser effect (NOE) values might be estimated.

By this stage of the modelling we should have all we need for the ensemble calculation: (1) the large pool of (semi-)random conformations, (2) the set of theoretically acquired biophysical parameters, and (3) their experimentally determined counterparts. The aspiration of the calculation procedures in this family of approaches is to select groups of conformers from the random pool that have theoretical values matching the observed experimental values. The software GAJOE, which is part of the software package EOM (Bernadó et al. 2007; Bernadó and Svergun 2012), selects conformers based only on SAXS data by iteratively fitting theoretical scattering curves to experimentally determined ones. The software ASTEROIDS (Salmon et al. 2010; Schneider et al. 2012) uses a genetic algorithm for selecting ensembles and iteratively repopulating the underlying potential energy landscapes by comparing theoretical parameters to available NMR and SAXS data. The algorithm converges to ensembles whose elements are unique, yet, on average, explain the experimental data equally well. Similarly, ENSEMBLE selects subsets of conformers based on a variety of NMR and SAXS data. Using these methods, the size of the final ensembles will be anywhere from between a few to several hundred conformations.

2.2 *Ensembles from Molecular Dynamics Simulations*

The starting point in these approaches (Chap. 2) is a single randomly oriented protein chain, as opposed to the large random pools of the previously described family of ensemble modelling procedures. The goal of the approach is to simulate the conformational sampling of this single random structure while taking experimental data into account. This class of simulations is called replica-exchange metadynamics (REMD) (Cavalli et al. 2013; Allison et al. 2009). Generally speaking, MD simulations aim to construct a plausible movement trajectory of a protein in a simulated environment. Such simulations are controlled by sophisticated functions, and usually cover a trajectory of 10–100 ns. However, the classical force fields used in simulations of structured proteins are inappropriate for IDPs, and the long-range structural changes an IDP might experience cannot be appropriately sampled in 100 ns (Cavalli et al. 2013). REMD simulations utilize a modified force field, introducing a penalty for deviating from the experimental measurements, and the values back-calculated from the structures sampled during the simulations, while also encouraging diverse sampling along pre-set variables, such as for example the radii of gyration. Additionally, multiple replica simulations are run in parallel; in other words, even though the time-scales for each replica are short, the combined outcome of the simulations offers a comprehensive sampling of the conformational space while remaining under the control of experimental data (Wu et al. 2009; Huang and Grzesiek 2010). It is worth noting that while the ensembles generated will be consistent with the experimental data used as restraints, there is no guarantee that they would remain consistent with cross-validation data not involved in the simulation. Theoretically, this could be achieved by sufficiently increasing the number of restraints that are used up to the point where specific long- or short-range structural information would become redundant.

2.3 *The Challenges of Ensemble Modelling*

Experimental observations from solution measurements provide information that is averaged on all the different conformations the protein chains have visited during the experiment; the selected theoretical conformers are therefore also required to fit the data on average, not one by one. As mentioned earlier, due to the immense number of conformations an IDP might sample, and also due to the limited amount of experimentally derived constraints used in the modelling, the resulting ensembles can only serve as discrete representations of highly complex probability distribution functions. In layman's terms, these methods give results that should be correct on average, although single conformers in the ensemble may not be biologically relevant or realistic.

Indeed, the greatest current challenge when dealing with ensemble models of IDPs is to determine if they provide accurate, realistic and biologically relevant representations of the range of conformations sampled by the proteins during their thermal fluctuations. The reliability of ensemble models surely lags far behind compared to that of the structural descriptions of structured proteins of well-defined folds.

The quality of an ensemble model depends strongly on the quality of the experimental data. Calculation procedures assume that the experimental observables are reliable, and therefore one will always get an output ensemble, even if the input experimental data is drastically inaccurate. This is especially true when using techniques, such as SAXS, which always yields output data, even in case of severe issues with the sample such as aggregation, degradation or impurity. The quality of the experimental data should therefore be rigorously checked.

The goal of the Protein Ensemble Database (Varadi et al. 2014) is to host not only the structural ensembles, but the underlying experimental data as well. We are convinced that making the data available will facilitate the development of new and more advanced calculation and validation procedures. Furthermore, the accessibility of the ensemble models may foster the establishment of a theoretical framework in which the function of disordered regions could be interpreted from ensemble descriptions.

3 The Structure of pE-DB

The data structure of the Protein Ensemble Database is implemented as a relational MySQL database. Relational databases record information in tables that are interconnected with each other, allowing complex and efficient searching and browsing. The pE-DB consists of a set of core tables and a dedicated module for each different supported experimental data type (Fig. 11.2). While the database is responsible for storing the relevant meta-information records, a dedicated file server hosts the protein ensembles in PDB format, the SAXS- and NMR-derived experimental data, the sequences in FASTA format, and a number of pre-generated plots and figures for each database entry. Each entry has a unique four-letter identifier, the pE-DB ID. These identifiers provide the means for connecting every piece of relevant meta-information to the referenced entry. These IDs can also be used for directly accessing the structural ensembles and the experimental data.

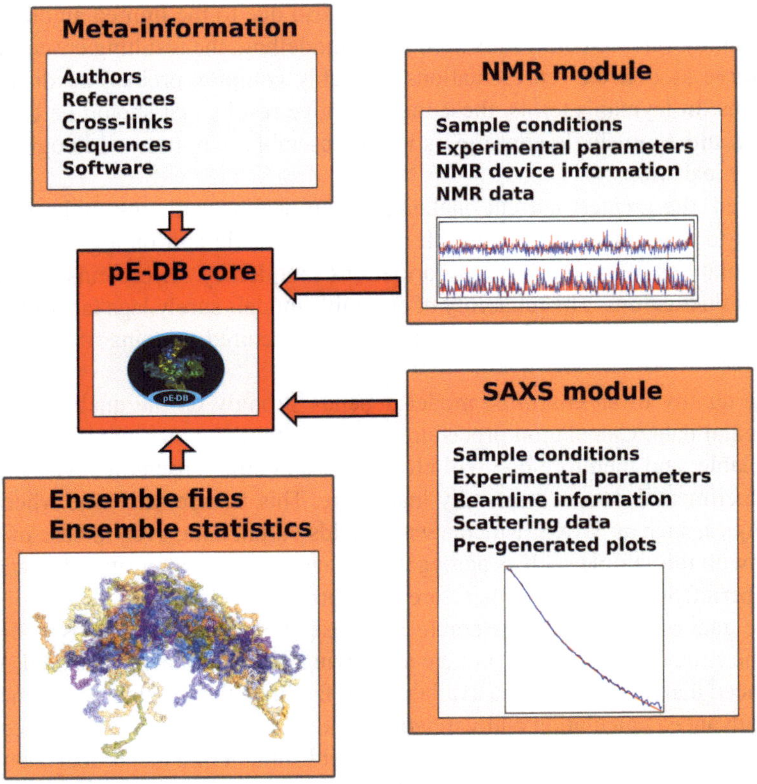

Fig. 11.2 The data structure of *pE-DB*. The *pE-DB* is organized in a modular manner, where the core module connects the various sub-parts. The currently supported data types (*NMR* and *SAXS*) both have their dedicated modules, and the ensemble files (in PDB format) are also stored separately. Finally, the meta-information module records the detailed information on the authors, on the proteins in the entries, the software used for the ensemble calculation, the relevant references and cross-links to other online resources

4 The Online User Interface

The most convenient way for users to interact with the database is by using the online user interface (http://pedb.vib.be). There are multiple intuitive ways for searching in the database, browsing the database entries by various criteria or downloading data in bulk. Advanced queries and direct SQL commands are also supported. Detailed information on every feature of the website is provided in the online user guide (http://pedb.vib.be/userguide.php).

Fig. 11.3 Browsing and searching features. The entries of the pE-DB can be conveniently browsed by *entries*, *proteins*, *types of data*, *authors* and *references*. These browsing buttons are on the *top* section of the screen, *right* on the *top* of the search field. Quick searches are possible by typing in a query term in this field, and fine tuning is possible by specifying the type of the search term (by default it will search every field in the database), and by selecting the data type of interest (by default it will return any type of data)

4.1 Browsing and Searching

The database offers browsing and searching tools in the top section of the main interface screen (Fig. 11.3). Browsing the entries by pE-DB IDs, proteins, author names, references and data types is supported. The number of entries per page can be adjusted on the browsing screen, and by clicking on any ID the user will be directed to the dedicated accession screen of the corresponding entry.

Searching in the database is possible by using either the fast and intuitive search window at the top section of the main screen or by using the 'Advanced search' tool. The default setting for the quick search option is to retrieve all the entries where the search term can be found, regardless of the underlying experimental data. Users may specify if the search term should be looked up only in a specific table, such as amongst the protein names, or the abstracts of the original publications. Valid search terms include identifiers from related databases such as UniProt (UniProt 2014) and DisProt (Sickmeier et al. 2007) IDs. Data types of interest can also be selected, allowing the retrieval of entries with only SAXS or NMR data, or requiring the combination of both data types.

The advanced search tool allows the use and combination of an arbitrary number of search terms. The type of search term (e.g. UniProt ID, author name, etc.) has to be specified. All the terms are connected with a Boolean 'AND' by default, but an 'OR' search can also be performed.

4.2 Data Retrieval

The pE-DB identifiers connect each ensemble model with their corresponding experimental data and their meta-information records. Consequently, every piece of information and data can be retrieved using the entry identifiers. Data can be downloaded for each entry either by navigating to its dedicated accession screen, or by using the bulk download tool of pE-DB found under the 'Download pE-DB' section on the left side menu. By providing a list of IDs, the sequences in FASTA format, ensembles in PDB format, NMR and SAXS experimental data in TEXT format, and the complete archives can be directly downloaded.

Furthermore, the complete database can be retrieved in flat SQL or tab-separated CSV formats along with every protein sequence, structural ensemble and the experimental data.

4.3 Accession Screen

The accession screen is where all the recorded information along with direct download links to the data and the ensembles is collected for the convenience of the user. The screen is divided into a number of sections, each of them dedicated to a different set of information (Fig. 11.4).

The 'General information' section provides a brief abstract of the data stored in the entry along with a description of the calculation protocol. Users can immediately see the method used for the calculation (random pool or MD simulation), the data type(s) used (SAXS and/or NMR) and whether there was any experimental validation of the structural ensemble. The latter information is important, since the database also hosts theoretical ensembles where the calculations were not compared against experimental evidence.

The 'Image gallery' section displays three sample conformations from the ensembles—one for the most compact conformer, one that has a radius of gyration closest to the average of the ensemble, and finally one that is the most extended conformer in the ensemble. Clicking on any of these figures directs the user to a JSmol visualization applet where the distribution of the radii of gyration is displayed along with each conformer. By selecting a single conformer, the users can immediately visualize the 3D representation in a fully customizable applet window (Fig. 11.5).

The 'Protein information' tab leads to a brief description of the protein(s) of the entry, the sequences, parameters (such as the molecular weights), and cross-links to other online repositories such as the UniProt (UniProt 2014), GenBank (Benson et al. 2013) or DisProt (Sickmeier et al. 2007) databases.

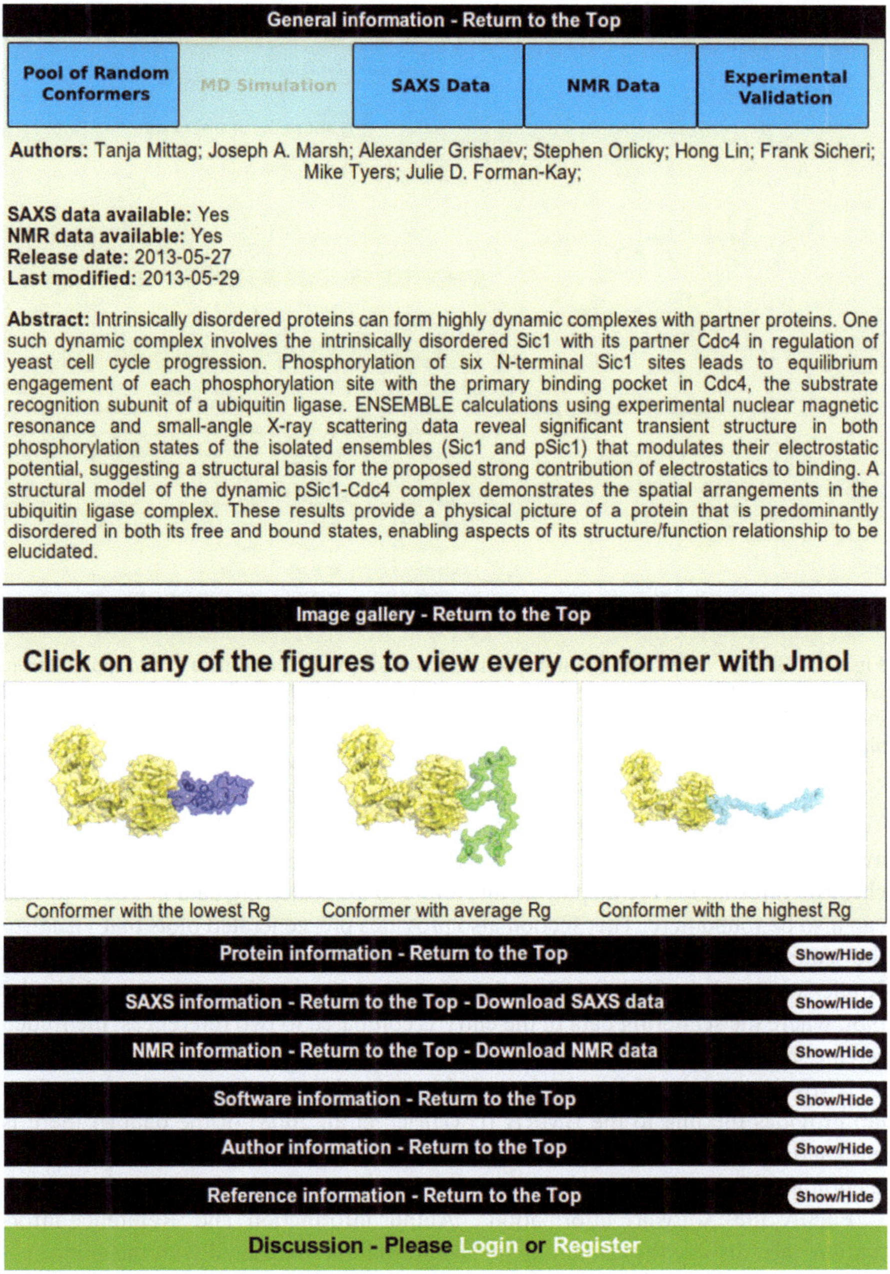

Fig. 11.4 The pE-DB entry screen. The main entry screens of every accession in the pE-DB have similar general outline. The *top* section provides *general information* about the entry, such as the type of calculation procedure or the experimental data used for the modelling, as well as a brief description of the entry in the '*Abstract*' section. The *image gallery* displays three examples of the various conformers found in the ensembles. Clicking on any of these will direct the user to a customizable JSmol applet. The sections below the image gallery show detailed information on the proteins, the experimental data, the software and authors, and finally the references. Each of these subsections can be expanded by clicking on the '*Show/Hide*' buttons

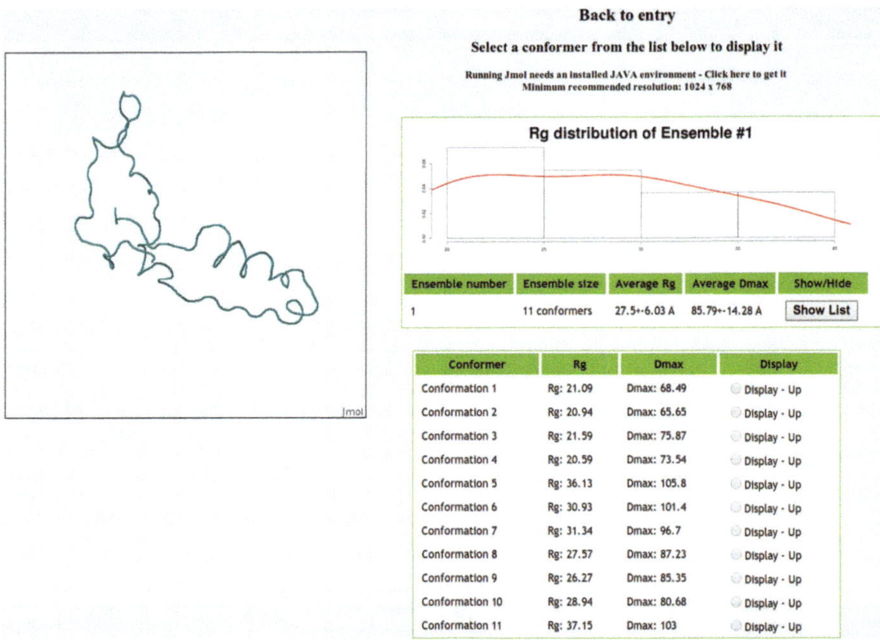

Fig. 11.5 Built-in visualization JSmol applet screen. Each pE-DB entry has a JSmol applet screen, where every conformer in each ensemble can be displayed in a fully customizable manner. General information on the *ensembles*, and the radius of gyration and Dmax values for each conformer are also provided here

When the ensembles rely on SAXS data, the 'SAXS information' section is available. Experimental parameters from sample components to beamline and synchrotron information are displayed, and a link to the scattering data in TEXT format can also be found here. This section also provides pre-generated plots that visualize the scattering data. The plots shown are the scattering curve, the Guinier plot, the Kratky plot and the P(r) function. Additionally, we provide a normalized Kratky plot, where the scattering data of the entry is compared to two references, the folded bovine serum albumin (BSA) and the unfolded tau protein.

NMR data (when applicable) is available at the 'NMR information' section. This tab provides the link to the BMRB ID (Ulrich et al. 2008) of the data (if there is one) and the experimental parameters. The data can also be downloaded from this section in TEXT format.

Finally, the 'Software information', 'Author information' and 'Reference information' tabs display the remaining meta-information associated with the entry.

A discussion forum where users can share their insights and doubts regarding the entry in question or the corresponding calculation procedures can be found at the bottom of each entry.

5 Data Submission

We encourage the research and development community to submit their own ensemble models to the Protein Ensemble Database in combination with the underlying experimental data. We also support the recalculation and evaluation of the ensembles that are already being hosted. Fortunately, since the release of the database in early 2014, we have already experienced a flow of submissions from research groups across the globe. Entries are being released only after the authors have published their study; the majority of submitted entries are therefore being kept on hold, invisible to the community and waiting to be released.

Data submission to pE-DB is a multi-step procedure that is initiated by contacting the pE-DB crew in the form of a pre-submission query that can be completed online under the 'Data submission' section of the user interface. The pre-submission request should provide a brief description of the ensembles, the experimental data types used, and the contact information of the authors. Following favourable evaluation of the request, we contact the authors and refer them to the online meta-information submission form. This form should record every piece of relevant information from sample conditions to personal and institute information. In parallel, we also require the submission of the ensembles in PDB format and the experimental data in TEXT format. Data submission is possible by either providing a download link or by using a secure FTP connection to our file server. Each submission is manually curated by experts in the field, and only ensembles that are based on high-quality experimental data are considered for deposition.

6 Towards the Functional Interpretation of Ensembles

It has recently become increasingly apparent that understanding the function of a protein is hindered by thinking in terms of static structures as opposed to a range of thermally accessible conformers (Tompa 2011). We are convinced that by developing and releasing the Protein Ensemble Database we are contributing to the field of structural biology with a new cornerstone in the evolution of ensemble descriptions for IDPs (Varadi et al. 2014). Whereas the Protein Data Bank (Berman et al. 2000) hosts the structures of proteins with well-defined folds, pE-DB stores the ensembles of flexible proteins that have previously been inaccessible.

These ensemble models currently lack the descriptive power of high-resolution crystal structures, and therefore their predictive power is more limited as well. Due to the immensely high degree of conformational freedom in IDPs, the models are completely dependent on the quality and amount of experimental data used during the calculation procedures. Theoretically, increasing the number of experimentally derived constraints will increase the quality of the ensemble models (Tompa and Varadi 2014).

Ensembles are often criticized, yet they have never been evaluated rigorously. By making the ensembles and the experimental observations available, researchers can access and cross-validate these models by using independent data, for example long-range information derived from FRET.

Without a doubt, the long-term objective of ensemble calculation is to achieve such descriptive power that the functions of IDPs can be reliably interpreted—and perhaps even predicted—from structural ensemble models. While we are clearly a long way from this goal, the availability of ensembles, protocols and experimental data will surely contribute to the development of next-generation ensemble calculation algorithms and the establishment of standardized protocols.

References

Allison JR, Varnai P, Dobson CM et al (2009) Determination of the free energy landscape of α-synuclein using spin label nuclear magnetic resonance measurements. J Am Chem Soc 131(51):18314–18326. doi:10.1021/ja904716h

Benson DA, Cavanaugh M, Clark K et al (2013) GenBank. Nucleic Acids Res 41(Database issue):D36–D42. doi:10.1093/nar/gks1195

Berman HM, Westbrook J, Feng Z et al (2000) The protein data bank. Nucleic Acids Res 28(1):235–242

Bernadó P, Svergun DI (2012) Structural analysis of intrinsically disordered proteins by small-angle X-ray scattering. Mol Biosyst 8(1):151–167. doi:10.1039/c1mb05275f

Bernadó P, Mylonas E, Petoukhov MV et al (2007) Structural characterization of flexible proteins using small-angle X-ray scattering. J Am Chem Soc 129(17):5656–5664. doi:10.1021/ja069124n

Canutescu AA, Shelenkov AA, Dunbrack RL Jr (2003) A graph-theory algorithm for rapid protein side-chain prediction. Protein Sci (A publication of the Protein Society) 12(9):2001–2014. doi:10.1110/ps.03154503

Cavalli A, Camilloni C, Vendruscolo M (2013) Molecular dynamics simulations with replica-averaged structural restraints generate structural ensembles according to the maximum entropy principle. Journal Chem Phys 138(9):094112. doi:10.1063/1.4793625

Chiti F, Dobson CM (2006) Protein misfolding, functional amyloid, and human disease. Annu Rev Biochem 75:333–366. doi:10.1146/annurev.biochem.75.101304.123901

Daughdrill GW, Kashtanov S, Stancik A et al (2012) Understanding the structural ensembles of a highly extended disordered protein. Mol Biosyst 8(1):308–319. doi:10.1039/c1mb05243h

Dunker AK, Silman I, Uversky VN et al (2008) Function and structure of inherently disordered proteins. Curr Opin Struct Biol 18(6):756–764. doi:10.1016/j.sbi.2008.10.002

Dyson HJ, Wright PE (2005) Intrinsically unstructured proteins and their functions. Nat Rev Mol Cell Biol 6(3):197–208. doi:10.1038/nrm1589

Eyal E, Najmanovich R, McConkey BJ et al (2004) Importance of solvent accessibility and contact surfaces in modeling side-chain conformations in proteins. J Comput Chem 25(5):712–724. doi:10.1002/jcc.10420

Feldman HJ, Hogue CW (2000) A fast method to sample real protein conformational space. Proteins 39(2):112–131

Feldman HJ, Hogue CW (2002) Probabilistic sampling of protein conformations: new hope for brute force? Proteins 46(1):8–23

Fisher CK, Stultz CM (2011) Constructing ensembles for intrinsically disordered proteins. Curr Opin Struct Biol 21(3):426–431. doi:10.1016/j.sbi.2011.04.001

Gsponer J, Babu MM (2009) The rules of disorder or why disorder rules. Prog Biophys Mol Biol 99(2/3):94–103. doi:10.1016/j.pbiomolbio.2009.03.001

Huang JR, Grzesiek S (2010) Ensemble calculations of unstructured proteins constrained by RDC and PRE data: a case study of urea-denatured ubiquitin. J Am Chem Soc 132(2):694–705. doi:10.1021/ja907974m

Jensen MR, Communie G, Ribeiro EA Jr et al (2011) Intrinsic disorder in measles virus nucleocapsids. Proc Natl Acad Sci U S A 108(24):9839–9844. doi:10.1073/pnas.1103270108

Jensen MR, Ruigrok RW, Blackledge M (2013) Describing intrinsically disordered proteins at atomic resolution by NMR. Curr Opin Struct Biol 23(3):426–435. doi:10.1016/j.sbi.2013.02.007

Krzeminski M, Marsh JA, Neale C et al (2013) Characterization of disordered proteins with ENSEMBLE. Bioinformatics 29(3):398–399. doi:10.1093/bioinformatics/bts701

Marsh JA, Forman-Kay JD (2011) Ensemble modeling of protein disordered states: experimental restraint contributions and validation. Proteins. doi:10.1002/prot.23220

Mertens HD, Piljic A, Schultz C et al (2012) Conformational analysis of a genetically encoded FRET biosensor by SAXS. Biophys J 102(12):2866–2875. doi:10.1016/j.bpj.2012.05.009

Mittag T, Marsh J, Grishaev A et al (2010) Structure/function implications in a dynamic complex of the intrinsically disordered Sic1 with the Cdc4 subunit of an SCF ubiquitin ligase. Structure 18(4):494–506. doi:10.1016/j.str.2010.01.020

Ozenne V, Bauer F, Salmon L et al (2012) Flexible-meccano: a tool for the generation of explicit ensemble descriptions of intrinsically disordered proteins and their associated experimental observables. Bioinformatics 28(11):1463–1470. doi:10.1093/bioinformatics/bts172

Salmon L, Nodet G, Ozenne V et al (2010) NMR characterization of long-range order in intrinsically disordered proteins. J Am Chem Soc 132(24):8407–8418. doi:10.1021/ja101645g

Schneider R, Huang JR, Yao M et al (2012) Towards a robust description of intrinsic protein disorder using nuclear magnetic resonance spectroscopy. Mol Biosyst 8(1):58–68. doi:10.1039/c1mb05291h

Sethi A, Anunciado D, Tian J et al (2013) Deducing conformational variability of intrinsically disordered proteins from infrared spectroscopy with Bayesian statistics. Chem Phys 422. doi:10.1016/j.chemphys.2013.05.005

Shen Y, Bax A (2007) Protein backbone chemical shifts predicted from searching a database for torsion angle and sequence homology. J Biomol NMR 38(4):289–302. doi:10.1007/s10858-007-9166-6

Sickmeier M, Hamilton JA, LeGall T et al (2007) DisProt: the database of disordered proteins. Nucleic Acids Res 35(Database issue):D786–D793. doi:10.1093/nar/gkl893

Tompa P (2002) Intrinsically unstructured proteins. Trends Biochem Sci 27(10):527–533

Tompa P (2011) Unstructural biology coming of age. Curr Opin Struct Biol 21(3):419–425. doi:10.1016/j.sbi.2011.03.012. (S0959-440X(11)00064-9 [pii])

Tompa P, Fuxreiter M (2008) Fuzzy complexes: polymorphism and structural disorder in protein-protein interactions. Trends Biochem Sci 33(1):2–8. doi:10.1016/j.tibs.2007.10.003

Tompa P, Varadi M (2014) Predicting the predictive power of IDP ensembles. Structure 22(2):177–178. doi:10.1016/j.str.2014.01.003

Ulrich EL, Akutsu H, Doreleijers JF et al (2008) BioMagResBank. Nucleic Acids Res 36(Database issue):D402–D408. doi:10.1093/nar/gkm957

UniProt C (2014) Activities at the universal protein resource (UniProt). Nucleic Acids Res 42(Database issue):D191–D198. doi:10.1093/nar/gkt1140

Uversky VN, Gillespie JR, Fink AL (2000) Why are "natively unfolded" proteins unstructured under physiologic conditions? Proteins 41(3):415–427

Varadi M, Kosol S, Lebrun P et al (2014) pE-DB: a database of structural ensembles of intrinsically disordered and of unfolded proteins. Nucleic Acids Res 42(Database issue):D326–D335. doi:10.1093/nar/gkt960

Wu KP, Weinstock DS, Narayanan C et al (2009) Structural reorganization of α-synuclein at low pH observed by NMR and REMD simulations. J Mol Biol 391(4):784–796. doi:10.1016/j.jmb.2009.06.063

Chapter 12
Order and Disorder in the Replicative Complex of Paramyxoviruses

Jenny Erales, David Blocquel, Johnny Habchi, Matilde Beltrandi, Antoine Gruet, Marion Dosnon, Christophe Bignon and Sonia Longhi

Abstract In this review we summarize available data showing the abundance of structural disorder within the nucleoprotein (N) and phosphoprotein (P) from three paramyxoviruses, namely the measles (MeV), Nipah (NiV) and Hendra (HeV) viruses. We provide a detailed description of the molecular mechanisms that govern the disorder-to-order transition that the intrinsically disordered C-terminal domain (N_{TAIL}) of their N proteins undergoes upon binding to the C-terminal X domain (XD) of the homologous P proteins. We also show that a significant flexibility persists within N_{TAIL}–XD complexes, which therefore provide illustrative examples of "fuzziness". The functional implications of structural disorder for viral transcription and replication are discussed in light of the ability of disordered regions to establish a complex molecular partnership and to confer a considerable reach to the elements of the replicative machinery.

Keywords Viral proteins · Nipah virus · Hedra virus · Measles virus · Intrinsic disorder · Folding upon binding

1 The Replicative Machinery of the Measles, Hendra and Nipah Viruses

The measles (MeV), Nipah (NiV) and Hendra (HeV) viruses are all members of the *Paramyxovirinae* sub-family within the *Paramyxoviridae* family of the *Mononegavirales* order. While MeV belongs to the *Morbillivirus* genus, NiV and HeV have been classified in the *Henipavirus* genus. *Paramyxoviridae* have a non-segmented, negative-stranded RNA genome that encodes six proteins: the nucleoprotein (N),

S. Longhi (✉) · J. Erales · D. Blocquel · J. Habchi · M. Beltrandi · A. Gruet · M. Dosnon · C. Bignon
Aix-Marseille Université, AFMB UMR 7257, 13288 Marseille, France
e-mail: Sonia.Longhi@afmb.univ-mrs.fr

CNRS, AFMB UMR 7257, 13288 Marseille, France

I. C. Felli, R. Pierattelli (eds.), *Intrinsically Disordered Proteins Studied by NMR Spectroscopy,* Advances in Experimental Medicine and Biology,
DOI 10.1007/978-3-319-20164-1_12

the phosphoprotein (P), the matrix protein, the F and H glycoproteins and the RNA-dependent RNA polymerase or "large" protein (L). The genome of *Paramyxoviridae* is not naked but encapsidated by multiple copies of the N protein, forming a helical nucleocapsid that serves as a template for both transcription and replication. These activities are ensured by the RNA-dependent RNA polymerase made of the L and P proteins, with P serving as an essential tethering factor between L and the nucleocapsid (Fig. 12.1). This ribonucleoprotein complex made of RNA and of the N, P and L proteins constitutes the replication machinery of *Paramyxoviridae.*

By virtue of their structural role in encapsidating the genome, paramyxoviral nucleoproteins are the most abundant viral proteins (Lamb and Kolakofsky 2001). Within MeV infected cells, N is found in a soluble, monomeric form (referred to as N°) and in a nucleocapsid assembled form. Following synthesis of the N protein, a chaperone is required to maintain this latter protein in a soluble and monomeric form. This role is played by the P protein, whose association simultaneously prevents illegitimate self-assembly of N (Huber et al. 1991; Spehner et al. 1997). This soluble N°-P complex is used as the substrate for the encapsidation of the nascent genomic RNA chain during replication (see Albertini et al. 2005; Blocquel et al. 2012a; Lamb and Kolakofsky 2001; Lamb and Parks 2007; Longhi and Canard 1999; Roux 2005 for reviews on transcription and replication). The assembled form of N also forms complexes with either isolated P or P bound to L, which are both essential to RNA synthesis by the viral polymerase (Buchholz et al. 1994; Ryan and Portner 1990).

Although L is supposed to possess all activities required for transcription and replication, including nucleotide polymerization, mRNA capping and polyadenylation, it is not active in the absence of P. The large size of *Paramyxoviridae* L, its low abundance in infected cells and the requirement of P for stability and/or prevention of spontaneous oligomerization (Chattopadhyay and Banerjee 2009) have rendered the molecular and structural characterization of *Paramyxoviridae* polymerase highly challenging. Indeed, so far no paramyxoviral polymerase has been purified to homogeneity, thereby explaining the scarcity of molecular data on this large, multifunctional enzyme. The only exceptions are the L/P complex from Rinderpest virus (RDV), which has been partially purified (Gopinath and Shaila 2008), and the Sendai virus (SeV) polymerase, which was shown to possess a methyltransferase activity in its C-terminal region (Ogino et al. 2005), in agreement with predictions (Ferron et al. 2002).

In the past decade we have gathered a wealth of bioinformatics and experimental evidence showing that paramyxoviruses N and P are enriched in intrinsically disordered regions (IDRs) (for reviews see Blocquel et al. 2012a; Habchi and Longhi 2012; Habchi et al. 2012; Longhi 2011; Longhi and Oglesbee 2010). Intrinsically disordered proteins (IDPs) and IDRs are ubiquitous proteins/regions lacking stable secondary and tertiary structures under physiological conditions of pH and salinity in the absence of their biological partner and thus exist as dynamic ensembles of conformers (for a recent review see Habchi et al. 2014). IDPs/IDRs are functional while being either fully or partly disordered. IDPs complement the functional repertoire of folded proteins, being able to interact with several partners and thereby

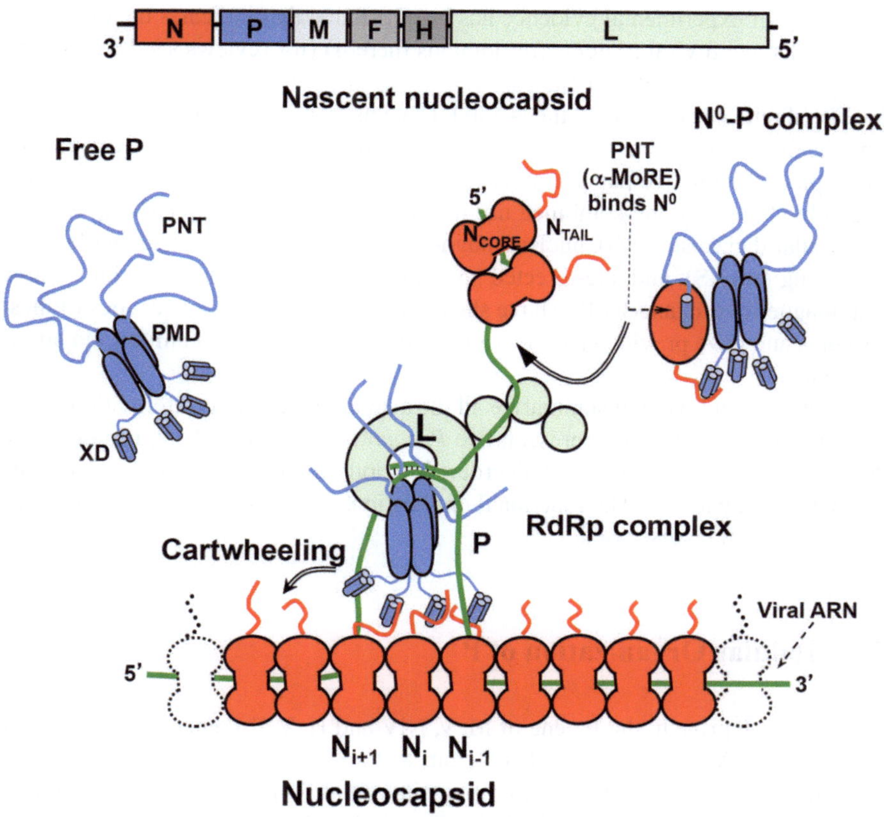

Fig. 12.1 *Schematic illustration of the Paramyxoviridae replicative complex*. The *top* of the figure represents the schematic organization of the genome encoding the nucleoprotein (*N*, *orange*), the phosphoprotein (*P*, *blue*), the matrix protein (*M*, *light grey*), the fusion protein (*F*, *grey*), the glycoprotein (*H*, *grey*) and the polymerase large subunit (*L*, *light green*). The *lower* part of the figure represents the scheme of the replicative complex with the RNA represented by a *green line*. The neo-synthetized RNA is shown already partially encapsidated. The *N* and *P* intrinsically disordered regions are symbolized by *lines*. The extended conformation of the disordered regions would allow the formation of a tripartite complex between *N*°, *P* and *L* required for nucleocapsid assembly. The P/L complex forms the RNA-dependent RNA polymerase (RdRp) complex, which cartwheels onto the nucleocapsid complex via the XD domain of *P*. *P* is shown as a tetramer to reflect the prevalence of this oligomeric state in paramyxoviral *P* proteins (see text). Modified from (Blocquel et al. 2012a)

exerting multiple biological functions (see Habchi et al. 2014 and references cited therein).

Bioinformatics studies have shown that viruses and eukaryotes have ten times more conserved disorder (roughly 1 %) than archaea and bacteria (0.1 %) (Chen et al. 2006), and have also indicated that viral proteins, and in particular proteins from RNA viruses, are enriched in short disordered regions (Tokuriki et al. 2009; Xue et al. 2012; Xue et al. 2010). Beyond these computational studies, a considerable body of experimental evidence has been collated that indicates the disordered nature of several viral proteins (or domains thereof) (for reviews see Uversky and Longhi 2012; Xue et al. 2014).

The abundance of IDRs in the N and P proteins from *Paramyxoviridae* members and the difficulty of obtaining homogenous polymers of N suitable for X-ray analysis explain the relative paucity of structural data obtained so far by X-ray crystallography. However, the combined use of other, more appropriate techniques, such as circular dichroism (CD), nuclear magnetic resonance (NMR), small angle X-ray scattering (SAXS), and site-directed spin labelling (SDSL) coupled with electron paramagnetic resonance (EPR) have shed light onto the molecular features of these proteins and have provided a quite accurate description of their conformational behaviour.

In this chapter we will summarize all the available molecular information on the N and P proteins of three representative *Paramyxoviridae* members, namely MeV, NiV and HeV, focusing on their disordered regions and the interactions they establish with their partners. The functional implications of disorder for transcription and replication will be discussed.

2 Modular Organization of P

Beyond the P protein, the P gene of MeV, NiV and HeV also encodes the C and V proteins. The V protein is translated from a P messenger resulting from co-transcriptional insertion of a G at the editing site of the P mRNA. The V protein thus shares the N-terminal module (MeV PNT, aa 1–230; NiV PNT, aa 1–406 and HeV PNT, aa 1–404) with the P protein and possesses a unique C-terminal, zinc-binding domain. Hence, the P protein consists of at least two domains: an N-terminal domain (PNT) common to both P and V, and a C-terminal domain (PCT) unique to the P protein (Fig. 12.2). The third protein, C, is encoded by an alternate open reading frame through ribosome initiation at an alternative translation codon.

Using various computational approaches (as described in Bourhis et al. 2007; Ferron et al. 2006; Lieutaud et al. 2013; Longhi et al. 2010), we have previously shown that the P proteins of *Paramyxovirinae* members have a modular organisation, being composed of alternating disordered and ordered regions (Habchi et al. 2010; Karlin et al. 2003). *Paramyxovirinae* PNT has been consistently predicted to be mostly disordered (Habchi et al. 2010; Karlin et al. 2003; Karlin et al. 2002b), being depleted in hydrophobic residues and enriched in so-called disorder-promoting residues (i.e. charged residues, along with glycine, serine and proline residues) (Campen et al. 2008). In the case of MeV, NiV and HeV, the disordered nature of PNT is also supported by several independent lines of experimental evidence: PNT was found to be highly sensitive to proteolysis, was shown to have a Stokes radius (R_S) higher than expected for a globular protein of the same size, and was shown to be disordered by both far-UV CD and NMR spectroscopy (Habchi et al. 2010; Karlin et al. 2003; Karlin et al. 2002b).

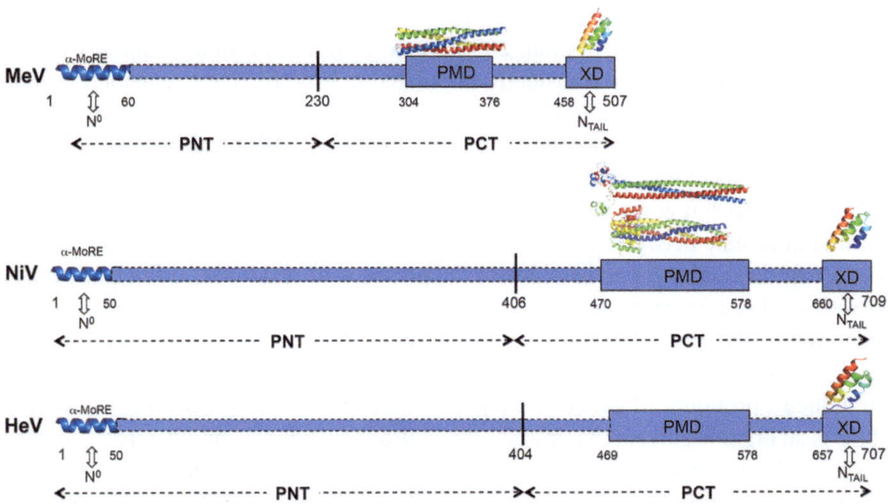

Fig. 12.2 *Modular organization of the MeV, NiV and HeV phosphoproteins*. Domain organization of P showing that it is composed of two moieties, PNT and PCT. Structured and disordered regions are represented as *large* or *narrow boxes*, respectively. PNT: N-terminal region of P; PCT: C-terminal region of P. PMD: P multimerisation domain; XD: X domain of P adopting a triple α-helical bundle. The α-MoRE partly preconfigured in solution and predicted/shown to adopt a stable α-helical conformation upon binding to N° is shown. Whenever available, the relevant crystal structures are shown above each domain. MeV PMD: PDB code 4BHV (Blocquel et al. 2014); MeV XD: PDB code 1OKS (Johansson et al. 2003); HeV XD: PDB code 4HEO (Communie et al. 2013b). The structural model of the NiV PMD trimeric form (Blocquel et al. 2013) and the structure of the tetrameric form as observed in crystals (Bruhn-Johannsen et al. 2014) are shown. The structural model of NiV XD is depicted (Habchi et al. 2011). All structures were drawn using Pymol (DeLano 2002)

Interestingly, a short (40–50 residues) ordered region is consistently predicted at the N-terminus of MeV, NiV and HeV P by all the disorder predictors implemented in the MeDor metaserver (Lieutaud et al. 2008) (Fig. 12.2). This N-terminal module with α-helical folding propensities corresponds to a conserved region amongst *Avulavirus*, *Henipavirus* and *Rubulavirus* members (Karlin et al. 2003), with that from *Rubulaviruses* having been shown to be involved in N°-binding (Watanabe et al. 1996). Using computational approaches, all *Paramyxovirinae* P proteins were found to share a short (11–16 residues) sequence motif within their first 40 residues (Karlin and Belshaw 2012). It has been proposed that this region would be conserved in all *Mononegavirales* phosphoproteins as a result of divergent evolution, and would be involved in binding to N° (Karlin and Belshaw 2012). In agreement, a similar N-terminal module, globally disordered yet containing transient α-helices (aa 1–60), has recently been identified and characterized in the vesicular stomatitis virus P protein (Leyrat et al. 2011a) and shown to fold upon binding to N° (Leyrat et al. 2011b). The N-terminal region of *Paramyxovirinae* P likely corresponds to an α-helical molecular recognition element (α-MoRE), where MoREs are short, order-prone regions within IDPs that have a certain propensity to bind to a partner and to undergo induced folding (i.e. a disorder-to-order transition) (Garner et al. 1999; Mohan et al. 2006; Oldfield et al. 2005; Vacic et al. 2007). In agreement with the predicted occurrence of a transiently populated α–MoRE, size exclusion chromatography (SEC) and dynamic light scattering (DLS) studies indeed unveiled that MeV, NiV and HeV PNT domains are not fully unfolded but rather conserve some degree of compactness typical of a premolten globule (PMG) conformation (Habchi et al. 2010; Karlin et al. 2002b). PMGs are characterized by an intermediate conformational state between a random coil and a molten globule and possess a certain degree of residual compactness due to the presence of residual and fluctuating secondary and/or tertiary structures (Dunker et al. 2001; Uversky 2002). The folding potential of the PNT domains was further confirmed using far-UV CD spectroscopy, where increasing concentrations of 2,2,2-trifluoroethanol (TFE) were shown to induce a pronounced gain of α-helicity (Habchi et al. 2010; Karlin et al. 2002b). TFE is an organic solvent that mimics the hydrophobic environment experienced by proteins during protein-protein interactions (Dahlman-Wright and McEwan 1996; Hua et al. 1998). The extent of residual compactness within the three PNT domains follows the order NiV PNT > HeV PNT > MeV PNT. Although it is plausible that the N-terminal region of PNT folds upon binding to N°, assessment of the effective folding-upon-binding abilities of this putative α–MoRE awaits the isolation and purification of a binding partner. Beyond N°, one such possible binding partner may be L and/or SNAP29, analogously to the closely related RDV and human parainfluenza type 3 virus, respectively (Ding et al. 2014; Sweetman et al. 2001).

PCT has a modular organization. It possesses a long disordered linker separating the P multimerisation domain, PMD, and the C-terminal X domain, XD (Habchi et al. 2010; Karlin et al. 2003) (Fig. 12.2). In the case of SeV, the disordered nature of this linker was experimentally confirmed by NMR studies (Bernadó et al. 2005; Houben et al. 2007a). In the case of MeV, indirect evidence in support of the disordered nature of this linker comes from the observation that spontaneous proteolytic cleavage occurs within this region (Longhi et al. 2003). In morbilliviruses and henipaviruses, an additional disordered region (referred to as a “spacer”) is predicted to occur upstream of PMD (Habchi et al. 2010; Karlin et al. 2003), with direct experimental confirmation of its disordered state having been provided in the case of MeV P (Communie et al. 2013a).

In line with both predictions (Habchi et al. 2010; Karlin et al. 2003) and spectroscopic studies (Habchi et al. 2011), the structures of MeV and HeV XD were shown to consist of a triple α-helical bundle (Communie et al. 2013b; Gely et al. 2010; Johansson et al. 2003; Kingston et al. 2004a) (Fig. 12.2). In MeV, NiV and HeV, XD is responsible for the interaction between P and the nucleocapsid (Habchi et al. 2012; Johansson et al. 2003; Kingston et al. 2004a). High-resolution structural data is also available for the X domains of the closely related SeV and mumps virus (MuV), the structures of which have been solved by nuclear magnetic resonance (NMR) and X-ray crystallography, respectively (Blanchard et al. 2004; Kingston et al. 2008). Contrary to all other paramyxoviral X domains investigated so far, MuV XD does not interact with the C-terminal region of N but rather establishes contacts with the structured N_{CORE} region of N (Kingston et al. 2004b). Interestingly, while in the majority of *Paramyxovirinae* members the C-terminal nucleocapsid-binding region of P adopts a stably folded, compact conformation, it is disordered in respiratory syncytial virus (RSV), a *Pneumovirinae* member (Llorente et al. 2006; Tran et al. 2007). In the same vein, recent experimental data unveiled that in the case of the P proteins from *Rubulavirus* members, intrinsic disorder further extends to their X domains, which were found to span a structural continuum ranging from stable to largely disordered in solution. They share, however, the ability to adopt, at least transiently, a common fold consisting of a triple α-helical bundle as in other *Paramyxovirinae* members (Kingston et al. 2008; Yegambaram et al. 2013).

Paramyxoviridae PMDs are predicted to adopt a coiled-coil structure (Habchi et al. 2010; Karlin et al. 2003; Llorente et al. 2006). The coiled-coil organization has been experimentally confirmed in the case of SeV (Tarbouriech et al. 2000), RDV (Rahaman et al. 2004), MeV (Blocquel et al. 2014; Communie et al. 2013a), MuV (Cox et al. 2013) and NiV (Bruhn-Johannsen et al. 2014). The oligomeric nature of *Paramyxoviridae* P proteins studied so far is in support of the so-called "cartwheeling" mechanism proposed for *Paramyxoviridae*, which posits that the polymerase complex cartwheels from one N monomer to another within the nucleocapsid in order to allow transcription and replication to take place (Kolakofsky et al. 2004). In spite of this common feature, however, the molecular information gathered so far has underscored considerable differences in the organization of P among various paramyxoviruses as detailed below.

PMDs from SeV (Tarbouriech et al. 2000) and MeV were shown to form a tetrameric coiled-coil (Blocquel et al. 2014; Communie et al. 2013a). The tetrameric coiled-coil organization of PMD has also been experimentally confirmed in the case of RDV (Rahaman et al. 2004), RSV (Llorente et al. 2006; Llorente et al. 2008) and MuV (Cox et al. 2013). In this latter case, however, the tetramer was found to consist of two sets of parallel helices in opposite orientation, i.e. to be a dimer of two antiparallel coiled-coil dimers (Cox et al. 2013), in striking contrast with all the *Paramyxoviridae* P proteins characterized so far, which were all shown to possess a parallel organization. Whether these differences in the organization are compatible with a common mechanism of transcription and replication remains to be experimentally established.

Strikingly, studies focused on NiV PMD yielded different results depending on whether the protein was studied in solution or by X-ray crystallography. Indeed, several independent biochemical and biophysical approaches, including SEC, SDS-PAGE, cross-linking, analytical ultracentrifugation and SAXS consistently converged to show that NiV PMD adopts a trimeric organization in solution (Blocquel et al. 2013). Notably, cross-linking experiments carried out by another group and making use of a different cross-linker (glutaraldehyde) revealed a tetrameric organization for NiV PMD (Salvamani et al. 2013). However it should be pointed out that the very high cross-linker concentrations used in those studies may have generated non-specific association. On the other hand, the crystallographic structure reported by Bruhn et al. unambiguously shows a tetrameric organization within the crystal (Bruhn-Johannsen et al. 2014). Figure 12.3a shows the trimeric model (Blocquel et al. 2013) and the crystallographic structure docked into the *ab initio*

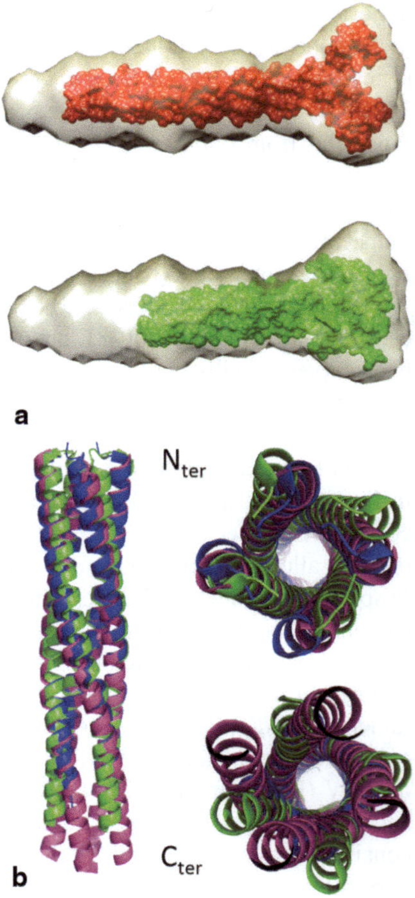

Fig. 12.3 *Structure of PMDs*. **a** SAXS envelope of NiV PMD (Blocquel et al. 2013) in which we docked either the trimeric form (Blocquel et al. 2013) (*top*) or the tetrameric form (PDB code 4N5B) (Bruhn-Johannsen et al. 2014) (*bottom*) of NiV PMD. **b** Superimposition among the two MeV PMD crystal structures solved by Blocquel et al. (Blocquel et al. 2013) (PDB codes 4BHV, shown in *green*, and 4C5Q, shown in *blue*) and the one solved by Communie et al. (Communie et al. 2013a) (PDB code 3ZDO, shown in *purple*). Modified from (Blocquel et al. 2014)

envelope of NiV PMD as obtained by SAXS. As shown in this figure, the trimeric model fits much better into the SAXS envelope, providing additional support for the occurrence of a trimeric form in solution. What could explain such a difference in the oligomeric state? We can speculate that the high local protein concentrations and/or the strong inter-molecular interactions within crystals might have biased the oligomeric state of the protein and promoted a tetrameric organization. That coiled-coils are able to modulate their oligomeric state according to the physico-chemical conditions (pH, temperature, etc.) or depending on whether they are located inside or outside the cell has already been reported (Dutta et al. 2001; Lupas and Gruber 2005). Of even more interest, the GCN4 leucine-zipper domain was shown to adopt different oligomeric states depending on the crystallization conditions, implying that the amino acid sequence does not specify a unique oligomeric state (Oshaben et al. 2012). It is also worth emphasising that conflicting experimental evidence is not unique to NiV PMD: indeed SeV PMD had also been shown to form trimers in solution (Curran 1998; Curran et al. 1995a) and to adopt a tetrameric coiled-coil conformation in the crystal (Tarbouriech et al. 2000). The experimental evidence pointing to a trimeric form of SeV P has perhaps been set aside too rapidly in light of the crystallographic data pointing to a tetrameric organization. However, the finding that both SeV and NiV PMD can form trimers in solution and tetramers in the crystal may reflect their intrinsic ability to adopt different oligomeric states that could be related to different functional forms of the P protein and to the different complexes (i.e. N-P, N°-P, P-L) that it can form within infected cells.

In the same vein, structural comparison among the different crystallographic structures of MeV PMD solved so far unveiled unexpected structural variations (Blocquel et al. 2014; Communie et al. 2013a). Although all the structures have a tetrameric coiled-coil organization, structural comparison unveiled considerable differences not only in the quaternary structure but also in the extent of disorder within the C-terminal region of the coiled-coil (Fig. 12.3b). The disordered nature of the C-terminal region is also supported by SAXS and SEC studies that show that MeV PMD exists as a dynamic equilibrium between two tetrameric forms of different compaction (Blocquel et al. 2014). As already discussed above for SeV and NiV PMD, the unexpected plasticity and flexibility of MeV PMD could be the first hint of the existence of different functional forms of the P protein, reflecting its multifunctional nature and pivotal role in the replicative cycle. These results also unveiled that the structure of coiled-coils can exhibit a certain degree of freedom, and that coiled-coils are less rigid than previously thought. They also bring awareness that conclusions about function and mechanism based on analysis of a single crystal structure of a dynamic protein can be easily biased, and they challenge to some extent the assumption according to which coiled-coil structures can be reliably predicted from the amino acid sequence (Blocquel et al. 2014).

In conclusion, the ability of SeV and NiV PMD to adopt different oligomeric states, together with the ability of MeV PMD to dynamically sample different forms that differ in the degree of compaction and extent of disorder, might be the basis for the ability of P to form different complexes critical for transcription and replication, with conformational changes possibly dictating the ability to form a transcriptase *versus* a replicase complex. Additional studies are necessary to obtain definite answers as to whether P oligomerization is strictly required for transcription and replication. Likewise, a detailed understanding of the role of disorder within PMD and of the functional impact of varying the P oligomeric state awaits future mutational studies.

3 Modular Organization of N

The nucleoprotein, N, is responsible for encapsidation of the viral genome (Lamb and Kolakofsky 2001). Not only does N protect viral RNA from degradation, but it also renders the latter competent for transcription and replication; indeed, the viral polymerase cannot transcribe nor replicate RNA when the latter is not encapsidated by the N protein within a helical nucleocapsid (Fig. 12.4a, b). In *Paramyxovirinae*,

N binds to exactly six nucleotides (Albertini et al. 2005), a property that dictates the so-called "rule of six", i.e. the requirement for the viral genome to be a multiple of six in order to ensure efficient transcription and replication (Kolakofsky et al. 1998).

Bioinformatics, deletion and electron microscopy studies have shown that *Paramyxovirinae* nucleoproteins are divided into two regions: a structured N-terminal moiety, N_{CORE} (aa 1–400 in MeV and aa 1–399 in henipaviruses), and a C-terminal domain, referred to as N_{TAIL} (aa 401–525 in MeV and aa 400–532 in henipaviruses) (Fig. 12.4c). While N_{CORE} contains all the regions necessary for self-assembly and RNA-binding, as well as for interaction with PNT within the N°-P complex, N_{TAIL} is responsible for interaction with XD (Bankamp et al. 1996; Buchholz et al. 1993; Curran et al. 1993; Karlin et al. 2002a; Kingston et al. 2004b; Liston et al. 1997; Myers et al. 1997; Myers et al. 1999).

N_{TAIL} domains from MeV, NiV and HeV possess features that are hallmarks of intrinsic disorder: (i) they are hyper-sensitive to proteolysis (Habchi et al. 2011; Karlin et al. 2002a), (ii) they cannot be visualized in cryo-electron microscopy reconstructions of nucleocapsids (Bhella et al. 2004), (iii) they have an amino acid

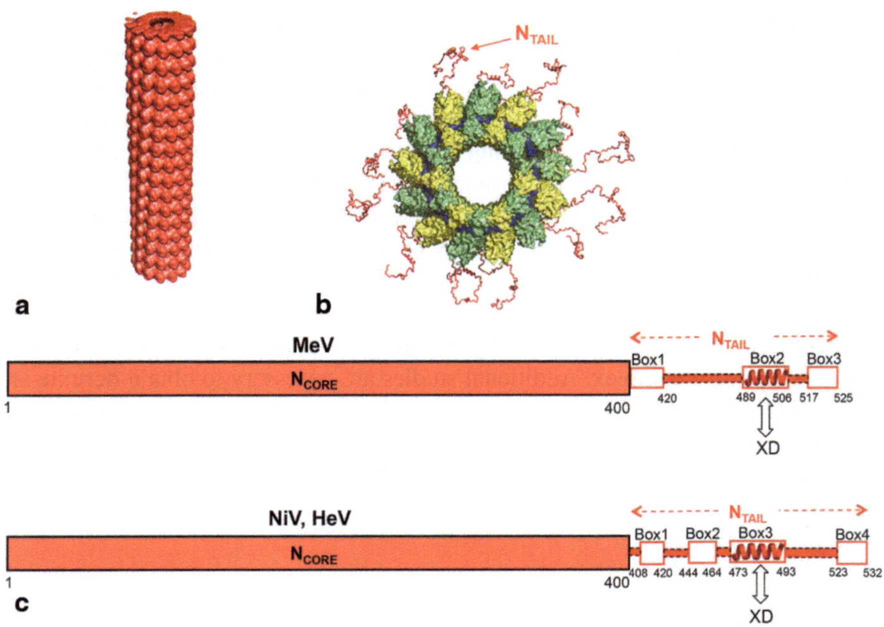

Fig. 12.4 *Organization of the MeV, NiV and HeV nucleoproteins.* **a** Cryo-electron microscopy reconstruction of the MeV nucleocapsid (Bhella 2007; Bhella et al. 2004). **b** Top view of the MeV nucleocapsid with the N_{TAIL} regions (in *red*) partly exposed at the surface. The viral RNA is represented in *blue*. Modified from (Ringkjøbing Jensen et al. 2011). **c** Modular organization of N from MeV and henipaviruses showing that N is composed of a folded domain, N_{CORE}, and a C-terminal disordered region, N_{TAIL}. The various boxes, corresponding to putative or experimentally proven MoREs, are shown, as is the α-MoRE (see *red helix*). The box interacting with the X domain of P is indicated by an *arrow*

sequence that is highly variable amongst phylogenetically related members (Habchi et al. 2010), and (iv) they are predicted to be mainly (if not fully) disordered by the secondary structure and disorder predictors implemented within the MeDor metaserver (Lieutaud et al. 2008). The disordered nature of these N_{TAIL} domains has been subsequently confirmed experimentally by both hydrodynamic and spectroscopic approaches, which have shown that they all belong to the PMG subfamily (Bourhis et al. 2004; Habchi et al. 2010; Longhi et al. 2003).

As for all N proteins of the *Mononegavirales* family, and with the sole exception of the N protein from the Borna disease virus (Rudolph et al. 2003), MeV and *Henipavirus* N proteins self-assemble to form large helical nucleocapsid-like particles with a broad size distribution when expressed in heterologous systems (Bhella et al. 2002; Communie et al. 2013b; Kerdiles et al. 2006; Spehner et al. 1991; Tan et al. 2004; Warnes et al. 1995). MeV nucleocapsids, as visualized by negative stain transmission electron microscopy (EM), have a typical herringbone-like appearance (Bhella et al. 2002; Bhella et al. 2004; Karlin et al. 2002a; Longhi et al. 2003; Schoehn et al. 2004). EM studies by two independent groups led to real-space helical reconstruction of MeV nucleocapsids (Bhella et al. 2004; Schoehn et al. 2004) (Fig. 12.4a). These studies showed that the removal of the disordered N_{TAIL} domain, which protrudes from the globular body of N_{CORE} and is at least partly exposed at the surface of the viral nucleocapsid (Heggeness et al. 1980, 1981; Karlin et al. 2002a), leads to increased nucleocapsid rigidity, with significant changes in both pitch and twist (see Bhella et al. 2004; Desfosses et al. 2011; Longhi et al. 2003; Schoehn et al. 2004).

So far, high-resolution structural data on *Paramyxoviridae* N is only available for RSV, whose N protein was crystallized in the form of N:RNA rings (Tawar et al. 2009). The nucleoprotein of RSV consists of two lobes and possesses an extended terminal arm that makes contacts with a neighbouring N monomer. The RNA is tightly packed between the two N lobes, being located on the external face of the N:RNA rings (Tawar et al. 2009). Using the structure of RSV N:RNA rings as a template, a model of MeV N:RNA has recently been built and docked within the electron density map of MeV nucleocapsids (Desfosses et al. 2011). Although the disordered N_{TAIL} domain could not be resolved in the reconstruction of the nucleocapsid, the fit suggests that N_{TAIL} would point toward the interior of the helical nucleocapsid (Desfosses et al. 2011). Thus, within the RSV nucleocapsid, the RNA is not accessible to the solvent, and has to be partially released from N to become accessible to the polymerase. Therefore, a conformational change must occur within N to allow exposure of the RNA. The disordered N_{TAIL} domain is thought to play a major role in this conformational change (see Sect. 11.7).

Although HeV, NiV and MeV N_{TAIL} domains are mostly disordered, bioinformatics analyses indicated the presence of short order-prone regions, possibly corresponding to MoREs (Bourhis et al. 2004; Habchi et al. 2010). In the case of MeV, one MoRE of α-helical nature was predicted to occur within one (i.e. Box2, aa 489–506) out of three regions (i.e. Box1-3) conserved within members of the *Morbillivirus* genus (Diallo et al. 1994) (Fig. 12.4c). While Box1 (aa 401–420) was shown to interact with a yet unidentified nucleoprotein receptor (NR) expressed at the surface of dendritic cells of lymphoid origin (Laine et al. 2003) as well as T and B lymphocytes (Laine et al. 2005), Box2 was shown to be the region responsible for interaction with XD (Blocquel et al. 2012b; Bourhis et al. 2004; Bourhis et al. 2005; Johansson et al. 2003). Analysis of the Cα chemical shifts of N_{TAIL} and of the mobility of spin labels grafted within Box2 showed that the α-MoRE of MeV N_{TAIL} is partly preconfigured as an α-helix in the absence of XD (Belle et al. 2008; Gely et al. 2010; Morin et al. 2006). More recently, an atomic-resolution ensemble description of the α-MoRE of MeV N_{TAIL} was obtained using recently developed tools designed to provide quantitative descriptions of conformational equilibria in IDPs on the basis of experimental NMR data (Bernadó et al. 2005; Jensen et al. 2008). By combining residual dipolar coupling (RDC) measurements and ensemble optimization methods (Bernadó et al. 2005; Jensen et al. 2008), the α-MoRE was shown to exist in a rapidly interconverting conformational equilibrium between an unfolded form and conformers containing four discrete α-helical elements situated around the interaction site (Ringkjøbing Jensen et al. 2011). Similar studies carried out on SeV N_{TAIL} unveiled a similar conformational behaviour, although in that case the α-MoRE was shown to sample an extended conformation and only three helical conformers (Jensen et al. 2010; Jensen et al. 2008).

Using various spectroscopic approaches, binding of XD was shown to trigger stable α-helical folding of the MoRE (Belle et al. 2008; Bischak et al. 2010; Bourhis et al. 2004; Bourhis et al. 2005; Gely et al. 2010; Johansson et al. 2003; Morin et al. 2006; Ringkjøbing Jensen et al. 2011).

The N_{TAIL} of henipaviruses differs from MeV N_{TAIL} in that the former possess four predicted MoREs (Habchi et al. 2010), with Box3 having been shown to be involved in interaction with XD (Habchi et al. 2011; Martinho et al. 2013) (Fig. 12.4c). Deletion studies (Blocquel et al. 2012c) with constructs bearing different combinations of the predicted MoREs shed light onto the structural state of the various boxes. In particular, they showed that Box3 is the counterpart of MeV Box2, being partially folded in solution and undergoing α-helical folding upon binding to XD (Habchi et al. 2011). Interestingly, SDSL EPR spectroscopy studies unveiled a considerable conformational heterogeneity within Box3 consistent with the occurrence of multiple helical conformers of different length (Martinho et al. 2013). In agreement, analysis of the Cα chemical shifts of the free form of HeV N_{TAIL} showed that Box3 is at least transiently populated as an α-helix (Communie et al. 2013b).

In henipaviruses, while the other boxes (e.g. Box1, Box2 and Box4) do not affect the ability of N_{TAIL} to interact with XD, they influence to some extent the α-helical folding of Box3 as well as the compaction properties of N_{TAIL}. Interestingly, subtle differences between NiV and HeV N_{TAIL} could also be observed (Blocquel et al. 2012c; Martinho et al. 2013). For example, a NiV variant devoid of Box1 and Box2 was found to possess a more extended conformation with respect to its HeV counterpart (Blocquel et al. 2012c). Since the two proteins have the same content in acidic residues, these observations suggest that other, more subtle sequence properties dictate the conformational behaviour of these proteins. Strikingly, the content in regular secondary structure was found not to be a major determinant of protein compaction, with N_{TAIL} proteins depleted in MoREs being nevertheless able to adopt a collapsed state. In the same vein, Box1, which was found to be an irregular-MoRE (I-MoRE), and hence to be devoid of α-helical propensities, was found to be a major determinant of protein compaction. The subtle differences observed between NiV and HeV N_{TAIL} domains have no significant effect on XD-binding abilities, with Box3 having been shown to be functionally interchangeable between the two viruses. Of interest, HeV N_{TAIL} was found to display an even better affinity for the heterologous X domain. However, there are no cross-interactions between proteins from HeV or NiV and proteins from MeV (Blocquel and Habchi *et al.*, unpublished data).

Beyond being disordered in isolation, the MeV and HeV N_{TAIL} domains were also shown to be disordered within full-length N proteins from nucleocapsid-like particles, as judged from NMR studies carried out on ^{15}N-labeled nucleocapsids (Communie et al. 2013b; Ringkjøbing Jensen et al. 2011) (Fig. 12.4b). In those studies, both MeV and HeV N_{TAIL} were found to retain their disordered state *in situ*, i.e. when appended to nucleocapsids. For both viruses, experimental evidence was obtained supporting a model in which the first 50 disordered amino acids of N_{TAIL} are conformationally restricted as the chain escapes from the inner channel to the outside of the nucleocapsid *via* the interstitial space between successive N_{CORE} helical turns (Communie et al. 2013b; Ringkjøbing Jensen et al. 2011). Notably, this model provides a plausible explanation for the increased rigidity of nucleocapsids in which the flexible N_{TAIL} region has been cleaved off. The inherent flexibility of intact nucleocapsids likely confers at least partial accessibility to the N-terminal region of $N_{TAIL,}$ thereby accounting for the ability of the Box1 region to bind to NR in the context of nucleocapsids released in the extracellular compartment (Laine et al. 2005; Laine et al. 2003). The flexibility of the N_{TAIL} region sandwiched between successive turns of the nucleocapsid may be the basis for variations in pitch and twist that may be related to switches between transcription and replication (Bhella 2007).

4 Molecular Mechanisms of N_{TAIL}-XD Complex Formation

The N_{TAIL} domains of MeV, NiV and HeV were shown to form a 1:1 stoichiometric complex with XD and to undergo induced folding upon binding to XD (Habchi et al. 2011; Johansson et al. 2003). Interestingly, while NMR titration experiments with ^{15}N-labeled N_{TAIL} indicated an α-helical transition within MeV N_{TAIL} upon addition of the homologous X domain, as judged from the appearance of new peaks in the α-helical region of the N_{TAIL} spectra (Bourhis et al. 2005; Gely et al. 2010; Habchi et al. 2011), no such peaks were observed in the HeV and NiV N_{TAIL} spectra even with saturating amounts of XD (Communie et al. 2013b; Baronti et al. 2015). This behaviour, which is often observed for IDPs undergoing folding-upon-binding events (Kiss et al. 2008; Mittag et al. 2008; Sue et al. 2008), supports an intermediate exchange regime among an ensemble of N_{TAIL} conformers at the XD surface, thus arguing for a considerable conformational heterogeneity in the bound form (see also Sect. 11.5) (Habchi et al. 2011).

In the case of MeV, a model of the interaction in which the α–MoRE of N_{TAIL} adopts an α–helical conformation and is embedded in a large hydrophobic cleft delimited by helices α2 and α3 of XD has been proposed (Johansson et al. 2003) and successively validated by Kingston and co-workers, who solved the crystal structure of a chimeric construct made of XD and the N_{TAIL}region encompassing residues 486–504 (Kingston et al. 2004a) (Fig. 12.5). Those studies unveiled that Box2 is tightly packed at the binding interface. The residues involved in the interaction of the two partners are mainly hydrophobic, involving Leu481, Leu484, Ile488, Phe497, Met500 and Ile504 from XD, and Ser491, Ala494, Leu495, Leu498 and Met501 from N_{TAIL} (Fig. 12.5). In addition, site directed mutagenesis studies revealed that Ala502 is also involved in the interaction with XD, as deduced from the 30-fold increase in the K_D observed with an N_{TAIL} variant bearing an Asp at this position (Shu et al. 2012). Recently, random mutagenesis studies confirmed the crucial role of N_{TAIL} residue Ser491 in complex formation (Gruet et al. 2013). Those studies also unveiled a previously unnoticed role for residue Arg497, whose side chain points out of the binding surface. In spite of its orientation towards the solvent, the side chain of Arg497 is at bonding distance from the OH group of Tyr480 of XD (see Fig. 12.5, right upper panel). Through generation and characterization of a "mirror" XD variant bearing the Y480F substitution, the crucial role of the Arg497-Tyr480 interaction in stabilizing the N_{TAIL}-XD complex was confirmed (Gruet et al. 2013).

Although previous studies suggested a possible implication of Box3 in binding of MeV N_{TAIL} to XD (Belle et al. 2008; Bourhis et al. 2005; Gely et al. 2010; Morin et al. 2006), recent isothermal titration calorimetry (ITC) studies ruled out a contribution of Box3 to binding (Blocquel et al. 2012b; Yegambaram and Kingston 2010).

While *Henipavirus* N_{TAIL}-XD complexes are characterized by an equilibrium dissociation constant (K_D) in the μM range (2 μM for NiV and 8 μM for HeV) (Habchi et al. 2011), we consistently found that the K_D of the MeV N_{TAIL}-XD complex is in the sub-micromolar range (Blocquel et al. 2012b; Bourhis et al. 2005; Shu et al. 2012). In contrast with our findings, the Kingston group reported a K_D in the micromolar range (15 μM) for the MeV N_{TAIL}-XD binding reaction (Yegambaram and Kingston 2010).

ITC studies revealed that *Henipavirus* N_{TAIL}-XD complexes are stable under NaCl concentrations as high as 1 M, suggesting that the interaction does not rely on polar contacts (Habchi et al. 2011), in line with an interaction driven by the burying of apolar residues of N_{TAIL} at the XD surface, as already observed in the case of MeV (Johansson et al. 2003; Kingston et al. 2004b).

While the crystal structure of HeV XD is available (Communie et al. 2013b), no structural data of the *Henipavirus* N_{TAIL}-XD complex is available so far. Considering the fact that far-UV CD spectroscopy revealed that binding of *Henipavirus* N_{TAIL} domains to XD results in an α-helical transition as in the case of MeV N_{TAIL} (Habchi et al. 2011), we modelled the more hydrophobic side of the amphipathic α-MoRE located within the Box3 region of *Henipavirus* N_{TAIL} at the hydrophobic surface delimited by helices α2 and α3 of XD using the MeV N_{TAIL}-XD structure as

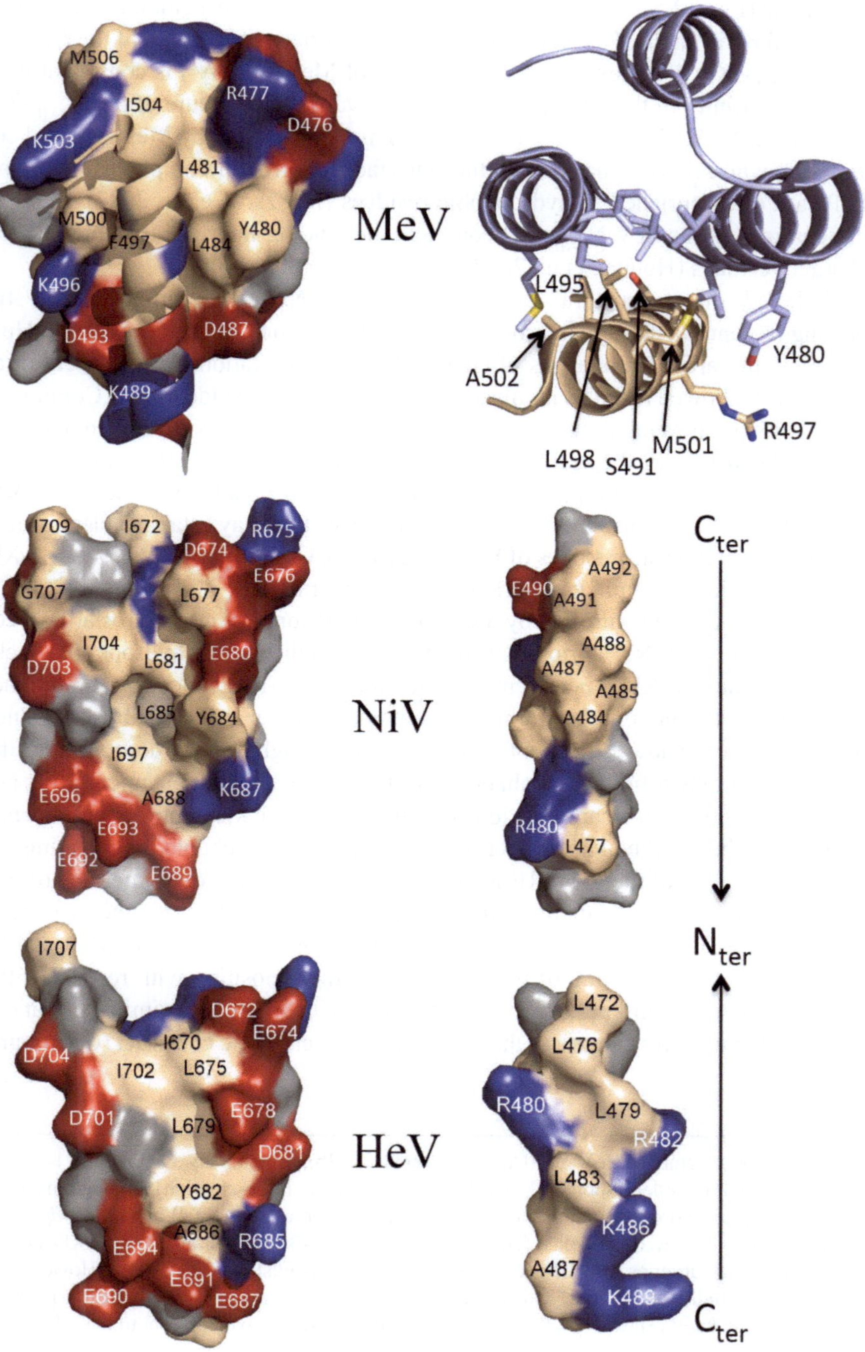

Fig. 12.5 *Structures of XD and of the interacting MoRE from N_{TAIL} in MeV, NiV and HeV*. The *top* of the figure shows the structure of the MeV Box2/XD complex (PDB code 1T6O, (Kingston et al. 2004a)) in surface representation (*left*) and ribbon representation (*right*) with the side chains of the residues involved in the interaction shown in sticks and in atom type colour. The *left-middle* and *bottom* panels represent (surface representation) the structural model of NiV (Habchi et al. 2011)

a template (Fig. 12.5). The two resulting models display a rather small interface area (439 Å^2 for NiV and 337 Å^2 for HeV). The lower buried surface area of the *Henipavirus* N_{TAIL}- XD complex as compared to that of MeV (634 Å^2) is consistent with the lower affinity of the binding reaction. The hydrophobic nature of the N_{TAIL}-XD interface in MeV, HeV and NiV is in agreement with findings by Meszaros and co-workers, who reported that the binding interfaces of protein complexes involving IDPs are often enriched in hydrophobic residues (Meszaros et al. 2007). In striking contrast, in the SeV N_{TAIL}-XD complex, the binding interface is dominated by charged residues (Houben et al. 2007b).

Although direct structural data on *Henipavirus* N_{TAIL}-XD complexes are still lacking, recent NMR studies provided the first clues on the structure of the HeV complex. In particular, analysis of chemical shift perturbations in reciprocal titration studies allowed residues involved in the interaction to be identified (Communie et al. 2013b). The availability of the crystal structure of HeV XD allowed mapping at the XD surface of the residues involved in binding to N_{TAIL} (Communie et al. 2013b). Although the binding interface is made of hydrophobic residues, the binding pocket of XD is surrounded by charged residues that may establish electrostatic interactions with basic residues of Box3 (Fig. 12.5). Contrary to the MeV N_{TAIL}-XD complex, which is dominated by hydrophobic interactions, in HeV the N_{TAIL}-XD interaction could be controlled by a combination of long-range electrostatic forces that correctly orient N_{TAIL} prior to accommodation in the narrow hydrophobic pocket at the surface of XD. The extent to which these long-range electrostatic interactions play a role in complex formation remains however to be established, since previous studies showed that salt concentrations as high as 1 M do not affect the N_{TAIL}-XD binding affinity (Habchi et al. 2011). Notably, neither chemical shifts nor electrostatic interactions are able to distinguish rotational symmetry about the axis of the N_{TAIL} helix, although two conformations are most probable, both having the hydrophobic face of the α-MoRE in contact with the hydrophobic interface of XD (Fig. 12.5). In both conformations, the two arginine residues flanking the hydrophobic face on HeV N_{TAIL} interact with acidic patches on the surface of XD. Notably, irrespective of the direction of the MoRE, its optimal position with respect to the observed chemical shift perturbations at the surface of XD is very similar to that observed in the crystal structure of the MeV N_{TAIL}-XD chimeric construct (Communie

---

and the crystal structure of HeV XD (Communie et al. 2013b), respectively. On the *right* is shown a helical structural model of both NiV and HeV Box3 (surface representation), oriented in such a way to show their hydrophobic side, which is supposed to interact with XD. When the *Henipavirus* XD/Box3 complexes were modelled (Habchi et al. 2011), Box3 was positioned at the XD surface in the same orientation as in the MeV XD/Box2 complex (Kingston et al. 2004a). However, as NMR titration studies on HeV XD and N_{TAIL} did not allow discrimination between the two possible orientations, we herein show HeV Box3 in the opposite orientation with respect to that of NiV. In the surface representations, all hydrophobic residues are represented in beige, while basic (Arg and Lys) and acidic residues (Asp and Glu) are shown in *blue* and *red*, respectively. The *lower* panel was modified from (Communie et al. 2013b)

et al. 2013b). Further studies will be necessary to draw final conclusions about the actual orientation of the α-MoRE at the XD surface of *Henipavirus* complexes.

From a mechanistic point of view, the finding that the α-MoRE of MeV N_{TAIL} is transiently populated as an α-helix might be taken as a hint suggesting that the molecular mechanism governing the folding coupled to binding of N_{TAIL} could rely on conformational selection (Tsai et al. 2001a; Tsai et al. 2001b). Two different, but not exclusive, binding mechanisms have been described in the literature for IDPs/IDRs (see (Habchi et al. 2014) and references cited therein). In the first one, called "folding after binding", the binding event takes place before folding (Shoemaker et al. 2000). In that case, the interaction with the partner is responsible for the gain of structure of the α-MoRE and for its stabilization. The second mechanism is conformer selection, where the partner binds to the pre-folded MoRE, thereby shifting the equilibrium of the conformational ensemble to the folded form. In MeV N_{TAIL}, complex formation seems to rely both on conformational selection and folding after binding. Indeed, the resonance behaviour of N_{TAIL} in titration experiments with unlabelled XD indicated a very poor fit to a two-state process, suggesting that binding may imply the formation of a binding intermediate in the form of a weak encounter complex, thereby implying a folding-after-binding mechanism. In further support of this hypothesis, recent data obtained by molecular dynamics simulations confirmed that binding preferentially occurs via an induced folding mechanism in spite of the partial pre-configuration of the α-MoRE (Wang et al. 2013). That binding coupled to folding events may rely on a mixed mechanism implying both induced folding (i.e. folding after binding) and conformational selection (i.e. folding before binding) has already been reported (Espinoza-Fonseca 2009).

The presence of a preconfigured MoRE with nevertheless a folding-after-binding mechanism is not a unique feature to MeV N_{TAIL}, having also been documented in the case of NiV (Baronti et al. 2015) and HeV N_{TAIL} (Communie et al. 2013b). In this latter case, quantitative analysis of peak intensities in the heteronuclear single quantum coherence (HSQC) spectra of N_{TAIL} at each XD titration point showed that the signal intensity decreases faster for the residues located at the extremities of the MoRE and for which a smaller amount of residual helical structure is observed in the isolated state of N_{TAIL}. This differential broadening suggests that XD binds to a short, central helix within the α-MoRE, and that this helix is subsequently extended via helical folding of the adjacent residues. Data therefore indicate that N_{TAIL} interacts with XD via a folding-upon-binding mechanism, with the folding event occurring on the micro- to millisecond time scale (Communie et al. 2013b).

5 Residual Flexibility within the N_{TAIL}-XD Complex

A low-resolution model of the MeV N_{TAIL}-XD complex obtained by SAXS studies showed that most of N_{TAIL} (residues 401–488) remains disordered within the complex (Bourhis et al. 2005). A recent study that made use of a combination of SDSL

EPR spectroscopy and modelling further supports a considerable residual flexibility in the bound form of MeV N_{TAIL} (Kavalenka et al. 2010). That study indeed showed that although the 505–525 region of N_{TAIL} becomes more rigid upon binding to XD as a result of the α-helical transition occurring within the neighbouring Box2 region (Belle et al. 2008), it nevertheless conserves a significant degree of freedom in the complex (Kavalenka et al. 2010). As such, the MeV N_{TAIL}-XD complex provides an illustrative example of "fuzziness", where this term has been coined by Tompa and Fuxreiter to designate the persistence of conspicuous regions of disorder within protein complexes implicating IDPs (Tompa and Fuxreiter 2008). *Henipavirus* N_{TAIL}-XD complexes were found to be similarly fuzzy. Indeed, the experimentally determined R_S of the NiV N_{TAIL}-XD complex (35.4±3.1 Å) suggests that binding to XD does not imply the formation of a compact complex (expected R_S of 22.3 Å) and that it rather retains a considerable flexibility (Habchi et al. 2011). In further support of the "fuzziness" within MeV, HeV and NiV N_{TAIL}-XD complexes, the many observable and relatively sharp NMR resonances that are nearly unaltered upon addition of XD provide evidence that these N_{TAIL} regions remain significantly disordered in the bound state (Gely et al. 2010; Habchi et al. 2011; Kingston et al. 2004a). Strikingly, the N_{TAIL}-XD complex from henipaviruses is even fuzzier, as judged from the vanishing of resonances of the MoREs at the beginning of titration with no reappearance even at saturation (Communie et al. 2013b); (Baronti et al. 2015). This observation suggests that even when bound to XD, the α-MoRE of both NiV and HeV N_{TAIL} remains highly dynamic, undergoing exchange between different conformers at the XD surface (Communie et al. 2013b).

What is the functional role of such fuzziness? We propose that the prevalently disordered nature of N_{TAIL} even after complex formation may serve as a platform for the capture of other binding partners. In agreement, in the case of MeV N_{TAIL}, Box1 has been shown to be responsible for the interaction with the cellular receptor NR (Laine et al. 2005; Laine et al. 2003; Laine et al. 2007), and Box3 was found to interact with the major inducible heat shock protein hsp70 (Couturier et al. 2010; Zhang et al. 2005). In MeV, viral transcription and replication are enhanced by hsp70, with this stimulation relying on an interaction with N_{TAIL} (Carsillo et al. 2006b; Oglesbee 2007; Oglesbee et al. 1993; Oglesbee et al. 1996; Vasconcelos et al. 1998a; Vasconcelos et al. 1998b; Zhang et al. 2005; Zhang et al. 2002). Two binding sites for hsp70 have been identified within N_{TAIL} (Zhang et al. 2005; Zhang et al. 2002); while α-MoRE provides a high-affinity binding site (K_D of 10 nM), a second low-affinity binding site is present within Box3 (Carsillo et al. 2006a; Zhang et al. 2002). Since hsp70 was shown to competitively inhibit the binding of XD to N_{TAIL} (Zhang et al. 2005), it has been proposed that hsp70 could enhance transcription and genome replication by reducing the stability of P-N_{TAIL} complexes, thereby promoting successive cycles of binding and release that are essential to polymerase movement along the nucleocapsid template (Bourhis et al. 2005; Zhang et al. 2005). The hsp70-dependent reduction of the stability of P-N_{TAIL} complexes would thus rely on competition between hsp70 and XD for binding to the α-MoRE of N_{TAIL}, with recruitment of hsp70 being ensured by both Box2 and Box3 (Zhang et al. 2005).

6 Evolvability of N_{TAIL}

In view of gaining additional insights into the molecular determinants that govern binding of MeV N_{TAIL} to XD, we used an approach that we termed "descriptive random mutagenesis" (Gruet et al. 2012). To that end, we generated a library of N_{TAIL} variants by error-prone polymerase chain reaction (PCR) and assessed how amino acid substitutions introduced at random within N_{TAIL} affect partner recognition. In contrast with directed evolution approaches, variants were picked at random in the absence of selection pressure and were characterized in terms of their sequence and binding abilities towards XD (Gruet et al. 2012). Their interaction strength towards XD was evaluated using a protein complementation assay based on split green fluorescent protein (GFP) reassembly (Magliery et al. 2005; Wilson et al. 2004). In this method, each partner is fused to a GFP moiety. The interaction between the two partners drives the re-assembly of the two GFP fragments and hence reconstitution of the fluorophore, thus producing a fluorescence signal. The stronger the interaction, the higher the fluorescence.

This approach not only identified determinants of N_{TAIL}-XD interaction that were in good agreement with previous work, but also provided new insights (Gruet et al. 2012). Among the 300 variants analysed, 224 encode full-length forms and a full coverage of the entire N_{TAIL} sequence was achieved (i.e. each amino acid of the N_{TAIL} sequence was found to be substituted at least once). The analysis of the library revealed that fluorescence was not correlated with the number of substitutions borne by the variants and rather depended on their location along the N_{TAIL} sequence. Most of the substitutions within N_{TAIL} were shown to affect its capacity to bind XD. While most of these substitutions decreased the interaction strength between N_{TAIL} and XD, some of them led to an increased interaction. Variants bearing substitutions within Box2 tend to display a reduced fluorescence, indicating that Box2 is poorly evolvable in terms of binding abilities toward XD. Variants bearing at least one substitution within Box2 (and irrespective of whether they brought additional substitutions elsewhere in the N_{TAIL} sequence or did not) were analysed and their average fluorescence was plotted as a function of the substitution position within N_{TAIL}. The resulting profile is quite well accounted for by the structure of the XD/Box2 complex, which shows that the most critical positions correspond to residues that have their side chains oriented towards the partner. Since many of the Box2 variants from the library also possess other substitutions elsewhere, five individual Box2 variants were generated with the specific aim of assessing the impact of the sole Box2 substitutions on the interaction. Results indicated that Box2 substitutions are on their own responsible for the observed decrease in the fluorescence. Importantly, the fluorescence drop, reflecting a decreased interaction, is not due to a decrease in the expression of the N_{TAIL} variants. Notably, this analysis also led to identification of an additional Box2 critical residue (Arg497) whose role in stabilizing the N_{TAIL}-XD complex had previously escaped detection (Gruet et al. 2012).

Notably, the analysis allowed the identification of five regulatory regions that dampen the interaction while being located outside the primary interaction site (i.e.

Box2). These sites, referred to as e-boxes (enhancer boxes), are located within the "fuzzy" N_{TAIL} region. The precise molecular mechanism by which e-boxes modulate the N_{TAIL}-XD interaction remains to be elucidated (Gruet et al. 2012).

Random mutagenesis also generated truncated variants resulting from the generation of a stop codon. In line with expectations, variants devoid of Box2 showed a dramatic drop in fluorescence, reflecting a loss of interaction. Unexpectedly however, variants that are only devoid of Box3 display an increased fluorescence, thus unveiling a possible inhibitory effect of Box3 on the interaction with XD.

In conclusion, this study unveiled that most of the N_{TAIL} sequence is sensitive to mutations and possesses a few regulatory sites located within fuzzy regions. The fuzziness of N_{TAIL} may therefore not only serve as a way to capture other binding partners but also to modulate the strength of interactions established by N_{TAIL}.

7 Functional Role of Structural Disorder within N and P in Terms of Transcription and Replication

In agreement with previous reports that underscored a relationship between disorder and protein interactivity (Dunker et al. 2005; Haynes et al. 2006; Uversky et al. 2005), the presence of the disordered N_{TAIL} region protruding at the surface of the viral nucleocapsid allows the establishment of a complex molecular partnership with a panel of structurally distinct cellular and viral partners. Indeed, in addition to P, MeV N_{TAIL} was also shown to interact with the M protein (Iwasaki et al. 2009) and various cellular proteins such as interferon regulatory factor 3 (IRF3) (Colombo et al. 2009; tenOever et al. 2002), hsp70 (Couturier et al. 2010; Zhang et al. 2005; Zhang et al. 2002), NR (Laine et al. 2005; Laine et al. 2003; Laine et al. 2007), the cell protein responsible for the nuclear export of N (Sato et al. 2006), peroxiredoxin 1 (Watanabe et al. 2011) and possibly components of the cell cytoskeleton (De and Banerjee 1999; Moyer et al. 1990). Among all these interactions, interaction with XD is critical as it allows the P/L complex to be recruited onto the nucleocaspid in order to allow transcription and replication to take place (Longhi 2007, 2009, 2011). Notably, a recent study by the Plemper group has challenged the previously accepted model according to which Box2 is strictly required to recruit the MeV polymerase complex; indeed, Box2 was found to be dispensable for MeV transcription and replication in the absence of the upstream N_{TAIL} region, which was found to act as a negative modulator (i.e. to prevent binding of the L-P complex to the nucleocapsid) (Krumm et al. 2013).

In *Paramyxovirinae*, the N_{TAIL}-XD interaction is also thought to trigger the opening of the nucleocapsid to provide access of the polymerase to the viral RNA. In agreement, EM studies showed that addition of XD triggers unwinding of MeV nucleocapsids (Bhella and Longhi, unpublished data). This dramatic conformational change is accompanied by an increased exposure of viral RNA to the solvent as indicated by its increased sensitivity to RNAse. In striking contrast with these findings, recent NMR studies have shown that addition of XD to HeV nucleocapsids does not trigger any major nucleocapsid rearrangement, as judged from the observation that the only affected residues are those belonging to the MoRE (Communie et al. 2013b). We can speculate that the expectedly necessary nucleocapsid unwinding requires either the full-length P protein, or the P-L complex and/or cellular cofactors. One such possible cellular cofactor could be hsp70, by analogy with previous studies that showed that hsp70-nucleocapsid complexes of the closely related canine distemper virus exhibit an expanded helical diameter, an increased fragility, and an enhanced exposure of the genomic RNA to nuclease degradation (Oglesbee et al. 1990; Oglesbee et al. 1989).

The induced folding of N_{TAIL} resulting from the interaction with P (and/or other physiological partners) could also exert an impact on the nucleocapsid conformation in such a way as to affect the structure of the replication promoter. Indeed, the replication promoter, located at the 3' end of the viral genome, is composed of two discontinuous elements that form a functional unit when juxtaposed on two successive helical turns (Tapparel et al. 1998). The switch between transcription and replication could be dictated by variations in the helical conformation of the nucleocapsid, which would result in a modification in the number of N monomers (and thus of nucleotides) per turn, thereby disrupting the replication promoter in favour of the transcription promoter (or *vice versa*). Morphological analyses, showing the occurrence of a large conformational flexibility within *Paramyxoviridae* nucleocapsids (Bhella et al. 2002; Bhella et al. 2004; Oglesbee et al. 1990; Oglesbee et al. 1989), tend to corroborate this hypothesis.

A tight N-P complex is predicted to hinder polymerase processivity, according to the cartwheeling mechanism, which posits that contacts between N_{TAIL} and XD have to be dynamically made/broken to allow the polymerase to progress along the nucleocapsid template in order to allow transcription and replication to take place. By contrast, in the so-called "jumping model" proposed in the case of rabies virus, the P protein would be permanently bound to the nucleocapsid template and the polymerase would jump between adjacent P dimers (Leyrat et al. 2010). Recent mutational studies that targeted the Box2 region of MeV N_{TAIL} unexpectedly showed that a reduced binding strength has no impact on the polymerase rate (Shu et al. 2012). This tolerance of the polymerase to N_{TAIL} substitutions is probably true only in a certain range of affinities, where in spite of a pronounced drop in the affinity towards XD, the N_{TAIL}-XD interaction remains strong enough to ensure recruitment of the polymerase. These results suggest that the accepted model whereby the N_{TAIL}-P_{XD} interaction has to be relatively weak to allow the polymerase to cartwheel on the nucleocapsid template needs to be revisited. A relatively labile complex can result from either an inherently lower affinity of the binding reaction, as in the case of henipaviruses (Habchi et al. 2011), whose N_{TAIL}-XD complexes are characterized by K_D values in the μM range, or from a tight complex whose strength is modulated by co-factors. Taking into account the ability of hsp70 to bind to the same N_{TAIL} sites as XD and thus to beat out the latter competitively (Zhang et al. 2005), it is tempting to speculate that the progression of the MeV polymerase complex along the template could be ensured by hsp70. In this model hsp70 would promote successive cycles of binding and release thanks to its destabilizing effect on the N_{TAIL}-

XD interaction. The prevalently disordered nature of N_{TAIL} even in the bound form would facilitate recruitment of hsp70, thus providing an easy means to modulate the N-P interaction strength, which would ultimately result in modulation of transcription and replication rates.

Similarly to N_{TAIL}, paramyxoviral PNT domains have been reported to interact with multiple partners, including both the unassembled and assembled forms of N (Chen et al. 2003; Curran et al. 1995b; Curran et al. 1994) and cellular proteins (Liston et al. 1995). In addition, in the context of the V and W proteins, PNT establishes many additional interactions with cell proteins, with these interactions playing a key role in the evasion of the interferon (IFN) response via both an antagonist activity of IFN signalling and inhibition of IFN induction (for reviews see Audsley and Moseley 2013; Fontana et al. 2008; Park et al. 2003; Ulane and Horvath 2002).

The presence of unstructured domains on both N and P would allow for coordinated interactions between the polymerase complex and a large surface area of the nucleocapsid template, including successive turns of the helix. Indeed, the maximal extension of MeV PNT as measured by SAXS is 40 nm (Longhi and Receveur-Bréchot, unpublished data). In comparison, one turn of the MeV nucleocapsid is 18 nm in diameter and 6 nm high (Bhella et al. 2002). PNT could thus easily stretch over several turns of the nucleocapsid, and since P is multimeric, N°-P might have a considerable extension. Likewise, the maximal extension of MeV N_{TAIL} in solution is 13 nm (Longhi et al. 2003). The very long reach of disordered regions could enable them to act as linkers and tether partners on large macromolecular assemblies, thereby acting as scaffolding engines as already described for intrinsically disordered scaffold proteins (Balazs et al. 2009; Cortese et al. 2008). A model can be proposed where, during replication, the extended conformation of PNT and N_{TAIL} would be key to allowing contact between the assembly substrate (N°-P) and the polymerase complex (L-P), thus leading to a tripartite N°-P-L complex. The plasticity of IDRs within N and P would therefore confer a considerable reach to the elements of the replication machinery.

8 Functional Advantages of Structural Disorder within Viral Proteins

Viruses have to quickly adapt to changes in their environment, and survive in both their host and the environment of their host. Because viruses are obligate intracellular parasites, they have an exquisitely close relationship with their host and their proteins have to interact with various components of the host including membranes, nucleic acids, and proteins. The lack of a rigid 3D structure gives IDPs/IDRs the necessary plasticity to establish various interactions with several partners at once. In the course of evolution, viruses have “learned” to hijack and manipulate host proteins for their benefit and to evade host defence mechanisms. A recent study by Davey and co-workers showed that viruses have achieved this ability through broad mimicry of host protein short linear motifs (SLiMs) (Davey et al. 2012), where the latter are embedded in disordered regions and play a variety of roles, including targeting host proteins for proteosomal degradation, cell signalling, directing proteins to the correct subcellular localization, deregulating cell cycle checkpoints, and altering transcription of host proteins (Davey et al. 2011). Importantly, binding to cell proteins through sites that mimic SLiMs also helps viral proteins to elude the host cell's immune system by rendering viral epitopes poorly recognizable by it (see Xue et al. 2014 and references therein cited).

Recent bioinformatics studies have also showed that viral proteins, and in particular proteins from RNA viruses, have a high content of disorder (Tokuriki et al. 2009; Xue et al. 2010). The authors proposed that beyond affording a broad partnership, the wide occurrence of disordered regions in viral proteins could also be related to the typical high mutation rates of RNA viruses. This represents a strategy for buffering the deleterious effects of mutations.

Disorder has also been reported to provide a means to tolerate insertions and/or deletions, and therefore to be abundant in regions with dual coding capacity (Jordan et al. 2000; Kovacs et al. 2010; Narechania et al. 2005; Rancurel et al. 2009; Romero et al. 2006). Consistent with this relationship, when we compared the modular organization of the P proteins within the *Paramyxovirinae* subfamily (Karlin et al. 2003), we noticed that a larger PNT domain in Henipaviruses accounts for the extra length of their P protein (Habchi et al. 2010).

In the same vein, PNT partially overlaps with the C protein (being encoded by the same RNA region), and the "spacer" region partially overlaps with the C-terminal domain of the V protein. The disordered nature of PNT and of the "spacer" region connecting PNT to PMD likely reflects a way of alleviating evolutionary constraints within overlapping reading frames. Disorder, which is encoded by a much wider portion of sequence space as compared to order, can indeed represent a strategy by which genes encoding overlapping reading frames can lessen evolutionary constraints imposed on their sequence by the overlap, allowing the encoded overlapping protein products to sample a wider sequence space without losing function.

Taking into account these considerations, as well as the correlation between overlapping genes and disorder and the typically high compaction of viral genomes that often contain overlapping reading frames, we have proposed that the main advantage of the abundance of disorder within viruses would reside in pleiotropy and genetic compaction (Xue et al. 2014). Indeed, disorder provides a solution to reduce both genome size and molecular crowding, where a single gene would (i) encode a single (regulatory) protein product that can establish multiple interactions via its disordered regions and hence exert multiple concomitant biological effects, and/or (ii) encode more than one product by means of overlapping reading frames. In fact, since disordered regions are less sensitive to structural constraints than ordered ones, the occurrence of disorder within one or both protein products encoded by an overlapping reading frame can represent a strategy to alleviate evolutionary constraints imposed by the overlap. As such, disorder would give viruses the ability to "handle" overlaps, thus further expanding the coding potential of viral genomes.

Acknowledgements We wish to thank all our co-workers who were involved in the studies herein summarized. This work was carried out with the financial support of the Agence Nationale de la Recherche, specific programs "Physico-Chimie du Vivant", ANR-08-PCVI-0020-01, and "ASTRID", ANR-11-ASTR-003-01. The work was also partly supported by the CNRS. D.B. was supported by a joint doctoral fellowship from the Direction Générale de l'Armement (DGA) and the CNRS. M.B. was partly supported by an Erasmus Master fellowship from the University of Milan and is presently supported by a Ph.D. fellowship from the French-Italian University. J.E. is supported by a post-doctoral fellowship from the Fondation pour la Recherche Médicale (FRM). The funders had no role in study design, data collection and analysis, decision to publish, or preparation of the manuscript.

References

Albertini AAV, Schoehn G, Ruigrok RW (2005) Structures impliquées dans la réplication et la transcription des virus Ã ARN non segmentés de sens négatif. Virologie 9:83–92

Audsley MD, Moseley GW (2013) Paramyxovirus evasion of innate immunity: diverse strategies for common targets. World J Virol 2:57–70

Balazs A, Csizmok V, Buday L et al (2009) High levels of structural disorder in scaffold proteins as exemplified by a novel neuronal protein, CASK-interactive protein1. FEBS J 276:3744–3756

Bankamp B, Horikami SM, Thompson PD et al (1996) Domains of the measles virus N protein required for binding to P protein and self-assembly. Virology 216:272–277

Baronti L, Erales J, Habchi J et al (2015) Dynamics of the intrinsically disordered C-terminal domain of the nipah virus nucleoprotein and interaction with the x domain of the phosphoprotein as unveiled by NMR spectroscopy. Chembiochem 6:268–276

Belle V, Rouger S, Costanzo S et al (2008) Mapping α-helical induced folding within the intrinsically disordered C-terminal domain of the measles virus nucleoprotein by site-directed spin-labeling EPR spectroscopy. Proteins Struct Funct Bioinform 73:973–988

Bernadó P, Blanchard L, Timmins P et al (2005) A structural model for unfolded proteins from residual dipolar couplings and small-angle x-ray scattering. Proc Natl Acad Sci U S A 102:17002–17007

Bhella D (2007) Measles virus nucleocapsid structure, conformational flexibility and the rule of six. In: Longhi S (ed) Measles virus nucleoprotein. Nova Publishers Inc., Hauppage

Bhella D, Ralph A, Murphy LB et al (2002) Significant differences in nucleocapsid morphology within the Paramyxoviridae. J Gen Virol 83:1831–1839

Bhella D, Ralph A, Yeo RP (2004) Conformational flexibility in recombinant measles virus nucleocapsids visualised by cryo-negative stain electron microscopy and real-space helical reconstruction. J Mol Biol 340:319–331

Bischak CG, Longhi S, Snead DM et al (2010) Probing structural transitions in the intrinsically disordered C-terminal domain of the measles virus nucleoprotein by vibrational spectroscopy of cyanylated cysteines. Biophys J 99:1676–1683

Blanchard L, Tarbouriech N, Blackledge M et al (2004) Structure and dynamics of the nucleocapsid-binding domain of the Sendai virus phosphoprotein in solution. Virology 319:201–211

Blocquel D, Bourhis JM, Eléouët JF et al (2012a) Transcription et réplication des Mononégavirales: une machine moléculaire originale. Virologie 16:225–257

Blocquel D, Habchi J, Costanzo S et al (2012b) Interaction between the C-terminal domains of measles virus nucleoprotein and phosphoprotein: A tight complex implying one binding site. Protein Sci 21:1577–1585

Blocquel D, Habchi J, Gruet A et al (2012c) Compaction and binding properties of the intrinsically disordered C-terminal domain of Henipavirus nucleoprotein as unveiled by deletion studies. Mol Biosyst 8:392–410

Blocquel D, Beltrandi M, Erales J et al (2013) Biochemical and structural studies of the oligomerization domain of the Nipah virus phosphoprotein: Evidence for an elongated coiled-coil homotrimer. Virology 446:162–172

Blocquel D, Habchi J, Durand E et al (2014) Coiled-coil deformations in crystal structures: the measles virus phosphoprotein multimerization domain as an illustrative example. Acta Cryst D 70:1589–1603

Bourhis J, Johansson K, Receveur-Bréchot V et al (2004) The C-terminal domain of measles virus nucleoprotein belongs to the class of intrinsically disordered proteins that fold upon binding to their physiological partner. Virus Res 99:157–167

Bourhis JM, Receveur-Bréchot V, Oglesbee M et al (2005) The intrinsically disordered C-terminal domain of the measles virus nucleoprotein interacts with the C-terminal domain of the phosphoprotein via two distinct sites and remains predominantly unfolded. Protein Sci 14:1975–1992

Bourhis JM, Canard B, Longhi S (2007) Predicting protein disorder and induced folding: from theoretical principles to practical applications. Curr Protein Pept Sci 8:135–149

Bruhn-Johannsen JF, Barnett K, Bibby J et al (2014) Crystal structure of the Nipah virus phosphoprotein tetramerization domain. J Virol 88:758–762

Buchholz CJ, Spehner D, Drillien R et al (1993) The conserved N-terminal region of Sendai virus nucleocapsid protein NP is required for nucleocapsid assembly. J Virol 67:5803–5812

Buchholz CJ, Retzler C, Homann HE et al (1994) The carboxy-terminal domain of Sendai virus nucleocapsid protein is involved in complex formation between phosphoprotein and nucleocapsid- like particles. Virology 204:770–776

Campen A, Williams RM, Brown CJ et al (2008) TOP-IDP-scale: a new amino acid scale measuring propensity for intrinsic disorder. Protein Pept Lett 15:956–963

Carsillo T, Traylor Z, Choi C et al (2006a) hsp72, a host determinant of measles virus neurovirulence. J Virol 80:11031–11039

Carsillo T, Zhang X, Vasconcelos D et al (2006b) A single codon in the nucleocapsid protein C terminus contributes to in vitro and in vivo fitness of Edmonston measles virus. J Virol 80:2904–2912

Chattopadhyay S, Banerjee AK (2009) Phosphoprotein, P of human parainfluenza virus type 3 prevents self-association of RNA-dependent RNA polymerase, L. Virology 383:226–236

Chen M, Cortay JC, Gerlier D (2003) Measles virus protein interactions in yeast: new findings and caveats. Virus Res 98:123–129

Chen JW, Romero P, Uversky VN et al (2006) Conservation of intrinsic disorder in protein domains and families: I. A database of conserved predicted disordered regions. J Proteome Res 5:879–887

Colombo M, Bourhis JM, Chamontin C et al (2009) The interaction between the measles virus nucleoprotein and the Interferon Regulator Factor 3 relies on a specific cellular environment. Virol J 6:59

Communie G, Crepin T, Maurin D et al (2013a) Structure of the tetramerization domain of measles virus phosphoprotein. J Virol 87:7166–7169

Communie G, Habchi J, Yabukarski F et al (2013b) Atomic resolution description of the interaction between the nucleoprotein and phosphoprotein of Hendra virus. PLoS Pathog 9:e1003631

Cortese MS, Uversky VN, Dunker AK (2008) Intrinsic disorder in scaffold proteins: getting more from less. Prog Biophys Mol Biol 98:85–106

Couturier M, Buccellato M, Costanzo S et al (2010) High Affinity Binding between Hsp70 and the C-Terminal Domain of the Measles Virus Nucleoprotein Requires an Hsp40 Co-Chaperone. J Mol Recognit 23:301–315

Cox R, Green TJ, Purushotham S et al (2013) Structural and functional characterization of the mumps virus phosphoprotein. J Virol 87:7558–7568

Curran J (1998) A role for the Sendai virus P protein trimer in RNA synthesis. J Virol 72:4274–4280

Curran J, Homann H, Buchholz C et al (1993) The hypervariable C-terminal tail of the Sendai paramyxovirus nucleocapsid protein is required for template function but not for RNA encapsidation. J Virol 67:4358–4364

Curran J, Pelet T, Kolakofsky D (1994) An acidic activation-like domain of the Sendai virus P protein is required for RNA synthesis and encapsidation. Virology 202:875–884

Curran J, Boeck R, Lin-Marq N et al (1995a) Paramyxovirus phosphoproteins form homotrimers as determined by an epitope dilution assay, via predicted coiled coils. Virology 214:139–149

Curran J, Marq JB, Kolakofsky D (1995b) An N-terminal domain of the Sendai paramyxovirus P protein acts as a chaperone for the NP protein during the nascent chain assembly step of genome replication. J Virol 69:849–855
Dahlman-Wright K, McEwan IJ (1996) Structural studies of mutant glucocorticoid receptor transactivation domains establish a link between transactivation activity in vivo and α-helix-forming potential in vitro. BioChemistry 35:1323–1327
Davey NE, Trave G, Gibson TJ (2011) How viruses hijack cell regulation. Trends Biochem Sci 36:159–169
Davey NE, Van Roey K, Weatheritt RJ et al (2012) Attributes of short linear motifs. Mol Biosyst 8:268–281
De BP, Banerjee AK (1999) Involvement of actin microfilaments in the transcription/replication of human parainfluenza virus type 3: possible role of actin in other viruses. Microsc Res Tech 47:114–123
DeLano WL (2002) The PyMOL molecular graphics system. Proteins: Struct Funct Bioinform 30:442–454
Desfosses A, Goret G, Farias Estrozi L et al (2011) Nucleoprotein-RNA orientation in the measles virus nucleocapsid by three-dimensional electron microscopy. J Virol 85:1391–1395
Diallo A, Barrett T, Barbron M et al (1994) Cloning of the nucleocapsid protein gene of peste-des-petits-ruminants virus: relationship to other morbilliviruses. J Gen Virol 75(1):233–237
Ding B, Zhang G, Yang X et al (2014) Phosphoprotein of human parainfluenza virus type 3 blocks autophagosome-lysosome fusion to increase virus production. Cell Host Microbe 15:564–577
Dunker AK, Lawson JD, Brown CJ et al (2001) Intrinsically disordered protein. J Mol Graph Model 19:26–59
Dunker AK, Cortese MS, Romero P et al (2005) Flexible nets. FEBS J 272:5129–5148
Dutta K, Alexandrov A, Huang H et al (2001) pH-induced folding of an apoptotic coiled coil. Protein Sci 10:2531–2540
Espinoza-Fonseca LM (2009) Reconciling binding mechanisms of intrinsically disordered proteins. Biochem Biophys Res Commun 382:479–482
Ferron F, Longhi S, Henrissat B et al (2002) Viral RNA-polymerases—a predicted 2'-O-ribose methyltransferase domain shared by all Mononegavirales. Trends Biochem Sci 27:222–224
Ferron F, Longhi S, Canard B et al (2006) A practical overview of protein disorder prediction methods. Proteins 65:1–14
Fontana JM, Bankamp B, Rota PA (2008) Inhibition of interferon induction and signaling by paramyxoviruses. Immunol Rev 225:46–67
Garner E, Romero P, Dunker AK et al (1999) Predicting binding regions within disordered proteins. Genome Inform Ser Workshop Genome Inform 10:41–50
Gely S, Lowry DF, Bernard C et al (2010) Solution structure of the C-terminal X domain of the measles virus phosphoprotein and interaction with the intrinsically disordered C-terminal domain of the nucleoprotein. J Mol Recognit 23:435–447
Gopinath M, Shaila MS (2008) Recombinant L and P protein complex of Rinderpest virus catalyses mRNA synthesis in vitro. Virus Res 135:150–154
Gruet A, Longhi S, Bignon C (2012) One-step generation of error-prone PCR libraries using Gateway(R) technology. Microb Cell Fact 11:14
Gruet A, Dosnon M, Vassena A et al (2013) Dissecting partner recognition by an intrinsically disordered protein using descriptive random mutagenesis. J Mol Biol 425:3495–3509
Habchi J, Longhi S (2012) Structural disorder within paramyxovirus nucleoproteins and phosphoproteins. Mol Biosyst 8:69–81
Habchi J, Mamelli L, Darbon H et al (2010) Structural disorder within Henipavirus nucleoprotein and phosphoprotein: from predictions to experimental assessment. PLoS ONE 5:e11684
Habchi J, Blangy S, Mamelli L et al (2011) Characterization of the interactions between the nucleoprotein and the phosphoprotein of Henipaviruses. J Biol Chem 286:13583–13602
Habchi J, Mamelli L, Longhi S (2012) Structural disorder within the nucleoprotein and phosphoprotein from measles, Nipah and Hendra viruses. In: Uversky VN, Longhi S (eds) Flexible viruses: structural disorder in viral proteins. Wiley, Hoboken, pp 47–94

Habchi J, Tompa P, Longhi S et al (2014) Introducing protein intrinsic disorder. Chem Rev 114:6561–6588

Haynes C, Oldfield CJ, Ji F et al (2006) Intrinsic disorder is a common feature of hub proteins from four eukaryotic interactomes. PLoS Comput Biol 2:e100

Heggeness MH, Scheid A, Choppin PW (1980) Conformation of the helical nucleocapsids of paramyxoviruses and vesicular stomatitis virus: reversible coiling and uncoiling induced by changes in salt concentration. Proc Natl Acad Sci U S A 77:2631–2635

Heggeness MH, Scheid A, Choppin PW (1981) The relationship of conformational changes in the Sendai virus nucleocapsid to proteolytic cleavage of the NP polypeptide. Virology 114:555–562

Houben K, Blanchard L, Blackledge M et al (2007a) Intrinsic dynamics of the partly unstructured PX domain from the Sendai virus RNA polymerase cofactor P. Biophys J 93:2830–2844

Houben K, Marion D, Tarbouriech N et al (2007b) Interaction of the C-terminal domains of sendai virus N and P proteins: comparison of polymerase-nucleocapsid interactions within the paramyxovirus family. J Virol 81:6807–6816

Hua QX, Jia WH, Bullock BP et al (1998) Transcriptional activator-coactivator recognition: nascent folding of a kinase-inducible transactivation domain predicts its structure on coactivator binding. BioChemistry 37:5858–5866

Huber M, Cattaneo R, Spielhofer P et al (1991) Measles virus phosphoprotein retains the nucleocapsid protein in the cytoplasm. Virology 185:299–308

Iwasaki M, Takeda M, Shirogane Y et al (2009) The matrix protein of measles virus regulates viral RNA synthesis and assembly by interacting with the nucleocapsid protein. J Virol 83:10374–10383

Jensen MR, Houben K, Lescop E et al (2008) Quantitative conformational analysis of partially folded proteins from residual dipolar couplings: application to the molecular recognition element of Sendai virus nucleoprotein. J Am Chem Soc 130:8055–8061

Jensen MR, Bernadó P, Houben K et al (2010) Structural disorder within sendai virus nucleoprotein and phosphoprotein: insight into the structural basis of molecular recognition. Protein Pept Lett 17:952–960

Johansson K, Bourhis JM, Campanacci V et al (2003) Crystal structure of the measles virus phosphoprotein domain responsible for the induced folding of the C-terminal domain of the nucleoprotein. J Biol Chem 278:44567–44573

Jordan IK, Sutter BA, McClure MA (2000) Molecular evolution of the Paramyxoviridae and Rhabdoviridae multiple-protein-encoding P gene. Mol Biol Evol 17:75–86

Karlin D, Belshaw R (2012) Detecting remote sequence homology in disordered proteins: discovery of conserved motifs in the N-termini of Mononegavirales phosphoproteins. PLoS ONE 7:e31719

Karlin D, Longhi S, Canard B (2002a) Substitution of two residues in the measles virus nucleoprotein results in an impaired self-association. Virology 302:420–432

Karlin D, Longhi S, Receveur V et al (2002b) The N-terminal domain of the phosphoprotein of morbilliviruses belongs to the natively unfolded class of proteins. Virology 296:251–262

Karlin D, Ferron F, Canard B et al (2003) Structural disorder and modular organization in Paramyxovirinae N and P. J Gen Virol 84:3239–3252

Kavalenka A, Urbancic I, Belle V et al (2010) Conformational analysis of the partially disordered measles virus NTAIL-XD complex by SDSL EPR spectroscopy. Biophys J 98:1055–1064

Kerdiles YM, Cherif B, Marie JC et al (2006) Immunomodulatory properties of morbillivirus nucleoproteins. Viral Immunol 19:324–334

Kingston RL, Hamel DJ, Gay LS et al (2004a) Structural basis for the attachment of a paramyxoviral polymerase to its template. Proc Natl Acad Sci U S A 101:8301–8306

Kingston RL, Walter AB, Gay LS (2004b) Characterization of nucleocapsid binding by the measles and the mumps virus phosphoprotein. J Virol 78:8630–8640

Kingston RL, Gay LS, Baase WS et al (2008) Structure of the nucleocapsid-binding domain from the mumps virus polymerase; an example of protein folding induced by crystallization. J Mol Biol 379:719–731

Kiss R, Bozoky Z, Kovacs D et al (2008) Calcium-induced tripartite binding of intrinsically disordered calpastatin to its cognate enzyme, calpain. FEBS Lett 582:2149–2154
Kolakofsky D, Pelet T, Garcin D et al (1998) Paramyxovirus RNA synthesis and the requirement for hexamer genome length: the rule of six revisited. J Virol 72:891–899
Kolakofsky D, Le Mercier P, Iseni F et al (2004) Viral DNA polymerase scanning and the gymnastics of Sendai virus RNA synthesis. Virology 318:463–473
Kovacs E, Tompa P, Liliom K et al (2010) Dual coding in alternative reading frames correlates with intrinsic protein disorder. Proc Natl Acad Sci U S A 107:5429–5434
Krumm SA, Takeda M, Plemper RK (2013) The measles virus nucleocapsid protein tail domain is dispensable for viral polymerase recruitment and activity. J Biol Chem 288:29943–29953
Laine D, Trescol-Biémont M, Longhi S et al (2003) Measles virus nucleoprotein binds to a novel cell surface receptor distinct from FcgRII via its C-terminal domain: role in MV-induced immunosuppression. J Virol 77:11332–11346
Laine D, Bourhis J, Longhi S et al (2005) Measles virus nucleoprotein induces cell proliferation arrest and apoptosis through NTAIL/NR and NCORE/FcgRIIB1 interactions, respectively. J Gen Virol 86:1771–1784
Laine D, Vidalain P, Gahnnam A et al (2007) Interaction of measles virus nucleoprotein with cell surface receptors: impact on cell biology and immune response. In: Longhi S (ed) Measles virus nucleoprotein. Nova Publishers Inc., Hauppage, pp 113–152
Lamb RA, Kolakofsky D (2001) Paramyxoviridae: the viruses and their replication. In: Fields BN, Knipe DM, Howley PM (eds) Fields virology, 4th edn. Lippincott-Raven, Philadelphia, pp 1305–1340
Lamb RA, Parks GD (2007) Paramyxoviridae: the viruses and their replication. In: Knipe DM, Howley PM (eds) Fields virology, 5th edn. Lippincott Williams & Wilkins, Philadelphia, pp 1450–1497
Leyrat C, Gerard FC, de Almeida Ribeiro E Jr et al (2010) Structural disorder in proteins of the rhabdoviridae replication complex. Protein Pept Lett 17:979–987
Leyrat C, Jensen MR, Ribeiro EA Jr et al (2011a) The N(0)-binding region of the vesicular stomatitis virus phosphoprotein is globally disordered but contains transient alpha-helices. Protein Sci 20:542–556
Leyrat C, Yabukarski F, Tarbouriech N et al (2011b) Structure of the vesicular stomatitis virus N(0)-P complex. PLoS Pathog 7:e1002248
Lieutaud P, Canard B, Longhi S (2008) MeDor: a metaserver for predicting protein disorder. BMC Genomics 9:S25
Lieutaud P, Ferron F, Habchi J et al (2013) Predicting protein disorder and induced folding: a practical approach. In: Dunn B (ed) Advances in protein and peptide sciences. Bentham Science Publishers, Sharjah, pp 441–492 (52)
Liston P, DiFlumeri C, Briedis DJ (1995) Protein interactions entered into by the measles virus P, V, and C proteins. Virus Res 38:241–259
Liston P, Batal R, DiFlumeri C et al (1997) Protein interaction domains of the measles virus nucleocapsid protein (NP). Arch Virol 142:305–321
Llorente MT, Barreno-Garcia B, Calero M et al (2006) Structural analysis of the human respiratory syncitial virus phosphoprotein: characterization of an a-helical domain involved in oligomerization. J Gen Virol 87:159–169
Llorente MT, Taylor IA, Lopez-Vinas E et al (2008) Structural properties of the human respiratory syncytial virus P protein: evidence for an elongated homotetrameric molecule that is the smallest orthologue within the family of paramyxovirus polymerase cofactors. Proteins 72:946–958
Longhi S (ed) (2007) Measles virus nucleoprotein. Nova Publishers Inc., Hauppage
Longhi S (2009) Nucleocapsid structure and function. Curr Top Microbiol Immunol 329:103–128
Longhi S (2011) Structural disorder within the measles virus nucleoprotein and phosphoprotein: functional implications for transcription and replication. In: Luo M (ed) Negative strand RNA virus. World Scientific Publishing, Singapore, pp 95–125
Longhi S, Canard B (1999) Mécanismes de transcription et de réplication des *Paramyxoviridae*. Virologie 3:227–240

Longhi S, Oglesbee M (2010) Structural disorder within the measles virus nucleoprotein and phosphoprotein. Protein Pept Lett 17:961–978

Longhi S, Receveur-Brechot V, Karlin D et al (2003) The C-terminal domain of the measles virus nucleoprotein is intrinsically disordered and folds upon binding to the C-terminal moiety of the phosphoprotein. J Biol Chem 278:18638–18648

Longhi S, Lieutaud P, Canard B (2010) Conformational disorder. Methods Mol Biol 609:307–325

Lupas AN, Gruber M (2005) The structure of alpha-helical coiled coils. Adv Protein Chem 70:37–78

Magliery TJ, Wilson CG, Pan W et al (2005) Detecting protein-protein interactions with a green fluorescent protein fragment reassembly trap: scope and mechanism. J Am Chem Soc 127:146–157

Martinho M, Habchi J, El Habre Z et al (2013) Assessing induced folding within the intrinsically disordered C-terminal domain of the Henipavirus nucleoproteins by site directed spin labeling EPR spectroscopy. J Biomol Struct Dyn 31:453–471

Meszaros B, Tompa P, Simon I et al (2007) Molecular principles of the interactions of disordered proteins. J Mol Biol 372:549–561

Mittag T, Orlicky S, Choy WY et al (2008) Dynamic equilibrium engagement of a polyvalent ligand with a single-site receptor. Proc Natl Acad Sci U S A 105:17772–17777

Mohan A, Oldfield CJ, Radivojac P et al (2006) Analysis of molecular recognition features (MoRFs). J Mol Biol 362:1043–1059

Morin B, Bourhis JM, Belle V et al (2006) Assessing induced folding of an intrinsically disordered protein by site-directed spin-labeling EPR spectroscopy. J Phys Chem B 110:20596–20608

Moyer SA, Baker SC, Horikami SM (1990) Host cell proteins required for measles virus reproduction. J Gen Virol 71:775–783

Myers TM, Pieters A, Moyer SA (1997) A highly conserved region of the Sendai virus nucleocapsid protein contributes to the NP-NP binding domain. Virology 229:322–335

Myers TM, Smallwood S, Moyer SA (1999) Identification of nucleocapsid protein residues required for Sendai virus nucleocapsid formation and genome replication. J Gen Virol 80: 1383–1391

Narechania A, Terai M, Burk RD (2005) Overlapping reading frames in closely related human papillomaviruses result in modular rates of selection within E2. J Gen Virol 86:1307–1313

Ogino T, Kobayashi M, Iwama M et al (2005) Sendai virus RNA-dependent RNA polymerase L protein catalyzes cap methylation of virus-specific mRNA. J Biol Chem 280: 4429–4435

Oglesbee M (2007) Nucleocapsid protein interactions with the major inducible heat shock protein. In: Longhi S (ed) Measles virus nucleoprotein. Nova Publishers Inc., Hauppage, pp 53–98

Oglesbee M, Tatalick L, Rice J et al (1989) Isolation and characterization of canine distemper virus nucleocapsid variants. J Gen Virol 70(9):2409–2419

Oglesbee M, Ringler S, Krakowka S (1990) Interaction of canine distemper virus nucleocapsid variants with 70K heat-shock proteins. J Gen Virol 71:1585–1590

Oglesbee MJ, Kenney H, Kenney T et al (1993) Enhanced production of morbillivirus gene-specific RNAs following induction of the cellular stress response in stable persistent infection. Virology 192:556–567

Oglesbee MJ, Liu Z, Kenney H et al (1996) The highly inducible member of the 70kDa family of heat shock proteins increases canine distemper virus polymerase activity. J Gen Virol 77:2125–2135

Oldfield CJ, Cheng Y, Cortese MS et al (2005) Coupled folding and binding with α-helix-forming molecular recognition elements. Biochemistry 44:12454–12470

Oshaben KM, Salari R, McCaslin DR et al (2012) The native GCN4 leucine-zipper domain does not uniquely specify a dimeric oligomerization state. BioChemistry 51:9581–9591

Park MS, Shaw ML, Munoz-Jordan J et al (2003) Newcastle disease virus (NDV)-based assay demonstrates interferon-antagonist activity for the NDV V protein and the Nipah virus V, W, and C proteins. J Virol 77:1501–1511

Rahaman A, Srinivasan N, Shamala N et al (2004) Phosphoprotein of the rinderpest virus forms a tetramer through a coiled coil region important for biological function. A structural insight. J Biol Chem 279:23606–23614
Rancurel C, Khosravi M, Dunker KA et al (2009) Overlapping genes produce proteins with unusual sequence properties and offer insight into de novo protein creation. J Virol 83:10719–10736
Ringkjøbing Jensen M, Communie G, Ribeiro ED Jr et al (2011) Intrinsic disorder in measles virus nucleocapsids. Proc Natl Acad Sci U S A 108:9839–9844
Romero PR, Zaidi S, Fang YY et al (2006) Alternative splicing in concert with protein intrinsic disorder enables increased functional diversity in multicellular organisms. Proc Natl Acad Sci U S A 103:8390–8395
Roux L (2005) Dans le génome des Paramyxovirinae, les promoteurs et leurs activités sont façonnés par la "règle de six". Virologie 9:19–34
Rudolph MG, Kraus I, Dickmanns A et al (2003) Crystal structure of the borna disease virus nucleoprotein. Structure (Camb) 11:1219–1226
Ryan KW, Portner A (1990) Separate domains of Sendai virus P protein are required for binding to viral nucleocapsids. Virology 174:515–521
Salvamani S, Goh Z, Ho K et al (2013) Oligomerization state of the multimerization domain of Nipah virus phosphoprotein. Process Biochem 48:1476–1480
Sato H, Masuda M, Miura R et al (2006) Morbillivirus nucleoprotein possesses a novel nuclear localization signal and a CRM1-independent nuclear export signal. Virology 352:121–130
Schoehn G, Mavrakis M, Albertini A et al (2004) The 12A structure of trypsin-treated measles virus N-RNA. J Mol Biol 339:301–312
Shoemaker BA, Portman JJ, Wolynes PG (2000) Speeding molecular recognition by using the folding funnel: the fly-casting mechanism. Proc Natl Acad Sci U S A 97:8868–8873
Shu Y, Habchi J, Costanzo S et al (2012) Plasticity in structural and functional interactions between the phosphoprotein and nucleoprotein of measles virus. J Biol Chem 287:11951–11967
Spehner D, Kirn A, Drillien R (1991) Assembly of nucleocapsidlike structures in animal cells infected with a vaccinia virus recombinant encoding the measles virus nucleoprotein. J Virol 65:6296–6300
Spehner D, Drillien R, Howley PM (1997) The assembly of the measles virus nucleoprotein into nucleocapsid-like particles is modulated by the phosphoprotein. Virology 232:260–268
Sue SC, Cervantes C, Komives EA et al (2008) Transfer of flexibility between ankyrin repeats in IkappaB* upon formation of the NF-kappaB complex. J Mol Biol 380:917–931
Sweetman DA, Miskin J, Baron MD (2001) Rinderpest virus C and V proteins interact with the major (L) component of the viral polymerase. Virology 281:193–204
Tan WS, Ong ST, Eshaghi M et al (2004) Solubility, immunogenicity and physical properties of the nucleocapsid protein of Nipah virus produced in Escherichia coli. J Med Virol 73:105–112
Tapparel C, Maurice D, Roux L (1998) The activity of Sendai virus genomic and antigenomic promoters requires a second element past the leader template regions: a motif (GNNNNN)3 is essential for replication. J Virol 72:3117–3128
Tarbouriech N, Curran J, Ruigrok RW et al (2000) Tetrameric coiled coil domain of Sendai virus phosphoprotein. Nat Struct Biol 7:777–781
Tawar RG, Duquerroy S, Vonrhein C et al (2009) 3D structure of a nucleocapsid-like nucleoprotein-RNA complex of respiratory syncytial virus. Science 326:1279–1283
tenOever BR, Servant MJ, Grandvaux N et al (2002) Recognition of the Measles Virus Nucleocapsid as a Mechanism of IRF-3 Activation. J Virol 76:3659–3669
Tokuriki N, Oldfield CJ, Uversky VN et al (2009) Do viral proteins possess unique biophysical features? Trends Biochem Sci 34:53–59
Tompa P, Fuxreiter M (2008) Fuzzy complexes: polymorphism and structural disorder in protein-protein interactions. Trends Biochem Sci 33:2–8
Tran TL, Castagne N, Bhella D et al (2007) The nine C-terminal amino acids of the respiratory syncytial virus protein P are necessary and sufficient for binding to ribonucleoprotein complexes in which six ribonucleotides are contacted per N protein protomer. J Gen Virol 88:196–206
Tsai CD, Ma B, Kumar S et al (2001a) Protein folding: binding of conformationally fluctuating building blocks via population selection. Crit Rev Biochem Mol Biol 36:399–433

Tsai CD, Ma B, Sham YY et al (2001b) Structured disorder and conformational selection. Proteins: structure. Funct Bioinform 44:418–427
Ulane CM, Horvath CM (2002) Paramyxoviruses SV5 and HPIV2 assemble STAT protein ubiquitin ligase complexes from cellular components. Virology 304:160–166
Uversky VN (2002) Natively unfolded proteins: a point where biology waits for physics. Protein Sci 11:739–756
Uversky VN, Longhi S (eds) (2012) Flexible viruses: structural disorder in viral proteins. Wiley, Hoboken
Uversky VN, Oldfield CJ, Dunker AK (2005) Showing your ID: intrinsic disorder as an ID for recognition, regulation and cell signaling. J Mol Recognit 18:343–384
Vacic V, Oldfield CJ, Mohan A et al (2007) Characterization of molecular recognition features, MoRFs, and their binding partners. J Proteome Res 6:2351–2366
Vasconcelos D, Norrby E, Oglesbee M (1998a) The cellular stress response increases measles virus-induced cytopathic effect. J Gen Virol 79:1769–1773
Vasconcelos DY, Cai XH, Oglesbee MJ (1998b) Constitutive overexpression of the major inducible 70kDa heat shock protein mediates large plaque formation by measles virus. J Gen Virol 79:2239–2247
Wang Y, Chu X, Longhi S et al (2013) Multiscaled exploration of coupled folding and binding of an intrinsically disordered molecular recognition element in measles virus nucleoprotein. Proc Natl Acad Sci U S A 110:E3743–E3752
Warnes A, Fooks AR, Dowsett AB et al (1995) Expression of the measles virus nucleoprotein gene in Escherichia coli and assembly of nucleocapsid-like structures. Gene 160:173–178
Watanabe N, Kawano M, Tsurudome M et al (1996) Identification of the sequences responsible for nuclear targeting of the V protein of human parainfluenza virus type 2. J Gen Virol 77:327–338
Watanabe A, Yoneda M, Ikeda F et al (2011) Peroxiredoxin 1 is required for efficient transcription and replication of measles virus. J Virol 85:2247–2253
Wilson CG, Magliery TJ, Regan L (2004) Detecting protein-protein interactions with GFP-fragment reassembly. Nat Methods 1:255–262
Xue B, Williams RW, Oldfield CJ et al (2010) Viral disorder or disordered viruses: do viral proteins possess unique features? Protein Pept Lett 17:932–951
Xue B, Dunker AK, Uversky VN (2012) Orderly order in protein intrinsic disorder distribution: disorder in 3500 proteomes from viruses and the three domains of life. J Biomol Struct Dyn 30:137–149
Xue B, Blocquel D, Habchi J et al (2014) Structural Disorder in Viral Proteins. Chem Rev 114:6880–6911
Yegambaram K, Kingston RL (2010) The feet of the measles virus polymerase bind the viral nucleocapsid protein at a single site. Protein Sci 19:893–899
Yegambaram K, Bulloch EM, Kingston RL (2013) Protein domain definition should allow for conditional disorder. Protein Sci 22:1502–1518
Zhang X, Glendening C, Linke H et al (2002) Identification and characterization of a regulatory domain on the carboxyl terminus of the measles virus nucleocapsid protein. J Virol 76:8737–8746
Zhang X, Bourhis JM, Longhi S et al (2005) Hsp72 recognizes a P binding motif in the measles virus N protein C-terminus. Virology 337:162–174

Chapter 13
Druggability of Intrinsically Disordered Proteins

Priyanka Joshi and Michele Vendruscolo

Abstract Although the proteins in all the current major classes considered to be druggable are folded in their native states, intrinsically disordered proteins (IDPs) are becoming attractive candidates for therapeutic intervention by small drug-like molecules. IDPs are challenging targets because they exist as ensembles of structures, thereby making them unsuitable for standard rational drug design approaches, which require the knowledge of the three-dimensional structure of the proteins to be drugged. As we review in this chapter, several different small molecule strategies are currently under investigation to target IDPs, including: (i) to stabilise IDPs in their natively disordered states, (ii) to inhibit interactions with ordered or disordered protein partners, and (iii) to induce allosteric inhibition. In this context, biophysical techniques, including in particular nuclear magnetic resonance (NMR) spectroscopy and small-angle X-ray scattering (SAXS) coupled with molecular dynamics simulations and chemoinformatics approaches, are increasingly used to characterize the structural ensembles of IDPs and the specific interactions that they make with their binding partners. By analysing the results of recent studies, we describe the main structural features that may render IDPs druggable, and describe techniques that can be used for drug discovery programs focused on IDPs.

Keywords Intrinsically disordered proteins · Drug design · Nuclear magnetic resonance spectroscopy · Small molecule library · Drug discovery

1 Introduction: IDPs as Therapeutic Targets

IDP Function and Dysfunction

As described in earlier chapters of this book, intrinsically disordered proteins (IDPs) play major roles in a wide range of biochemical processes in living organisms. A range of recent studies has revealed that the functional diversity provided by disordered regions complements that of ordered regions of proteins (Iakoucheva et al.

M. Vendruscolo (✉) · P. Joshi
Department of Chemistry, University of Cambridge, Cambridge, CB2 1EW, UK
e-mail: mv245@cam.ac.uk

I. C. Felli, R. Pierattelli (eds.), *Intrinsically Disordered Proteins Studied by NMR Spectroscopy,* Advances in Experimental Medicine and Biology,
DOI 10.1007/978-3-319-20164-1_13

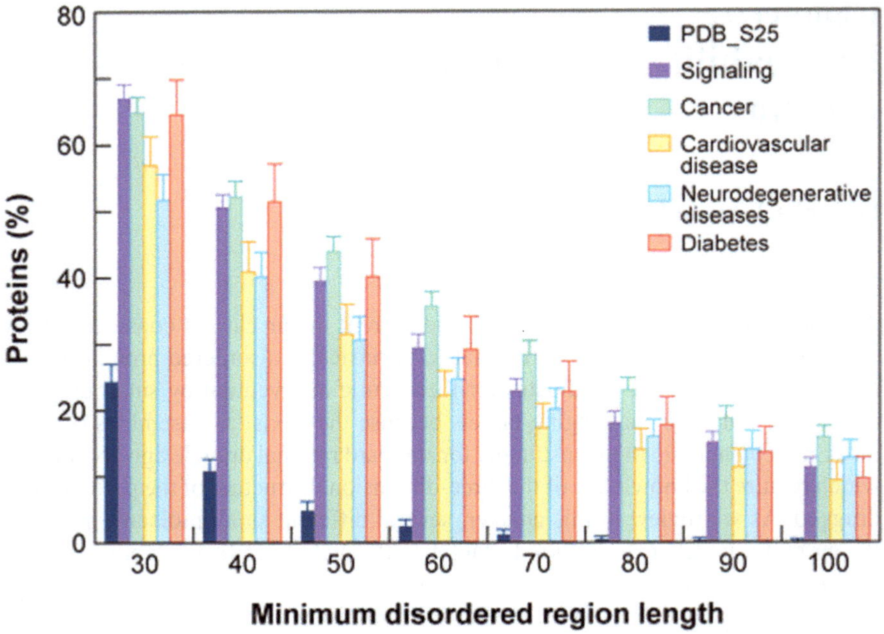

Fig. 13.1 Presence of intrinsic disorder in disease-associated proteins. The histogram presents the percentage of disease-associated proteins with more than 30 to more than 100 consecutive residues predicted to be disordered, with error bars representing 95 % confidence intervals; for comparison, data for signalling and ordered proteins are also shown. Reprinted from (Uversky et al. 2008)

2002; Dyson and Wright 2005; Tompa 2012; Uversky 2013; Dunker et al. 2002; van der Lee et al. 2014; Babu et al. 2012), in particular in terms of key cellular functions such as signaling and regulation (Iakoucheva et al. 2002; Dyson and Wright 2005; Tompa 2012; Uversky 2013; Dunker et al. 2002; Uversky et al. 2008, 2009; Knowles et al. 2014; Babu et al. 2012; van der Lee et al. 2014). The high flexibility and lack of stable secondary and tertiary structures allow IDPs to interact with multiple partners, often placing them at the hubs of protein-protein interaction networks (Mészáros et al. 2011; Malaney et al. 2013; Dunker et al. 2005). It has also been realized that the failure of the regulatory processes responsible for the correct behaviour of IDPs, which can be referred to as 'IDP homeostasis', is associated with a variety of pathological conditions (Uversky et al. 2008, 2009; Knowles et al. 2014). Indeed, intrinsic disorder is often observed in peptides and proteins implicated in a series of human conditions including cancer, cardiovascular diseases and neurodegenerative disorders (Uversky et al. 2008, 2009; Knowles et al. 2014) (Fig. 13.1).

IDPs Are Implicated in Misfolding Diseases

Among all the diseases associated with IDPs, great attention has recently been devoted to misfolding diseases (Chiti and Dobson 2006; Dobson 2001; Knowles et al. 2014). Most of these diseases originate from the conversion of specific proteins from their soluble functional states into stable, highly ordered filamentous protein aggregates, known as amyloid fibrils, which accumulate in a variety of organs and tissues (Chiti and Dobson 2006; Dobson 2001; Knowles et al. 2014). Amyloid fibrils display common properties including a characteristic cross-β structure in which continuous β-sheets are formed with β-strands running perpendicularly to the long axis of the fibrils (Fitzpatrick et al. 2013; Eisenberg and Jucker 2012; Knowles et al. 2014). Although amyloid fibrils from different diseases are structurally and morphologically similar to each other, the polypeptide chains that constitute the fibrils are diverse and the native conformation may be rich in β-sheets and α-helices, or they may be natively unfolded (Fitzpatrick et al. 2013; Eisenberg and Jucker 2012; Knowles et al. 2014). The conformational diversity of the native states of amyloidogenic proteins, as opposed to the close structural similarity of the resultant amyloid fibrils, implies that substantial structural rearrangements need to occur for fibril formation to happen. As IDPs are devoid of stable structure, the primary step of their fibrillogenesis requires the stabilization of monomeric or oligomeric partially folded conformations. Such partially folded conformations are characterised by the presence of specific intermolecular interactions, including electrostatic attraction, hydrogen bonding, and hydrophobic contacts, which promote the assembly process. Therefore, a possible strategy to target aggregation-prone IDPs implicated in misfolding diseases is to stabilize the protein in conformations that do not readily enable the formation of higher order species. Indeed, recent studies are beginning to suggest that such approaches may be promising (Zhu et al. 2013; Toth et al. 2014; Dunker and Uversky 2010; Metallo 2010; Rezaei-Ghaleh et al. 2012; Uversky 2012).

Therapeutic Targeting of IDPs
The foremost reason to target IDPs pharmacologically is that they are closely associated with a wide range of diseases. Therapeutic intervention strategies aimed at maintaining the normal functionality of these proteins require the characterization of their structures and functional mechanisms, as indeed is beginning to be done (Iakoucheva et al. 2002; Dyson and Wright 2005; Tompa 2012; Uversky 2013; Dunker et al. 2002; Uversky et al. 2008, 2009; Knowles et al. 2014). This objective may be achieved by a range of different strategies, including: (i) the stabilization of IDPs in their natively disordered states, for instance in the case of aggregation-prone proteins, (ii) the inhibition of interactions with molecular partners, in cases where a small molecule perturbs the binding interface, and (iii) the regulation of the behaviour of IDPs by allosteric effects, for instance when the binding of a small molecule to an ordered region causes the disordered region to become ordered.

Most drugs target enzymes or cell surface receptors by modulating their functions, where small molecules can mimic the interactions made by their natural substrates (Imming et al. 2006). Even though enzymes possess a certain degree of flexibility to be able to act on a variety of substrates, their structures tend to fluctuate around equilibrium positions, making it easier to identify binding pockets and subsequently tailor drugs to fit in them. By contrast, IDPs explore large ensembles of structures, which exhitbit large conformational fluctuations and no evidence of permanent binding pockets. This type of conformational features does not present suitable cavities for small drug-like molecules to form stable interactions (Dunker and Uversky 2010; Metallo 2010; Toth et al. 2014; Zhu et al. 2013).

Quite generally, the binding of a small molecule to a disordered protein may seem counterintuitive because of the large difference expected between the entropic loss and the enthalpic gain upon binding (Metallo 2010; Toth et al. 2014; Zhu et al. 2013). As we will describe in more detail in the remainder of this chapter, however, some intrinsically disordered proteins have been shown to be capable of forming adaptable, specific interfaces for small molecule binding (Metallo 2010; Toth et al. 2014; Zhu et al. 2013).

Druggability of IDPs

The ability to find compounds that bind to a target protein does not by itself imply that this protein is fully druggable, as the compounds should also be orally bioavailable (Hopkins and Groom 2002). Lipinski proposed a set of five criteria to readily estimate bioavailability, the so-called 'rule of five', that includes restrictions on the molecular weight (<500 Da), number of hydrogen bond acceptors (<10), number of hydrogen bond donors (<5) and the octanol-water partition coefficient (<5) of the potential lead compound (Lipinski 2004). These rules should be in some way combined with the primary demand for the design of small molecule drugs, that they should be able to form relatively strong interactions with the protein to overcome the entropy loss upon binding. IDPs are in this context particularly challenging targets, as their mobility implies that, compared to structured proteins, their interactions with small molecules are weaker and more transient, and the entropic loss greater.

As noted above, the high abundance and peculiar roles of IDPs in protein-protein interactions (PPIs) makes them desirable drug targets. However, attempts at developing small molecule drugs that block protein-protein interactions have generally been challenging (Arkin and Wells 2004; Wells and McClendon 2007; Hopkins and Groom 2002). Investigations on the reasons that make PPIs difficult drug targets are beginning to reveal some of the molecular mechanisms responsible for these problems. It has been shown that PPIs display complex binding surfaces, have discontinuous epitopes or multiple continuous epitopes, are devoid of groves or pockets, exhibit interaction regions that are often as large as 1500–3000 Å^2, and have energies that are not evenly distributed over a large contact area but over smaller regions called hotspots (Jones and Thornton 1996; Conte et al. 1999; Wells and McClendon 2007).

Specialized libraries for IDPs have not been extensively adopted to date, and the properties of compounds that bind IDPs have not been intensively investigated. Compounds that bind IDPs tend to have properties that make them different from those from more conventional target classes, as in particular they are larger and more three-dimensional. In the absence of a more complete understanding of these properties, the question regarding the set of compounds that could adequately provide coverage of the chemical space specifically for IDPs is still open. To this end, the construction of libraries of compounds specific for IDPs would be highly useful, as it would lead to a better and faster design of inhibitor libraries and promote more effective efforts towards drug discovery programs for IDPs.

2 Strategies for Drug Discovery for IDPs

Strategies for therapeutic intervention for IDPs involve a wide range of objectives, including:

I. To maintain IDPs in their natively disordered states;
II. To inhibit interactions with ordered or disordered partners;
III. To induce allosteric inhibition.

Below we discuss some examples that illustrate how these strategies have begun to be implemented in recent studies.

2.1 Case Study 1. Targeting the p53-Mdm2 Interaction

p53 protein is a transcription factor at the centre of a large signalling network that targets genes involved in cell cycle regulation, apoptosis and DNA repair (Muller and Vousden 2014). For this reason, p53 dysregulation is a major factor in cancer development (Muller and Vousden 2014). The interactions made by p53 can involve: (i) the N-terminal domain (the transcription domain), (ii) the C-terminal domain (the regulatory domain), and (iii) the DNA binding domain (DBD). Of these domains only the DBD is structured, while the N- and C-terminal domains are intrinsically disordered (Dawson et al. 2003; Oldfield et al. 2008). Thus, p53 extensively uses disordered regions to mediate and modulate interactions with other proteins, as about 70% of the interactions of p53 are of this type (Oldfield et al. 2008; Uversky et al. 2008).

Among the proteins that regulate p53, Mdm2 plays a major role by inactivating it through binding to its transcription activation domain (Kubbutat et al. 1997). X-ray and NMR studies have revealed that the Mdm2 binding region of p53 near the N-terminus, which is highly flexible when unbound, forms a α-helical structure that binds into a deep groove on the surface of Mdm2 (Chène 2004; Michelsen et al. 2012). Several peptides and small molecules have been designed to inhibit this interaction (Wang et al. 2011; Fry et al. 2013). Predictions made using the molecular recognition features (MoRFs) method (van der Lee et al. 2014), as well as the presence of hydrophobic clusters and other distinctive structural features of the p53-Mdm2 complex, have rationalised why this region appears to be a promising drug target (Cheng et al. 2006), thus providing support to the notion that a protein-protein interaction involving one disordered and one structured partner is likely to be druggable (Uversky 2012). Analysis of the MoRF dataset (Habchi et al. 2014; Cheng et al. 2006) has suggested that for example in some cases the disordered region of a protein can form an α-helix with a hydrophobic face that fits into the groove of the ordered protein. In the disorder-to-order transition, the binding energy should balance off the high entropy of the unfolded state. This interaction may thus be weaker than that between two ordered proteins, and thus potentially more easily targeted by small molecule inhibitors.

2.2 Case Study 2. Targeting the c-Myc-Max Interaction

The oncoprotein c-Myc is a transcription factor implicated in a broad range of human cancers (Dang 1999). Its activity is dependent on heterodimerisation with the disordered protein Max (Blackwood and Eisenman 1991). Both c-Myc and Max are highly disordered in their free forms and undergo mutual coupled binding and folding to form the heterodimer complex via a basic helix-loop-helix leucine zipper domain present in both proteins (Hammoudeh et al. 2009; Harvey et al. 2012; Michel and Cuchillo 2012). This interaction has been reported to be druggable by small molecules that bind to the monomeric and disordered c-Myc (Hammoudeh et al. 2009; Harvey et al. 2012; Michel and Cuchillo 2012). NMR spectroscopy revealed that these molecules bind to a disordered site in the c-Myc monomer located at the interface between the helix-loop-helix and leucine zipper in the c-Myc-Max complex (Hammoudeh et al. 2009). c-Myc is flexible in both the free and small molecule-bound states with only secondary structural elements being transiently populated, thus representing a clear example of the strategy of using small molecules to inhibit interactions between disordered partners. This case study highlights the importance of the specificity of the interactions between disordered proteins and small molecules, and provide insights into the types of small molecules that could be capable of binding highly dynamic targets.

2.3 Case Study 3. Targeting the Aβ Peptide

The aggregation of the Aβ peptide into oligomeric assemblies and ordered fibrils is associated with Alzheimer's disease (Chiti and Dobson 2006; Haass and Selkoe 2007; Knowles et al. 2014). For this reason, this peptide has been the subject of intense research and an increasing number of small molecules have been reported to interfere with its aggregation process (Necula et al. 2007; McKoy et al. 2012; Hawkes et al. 2009; Porat et al. 2006; Wang et al. 2014; Yamin et al. 2009). The existence of various metastable structures of this peptide, however, poses tremendous challenges in developing strategies to avoid its aggregation. Its two main isoforms, Aβ40 and Aβ42, are highly disordered, with the latter having a greater pathological relevance due to its more aggressive aggregation behaviour. Although no predominant binding modes have been identified, many of the small molecules that have been reported to interact with the Aβ peptide appear to do so at the central portion of the peptide (residues 16–22). A recent mapping of the free energy landscape of this peptide (Zhu et al. 2013) (Fig. 13.2) has suggested that the presence of these small molecules may affect the populations of the conformational substates (Zhu et al. 2013). These findings suggest the intriguing possibility that the binding of small molecules to the Aβ peptide may change its behaviour by shifting the statistical weights of multiple substates.

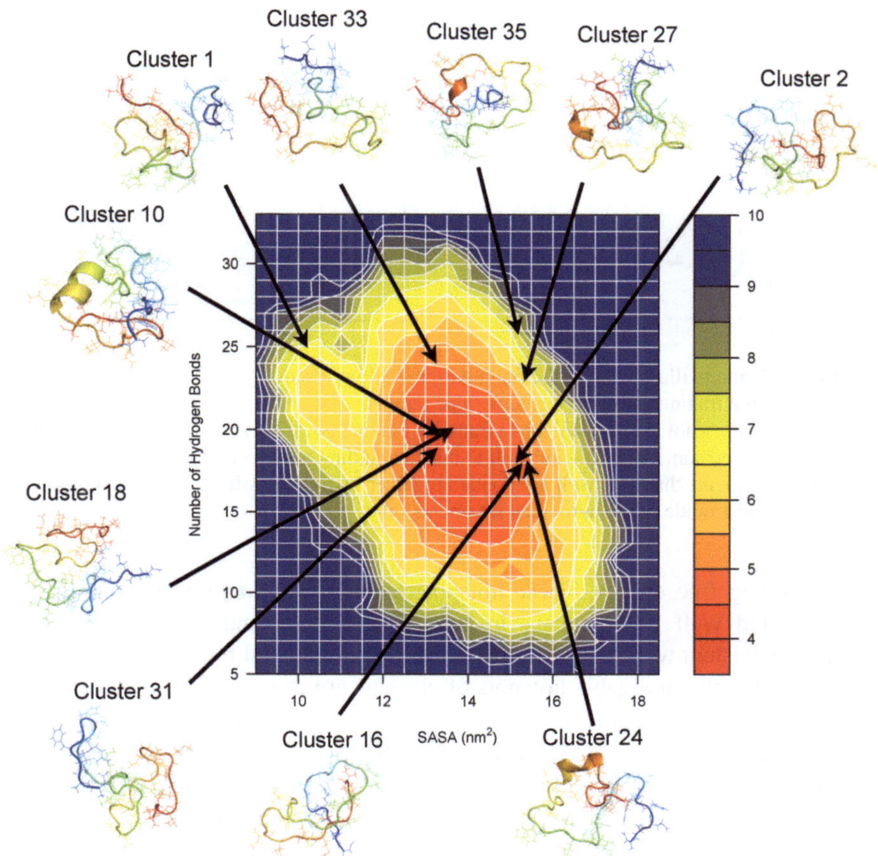

Fig. 13.2 Representative free energy landscape of an IDP. The free energy landscape of the Aβ peptide is shown as a function of the number of hydrogen bonds (backbone-backbone, backbone-sidechain, and sidechain-sidechain) and of the solvent-exposed surface area of hydrophobic residues. The most populated clusters of structures, of which some examples are shown, are found in different regions of the free energy landscape. Reprinted from (Zhu et al. 2013)

3 Fragment-based Drug Discovery for IDPs

So far, owing to the heterogeneous structural characteristics of IDPs, it has been challenging to identify possible initial drug candidates (or 'hits') through high-throughput screening of compounds. A promising approach to overcome this problem is based on fragment-based drug design (FBDD), an approach that allows smaller starting structures to be identified and then subsequently grown into small-drug like molecules (Warr 2009; Hajduk and Greer 2007; Congreve et al. 2008; Murray and Rees 2009; Murray and Blundell 2010). Druggable epitopes are identified in an unbound protein structure using such fragments. This approach is based on the observation that a few adjacent residues often make a significant contribution

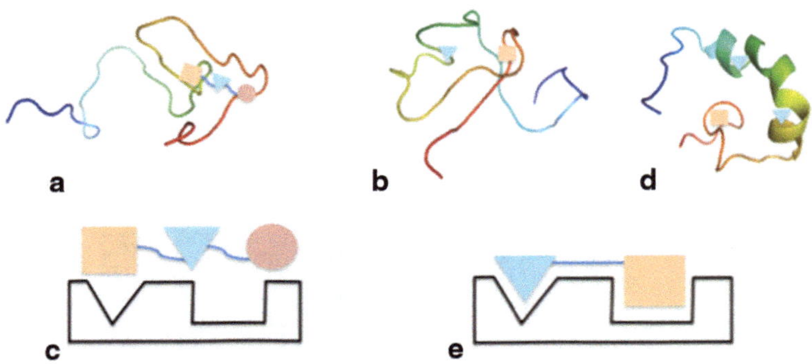

Fig. 13.3 Schematic illustration of the fragment-based drug discovery strategy. A small molecule made up of three fragments (*orange square, blue triangle and pink circle*) is bound weakly to an IDP (**a**) since it does not fit very well into a transient binding pocket (**c**). The fragments are considered separately and some of them bind well to distinct binding pockets in different conformations (**b**, **d**). The fragments that bind well are bonded together by a linker whose length and properties match the binding mode of the individual fragments (**e**)

to the binding free energy. These hotspot regions (Wells and McClendon 2007; Clackson and Well 1995) consist in large part of polar and conserved residues, which is consistent with their roles in binding. Experimental fragment screens have confirmed that the druggable hotspots of proteins are characterized by their ability to bind a variety of fragments and that the number of different probe molecules observed to bind to a particular site predicts the potential importance of the site, as well as its overall druggability (Kozakov et al. 2011). A set of fragment probes containing diverse functional groups and shapes, which enable them to bind to a variety of hotspots and binding sites, has recently been used to identify druggable sites in Aβ and α-synuclein (Zhu et al. 2013; Toth et al. 2014). In the case of Aβ, the identification of hotspots and binding pockets was carried out using a computational fragment probe mapping methodology (Zhu et al. 2013).

Unlike the case of structured proteins, as for example G protein-coupled receptors (GPCRs) and kinases, dedicated small molecule libraries are not yet available for IDPs. The development of such libraries may be based on the idea that IDP sequences present peculiar features that enable them to assume heterogeneous structures. Therefore, for individual IDPs, privileged small molecule scaffolds can in principle be identified for drug discovery. The term ‘privileged scaffold’ stems from the notion that multiple molecules of similar scaffolds have similar bioactivities (Welsch et al. 2010). Further, these privileged scaffolds can be used to identify compounds that are biologically active towards IDPs by building chemical libraries. This ligand-based approach could be used together with structure-based methods where an ensemble of structures are probed by multiple fragments to identify hotspots comprised of conserved residues across the ensemble. Subsequently, the fragment ligands could be linked to achieve greater binding (Fig. 13.3). In our group we are currently pursuing a strategy in which fragments from known inhibitors of Aβ are used to identify new molecules that could be potential inhibitors of aggregation. This approach could also be used to identify those scaffolds that contribute significantly towards the free energy of binding to the monomeric peptide, then further expanding them with appropriate linkers and functional groups to optimize binding and bioavailability.

Quite generally, the FBDD approach is most promising for drug design, given that a few fragments can sample vast amounts of chemical space, reducing the number of compounds for screening (Sun et al. 2011). Fragments bound at various regions of IDPs can be linked together via an appropriate linker (Fig. 13.3). In most cases, fragment-sized compounds need hydrogen bonds to achieve detectable binding. This enthalpic gain compensates for the entropic loss upon binding of the small molecules, and overall lowers the free energy of the protein upon binding. This effect is similar to the mimicking of the interacting partners of IDPs by these molecules.

4 Techniques Used to Study IDP-Small Molecule Interactions

Any drug discovery programme should include a quantitative assessment of the interactions of therapeutic targets with drug candidates. In the case of IDPs such interactions can be characterized in great detail by a wide range of experimental techniques, including NMR spectroscopy and small-angle X-ray scattering (SAXS), particularly if complemented with advanced computational methods.

4.1 Nuclear Magnetic Resonance (NMR) Spectroscopy

4.1.1 NMR Spectroscopy for Drug Development for IDPs

As described in earlier chapters of this book (Chaps. 3–5), NMR spectroscopy is a particularly suitable technique to study the structure and dynamics of IDPs. In the context of drug development, already in the early 1970s NMR spectroscopy was used to study ligand-protein transient interactions, mainly to obtain information from the ligand point of view, by observing perturbations of different NMR parameters including nuclear relaxation time, chemical shifts, inter and intra-molecular nuclear Overhauser effects (NOEs) and intermolecular spin diffusion (Pellecchia et al. 2002). NMR spectroscopy has thus played a major role in the development of rational drug design approaches based on the knowledge of the three-dimensional structures of potential pharmacological targets and their complexes with drug-like molecules (Pellecchia et al. 2008). In the last 25 years, an extended arsenal of NMR methodologies has become available at different stages of the drug discovery process, including target identification, screening compounds, hit validation and lead-compound characterization, structural and mechanistic characterization of ligand-receptor interactions to design and assess target druggability, pharmacophore identification, and potentially structure based drug design (Hajduk et al. 1999; Pellecchia et al. 2002; Renaud and Delsuc 2009). More recently, NMR spectroscopy has demonstrated its utility in FBDD strategies, thus providing an alternative to conventional high-throughput screening, or to support hit-to-lead optimization for a particular drug target (Klages et al. 2007). NMR spectroscopy can also be used to determine low-resolution structures of target-ligand complexes for IDPs or membrane proteins that are not amenable to crystallographic approaches (Pellecchia et al. 2008). Several NMR-based strategies have been developed for FBDD applications, ranging from the traditional chemical shift mapping to ligand-based techniques that monitor changes in ligand nuclear spin relaxation properties upon binding, to measurements of diffusion (Klages et al. 2007). Some of these approaches are better suited to screen and/or to validate hits coming from high-throughput screening (HTS) campaigns, whereas others are better suited to guide hit optimization into more potent, selective drug-like compounds.

4.1.2 Using Chemical-Shift Mapping for Ligand Binding, Screening and Validation

Applications of NMR-based approaches have been extended to 'non-traditional' targets such as protein-protein interactions and IDPs. Chemical-shift mapping is one of the most robust, reliable and reproducible ligand binding assays available, and is thus the most utilized approach for ligand binding studies (Hajduk et al. 1999; Pellecchia et al. 2002). This technique exploits the differences in chemical shifts between free and bound protein targets in ^{15}N-^{1}H and ^{13}C-^{1}H two-dimensional correlation spectra of the target upon titration of a ligand or a mixture of ligands. When combined with resonance assignments, at least for proteins with a molecular weight under 30–40 kDa, this approach can give valuable structural information on the site of binding. This method can be extended to larger macromolecular targets in which an amino acid type has been selectively labelled to reduce spectral complexity, thus extending its applicability to targets greater than 100 kDa (Pellecchia et al. 2002). Similar to ordered targets, when the structures of the IDPs have been previously determined and characterized by NMR spectroscopy, in some cases it should be possible to rapidly derive ligand-protein distances via NOE-type experiments that allow sufficiently more precise determination of the ligand binding mode. One of the advantages of this method is that compounds that bind to a given protein can be found and characterized without the need to develop more complex assays (Pellecchia et al. 2008). This aspect is particularly useful in the case of IDPs, where assay development is challenging given the heterogeneity that exists among the different species. Unlike X-ray crystallography, however, this approach does not readily provide information on the dissociation constants of the complexes, nor can it be easily used to monitor ligand binding. Moreover, the amount of protein that is required for individual NMR experiments is relatively high as compared to other techniques, making it expensive for the testing of large libraries of compounds. Hence, these assays have found widespread use in hit validation and in FBDD, in which smaller libraries are used. Taken together, the results obtained thus far indicate that strategies based on the use of FBDD, NMR and computational methods provide an attractive strategy for targeting IDPs for drug design.

Instead of studying the protein-ligand complex upon binding, ligand-based methods could be used to observe the perturbations induced by sub-stoichiometric amounts of targets on the NMR spectra of the ligands. Examples of these approaches include the saturation transfer difference (STD) or T1ρ measurements. For IDPs, whose structures are difficult to calculate (as explained in Chaps. 2–5), these 'transferred' ligand-based methods could prove valuable. These approaches, however, are less informative than chemical-shift mapping and are used primarily for screening and for validating ligand binding. This information could be very useful in pharmacophore-based design (Pellecchia et al. 2008).

Hybrid approaches based on the combination of NMR computational docking and screening methods could also be used. In these methods, NMR restraints such as internuclear distance information, chemical shift mapping and residual dipolar couplings are used to calculate an ensemble of conformations of IDPs through molecular dynamics simulations (Vendruscolo 2007), even without full structure determination. Selective conformations based on low energy, compaction, or other structural indicators can then be subjected to virtual screening via docking wherein the top ranked compounds are experimentally verified using NMR spectroscopy. Introducing the docking approach reduces the number of test compounds in a library and saves on running an expensive and time-consuming high-throughput screening experiment. Similarly, a fragment-based library based on these docked compounds could also be used via the ligand-based methods to identify pharmacophores for IDPs.

4.1.3 Using NMR Spectroscopy for Hit and Lead Optimization

Some of the NMR approaches, such as the structure-activity relationships ('SAR by NMR') (Shuker et al. 1996), have proven very useful in deriving high-affinity ligands for challenging targets for which other approaches have failed to produce viable leads. In targets that have been previously classified as challenging and 'undruggable' or for which there are few alternatives-such as IDPs-NMR information is clearly very useful. The observation that some targets yield more free energy of binding per atom for the initial binding fragments provides a good indicator for assessing the druggability of the target and identifying potential hotspots on the surface of the protein (Pellecchia et al. 2008). These hotspots can also be identified using computational algorithms, such as for example FTMap (Ivetac and McCammon 2012), as shown in a recent study on the Aβ peptide (Zhu et al. 2013).

4.2 Small-Angle X-ray Scattering

As described earlier in this book (Chap. 8), SAXS methods can be effectively employed to study systems with conformational polydispersity, i.e. completely or partially disordered macromolecules, including multi-domain proteins with flexible linkers and IDPs. SAXS is a label-free biophysical method that is particularly suitable in the study of the overall structure and structural transitions of biological macromolecules in solution (Chap. 8). It is a powerful, although low-resolution, tool for the quantitative analysis of flexible systems such as IDPs, and is highly complementary to the high-resolution methods X-ray crystallography and NMR spectroscopy.

Advances in instrumentation at synchrotron facilities, including data collection times within seconds, charge-coupled device detectors, automated sample changers, and integrated robotic control software, have enabled the collection of scattering profiles for high-throughput studies within seconds, making data collection on IDPs relatively accessible (Bernado and Svergun 2012). This progress is very significant for drug discovery programs aimed at IDPs, as structural effects induced by small molecules can be studied using a medium-throughput screen. Analysis of the conformationally heterogeneous ensembles provides quantitative information about flexibility and insights into the structural features (Bernadó et al. 2007). This type of information is extremely useful for drug discovery pipelines, where after an *in silico* hit identification, the goal is to identify lead compounds. As SAXS methods give relatively low-resolution information (Blobel et al. 2009) but well-defined evidence for overall structural changes, they are very useful tools for studying the effects of small molecules on IDPs.

In peptide and protein systems that undergo aggregation, such as for example the Aβ peptide, possible therapeutic interventions can be aimed at targeting the flexible monomeric species to stabilize them without forming higher order oligomers (Toth et al. 2014; Zhu et al. 2013). Compounds interacting with peptides and proteins at the monomeric level are expected to have three kinds of effects: (i) to inhibit aggregation, (ii) to accelerate aggregation (iii) or to leave aggregation unaltered. These effects can be monitored through time-dependent scattering measurements. In particular, Kratky plots collected at different stages of the aggregation process give the scattering profiles characteristic of either an ordered protein or a disordered protein (Bernado and Svergun 2012). The effects of the molecule on the protein can be observed by studying the Kratky plots, which are characteristically different for ordered, disordered and multidomain proteins. Alternatively, for an IDP, the radius of gyration could be monitored as a function of time. Since flexible proteins normally have larger radius of gyration values than fully folded ones, the addition of an effector compound often leads to a decrease in the radius of gyration. More generally, any deviation from the normal trend of the aggregation behaviour of an IDP suggests that the molecule has an effect. Such data can be used to build low-resolution models of the structural species in cases where structural species are not very well described, as for example the Aβ peptide, where structures are challenging to obtain. Additionally, SAXS data could also be used in combination with molecular simulations to study the dynamics of these interactions or to obtain models of these structural species.

4.3 Chemical Kinetics

Chemical kinetics is a powerful approach for the quantitative analysis of the rates at which biochemical reactions take place (Knowles et al. 2009). This method offers the possibility of detecting and analysing even very weak binding events and their effects, and thus shows promise for studying the interactions of small molecules with IDPs (Arosio et al. 2014). The inhibitory effects of different compounds on protein aggregation have been evaluated by monitoring the kinetics of aggregation *in vitro*, in particular by means of thioflavin T (ThT) fluorescence based assays (Cohen et al. 2013; Arosio et al. 2014; Meisl et al. 2014). Potential drugs are based on their capacity to delay or arrest fibril formation as monitored in this way (Arosio et al. 2014). This approach has the advantage of rapidly providing potential information on specific compounds, and it can define the specific mechanism by which fibril formation is inhibited by enabling the characterization of the individual microscopic processes underlying the aggregation process, including primary nucleation, elongation, fragmentation and secondary nucleation (Arosio et al. 2014). By using this method, it has recently been possible to reveal that the aggregation of Aβ42 into toxic oligomers and fibrils is primarily caused by secondary nucleation events in which the presence of existing aggregates catalyses the formation of new ones (Cohen et al. 2013). Compounds that prevent toxic oligomers from forming are thus likely to be promising drug candidates in this case (Cohen et al. 2013; Arosio et al. 2014). For IDPs, it is clear that the understanding of the mechanism of inhibition of any potential drug is a key requirement to achieve the desired targeting (Arosio et al. 2014).

4.4 Molecular Dynamics Simulations

As discussed in Chap. 2 of this book, molecular dynamics simulations, in particular when used in combination with advanced sampling techniques and information from experimental data, can provide highly quantitative information on the dynamics of a small molecule upon binding. When the objective is to stabilize the monomeric native states of IDPs, the major goal of these efforts should be to map their free energy landscapes (Vendruscolo and Dobson 2005). Unlike ordered proteins, which have fully-folded native structures that populate well-defined free energy minima, IDPs are characterized by numerous local free energy minima without any stable well-folded conformations. The target of the structure-based drug discovery strategy, therefore, rather than being an individual conformation, should be the variety of low free energy states populated by IDPs. The interaction with a particular binding partner is likely to affect the whole free energy landscape of an IDP, making some minima deeper and some barriers higher (Fig. 13.4). One possibility is to follow a 'double-hit' strategy in which the first step is to identify a small molecule that causes a conformational change in the target IDP by creating a binding pocket, and the second step is to find a molecule capable of recognizing this binding pocket.

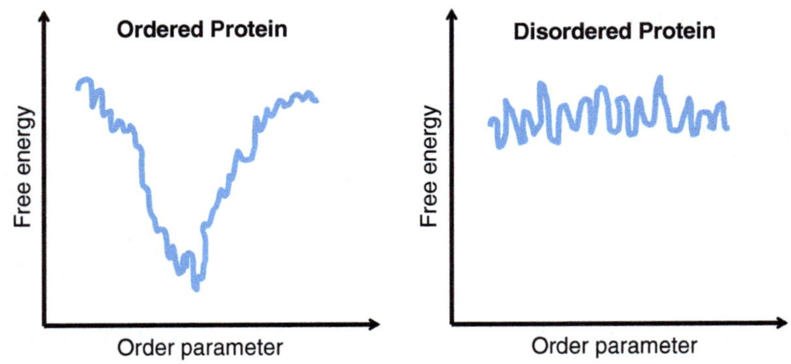

Fig. 13.4 Schematic illustration of the binding of small molecules to target IDPs. In ordered proteins, the free energy landscape is characterised by a well-defined global minimum, which contains structured binding pockets that can bind small molecules with high affinity. In IDPs, the free energy landscape exhibits multiple minima populated by conformations with transient binding pockets and generally low affinity for small molecules

To implement this strategy, it would be very helpful to characterize the free energy landscape of the target IDP in order to gauge the effects of the lead compounds of interest. From an ensemble of conformations, structures of low free energies could be further stabilized by exploiting chemical features suitable to form significant enthalpic interactions with the target IDP.

4.5 Chemoinformatics

Chemoinformatic approaches exploit the knowledge base of the existing chemical space (Dobson 2004) of small molecules. Although the chemical space for IDPs has not yet been fully mapped, the existence of specific physicochemical properties of IDPs suggests that there may be chemical and functional groups particularly favourable for IDP binding. Following this approach, databases such as ChEMBL, PubChem, DrugBank, and ZINC (Gaulton et al. 2012; Wishart et al. 2006; Irwin and Shoichet 2004; Wang et al. 2009) can be exploited to search for compounds that could be repurposed towards targeting IDP ensembles (Varadi et al. 2014). It will be exciting to see whether advances will be made in the future through this strategy.

5 Conclusions

In this chapter we have described how IDPs represent highly challenging, and yet crucial, drug targets. Traditional drug discovery strategies are not ideally suited for these peptides and proteins, as they are structurally heterogeneous and usually devoid of clear binding sites, and it is therefore difficult to design small molecules that bind IDPs with high affinity and high specificity. Despite these problems, we have discussed how recent developments in biophysical techniques, including NMR spectroscopy, SAXS, chemical kinetics and molecular dynamics simulations, have provided initial evidence that IDPs may be druggable. In perspective, given the small but fair amount of success stories so far, we can anticipate that IDPs will be included in the list of targets in drug discovery initiatives in the future.

References

Arkin MR, Wells JA (2004) Small-molecule inhibitors of protein-protein interactions: progressing towards the dream. Nat Rev Drug Discov 3(4):301–317

Arosio P, Vendruscolo M, Dobson CM et al (2014) Chemical kinetics for drug discovery to combat protein aggregation diseases. Trends Pharmacol Sci 35(3):127–135

Babu MM, Kriwacki RW, Pappu RV (2012) Versatility from protein disorder. Science 337(6101):1460–1461

Bernadó P, Svergun DI (2012) Structural analysis of intrinsically disordered proteins by small-angle X-ray scattering. Mol BioSys 8(1):151–167

Bernadó P, Mylonas E, Petoukhov MV et al (2007) Structural characterization of flexible proteins using small-angle X-ray scattering. J Am Chem Soc 129(17):5656–5664

Blackwood EM, Eisenman RN (1991) Max: a helix-loop-helix zipper protein that forms a sequence-specific DNA-binding complex with Myc. Science 251(4998):1211–1217

Blobel J, Bernadó P, Svergun DI et al (2009) Low-resolution structures of transient protein-protein complexes using small-angle X-ray scattering. J Am Chem Soc 131(12):4378–4386

Chène P (2004) Inhibition of the p53-MDM2 interaction: targeting a protein-protein interface. Mol Cancer Res 2(1):20–28

Cheng Y, LeGall T, Oldfield CJ et al (2006) Rational drug design via intrinsically disordered protein. Trends Biotechnol 24(10):435–442

Chiti F, Dobson CM (2006) Protein misfolding, functional amyloid, and human disease. Annu Rev Biochem 75:333–366

Clackson T, Well JA (1995) A hotspot of binding energy in a hormone-receptor interface. Science 267(5196):383–386

Cohen SIA, Linse S, Luheshi LM et al (2013) Proliferation of amyloid-β42 aggregates occurs through a secondary nucleation mechanism. Proc Natl Acad Sci U S A 110(24):9758–9763

Congreve M, Chessari G, Tisi D et al (2008) Recent developments in fragment-based drug discovery. J Med Chem 51(13):3661–3680

Conte LL, Chothia C, Janin J (1999) The atomic structure of protein-protein recognition sites. J Mol Biol 285(5):2177–2198

Dang CV (1999) c-Myc target genes involved in cell growth, apoptosis, and metabolism. Mol Cell Biol 19(1):1–11

Dawson R, Müller L, Dehner A et al (2003) The N-terminal domain of p53 is natively unfolded. J Mol Biol 332(5):1131–1141

Dobson CM (2001) The structural basis of protein folding and its links with human disease. Philos Trans R Soc B 356(1406):133–145

Dobson CM (2004) Chemical space and biology. Nature 432(7019):824–828

Dunker AK, Uversky VN (2010) Drugs for 'protein clouds': targeting intrinsically disordered transcription factors. Curr Op Pharmacol 10(6):782–788

Dunker AK, Brown CJ, Lawson JD et al (2002) Intrinsic disorder and protein function. Biochemistry 41(21):6573–6582

Dunker AK, Cortese MS, Romero P et al (2005) Flexible nets. The roles of intrinsic disorder in protein interaction networks. FEBS J 272(20):5129–5148

Dyson HJ, Wright PE (2005) Intrinsically unstructured proteins and their functions. Nat Rev Mol Cell Biol 6(3):197–208
Eisenberg D, Jucker M (2012) The amyloid state of proteins in human diseases. Cell 148(6):1188–1203
Fitzpatrick AWP, Debelouchina GT, Bayro MJ et al (2013) Atomic structure and hierarchical assembly of a cross-β amyloid fibril. Proc Natl Acad Sci U S A 110(14):5468–5473
Fry DC, Wartchow C, Graves B et al (2013) Deconstruction of a nutlin: dissecting the binding determinants of a potent protein-protein interaction inhibitor. ACS Med Chem Lett 4(7):660–665
Gaulton A, Bellis LJ, Bento AP et al (2012) ChEMBL: a large-scale bioactivity database for drug discovery. Nucl Acids Res 40(Database issue):D1100–D1107
Haass C, Selkoe DJ (2007) Soluble protein oligomers in neurodegeneration: lessons from the Alzheimer's amyloid β-peptide. Nat Rev Mol Cell Biol 8(2):101–112
Habchi J, Tompa P, Longhi S et al (2014) Introducing protein intrinsic disorder. Chem Rev 114(13):6561–6588
Hajduk PJ, Greer J (2007) A decade of fragment-based drug design: strategic advances and lessons learned. Nat Rev Drug Discov 6(3):211–219
Hajduk PJ, Meadows RP, Fesik SW (1999) NMR-based screening in drug discovery. Q Rev Bioph 32(03):211–240
Hammoudeh DI, Follis AV, Prochownik EV et al (2009) Multiple independent binding sites for small-molecule inhibitors on the oncoprotein c-Myc. J Am Chem Soc 131(21):7390–7401
Harvey SR, Porrini M, Stachl C et al (2012) Small-molecule inhibition of c-MYC: MAX leucine zipper formation is revealed by ion mobility mass spectrometry. J Am Chem Soc 134(47):19384–19392
Hawkes CA, Ng V, McLaurin J (2009) Small molecule inhibitors of Aβ aggregation and neurotoxicity. Drug Develop Res 70(2):111–124
Hopkins AL, Groom CR (2002) The druggable genome. Nat Rev Drug Discov 1(9):727–730
Iakoucheva LM, Brown CJ, Lawson JD et al (2002) Intrinsic disorder in cell-signaling and cancer-associated proteins. J Mol Biol 323(3):573–584
Imming P, Sinning C, Meyer A (2006) Drugs, their targets and the nature and number of drug targets. Nat Rev Drug Discov 5:821–834
Irwin JJ, Shoichet BK (2004) ZINC—a free database of commercially available compounds for virtual screening. J Chem Inf Model 45(1):177–182
Ivetac A, McCammon JA (2012) A molecular dynamics ensemble-based approach for the mapping of druggable binding sites. In: Baron R (ed) Computational drug discovery and design, vol 819. Methods in molecular biology. Springer New York, New York, pp 3–12
Jones S, Thornton JM (1996) Principles of protein-protein interactions. Proc Natl Acad Sci U S A 93(1):13–20
Klages J, Coles M, Kessler H (2007) NMR-based screening: a powerful tool in fragment-based drug discovery. Analyst 132(7):692
Knowles TP, Waudby CA, Devlin GL et al (2009) An analytical solution to the kinetics of breakable filament assembly. Science 326(5959):1533–1537
Knowles TP, Vendruscolo M, Dobson CM (2014) The amyloid state and its association with protein misfolding diseases. Nat Rev Mol Cell Biol 15(6):384–396
Kozakov D, Hall DR, Chuang GY et al (2011) Structural conservation of druggable hotspots in protein-protein interfaces. Proc Natl Acad Sci U S A 108(33):13528–13533
Kubbutat MH, Jones SN, Vousden KH (1997) Regulation of p53 stability by Mdm2. Nature 387(6630):299–303
Lipinski CA (2004) Lead- and drug-like compounds: the rule-of-five revolution. Drug Discov Today 1(4):337–341
Malaney P, Pathak RR, Xue B et al (2013) Intrinsic disorder in PTEN and its interactome confers structural plasticity and functional versatility. Sci Rep 3:2035
McKoy AF, Chen J, Schupbach T et al (2012) A novel inhibitor of amyloid β (Aβ) peptide aggregation: from high throughput screening to efficacy in an animal model of Alzheimer disease. J Biol Chem 287(46):38992–39000

Meisl G, Yang X, Hellstrand E et al (2014) Differences in nucleation behavior underlie the contrasting aggregation kinetics of the Aβ40 and Aβ42 peptides. Proc Natl Acad Sci U S A 111(26):9384–9389

Mészáros B, Simon I, Dosztányi Z (2011) The expanding view of protein–protein interactions: complexes involving intrinsically disordered proteins. Phys Biol 8(3):035003

Metallo SJ (2010) Intrinsically disordered proteins are potential drug targets. Curr Op Chem Biol 14(4):481–488

Michel J, Cuchillo R (2012) The impact of small molecule binding on the energy landscape of the intrinsically disordered protein c-Myc. PloS ONE 7(7):e41070

Michelsen K, Jordan JB, Lewis J et al (2012) Ordering of the N-terminus of human MDM2 by small molecule inhibitors. J Am Chem Soc 134(41):17059–17067

Muller PA, Vousden KH (2014) Mutant p53 in cancer: new functions and therapeutic opportunities. Cancer Cell 25(3):304–317

Murray CW, Blundell TL (2010) Structural biology in fragment-based drug design. Curr Op Struct Biol 20(4):497–507

Murray CW, Rees DC (2009) The rise of fragment-based drug discovery. Nat Chem 1(3):187–192

Necula M, Kayed R, Milton S et al (2007) Small molecule inhibitors of aggregation indicate that amyloid beta oligomerization and fibrillization pathways are independent and distinct. J Biol Chem 282(14):10311–10324

Oldfield CJ, Meng J, Yang JY et al (2008) Flexible nets: disorder and induced fit in the associations of p53 and 14-3-3 with their partners. BMC Genomics 9(Suppl 1):S1

Pellecchia M, Sem DS, Wuthrich K (2002) NMR in drug discovery. Nat Rev Drug Discov 1(3):211–219

Pellecchia M, Bertini I, Cowburn D et al (2008) Perspectives on NMR in drug discovery: a technique comes of age. Nat Rev Drug Discov 7(9):738–745

Porat Y, Abramowitz A, Gazit E (2006) Inhibition of amyloid fibril formation by polyphenols: structural similarity and aromatic interactions as a common inhibition mechanism. Chem Biol Drug Des 67(1):27–37

Renaud JP, Delsuc MA (2009) Biophysical techniques for ligand screening and drug design. Curr Op Pharmacol 9(5):622–628

Rezaei-Ghaleh N, Blackledge M, Zweckstetter M (2012) Intrinsically disordered proteins: from sequence and conformational properties toward drug discovery. ChemBioChem 13(7):930–950

Shuker SB, Hajduk PJ, Meadows RP et al (1996) Discovering high-affinity ligands for proteins: SAR by NMR. Science 274(5292):1531–1534

Sun C, Petros AM, Hajduk PJ (2011) Fragment-based lead discovery: challenges and opportunities. J Comput-Aided Mol Des 25(7):607–610

Tompa P (2012) Intrinsically disordered proteins: a 10-year recap. Trends Biochem Sci 37(12):509–516

Toth G, Gardai SJ, Zago W et al (2014) Targeting the intrinsically disordered structural ensemble of α-synuclein by small molecules as a potential therapeutic strategy for Parkinson's disease. PloS ONE 9(2):e87133

Uversky VN (2012) Intrinsically disordered proteins and novel strategies for drug discovery. Expert Opin Drug Discov 7(6):475–488

Uversky VN (2013) Unusual biophysics of intrinsically disordered proteins. Biochim Bioph Acta 1834(5):932–951

Uversky VN, Oldfield CJ, Dunker AK (2008) Intrinsically disordered proteins in human diseases: introducing the D_2 concept. Annu Rev Bioph 37:215–246

Uversky VN, Oldfield CJ, Midic U et al (2009) Unfoldomics of human diseases: linking protein intrinsic disorder with diseases. BMC Genomics 10(Suppl 1):S7

van der Lee R, Buljan M, Lang B et al (2014) Classification of intrinsically disordered regions and proteins. Chem Rev 114:6589–6631

Varadi M, Kosol S, Lebrun P et al (2014) pE-DB: a database of structural ensembles of intrinsically disordered and of unfolded proteins. Nucl Acids Res 42(D1):D326–D335

Vendruscolo M (2007) Determination of conformationally heterogeneous states of proteins. Curr Op Struct Biol 17(1):15–20

Vendruscolo M, Dobson CM (2005) Towards complete descriptions of the free–energy landscapes of proteins. Philos Trans R Soc, A 363(1827):433–452

Wang Y, Xiao J, Suzek TO et al (2009) PubChem: a public information system for analyzing bioactivities of small molecules. Nucl Acids Res 37(suppl 2):W623–W633

Wang J, Cao Z, Zhao L et al (2011) Novel strategies for drug discovery based on Intrinsically Disordered Proteins (IDPs). Int J Mol Sci 12(5):3205–3219

Wang Q, Liang G, Zhang M et al (2014) De novo design of self-assembled hexapeptides as β-Amyloid (Aβ) peptide inhibitors. ACS Chem Neurosci 5(10):972–81

Warr WA (2009) Fragment-based drug discovery. J Comput-Aided Mol Des 23(8):453–458

Wells JA, McClendon CL (2007) Reaching for high-hanging fruit in drug discovery at protein-protein interfaces. Nature 450(7172):1001–1009

Welsch ME, Snyder SA, Stockwell BR (2010) Privileged scaffolds for library design and drug discovery. Curr Op Chem Biol 14(3):347–361

Wishart DS, Knox C, Guo AC et al (2006) DrugBank: a comprehensive resource for in silico drug discovery and exploration. Nucl Acids Res 34(Database issue):D668–D672

Yamin G, Ruchala P, Teplow DB (2009) A peptide hairpin inhibitor of amyloid beta-protein oligomerization and fibrillogenesis. BioChemistry 48(48):11329–11331

Zhu M, De Simone A, Schenk D et al (2013) Identification of small-molecule binding pockets in the soluble monomeric form of the Aβ42 peptide. J Chem Phys 139(3):035101

Chapter 14
Beta Amyloid Hallmarks: From Intrinsically Disordered Proteins to Alzheimer's Disease

Magdalena Korsak and Tatiana Kozyreva

Abstract Beta amyloid protein (Aβ) is one of the intrinsically disordered proteins associated with neurodegenerative diseases like Parkinson's, prion disease and Alzheimer's disease (AD) in particular. Although the direct involvement of Aβ peptides in AD is well documented and their aggregative ability is closely related to their neurotoxicity, the precise mechanism of the neurotoxic effects of Aβ peptides remains unclear. There is still a significant gap between the site-specific structural information and the complex structural diversity of Aβ amyloids. The description of the structural polymorphisms of Aβ amyloids can provide valuable information of the molecular basis of AD onset-progress and is essential for comprehension of the Aβ aggregation pathways, in particular its structural evolution. In this review we tried to illustrate the emerging trend of defining several human neurodegenerative disorders as syndromes of protein folding and oligomerization through the example of AD.

Keywords Neurotoxicity · Alzheimer's disease · Aβ amyloids · Aggregation

> You have to begin to lose your memory, if only in bits and pieces, to realize that memory is what makes our lives. Life without memory is no life at all, just as an intelligence without the possibility of expression is not really an intelligence. Our memory is our coherence, our reason, our feeling, even our action. Without it, we are nothing.
>
> *Luis Bunuel*

T. Kozyreva (✉) · M. Korsak
Giotto Biotech, Via Madonna del Piano 6, 50019 Sesto Fiorentino, Italy
e-mail: kozyreva@giottobiotech.com

M. Korsak
e-mail: magdalena.korsak1@gmail.com

I. C. Felli, R. Pierattelli (eds.), *Intrinsically Disordered Proteins Studied by NMR Spectroscopy,* Advances in Experimental Medicine and Biology,
DOI 10.1007/978-3-319-20164-1_14

1 Introduction

Over 600 disorders afflict the nervous system[1]. Neurodegenerative diseases are defined as hereditary and sporadic conditions that are characterized by progressive nervous system dysfunction. These disorders are often associated with atrophy of the affected central or peripheral structures of the nervous system. They include diseases such as Alzheimer's disease (AD) and other dementias, brain cancer, degenerative nerve diseases, encephalitis, epilepsy, genetic brain disorders, head and brain malformations, hydrocephalus, stroke, Parkinson's disease, multiple sclerosis, amyotrophic lateral sclerosis (ALS or Lou Gehrig's Disease), Huntington's disease, and prion diseases.

Neuroscientific research has enjoyed rapid progress fuelled by technologically sophisticated, multidisciplinary approaches. If we consider the vast amount of literature published on the subject of neurodegeneration just in the last years, we could notice a substantial progress in understanding the molecular basis of these terrifying diseases, but what we know is still a drop in the bucket, and the main question remains: what causes neurodegeneration and how can we cope with it?

In this short review we illustrate the emerging trend of defining several human neurodegenerative disorders as syndromes of protein folding and oligomerization through the example of AD. AD is recognized as a major public health problem in developed countries, and knowledge of its causes and mechanisms has grown enormously in the past decade. This insidious and devastating brain degeneration that robs its victims of their memory, reasoning, abstraction, and language abilities affects one in four individuals over 85 years of age. According to the World Alzheimer Report, the disease is expected to affect 115.4 million people by the year 2050[2]. From an economic point of view, AD is one of the most expensive diseases because it usually spans many years and patients require intense daily care.

Symptoms associated with the illness include cognitive dysfunction and neuronal death in the brain, both of which are indicative of a significant progressive neurodegenerative disease (Walsh and Selkoe 2004). No effective treatments for AD are currently available and not even its pathogenesis is fully understood. The current limited number of treatments for Alzheimer's disease merely address symptoms rather than the root cause. The so-called "amyloid hypothesis" has come to dominate explanations for the damage that occurs to the brain. The presence of amyloid plaques and congophilic angiopathy in the brain cortex and hippocampus is considered to be a major pathological feature of AD (Evin and Weidemann 2002). Almost nine decades passed from the moment when Alois Alzheimer peered through a microscope at the brain of his first patient and prophetically wrote, "scattered through the entire cortex ... one found miliary foci that were caused by the deposition of a peculiar substance ..." (Selkoe 1994) and when neuroscientists

[1] European Commission website 2014. Neurodegenerative Disorders. Accessed 26 October 2014. http://ec.europa.eu/health/major_chronic_diseases/diseases/brain_neurological/index_en.htm.

[2] Alzheimer's Disease International website 2014. World Alzheimer's Reports. Accessed 26 October 2014. http://www.alz.co.uk/research/world-report.

isolated and chemically characterized the nature of the amyloid material found in the senile plaque, revealing the Aβ protein (Glenner and Wong 1984; Masters et al. 1985a) and its precursor APP (amyloid precursor protein) (Kang et al. 1987; Evin and Weidemann 2002). They observed that amyloid peptides constitute ~90% of the plaque material (Glenner and Wong 1984; Masters et al. 1985b; Kang et al. 1987), while the remaining 10% of amyloid plaques are composed of proteins from the apolipoprotein E class, lipids from membranes of degenerated portions of the intercommunicated nerve extensions called axons, metal ions such as Cu, Zn, Fe and Al, and traces of other components from the extracellular liquid. Transmission electron microscopy of amyloid plaques revealed numerous unbranched filaments, representing amyloid fibrils, surrounded by amorphous aggregates of diffuse amyloid. Amyloid plaques fall into two broad morphological categories: diffuse and neuritic. Both plaque types are detectable with anti-Aβ antibodies, but only neuritic plaques are prominently stained by sheet-binding dyes such as Congo red and thioflavin S. Neuritic dystrophies are swollen and distorted processes of axonal or dendritic origin that radiate from the core of a neuritic plaque. They are detectable with antibodies against APP, phospho-tau, neurofilaments and ubiquitin, indicating a disruption of protein transport and attempts to degrade this blockage (Dickson et al. 1999). Progressive neuritic plaque deposition is a hallmark of AD. Neuritic plaque formation commonly begins in the neocortex and later affects the hippocampus and amygdala. By the end stage of the disease, neuritic plaques are present in the brainstem and other subcortical structures (Thal et al. 2002). It has been suggested that the presence and a substantial increase in diffuse plaque is associated with the preclinical stages of AD (Knopman et al. 2003; Vlassenko et al. 2011). The surprising and totally unexpected observation that Aβ is produced constantly throughout life as a physiologically normal metabolite generated in healthy people changed the common concept of AD (Hardy and Selkoe 2002). Plaques have also been found in cognitively normal individuals, and plaque burden does not correlate with memory decline. Moreover, successful removal of amyloid plaques by immunotherapy fails to improve cognition (Haass 2010). Additionally, Dobson and co-workers have shown that virtually any protein can be induced into forming amyloid, suggesting that amyloid is a primordial, highly stable polypeptide form that had to be overcome by evolution in order to create functional globular proteins (Dobson and Misfolding 2003). Soluble Aβ oligomers are now believed to act as neurotoxic entities rather than amyloid plaques.

It appears that the soluble Aβ monomers require conversion to a largely β-pleated sheet conformation and subsequent aggregation before they can confer neurotoxicity *in vitro* (Lorenzo and Yankner 1994; Howlett et al. 1995). The production of the toxic species—soluble Aβ oligomers—and their subsequent ability to cause neuronal injury depends on the precision of an intramembranous proteolytic cleavage of APP (Haass and Selkoe 2007). Proteolytic processing of APP protein involves three types of proteases.

APP is a member of an evolutionarily conserved gene family with two mammalian homologs, amyloid precursor-like proteins (APLP) 1 and 2 (Wasco et al. 1992, 1993). These proteins contain highly similar sequences in their ectodomains and intracellular carboxy-termini, but the transmembrane region comprising the Aβ peptide is unique to APP (Bayer et al. 1999). Although its primary physiological function remains unclear, APP has been implicated in a variety of processes such as intracellular signalling, synapse adhesion, trophic support, axon remodelling and apoptosis (Zheng and Koo 2011).

APP is ubiquitously expressed. There are three major APP isoforms resulting from alternative splicing of its 18 exon gene: APP695, APP751 and APP770 (Yoshikai et al. 1990). APP751 and APP770 are the main transcripts found in non-neuronal tissue. APP695 is the most abundant isoform in the brain, where its expression is primarily limited to neurons. Brain region-specific variation in APP695 expression occurs in both mouse and human, with the highest transcript levels found in the cortex, hippocampus and cerebellum (Sola et al. 1993).

APP is processed via two major pathways that utilize different enzymes and result in distinct cleavage products. The non-amyloidogenic pathway precludes the formation of Aβ due to constitutive α-secretase-mediated cleavage in the middle of the Aβ domain (Esch et al. 1990). It was initially proposed that a zinc-dependent, transmembrane protease served as α-secretase (Roberts et al. 1994). Three members of the A disintegrin and metalloproteinase (ADAM) family were later found to possess α-secretase activity: ADAM-10, ADAM-17, and ADAM-9 (Lammich et al. 1999). More recent evidence, however, suggests that ADAM-10 serves as the primary α-secretase in neurons (Kuhn et al. 2010). Alpha-cleavage of APP occurs mainly at the plasma membrane, releasing a soluble α-APP fragment (sAPP α) into the lumen/extracellular space and creating a membrane-bound, 83-residue C-terminal fragment (C83) (Sisodia 1992). Subsequent intramembranous cleavage of C83 by γ-secretase liberates a soluble, 3 kDa fragment (p3) and the APP intracellular domain (AICD) (Zheng and Koo 2011). The p3 fragment is rapidly degraded, while AICD may act as a transcriptional regulator (Haass and Selkoe 2007).

The amyloidogenic processing of APP primarily occurs in the endocytic pathway. β-secretase initiates the sequence of amyloidogenic cleavage events. Cleavage of APP at the β-site generates a soluble amino-terminal fragment (sAPP β) and a membrane-associated, 99-residue C-terminal fragment (C99). γ-secretase then performs a stepwise, intramembrane cleavage of the C99 fragment, liberating Aβ and AICD. Aβ peptides range from 37 to 43 amino acids in length; however, under physiological conditions, the majority of Aβ produced is 40 amino acids long (Aβ40). The 42 amino acid variant (Aβ42) normally only comprises a minor fraction of the total Aβ. Nevertheless Aβ42 has a more hydrophobic nature, and its aggregative ability and neurotoxicity are therefore much greater than those of Aβ40; both of these forms (Aβ40 and Aβ42) are therefore targets of intense study (Haass 2010). β-site cleaving enzyme 1 (BACE1) was identified as the enzyme responsible for APP β–cleavage. BACE1 is a type 1 membrane-bound aspartyl protease with its active site facing the lumen. It is capable of cleaving APP at two positions: the aspartate at position 1 of the Aβ sequence or the glutamate at position 11. BACE1 is found in a variety of tissues, but is predominantly expressed in neurons (Sinha et al. 1999). Intracellularly, BACE1 mainly localizes to the trans-Golgi network and endosomes. However, BACE1 is also trafficked between the Golgi and the plasma membrane, where

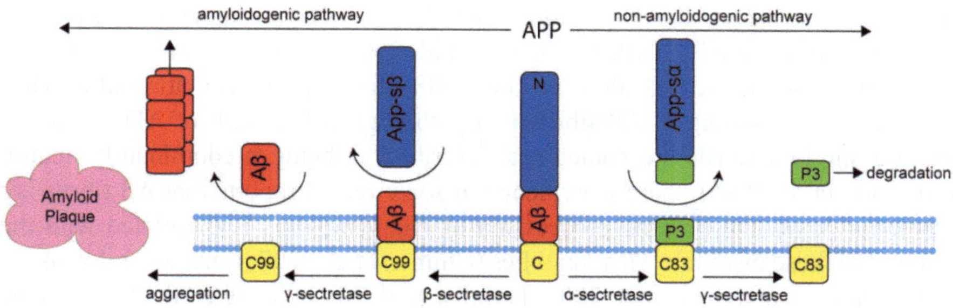

Fig. 14.1 APP metabolism by the secretase enzymes. (Barrantes et al. 2010)

it is enriched in lipid rafts. From the plasma membrane, BACE1 is internalized and sorted into endosomes or recycled to the trans-Golgi network (Walter et al. 2001). It serves as the primary β-secretase and is the rate-limiting enzyme in Aβ production. There are currently four known components of γ-secretase: presenilin (PS1 or PS2), nicastrin, anterior pharynx defective 1 (APH1) and presenilin enhancer 2 (PEN-2). These proteins assemble into the γ–secretase complex while cycling through the endoplasmic reticulum/Golgi (Edbauer et al. 2003). Once mature, γ-secretase is primarily found at the plasma membrane and in the endosomal/lysosomal system. Although PS, nicastrin, APH1 and PEN-2 are all required for γ-secretase activity, PS contains the catalytic active site needed for γ-cleavage of APP (Edbauer et al. 2003) (Fig. 14.1).

1.1 Polymorphism Widely Observed for Aβ Amyloidal Aggregates in vitro

Aβ molecules can spontaneously self-aggregate *in vitro* into different species (Jan et al. 2010) under distinct conditions. Aβ peptides can form soluble oligomers and protofibrils, which could be intermediates of a fibrillation process (Fig. 14.2) (Benilova et al. 2012). Studies have revealed the high neurotoxicity of these species

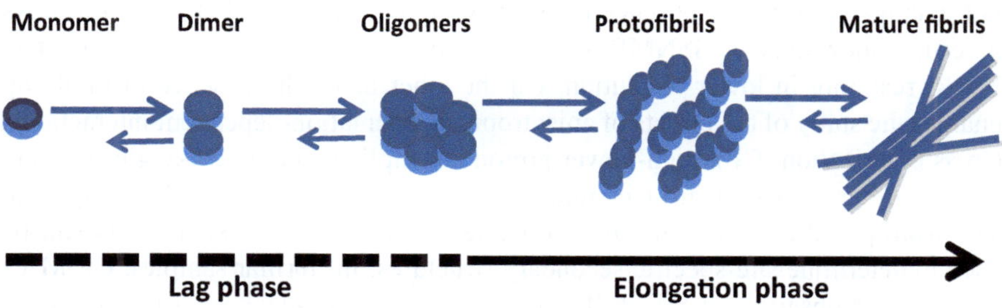

Fig. 14.2 A putative schematic of aggregation of Aβ with two kinetic phases. In the lag phase, monomers slowly form oligomers (*dashed lines*). In the elongation phase, oligomers promote fibril formation via protofibrils (*straight line*). (Kumar et al. 2011)

and their close links with AD (Benilova et al. 2012; Chimon et al. 2007; Hoshi et al. 2003; Krafft and Klein 2010; Kirkitadze et al. 2002; Lambert et al. 1998; Lesne´ et al. 2006; Noguchi et al. 2009; Selkoe, 2008). Instead, the formation of amyloidal deposits consisting of Aβ fibrils is a pathological hallmark of AD. Although mature amyloid fibrils are sometimes described as being predominantly neutral (Aksenov et al. 1996), there is evidence in some recent reports that Aβ fibrils are also neurotoxic and that the progression of AD symptoms is correlated with the amount of these insoluble Aβ assemblies (Chimon et al. 2007; Lorenzo and Yankner 1994; Meyer-Luehmann et al. 2008; Petkova et al. 2005; Qiang et al. 2012; Selkoe et al. 2004; Walsh et al. 1999).

Besides the different types of aggregates, the morphology of the same type of Aβ assemblies can also vary significantly depending on different aggregation conditions. This phenomenon is typically called polymorphism and has also been found in samples derived from AD patient brain tissues (Paravastu et al. 2009; Lu et al. 2013). Furthermore, it has been shown that different morphologies of Aβ fibrils can cause changes in neurotoxicity (Petkova et al. 2005).

1.2 Overview of Functional Polymorphism and Structural Models of Amyloidal Fibrils

The polymorphism of Aβ fibrils has been determined *in vitro* by numerous different structural studies such as cryo-electron microscopy (cryo-EM) (Fändrich et al. 2011; Meinhardt et al. 2009), which revealed a large spectrum of Aβ40 fibril polymorphisms, and solid-state nuclear magnetic resonance (SSNMR) (Benzinger et al. 1998; Bertini et al. 2011a; Lansbury et al. 1995; Paravastu et al. 2008; Petkova et al. 2002, 2005, 2006; Qiang et al. 2011).

SSNMR is one of the best techniques for obtaining atomic resolution structures of amyloid fibrils sufficient for the development of full molecular models (Tycko 2010; Wasmer et al. 2008), as X-ray crystallography is limited to small amyloidogenic peptides and cryo-EM is hindered by its relatively low spatial resolution. SSNMR has allowed for substantial advancement in understanding the structure of amyloid fibrils. Dipole-dipole couplings and chemical shift anisotropies are not averaged due to the absence of isotropic tumbling in solid-state samples. As a result, linewidths in SSNMR spectra are broadened relative to solution-state NMR, resulting in lower resolution. On the other hand, the absence of tumbling enables the study of the effects of anisotropic or orientation-dependent interactions. Cross polarization (CP), high-power proton decoupling, and magic-angle spinning (MAS) are used as standard techniques to obtain high-resolution SSNMR spectra. The isotropic chemical shift values obtained from CP-MAS experiments can be used to determine site-specific secondary structures. In fibrillar samples, CP-MAS is useful for observing rigid fibrillar parts, while dipolar-dephasing MAS is used to detect soluble components of mobile parts (Naito et al. 2004). Rotational resonance is used for determining homonuclear-internuclear distances, and rotational echo double resonance (REDOR) is used to determine heteronuclear-internuclear distances (Tycko and Ishii 2003). Other techniques, such as radiofrequency-driven recoupling and dipolar-assisted rotational resonance, are useful for obtaining folding information on amyloid fibrils (Balbach et al. 2002).

The only restriction of high-resolution studies of Aβ assemblies by SSNMR is the requirement for highly ordered and homogeneous samples. However, in amyloid systems it is rather common that several differently shaped aggregates coexist in a mixture, making preparation of the samples very difficult. This limitation can be overcome by programmed isotopic labelling schemes or sequence truncation, but each of these methods hampers complete structural analysis by SSNMR. Another strategy to overcome this restraint is the so-called seeding procedure, which leads not only to an improvement in the homogeneity of Aβ fibrils, but also permits insights into the molecular structures of amyloid fibrils developing in human tissue. This tactic is feasible after reconstruction of isotope enriched *in vivo* fibrils using as seeds brain tissues from patients as seeds with AD (Paravastu et al. 2009). This approach is made possible by the fact that *in vitro* studies have shown that the seeding procedure allows fibrils to be obtained that exactly retain the same molecular structures, corresponding to the seeds from the brain of patients with AD (Paravastu et al. 2008; Petkova et al. 2005). This strategy also has many drawbacks because the sample obtained in this way might contain significant contamination from many tissue components such as lipids. However, it allows the restoration of pathological samples from AD patients and the obtainment of high-resolution structural characterizations of these assemblies.

As heterogeneity also depends heavily upon different aggregation conditions, several different factors must be taken into consideration. One important key factor is the purity of the starting materials. The presence of pre-existing aggregates can result in a lack of reproducibility, as well as in changes of kinetics, fibrillogenesis processes, and neurotoxic activity (Fezoui et al. 2000). Several different protocols have been developed in order to solve this issue, resulting in standardized aggregate-free Aβ peptide samples (Broersen et al. 2011, Jao et al. 1997, Fezoui et al. 2000). Disaggregation of the Aβ assemblies involved the use of the structure-breaking organic solvents hexafluoroisopropanol (HFIP) and dimethyl sulfoxide (DMSO), trifluoroacetic acid (TFA) pre-treatment, and pre-dissolution of the peptide in a dilute base solution (e.g. NaOH). Each of these approaches allows aggregate-free material to be obtained. Another significant aspect is represented by the conditions of the fibrillation protocols such as concentration, pH, ionic strength, and temperature. The incubation of the sample in quiescent conditions leads to fibrils with the "twisted pairs" morphology (Paravastu et al. 2008), while gentle agitation results in striated ribbons (Petkova et al. 2006) or "flat" striated bundles (Bertini et al. 2011b). The selection of the fibrillation strategy is essential for obtaining high quality samples for high resolution SSNMR.

Aβ40 amyloid fibrils have been widely investigated and several structural models have recently been proposed using SSNMR (Bertini et al. 2011a; Paravastu et al. 2008; Petkova et al. 2002, 2006), A structural model of Aβ42 fibrils was proposed based on solution NMR and mutagenesis (Lührs et al. 2005).

In all these fibrillar assemblies Aβ molecules are densely packed in extended β-sheets (ladders) but exhibit polydispersed morphology and differ from one another in length and bundle width. Mature fibrils are characterized by a specific filamentous structure and usually contain 2–6 protofilaments that are more than 1 μm long and 8–12 nm in diameter. They display a cross-β X-ray fibre diffraction pattern as well as stainability with Congo red, resulting in green birefringence; they also bind Thioflavin-T (Merz et al. 1983; Petkova et al. 2005; Serpell 2000). It is widely accepted that amyloid fibrils are insoluble deposits, but recent studies have shown that many biochemical factors, e.g. biological lipids, are able to efficiently revert the fibrillation process and convert these insoluble assemblies into soluble, highly toxic intermediate species that retain the same biochemical and biophysical properties (Martins et al. 2008).

The distinct polymorphism of fibrillar assemblies is reflected in variations in overall structural symmetry and differences in specific aspects of structural elements and various residue sites, as well as in the nature of β-sheet structures including topologies of the β1-turn-β2 motif and inter-protofilament contacts (Ahmed et al. 2010; Bertini et al. 2011b; Lührs et al. 2005; Paravastu et al. 2008; Petkova et al. 2006) (Fig. 14.3). Recent studies have also shown that, in Aβ fibrils carrying mutants, both parallel and anti-parallel registry can occur (Qiang et al. 2011).

Although the structural polymorphism of Aβ40 fibrils was initially described as taking place mainly at the supramolecular level (Paravastu et al. 2008), a recent structural model of mature Aβ fibrils (Bertini et al. 2011b) suggests that polymorphism can already originate at the level of the zipper of the β1-turn-β2 motif, the inter-protofilament interface and the N-terminal conformation. The N-terminal part of the peptide in the fibril model can adopt β_N-strand conformation, but there is also evidence in the literature that a disordered conformation on the N-terminal part in some other Aβ fibrils is also possible (Paravastu et al. 2008; Petkova et al. 2006; Sachse et al. 2008, 2010). It can therefore be concluded that the N-terminal part of Aβ40 peptides can adopt distinct conformations and thereby also contribute to structural diversity.

It is reasonable to speculate that the folding of the monomer and supramolecular packing (within and among protofilaments) are linked or even define the morphology of amyloid fibrils through structural duplication/propagation along the fibril axis similar to the growth of 1D nanomaterials (Bertini et al. 2011b).

1.3 Review of the Methods for Immobilizing and Investigating Amyloidal Intermediate Species

Since various studies had reported that the neurotoxicity of Aβ peptides might be ascribed to pre-fibrillar assemblies (i.e. not necessarily fibrils (Benilova et al. 2012; Chimon et al. 2007; Hoshi et al. 2003; Kirkitadze et al. 2002; Krafft and Klein 2010; Lambert et al. 1998; Lesne′ et al. 2006; Noguchi et al. 2009; Selkoe 2008)), the high-resolution structural characterisation of these soluble species has become an

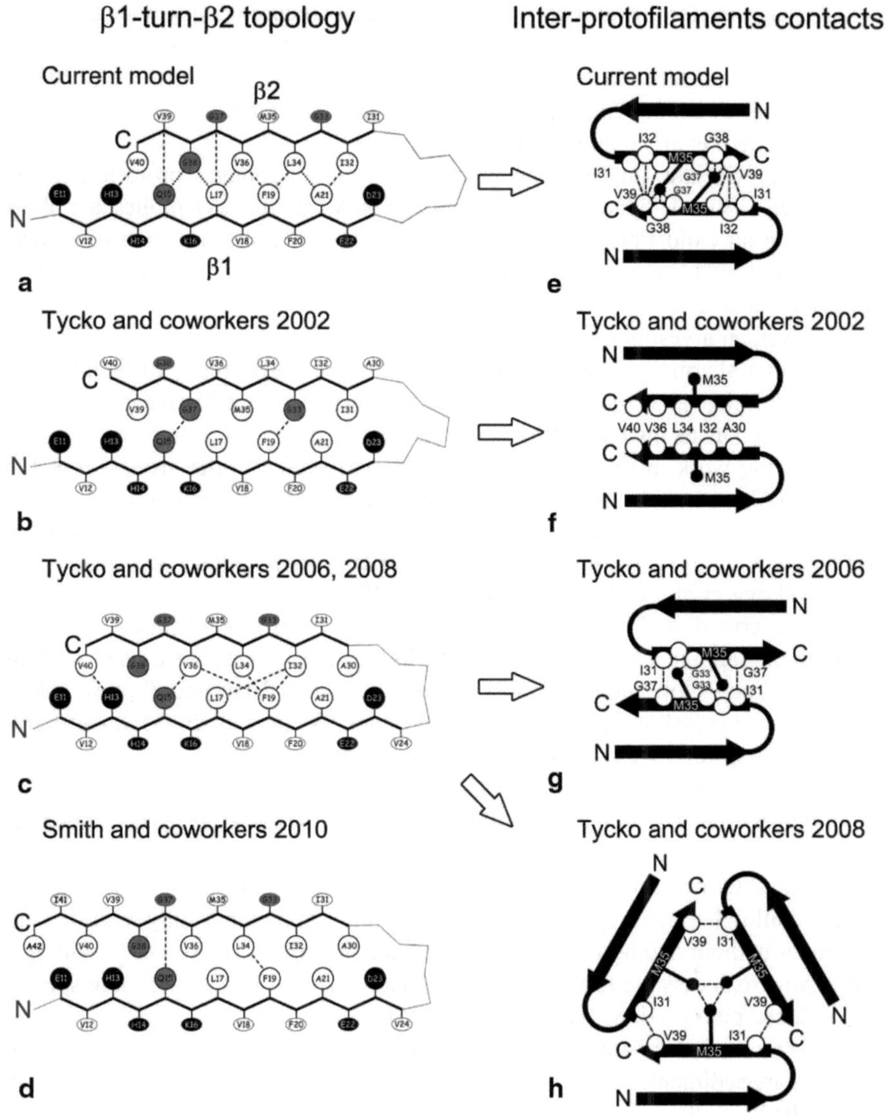

Fig. 14.3 Different topologies of β1-turn-β2 motif (*left column*) and inter-protofilament contacts (*right column*) in various SSNMR-derived structural models of Aβ fibrils (**a** Bertini et al. 2011b; **b** Petkova et al. 2002; **c** Petkova et al. 2006; Paravastu et al. 2008; **d** Ahmed et al. 2010). Adapted from Bertini et al. 2011b

overarching objective for understanding Aβ aggregation pathways and the complex molecular mechanism of AD (Benilova et al. 2012).

Several *in vitro* SSNMR studies established an initial high-resolution insight into certain non-fibrillar (Ahmed et al. 2010; Chimon and Yoshitaka 2005; Chimon et al. 2007; Lopez del Amo et al. 2012) and protofibrillar (Qiang et al. 2012; Scheidt et al.

2011) Aβ aggregates. Several other experimental and theoretical methods have also been used to obtain residue-specific information on prefibrillar Aβ aggregates and on structural persistence in the monomer (Bernstein et al. 2009; Bertini et al. 2013a; Danielsson et al. 2006; Fändrich 2012; Fawzi et al. 2011; Gallion 2012; Haupt et al. 2012; Kheterpal et al. 2006; Pan et al. 2011).

These prefibrillar intermediate assemblies, e.g. oligomers, protofibrils, and Aβ-derived diffusible ligands (ADDLs), as well as Aβ annular assemblies, are likely involved in amyloid fibril formation. All of these assemblies are rich in β-sheet structure and bind Congo red and Thioflavin-T, although more weakly than mature fibrils (Jan et al. 2010).

However, an investigation of these prefibrillar deposits faces many obstacles. As they are often thermally unstable compared to mature fibrils, many different methods to immobilize and study these species have been developed in recent years and each strategy has its own advantages and disadvantages.

Pioneering work in this direction has been performed by the groups of Smith and Ishii (Ahmed et al. 2010; Chimon and Yoshitaka 2005; Chimon et al. 2007), who focused on the Aβ oligomeric form. Taking advantage of the fact that various proteins including Aβ retain their structures after lyophilisation (Benzinger et al. 1998; Petkova et al. 2002; Studelska et al. 1997), many groups trap the thermally labile intermediate by freeze-trapping and subsequent lyophilisation. Different methods of trapping oligomeric and protofibrillar Aβ species were used such as filtration through low molecular weight cut off filters (Bitan and Teplow 2005), photo-induced crosslinking of unmodified proteins (Bitan et al. 2003), organic solvents (Haupt et al. 2012), density gradient centrifugation (Ward et al. 2000), size exclusion chromatography (SEC) (Bitan et al. 2003; Hartley et al. 1999; Jan et al. 2010; Walsh et al. 1997) and interaction partners (Bieschke et al. 2012; Lopez del Amo et al. 2012; Scheidt et al. 2011).

A method termed sedimented solute NMR (sedNMR) has recently been developed and allows Aβ aggregates to be collected and trapped in a fully hydrated environment without adding cosolvents or interaction partners, therefore providing a unique way to access the formation kinetics and structural features of these species with reduced perturbations (Bertini et al. 2013b).

In this method, solid-state NMR (SSNMR) experiments are used to observe proteins that are sedimented from solution using an ultracentrifugal field (Bertini et al. 2011a, 2012a, b, 2013a; Gardiennet et al. 2012; Polenova 2011; Ravera et al. 2013). The application of SSNMR is possible, as it has been reported (Ahmed et al. 2010; Bernstein et al. 2009; Fawzi et al. 2011; Kirkitadze et al. 2002; Lee et al. 2011) that Aβ peptides in aqueous solutions spontaneously form soluble aggregates of high molecular weight (50–200 kDa) that are large enough to sediment and thus become visible by SSNMR.

Sedimented solute NMR relies on the fact that sedimented macromolecules may be rotationally impaired by self-crowding, thus giving rise to solid-state NMR spectra. The sedimentation of macromolecules into this type of solid-like phase can be achieved in two different ways. One way is direct *in situ* sedimentation by magic angle spinning (MAS) of the NMR rotor (MAS-induced sedimentation), which acts as an ultracentrifuge (Bertini et al. 2011a) (Fig. 14.4a). *In situ* sedNMR can be used to address the kinetics of formation of soluble Aβ assemblies by simultaneously monitoring the disappearance of the monomer and the appearance of the oligomers. Another method is *ex situ* sedimentation by common ultracentrifuge (UC-induced sedimentation) (Bertini et al. 2012a; Gardiennet et al. 2012) with the help of devices designed to pack NMR rotors with precipitates or microcrystals (Bockmann et al. 2009) (Fig. 14.4b). *Ex situ* sedNMR allows one to select different oligomeric species by changing experimental conditions such us ultracentrifugation frequency or time and the initial Aβ peptide concentration.

Sedimentation through ultracentrifugation, either by magic angle spinning (*in situ*) or preparative ultracentrifugation (*ex situ*), can be used to immobilize and characterize oligomeric species, measure their formation kinetics, selectively sediment some of these species by their different molecular weights, and reveal the atomic-level structural features of soluble Aβ assemblies.

The collective data obtained from all of these methods demonstrate probable pathways from these fibril precursors to terminal fibrillar states and also provide

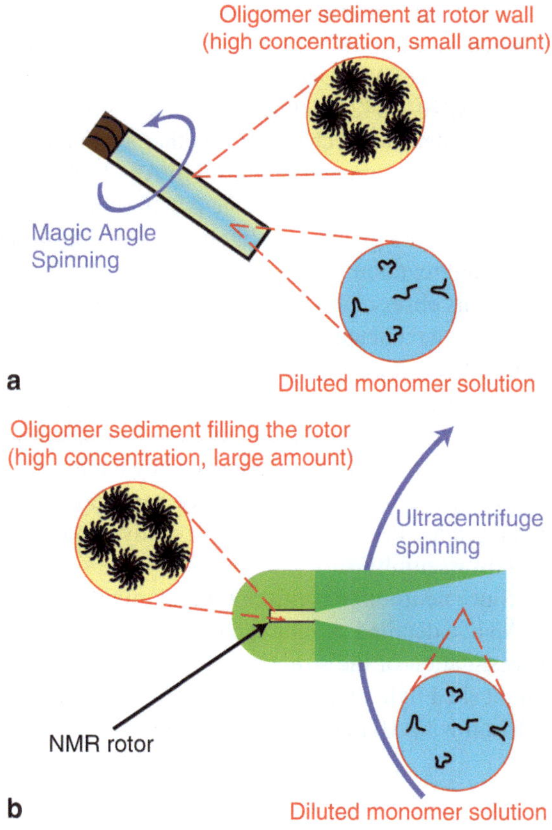

Fig. 14.4 Representation of the sedimentation process. In magic angle spinning-induced (*in situ*) sedimentation (*top*), the sediment is created in a thin layer at the rotor walls (the width of the sediment layer is greatly exaggerated). Sedimentation induced by preparative ultracentrifugation (*ex situ*) (*bottom*) can be used to effectively fill the rotor with sediment. Adapted from Bertini et al. 2013a)

evidence that these prefibrillar assemblies already contain β-strand structures but with different supramolecular organizations and reduced structural order and periodic symmetry, which might define the structures of the multiple conformers in fibrils in amyloid misfolding. However, some authors have suggested that certain oligomers that do not further aggregate to amyloid fibres and that have different secondary structures exist among different Aβ intermediate species.

1.4 Association Between Neurotoxicity and Different Ratios of Aβ40 to Aβ42

Aside from the fact that neurotoxicity can be induced by intermediate deposits, recent studies have established that the ratio of Aβ40 to Aβ42 is an important factor for providing stability to intermediate, neurotoxic species and affecting aggregation kinetics (Frost et al. 2003; Herzig et al. 2004; Jan et al. 2008, 2010; Kim et al. 2007; Kuperstein et al. 2010; Pauwels et al. 2012; Snider et al 2005; Wang et al. 2006; Yan and Wang, 2007; Yoshiike et al. 2003; Younkin 1995; Zou et al. 2003).

Aβ42 and Aβ40 alloforms co-exist in a molar ratio of 1:9 under normal physiological conditions in the brain. In patients with familial AD this ratio is shifted to a higher level of Aβ42, corresponding to a ratio of 3:7. Investigations of the properties of the Aβ40/Aβ42 mixture by different groups have clearly shown that these two species interact and change each other's dynamic behaviour. Even minor alterations in the relative amount of the Aβ40/Aβ42 ratio dramatically affect the biophysical and biological properties of the Aβ mixtures reflected in their aggregation kinetics by altering the pattern of oligomer formation. It has been shown that Aβ40 delays Aβ42 aggregation, while Aβ42 has an opposite effect and induces Aβ40 aggregation. This observation was also confirmed by *in vivo* studies, which have shown that a higher percentage of Aβ40 peptides in the brain might be protective (Kim et al. 2007; Wang et al. 2006).

Although it is generally assumed that alterations in ratios of the Aβ40 and Aβ42 mixture can stabilize distinct intermediate species associated with toxicity, the possibility that neurotoxicity is induced by a number of different conformations should not be neglected. This possibility can be explained by the fact that those toxic intermediate deposits might exist in dynamic equilibrium through their assembly and disassembly.

Since neurotoxic conformation(s) might by induced by a particular ratio of Aβ40 to Aβ42 a detailed structural characterization of Aβ40:Aβ42 mixed fibrils is needed. Investigation of the structure of Aβ40:Aβ42 mixed fibrils in different molar ratios might shed light on the processes related to the onset of the disease and on their correlation with Aβ peptides reciprocal ratios.

1.5 Summary of Expression and Purification Procedures for the Preparation of Aβ Peptides

Below we describe the methods of sample preparation of different kinds of Aβ peptides for NMR investigations.

The AβM40 and AβM42 peptides without any fusion tags, as well as Aβ40 and Aβ42 without the starting methionine fused with an N-terminal hexahistidine affinity tag were expressed as inclusion bodies in *E. coli*. Although inclusion bodies are usually undesirable due to problems with protein refolding, they are a good alternative in the case of intrinsically disordered proteins because their separation from the cell lysate is an easy and efficient protein purification method.

1.6 Aβ Peptides With MET

The cDNA of AβM40/AβM42 was cloned in the pET3a vector using the NdeI and BamHI restriction enzymes. The peptides were expressed in the BL21(DE3)pLys *E. coli* strain. The presence of an exogenous N-terminal methionine, due to the translation start codon, as reported in the literature (Walsh et al. 2009), does not affect the fibrillation kinetics or morphology of the fibrils formed by AβM40 or AβM42.

Many papers report the expression conditions for Aβ peptide production as follows: 8 to 16 h of incubation after induction with the addition of 0.8–1 mM of isopropyl β-D-1-thiogalactopyranoside (IPTG) and a temperature of 25–27 °C (Long et al. 2011; Zhang et al. 2009). In order to increase the protein yield, especially in terms of labelled protein production, we tried to increase the variable cell culture parameters as much as possible, boosting protein expression as an insoluble fraction. Expression of the fusion constructs under control of the T7 promoter/lac operator in *E. coli* BL21(DE3) cells provides high yields of the fusion proteins, which accumulate in inclusion bodies.

In the case of AβM40/AβM42 the growth was performed using the Marley method (Marley et al. 2001). The cells transformed with the AβM40/AβM42 expression plasmid were predominately grown in rich medium at 37 °C until OD600 reached 0.6, then centrifuged and exchanged into an isotopically defined minimal media enriched with $(^{15}NH_4)_2SO_4$ (1 g/L) and (^{13}C) glucose (4 g/L). The peptide expression was induced with 1.2 mM IPTG and the cells were harvested after 4 h of incubation at 39 °C.

The peptides were purified as reported (Bertini et al. 2011b; Hellstrand et al. 2010; Jan et al. 2010; Walsh et al. 1997; 2009) with some modifications using a combination of anion exchange and size exclusion chromatography. All the manipulations were performed at slightly alkaline pH in order to avoid the formation of structural contaminants produced by isoelectric precipitation.

There are some reports in the literature regarding possible aggregation and as a consequence a decrease in protein yield in pre-packed purification columns versus free resin (Walsh et al. 2009; Zhang et al. 2009). We also observed that purification of AβM40 and AβM42 by ion exchange chromatography in column mode (diethylaminoethyl (DEAE) cellulose column) led to much lower yields of monomeric peptide than in batch mode because of protein precipitation and aggregation.

The inclusion bodies were first solubilized with 8 M urea and then purified by ion exchange chromatography performed in batch mode. For this purpose DE52 resin on a Büchner funnel with filter paper on a vacuum glass bottle was used. Elution was performed using different concentrations of NaCl buffer. The protein was eluted with a 125 mM concentration of the salt. In the case of AβM42, the protein was also present in 20 mM, 150 mM, 200 mM and 1 M fractions, probably due to progressive sample aggregation. All the obtained fractions of diluted protein were concentrated to the final volume using an Amicon device. We examined a number of methods to concentrate the Aβ solution. Although several different methods within the 3 kDa molecular mass cut-off for centrifugal devices proved useful, the Amicon device was the best solution for highly concentrated proteins.

This two-step purification allows a highly pure product to be obtained with a yield of about 10 mg of AβM40 and 5–10 mg of AβM42 per litre of culture.

1.7 Aβ Peptides Without Met

Aβ40 and Aβ42 were produced in *E. coli* cytoplasm as fusion proteins with N-terminal hexahistidine affinity tags. The fusion construct consists of a soluble polypeptide segment comprised of 19 repeats of the tetrapeptide sequence NANP, a TEV protease cleavage site (sequence ENLYFQ), dipeptide linker sequences and the Aβ sequence.

We applied a similar approach for the expression and purification of Aβ peptides (Aβ40 and Aβ42) without methionine. Cells were grown in rich medium (lysogeny broth (LB) or Terrific Broth (TB)) until they reached high OD values. Peptide expression was induced with 1–1.2 mM IPTG and cells were harvested after 4 h incubation at 39 °C. After the removal of the soluble proteins the inclusion bodies were solubilized with 20 mM TRIS pH 8, 8 M and purified by affinity chromatography using a nickel chelating (His-Trap) column under denaturating conditions, followed by digestion with AcTEV protease. In order to avoid protein aggregation and improve the separation of cut/uncut protein, 8 M urea or 6 M guanidine hydrochloride was introduced in two affinity column steps.

As a final purification step for both peptide families (Aβ40,42; AβM40,42), a gel filtration in 50 mM ammonium acetate pH 8.5 using a Superdex 75 26/60 column was used. It is important to keep the pH of the solution over 4–7, the pH range where aggregation is maximized.

The aggregation of Aβ peptides is strongly influenced by the presence of structural and chemical impurities; therefore, before proceeding with SEC, all samples were denatured using 6 M guanidine hydrochloride as described previously (Walsh et al. 1997).

The protein obtained was dialysed in water and lyophilised.

In the case of Aβ42 peptides and even more so with their mutants, due to problems with tag digestion, we have to adopt a different purification procedure, introducing organic solvents.

The inclusion bodies were solubilized with 6 M guanidinium chloride and the protein was purified by metal chelating affinity chromatography on Ni2+-nitrilotriacetic acid agarose in the presence of 6 M guanidinium chloride, thus lowering the pH. The fusion proteins were further purified via reversed phase high-performance liquid chromatography (RP-HPLC) using a semi-preparative Zorbax SB300 C8 column (Agilent), lyophilized from aqueous acetonitrile and directly used for TEV protease cleavage. A cleavage efficiency of about 70% was achieved after incubation at pH 8.0 and 4 °C for 16 h at a protein concentration of 100 μM in the presence of 5 μM TEV protease. The cleavage mixture was subsequently applied to RP-HPLC, which allowed quantitative separation of the hydrophobic Aβ42 peptide from the other, more hydrophilic components in the cleavage reaction. The final yield of purified Aβ40 was 20 mg/L of culture.

Another problem encountered was associated with solubilisation of lyophilised protein in water. Because the Aβ peptides undergo time- and concentration-dependent aggregation in acetonitrile-water (Shen and Murphy 1995), the dry, purified peptides adopt different structures and aggregation states (Soto et al. 1995). Depending on the peptide batch and the particular aggregation conditions, considerable discrepancies exist across different laboratories as well as within the same laboratory over time. In addition, numerous studies have established that neurotoxicity and the kinetics of aggregation are directly related to assembly state in solution. Therefore, direct solubilisation of the Aβ peptides in aqueous media should always be avoided because it generates batch-dependent mixtures of aggregates and structures (Hou et al. 2004).

A couple of procedures have been proposed to disaggregate the Aβ and generate a monomeric random coil structure: pre-dissolution of the peptide in dilute base solution, and pre-dissolution in TFA and HFIP solvents (Jao et al. 1997). Finally, we applied a protocol that has already been reported (Broersen et al. 2011) for the solubilisation of Aβ peptide that involves sequential solubilisation using the structure-breaking organic solvents HFIP and DMSO followed by column purification and results in standardized aggregate-free Aβ peptide. As was reported in pure DMSO, Aβ appears to be monomeric and lacks any β-sheet character (Shen and Murphy 1995).

2 Conclusions

Although the direct involvement of Aβ peptides in AD is well documented, the toxic Aβ species and the precise mechanism of its neurotoxicity remain unclear. Moreover, there is still a significant gap between site-specific structural information and the complex structural diversity of Aβ amyloids. A detailed structural and functional characterization of fibrillar assemblies as well as the various prefibrillar intermediates is crucial for understanding Aβ aggregation pathways and identifyingtoxic Aβ species. The recognition of the real culprit for AD onset is fundamental for the design of new, effective therapeutic strategies targeted at preventing the formation or impairing the activity of toxic Aβ assemblies involved in AD.

References

Ahmed M, Davis J, Aucoin D et al (2010) Structural conversion of neurotoxic amyloid-β(1–42) oligomers to fibrils. Nat Struct Mol Biol 17(5):561–567

Aksenov MY, Aksenova MV, Butterfield DA et al (1996) Glutaminesynthetase-induced enhancement of β-amyloid peptide Aβ (1–40) neurotoxicity accompanied by abrogation of fibril formation and Aβ fragmentation. J Neurochem 66:2050–2056

Balbach JJ, Petkova AT, Oyler NA et al (2002) Supramolecular Structure in Full-Length Alzheimer's β-Amyloid Fibrils: Evidence for a Parallel b -Sheet Organization from Solid-State Nuclear Magnetic Resonance. Biophys J 83:1205–1216

Barrantes FJ, Borroni V, Vallés S (2010) Neuronal nicotinic acetylcholine receptor–cholesterol crosstalk in Alzheimer's disease. FEBS Lett 584(9):1856–1863

Bayer TA, Cappai R, Masters CL et al (1999) It all sticks together the APP-related family of proteins and Alzheimer's disease. Mol Psychiatry 4(6):524–528

Benilova I, Karran E, De Strooper B (2012) The toxic Aβ oligomer and Alzheimer's disease: an emperor in need of clothes. Nat Neurosci 15(3):349–357

Benzinger TLS, Gregory DM, Burkoth TS et al (1998) Propagating structure of Alzheimer's β-amyloid(10–35) is parallel β-sheet with residues in exact register. Proc Natl Acad Sci U S A 95(23):13407–13412

Bernstein SL, Dupuis NF, Lazo ND et al (2009) Amyloid-β protein oligomerization and the importance of tetramers and dodecamers in the aetiology of Alzheimer's disease. Nat Chem 1(4):326–331

Bertini I, Luchinat C, Parigi G et al (2011a) Solid-state NMR of proteins sedimented by ultracentrifugation. Proc Natl Acad Sci U S A 108(26):10396–10399

Bertini I, Gonnelli L, Luchinat C et al (2011b) A new structural model of Aβ40 Fibrils. J Am Chem Soc U S A 133(40):16013–16022

Bertini I, Engelke F, Gonnelli L et al (2012a) On the use of ultracentrifugal devices for sedimented solute NMR. J Biomol NMR 54(2):123–127

Bertini I, Engelke F, Luchinat C et al (2012b) NMR properties of sedimented solutes. Phys Chem Chem Phys 14(2):439–447

Bertini I, Gallo G, Korsak M et al (2013a) Formation kinetics and structural features of β-amyloid aggregates by sedimented solute NMR. Eur J Chem Biol 14(14):1891–1897

Bertini I, Luchinat C, Parigi G et al (2013b) SedNMR: on the edge between solution and solid-state NMR. Acc Cheml Res 46(9):2059–2069

Bieschke J, Herbst M, Wiglenda T et al (2012) Small-molecule conversion of toxic oligomers to nontoxic β-sheet-rich amyloid fibrils. Nat Chem Biol 8(1):93–101

Bitan G, Teplow DB (2005) Preparation of aggregate-free, low molecular weight amyloid β for assembly and toxicity assays. Methods Mol Biol 299:3–9

Bitan G, Kirkitadze MD, Lomakin A et al (2003) Amyloid β-protein (Abeta) assembly: Abeta 40 and Abeta 42 oligomerize through distinct pathways. Proc Natl Acad Sci U S A 100(1):330–335

Bockmann A, Gardiennet C, Verel R et al (2009) Characterization of different water pools in solid-state NMR protein samples. J Biomol NMR 45(3):319–327

Broersen K, Jonckheere W, Rozenski J et al (2011) A Standardized and biocompatible preparation of aggregate-free amyloid β peptide for biophysical and biological studies of Alzheimer's disease. Protein Eng Des Sel PEDS 24(9):743–750

Chimon S, Yoshitaka I (2005) Capturing intermediate structures of Alzheimer's β-amyloid, Aβ1–40, by solid state NMR spectroscopy. J Am Chem Soc U S A 127(39):13472–13473
Chimon S, Shaibat MA, Jones CR et al (2007) Evidence of fibril-like b-sheet structures in a neurotoxic amyloid intermediate of Alzheimer's β-amyloid. Nat Struct Mol Biol 14:1157–1164
Danielsson J, Andersson A, Jarvet J et al (2006) ^{15}N relaxation study of the amyloid beta-peptide: structural propensities and persistence length. Magn Reson Chem 44:S114–S121
Dickson TC, King CE, McCormack GH et al (1999) Neurochemical diversity of dystrophic neurites in the early and late stages of Alzheimer's disease. Exp Neurol 1:100–110
Dobson CM, Misfolding PF (2003) Nature 426:884–890
Esch FS, Keim PS, Beattie EC et al (1990) Cleavage of amyloid β peptide during constitutive processing of its precursor. Science (New York N.Y.) 248(4959):1122–1124
Evin G, Weidemann A (2002) Biogenesis and metabolism of Alzheimer's disease Aβ amyloid peptides. Peptides 23:1285–1297
Fändrich M (2012) Oligomeric intermediates in amyloid formation: structure determination and mechanisms of toxicity. J Mol Biol 421(4–5):427–440
Fändrich M, Schmidt M, Grigorieff N (2011) Recent progress in understanding Alzheimer's β-amyloid structures. Trends Biochem Sci 36:338–345
Fawzi NL, Ying J, Ghirlando R et al (2011) Atomic-resolution dynamics on the surface of amyloid-β protofibrils probed by solution NMR. Nature 480(7376):268–272
Fezoui Y, Hartley DM, Harper JD et al (2000) An improved method of preparing the amyloid β-protein for fibrillogenesis and neurotoxicity experiments. Amyloid 7(3):166–178
Frost D, Gorman PM, Yip CM et al (2003) Co-incorporation of Aβ 40 and Aβ 42 to form mixed pre-fibrillar aggregates. Eur J Biochem 270(4):654–663
Gallion SL (2012) Modeling amyloid-β as homogeneous dodecamers and in complex with cellular prion protein. PLoS ONE 7(11):e49375
Gardiennet C, Schutz AK, Hunkeler A et al (2012) A sedimented sample of a 59 kDa dodecameric helicase yields high-resolution solid-state NMR spectra. Angew Chem Int Ed 51(31):7855–7858
Glenner GG, Wong CW (1984) Alzheimer's disease: initial report of the purification and characterization of a novel cerebrovascular amyloid protein. Biochem Biophys Res Commun 120(3):885–890
Haass C (2010) Initiation and propagation of neurodegeneration. Nat Med 16(11):1201–1204
Haass C, Selkoe DJ (2007) Soluble protein oligomers in neurodegeneration: lessons from the Alzheimer's amyloid β-peptide. Nat Rev Mol Cell Biol 8(2):101–112
Hardy J, Selkoe DJ (2002) The amyloid hypothesis of Alzheimer's disease: progress and problems on the road to therapeutics. Science 297:353–356
Hartley DM, Walsh DM, Chian P et al (1999) Protofibrillar intermediates of amyloid β-protein induce acute electrophysiological changes and progressive neurotoxicity in cortical neurons. J Neurosci 19(20):8876–8884
Haupt C, Leppert J, Ronicke R et al (2012) Structural basis of β-amyloid-dependent synaptic dysfunctions. Angew Chem Int Ed 51(7):1576–1579
Hellstrand E, Barry B, Dominic MW et al (2010) Amyloid β-Protein aggregation produces highly reproducible kinetic data and occurs by a two-phase process. ACS Chem Neurosci 1(1):13–18
Herzig MC, Winkler DT, Burgermeister P et al (2004) Aβ is targeted to the vasculature in a mouse model of hereditary cerebral hemorrhage with amyloidosis. Nat Neurosci 7(9):954–960
Hoshi M, Sato M, Sato M et al (2003) Spherical aggregates of (b-amyloid (amylospheroid) show high neurotoxicity and activate tau protein kinase I/glycogen synthase kinase-3b. Proc Natl Acad Sci U S A 100:6370–6375
Hou L, Shao H, Zhang Y et al (2004) Solution NMR studies of the aβ(1–40) and aβ(1–42) peptides establish that the Met35 oxidation state affects the mechanism of amyloid formation. J Am Chem Soc 126(7):1992–2005
Howlett DR, Jennings KH, Lee DC et al (1995) Aggregation state and neurotoxic properties of Alzheimer β-amyloid peptide. Neurodegeneration 4(1):23–32

Jan A, Gokce O, Luthi-Carter R et al (2008) The ratio of monomeric to aggregated forms of Aβ40 and Aβ42 is an important determinant of amyloid-β aggregation, fibrillogenesis, and toxicity. J Biol Chem 283(42):28176–28189

Jan A, Hartley DM, Lashuel HA (2010) Preparation and characterization of toxic aβ aggregates for structural and functional studies in Alzheimer's disease research. Nat Protoc 5(6): 1186–1209

Jao S-C, Kan M, Talafous J et al (1997) Trifluoroacetic acid pretreatment reproducibly disaggregates the amyloid β-peptide. Amyloid Internatl J Exp Clin Invest 4(4):240–252

Kang J, Lemaire HG, Unterbeck A et al (1987) The precursor of Alzheimer's disease amyloid A4 protein resembles a cell surface receptor. Nature 325:733–736

Kheterpal I, Chen M, Cook KD et al (2006) Structural differences in Abeta amyloid protofibrils and fibrils mapped by hydrogen exchange–mass spectrometry with on-line proteolytic fragmentation. J Mol Biol 361(4):785–795

Kim J, Onstead L, Randle S et al (2007) Aβ40 inhibits amyloid deposition in vivo. J Neurosci 27(3):627–633

Kirkitadze MD, Bitan G, Teplow DB (2002) Paradigm shifts in Alzheimer's disease and other neurodegenerative disorders: the emerging role of oligomeric assemblies. J Neurosci Res 69:567–577

Knopman DS, Parisi JE, Salviati A et al (2003) Neuropathology of cognitively normal elderly. J Neuropathol Exp Neurol 62:1087–1095

Krafft GA, Klein WL (2010) ADDLs and the signaling web that leads to Alzheimer's disease. Neuropharmacology 59(4–5):230–242

Kuhn P-H, Wang H, Dislich B et al (2010) ADAM10 is the physiologically relevant, constitutive α-secretase of the amyloid precursor protein in primary neurons. EMBO J, 29(17):3020–3027

Kumar S, Rezaei-Ghaleh N, Terwel D et al (2011) Extracellular phosphorylation of the amyloid β-peptide promotes formation of toxic aggregates during the pathogenesis of Alzheimer's disease. Eur Mol Biol Organ J 30(11):2255–2265

Kuperstein I, Broersen K, Benilova I et al (2010) Neurotoxicity of Alzheimer's disease Aβ peptides is induced by small changes in the Aβ42 to Aβ40 ratio. Eur Mol Biol Organ J 29(19):3408–3420

Lambert MP, Barlow AK, Chromy BA et al (1998) Diffusible, nonfibrillar ligands derived from Aβ1–42 are potent central nervous system neurotoxins. Proc Natl Acad Sci U S A 95:6448–6453

Lammich S, Kojro E, Postina R et al (1999) Constitutive and regulated α-secretase cleavage of Alzheimer's amyloid precursor protein by a disintegrin metalloprotease. Proc Natl Acad Sci U S A 96(7):3922–3927

Lansbury PT, Costa PR, Griffiths JM et al (1995) Structural model for the β-amyloid fibril based on interstrand alignment of an antiparallel-sheet comprising a C-terminal peptide. Nat Struct Mol Biol 2(11):990–998

Lee J, Culyba EK, Powers ET et al (2011) Amyloid-β forms fibrils by nucleated conformational conversion of oligomers. Nat Chem Biol 7(9):602–609

Lesné S, Koh MT, Kotilinek L et al (2006) A specific amyloid-β protein assembly in the brainimpairs memory. Nature 440(7082):352–357

Long F, Cho W, Ishii Y (2011) Expression and purification of ^{15}N- and ^{13}C-Isotope labeled 40-residue human Alzheimer's B-amyloid peptide for NMR-based structural analysis. Protein Expr Purif 79(1):16–24

Lopez del Amo JM, Fink U, Dasari M et al (2012) Structural properties of EGCG-induced, nontoxic Alzheimer's disease Aβ oligomers. J Mol Biol 421(4–5):517–524

Lorenzo A, Yankner BA (1994) Beta-amyloid neurotoxicity requires fibril formation and is inhibited by Congo red. Proc Natl Acad Sci U S A 91(25):12243–12247

Lu J-X, Qiang W, Yau W-M et al (2013) Molecular structure of β-amyloid fibrils in Alzheimer's disease brain tissue. Cell 154(6):1257–1268

Lührs T, Ritter C, Adrian M et al (2005) 3D structure of Alzheimer's amyloid β (1–42) fibrils. Proc Natl Acad Sci U S A 102(48):17342–17347

Marley J, Lu M, Bracken C (2001) A method for efficient isotopic labeling of recombinant proteins. J Biomol NMR 20(1):71–75

Martins IC, Kuperstein I, Wilkinson H et al (2008) Lipids revert inert Aβ amyloid fibrils to neurotoxic protofibrils that affect learning in mice. Eur Mol Biol Organ J 27(1):224–233

Masters CL, Simms G, Weinman NA et al (1985a) Amyloid plaque core protein in Alzheimer disease and Down syndrome. Proc Natl Acad Sci U S A 82:4245–4249

Masters CL, Multhaup G, Simms G et al (1985b) Neuronal origin of a cerebral amyloid: neurofibrillary tangles of Alzheimer's disease contain the same protein as the amyloid of plaque cores and blood vessels. EMBO J 4(11):2757–2763

Meinhardt J, Sachse C, Hortschansky P et al (2009) Aβ(1–40) fibril polymorphism implies diverse interaction patterns in amyloid fibrils. J Mol Biol 386(3):869–877

Merz PA, Wisniewski HM, Somerville RA et al (1983) Ultrastructural morphology of amyloid fibrils from neuritic and amyloid plaques. Acta Neuropathol 60(1–2):113–124

Meyer-Luehmann M, Spires-Jones TL, Prada C et al (2008) Rapid appearance and local toxicity of amyloid-β plaques in a mouse model of Alzheimer's disease. Nature 451:720–77U5

Naito A, Kamihira M, Inoue R et al (2004) Structural diversity of amyloid fibril formed inhuman calcitonin as revealed by site-directed ^{13}C solid-state NMR spectroscopy. Magn Reson Chem 42:247–257

Noguchi A, Matsumura S, Dezawa M et al (2009) Isolation and characterization of patient-derived, toxic, high mass amyloid β-protein (Aβ) assembly from Alzheimer disease brains. J Biol Chem 284(47):32895–32905

Pan J, Han J, Borchers CH et al (2011) Conformer-specific hydrogen exchange analysis of Ab(1–42) oligomers by top-down electron capture dissociation mass spectrometry. Anal Chem 83(13):5386–5393

Paravastu AK, Leapman RD, Yau W-M et al (2008) Molecular structural basis for polymorphism in Alzheimer's β-amyloid fibrils. Proc Natl Acad Sci U S A 105(47):18349–18354

Paravastu AK, Quhwash I, Leapman RD et al (2009) Seeded growth of beta-amyloid fibrils from Alzheimer's brain-derived fibrils produces a distinct fibril structure. Proc Natl Acad Sci U S A 106(18):7443–7448

Pauwels K, Williams TL, Morris KL et al (2012) Structural basis for increased toxicity of pathological aβ42:aβ40 ratios in Alzheimer disease. J Biol Chem 287(8):5650–5660

Petkova AT, Ishii Y, Balbach JJ et al (2002) A structural model for Alzheimer's β-amyloid fibrils based on experimental constraints from solid state NMR. Proc Natl Acad Sci U S A 99(26):16742–16747

Petkova AT, Leapman RD, Guo ZH et al (2005) Self-propagating, molecular-level polymorphism in Alzheimer's β-amyloid fibrils. Science 307(5707):262–265

Petkova AT, Yau W-M, Tycko R (2006) Experimental constraints onquaternary structure in Alzheimer's β-amyloid fibrils. BioChemistry 45(2):498–512

Polenova T (2011) Protein NMR spectroscopy: spinning into focus. Nat Chem 3(10):759–760

Qiang W, Yau W-M, Tycko R (2011) Structural evolution of Iowa mutant β-amyloid fibrils from polymorphic to homogeneous states under repeated seeded growth. J Am Chem Soc 133(11):4018–4029

Qiang W, Yau W-M, Luo Y et al (2012) Antiparallel β-sheet architecture in Iowa-mutant β-amyloid fibrils. J Am Chem Soc 109(12):4443–4448

Ravera E, Corzilius B, Michaelis VK et al (2013) Dynamic nuclear polarization of sedimented solutes. J Am Chem Soc U S A 135(5):1641–1644

Roberts SB, Ripellino JA, Ingalls KM et al (1994) Nonamyloidogenic cleavage of the β-amyloid precursor protein by an integral membrane metalloendopeptidase. J Biol Chem 269(4):3111–3116

Sachse C, Fändrich M, Grigorieff N (2008) Paired beta-sheet structure of an Aβ(1–40) amyloid fibril revealed by electron microscopy. Proc Natl Acad Sci U S A 105(21):7462–7466

Sachse C, Grigorieff N, Fändrich M (2010) Nanoscale flexibility parameters of Alzheimer amyloid fibrils determined by electron cryo-microscopy. Angew Chem Int Ed 49(7):1321–1323

Scheidt HA, Morgado I, Rothemund S et al (2011) Solid-state NMR spectroscopic investigation of Aβ protofibrils: implication of aβ-sheet remodeling upon maturation into terminal amyloid fibrils. Angew Chem Int Ed 50(12):2837–2840

Selkoe DJMD (1994) Alzheimer's disease: a central role for amyloid. J Neuropathol Exp Neurol 53(5):438–447
Selkoe DJ (2004) Cell biology of protein misfolding: the examples of Alzheimer's and Parkinson's diseases. Nat Cell Biol 6(11):1054–1061
Selkoe DJ (2008) Soluble oligomers of the amyloid β-protein impair synaptic plasticity and behavior. Behav Brain Res 192(1):106–113
Serpell LC et al (2000) Alzheimer's amyloid fibrils: structure and assembly. Biochim Biophys Acta 1502(1):16–30
Shen CL, Murphy RM (1995) Solvent effects on self-assembly of β-amyloid peptide. Biophys J 69(2):640–651
Sinha S, Anderson JP, Barbour R et al (1999) Purification and cloning of amyloid precursor protein β-secretase from human brain. Nature 402(6761):537–540
Sisodia SS (1992) β-amyloid precursor protein cleavage by a membrane-bound protease. Proc Natl Acad Sci U S A 89(13):6075–6079
Sola C, Mengod G, Probst A et al (1993) Di_erential regional and cellular distribution of β-amyloid precursor protein messenger RNAs containing and lacking the Kunitz protease inhibitor domain in the brain of human, rat and mouse. Neuroscience 53(1):267–295
Snider BJ, Norton J, Coats MA et al (2005) Novel presenilin 1 mutation (S170F) causing Alzheimer disease with Lewy bodies in the third decade of life. Arch Neurol 62:1821–1830
Soto C, Castaño EM, Kumar RA et al (1995) Fibrillogenesis of synthetic amyloid-B peptides is dependent on their initial secondary structure. Neurosci Lett 200(2):105–108
Studelska DR, McDowell LM, Espe MP et al (1997) Slowed enzymatic turnover allows characterization of intermediates by solid-state NMR. BioChemistry 36(50):15555–15560
Thal DR, Rüb U, Orantes M et al (2002) Phases of Aβ-deposition in the human brain and its relevance for the development of AD. Neurology 58:1791–1800
Tycko R (2010) Solid-state NMR studies of amyloid fibril structure. Annu Rev Phys Chem 62:279–299
Tycko R, Ishii Y (2003) Constraints on supramolecular structure in amyloid fibrils from two-dimensional solid-state NMR spectroscopy with uniform isotopic labeling. J Am Chem Soc 125:6606–6607
Vlassenko AG, Mintun MA, Xiong C et al (2011) Amyloid-β plaque growth in cognitively normal adults: longitudinal [11C]Pittsburgh compound B data. Annals of neurology,70(5):857–861
Walsh DM, Lomakin A, Benedek GB et al (1997) Amyloid β-protein fibrillogenesis detection of a protofibrilar intermediate. J Biol Chem 272(35):22364–22372
Walsh DM, Hartley DM, Kusumoto Y et al (1999) Amyloid β-protein fibrillogenesis. Structure and biological activity of protofibrillar intermediates. J Biol Chem 274(36):25945–25952
Walsh DM, Thulin E, Minogue AM et al (2009) A facile method for expression and purification of the Alzheimer's disease associated amyloid β peptide. FEBS J 276(5):1266–1281
Walter J, Fluhrer R, Hartung B et al (2001) Phosphorylation regulates intracellular traficking of β-secretase. J Biol Chem 276(18):14634–14641
Wang R, Wang B, He W et al (2006) Wild-type presenilin 1 protects against Alzheimer disease mutation-induced amyloid pathology. J Biol Chem 281(22):15330–15336
Ward RV, Jennings KH, Jepras R et al (2000) Fractionation and characterization of oligomeric, protofibrillar and fibrillar forms of β-amyloid peptide. Biochem J 348(Pt 1):137–144
Wasco W, Bupp K, Magendantz M et al (1992) Identification of a mouse brain cDNA that encodes a protein related to the Alzheimer disease-associated amyloid β protein precursor. Proc Natl Acad Sci U S A 89(22):10758–10762
Wasco W, Gurubhagavatula S, Paradis MD et al (1993) Isolation and characterization of APLP2 encoding a homologue of the Alzheimer's associated amyloid β protein precursor. Nat Genet 5(1):95–100
Wasmer C, Lange A, Van Melckebeke H et al (2008) Amyloid fibrils of the HET-s(218–289) prion form a β solenoid with a triangular hydrophobic core. Science 319(5869):1523–1526
Yan Y, Wang C (2007) Aβ40 protects non-toxic Aβ42 monomer from aggregation. J Mol Biol 369(4):909–916

Yoshiike Y, Chui D-H, Akagi T et al (2003) Specific compositions of amyloid-beta peptides as the determinant of toxic β-aggregation. J Biol Chem 278(26):23648–23655
Yoshikai S, Sasaki H, Doh-ura K et al (1990) Genomic organization of the human amyloid β-protein precursor gene. Gene 87(2):257–263
Younkin SG (1995) Evidence that Aβ 42 is the real culprit in Alzheimer's disease. Ann Neurol 37(3):287–288
Zhang L, Yu H, Song C et al (2009) Expression, purification, and characterization of recombinant human β-Amyloid42 peptide in escherichia coli. Protein Exp Purif 64(1):55–62
Zheng H, Koo EH (2011) Biology and pathophysiology of the amyloid precursor protein. Mol Neurodegener 6(1):27
Zou K, Kim D, Kakio A et al (2003) Amyloid β-protein (Aβ)1–40 protects neurons from damage induced by Aβ1–42 in culture and in rat brain. J Neurochem 87(3):609–619

MIX
Papier aus verantwortungsvollen Quellen
Paper from responsible sources
FSC® C105338

If you have any concerns about our products,
you can contact us on
ProductSafety@springernature.com

In case Publisher is established outside the EU,
the EU authorized representative is:
Springer Nature Customer Service Center GmbH
Europaplatz 3, 69115 Heidelberg, Germany

Printed by Libri Plureos GmbH
in Hamburg, Germany